Dominic Breit, Eduard Feireisl, Martina Hofmanová

Stochastically Forced Compressible Fluid Flows

De Gruyter Series in Applied and Numerical Mathematics

Volume 3

Dominic Breit, Eduard Feireisl,
Martina Hofmanová

Stochastically Forced Compressible Fluid Flows

DE GRUYTER

Mathematics Subject Classification 2010
Primary: 60H15, 35Q30, 76N10; Secondary: 35R60

Authors
Dr Dominic Breit
Heriot-Watt University
School of Mathematical and Computer Sciences
Edinburgh EH14 4AS
UK
d.breit@hw.ac.uk

Prof. Dr Eduard Feireisl
Institute of Mathematics AS CR
Žitná 25
115 67 Praha 1
Czech Republic
feireisl@math.cas.cz

Prof. Dr Martina Hofmanová
Universität Bielefeld
Fakultät für Mathematik
Universitätsstr. 25
33615 Bielefeld
Germany
hofmanova@math.uni-bielefeld.de

ISBN 978-3-11-049050-3
e-ISBN (PDF) 978-3-11-049255-2
e-ISBN (EPUB) 978-3-11-049076-3
Set-ISBN 978-3-11-049256-9
ISSN 2512-1820

Library of Congress Cataloging-in-Publication Data
A CIP catalog record for this book has been applied for at the Library of Congress.

Bibliographic information published by the Deutsche Nationalbibliothek
The Deutsche Nationalbibliothek lists this publication in the Deutsche Nationalbibliografie; detailed bibliographic data are available on the Internet at http://dnb.dnb.de.

Typesetting: VTeX UAB, Lithuania
Printing and binding: CPI books GmbH, Leck
♾ Printed on acid-free paper
Printed in Germany

www.degruyter.com

Acknowledgements

The research leading to the results in this book has been initiated in November 2013 when M.H. was visiting D.B. in Munich. She would like to thank the Mathematical Institute at LMU Munich, in particular the research group around L. Diening, for the hospitality. Many further results have been obtained during the numerous visits of D.B. and M.H. in Prague. Thanks go to the Institute of Mathematics of the Academy of Sciences of the Czech Republic for their hospitality. Large parts of the research have also been carried out during several visits of D.B. in Berlin. He wishes to express gratitude to the Mathematical Institute at TU Berlin for the stimulating research environment. Let us finally mention the support we had from the Edinburgh Mathematical Society for a visit of M.H. at Heriot-Watt University.
The research of E.F. leading to the results of this book has received funding from the European Research Council under the European Union's Seventh Framework Programme (FP7/2007-2013)/ERC Grant Agreement 320078. The Institute of Mathematics of the Academy of Sciences of the Czech Republic is supported by RVO:67985840.

For fruitful discussions we thank to Bohdan Maslowski.

Dominic Breit — Edinburgh
Eduard Feireisl — Prague
Martina Hofmanová — Bielefeld

August 2017

https://doi.org/10.1515/9783110492552-201

Notation

$\aleph_0$	Cardinality of the set of integers
c, C	Generic constants which differ from line to line
$a \lesssim b$	$a \leq cb$
B^c	Complement of a set B
$\mathcal{O}$	A domain – an open connected subset of $\mathbb{R}^N$
$\lvert\mathcal{O}\rvert$	Lebesgue measure of the domain $\mathcal{O}$
$\mathfrak{L}$	1-dimensional Lebesgue measure
$\mathfrak{L}^N$	N-dimensional Lebesgue measure
$\mathfrak{B}$	Borel σ-algebra
$\mathbb{T}^N$	N-dimensional flat torus $([-1,1]\vert_{\{-1,1\}})^N$
$\mathbb{T}_L^{N+1}$	Space-time $(N+1)$-dimensional torus $([-L,L]\vert_{\{-L,L\}}) \times \mathbb{T}^N$
$\mathbb{R}^{n\times N}$	Space of $n \times N$ matrices over $\mathbb{R}$
$\mathbb{A} : \mathbb{B}$	Scalar product $\sum_{ij} A_{ij}B_{ij}$ between two matrices $\mathbb{A}$, $\mathbb{B}$
$\mathbb{I}$	Identity matrix $(\delta_{ij})_{i,j=1}^N$ in $\mathbb{R}^{N\times N}$
B_b	Bounded Borel measurable functions
C	Continuous functions
C_c	Continuous functions with compact support
C_0	Continuous functions vanishing at infinity
C_b	Bounded continuous functions
C^α	α-Hölder continuous functions
C^k	k-times continuously differentiable functions
C_c^k	C^k-functions with compact support
$C^{k,\alpha}$	k-times continuously differentiable functions with α-Hölder continuous derivatives
C^∞	∞-times continuously differentiable functions
$C_c^\infty/\mathcal{D}$	C^∞-functions with compact support
$\mathcal{D}'$	Dual of C_c^∞
C_{div}^∞	C^∞-functions with vanishing divergence
$\mathcal{D}'_{\mathrm{div}}$	Dual of C_{div}^∞
L^p	Lebesgue space of p-integrable functions
L_{loc}^p	Lebesgue space of locally p-integrable functions
L_{div}^p	L^p-functions with vanishing divergence
p'	Dual exponent of p: $p' = p/(p-1)$
$W^{k,p}$	Sobolev functions with differentiability k and integrability p
$W_{\mathrm{div}}^{k,p}$	$W^{k,p}$-functions with vanishing divergence
$W^{-k,p}$	Dual space of $W^{k,p'}$
$(e_{\mathbf{m}})_{\mathbf{m}\in\mathbb{Z}^N}$	Trigonometric polynomials on $\mathbb{T}^N$
$\mathcal{M}_b$	Bounded signed measures
$\mathcal{M}_b^+$	Non-negative bounded measures

https://doi.org/10.1515/9783110492552-202

$\mathcal{M}_R^+$	Non-negative Radon measures
Δ^{-1}	Solution operator to the Laplace equation
$\mathcal{P}_H\mathbf{v}$	Helmholtz projection $\mathbf{v} - \nabla\Delta^{-1}\operatorname{div}\mathbf{v}$ of a function $\mathbf{v} : \mathbb{R}^N \to \mathbb{R}^N$ ($\mathbb{T}^N \to \mathbb{T}^N$)
$\mathcal{Q}\mathbf{v}$	Gradient part $\nabla\Delta^{-1}\operatorname{div}\mathbf{v}$ of a function $\mathbf{v} : \mathbb{R}^N \to \mathbb{R}^N$ ($\mathbb{T}^N \to \mathbb{T}^N$)
X^*	Dual space of X
$\\|\cdot\\|_X$	Norm on X
$\langle\cdot,\cdot\rangle_X$	Inner product on X
$\langle\cdot,\cdot\rangle_{X^*,X}$	Duality pairing between X^* and X
$\rightharpoonup$	Weak convergence
$\overset{*}{\rightharpoonup}$	Weak-$*$ convergence
$\overset{d}{\to}$	Convergence in law
$L^p(0,T;X)$	Bochner space of X-valued p-integrable functions
$L^p_{\mathrm{loc}}(0,\infty;X)$	Bochner space of X-valued locally p-integrable functions
$C([0,T];X)$	Continuous functions with values in X
$C_{\mathrm{loc}}([0,\infty);X)$	Locally continuous functions with values in X
$C^\alpha([0,T];X)$	α-Hölder continuous functions with values in X
$C_w([0,T];X)$	Weakly continuous functions with values in X
$W^{k,p}(0,T;X)$	k-times weakly differentiable functions with values in X and integrability p
$(\Omega,\mathfrak{F},\mathbb{P})$	Probability space with sample space Ω, σ-algebra $\mathfrak{F}$, and probability measure $\mathbb{P}$
$(\mathfrak{F}_t)_{t\geq 0}$	Filtration
$(\sigma_t[\mathbf{U}])_{t\geq 0}$	Canonical filtration/history of a stochastic process/random distribution $\mathbf{U}$
$(\Omega,\mathfrak{F},(\mathfrak{F}_t)_{t\geq 0},\mathbb{P})$	Filtered probability space with filtration $(\mathfrak{F}_t)_{t\geq 0}$
$([0,1],\overline{\mathfrak{B}([0,1])},\mathfrak{L})$	Standard probability space
$\mathbb{E}$	Expectation
$\mathbb{E}[\cdot\|\mathfrak{F}]$	Conditional expectation given $\mathfrak{F}$
$\mathcal{L}[\cdot]$	Law of a random variable
$\mathcal{L}_X[\cdot]$	Law of a random variable on the space X
$\overset{d}{\sim}$	Equality in law
$L^p_{\mathrm{prog}}(\Omega\times[0,T];X)$	L^p-integrable progressively measurable X-valued random variable
$L(\mathfrak{U},H)$	Continuous linear operators from $\mathfrak{U}\to H$
$L_2(\mathfrak{U},H)$	Hilbert–Schmidt operators from $\mathfrak{U}\to H$
$(e_k)_{k\in\mathbb{N}}$	Complete orthonormal system in $\mathfrak{U}$
$W=\sum_{k=1}^\infty e_k W_k$	Cylindrical Wiener process in $\mathfrak{U}$
$\langle\langle\mathbf{U}\rangle\rangle$	Quadratic variation of the stochastic process $\mathbf{U}$
$\langle\langle\mathbf{U},\mathbf{V}\rangle\rangle$	Cross variation of stochastic processes $\mathbf{U}$ and $\mathbf{V}$

Contents

Part I: **Preliminary results**

1 Elements of functional analysis

We will exclusively use functions $v = v(t, x)$ with the time $t \in I$ and the space variable $x \in \mathcal{O} \subset \mathbb{R}^N$, where I is an interval and $\mathcal{O}$ denotes a *domain* – an open connected subset of $\mathbb{R}^N$. Sometimes, it will be convenient to separate the time and space variables and consider $v = v(t, \cdot)$ as a mapping ranging in a suitable topological space X of functions depending on the x-variable. To avoid problems related to the presence of a kinematic boundary in the equation of fluid mechanics, we mostly focus on functions that are space periodic, meaning the spatial domain $\mathcal{O}$ is identified with the flat torus $\mathbb{T}^N$, given by

$$\mathbb{T}^N = \left([-1,1]|_{\{-1,1\}}\right)^N.$$

The length of period 2 is taken only for the sake of convenience. All results stated in this book have been obtained for a general torus given by

$$\Pi_{i=1}^N [a_i, b_i]|_{\{a_i, b_i\}}.$$

If not otherwise stated, all functions (or vector-valued functions) are real-valued.

1.1 Continuous functions, measures

For a topological space X, the symbol $C(X)$ denotes the space of continuous functions on X, $C_c(X)$ is the space of all continuous functions compactly supported in X, and $C_b(X)$ is the space of all bounded continuous functions on X.

If K is compact, $C(K)$ is a Banach space with the norm

$$\|v\|_{C(K)} = \sup_{y \in K} |v(y)|, \quad v \in C(K).$$

For $X \subset \mathbb{R}^N$ or $X \subset \mathbb{R}$ we simply write $\|\cdot\|_{C_x}$ and $\|\cdot\|_{C_t}$. Similarly, for functions $v : K \to Y$ ranging in a metric space Y with metric d_Y, we define a metric on $C(K; Y)$ as

$$d_{C(K;Y)}[v, w] = \sup_{y \in K} d_Y[v(y), w(y)], \quad v, w \in C(K; Y).$$

If there is no danger of confusion, we write $C(K)$ instead of $C(K; \mathbb{R}^M)$.

The following result is known as the *Arzelà–Ascoli theorem*; see Kelley [Kel55, Chapter 7, Theorem 17].

Theorem 1.1.1. *Let $K \subset \mathbb{R}^N$ be compact and Y a compact topological metric space endowed with a metric d_Y. Let $(v_n)_{n \in \mathbb{N}}$ be a sequence of functions in $C(K; Y)$ that is equi-*

https://doi.org/10.1515/9783110492552-001

continuous, meaning that, for any $\varepsilon > 0$, *there is a* $\delta > 0$ *such that*

$$d_Y[v_n(y), v_n(z)] \le \varepsilon \text{ provided } |y - z| < \delta \text{ independently of } n \in \mathbb{N}.$$

Then $(v_n)_{n\in\mathbb{N}}$ *is precompact in* $C(K;Y)$, *that is, there exist a subsequence (not relabeled) and a function* $v \in C(K;Y)$ *such that*

$$\sup_{y\in K} d_Y[v_n(y), v(y)] \to 0 \quad \text{as } n \to \infty.$$

Next we recall the *Stone–Weierstrass theorem*; see Cullen [Cul68].

Theorem 1.1.2. *Suppose* K *is a compact Hausdorff space and* $\mathcal{A}$ *is a subalgebra of* $C(K;\mathbb{R})$ *which contains a non-zero constant function. Then* $\mathcal{A}$ *is dense in* $C(K;\mathbb{R})$ *if and only if it separates points.*

Remark 1.1.3. A set of continuous functions $\mathcal{A}$ on K separates points if, for $x, y \in K$, $x \neq y$, there is $f \in \mathcal{A}$ such that $f(x) \neq f(y)$. Note that a topological space is Hausdorff if, for any two points $x \neq y$, there are open sets U_x, U_y, $x \in U_x$, $y \in U_y$, $U_x \cap U_y = \emptyset$. In particular, any topological space in which $C(X;\mathbb{R})$ separates points is Hausdorff and the "if" part of Theorem 1.1.2 holds without the explicit requirement K to be Hausdorff.

A function vanishes at infinity if, for any $\varepsilon > 0$, there is a compact $K_\varepsilon \subset X$ such that $|f(x)| < \varepsilon$ for $x \notin K_\varepsilon$. The space of continuous functions vanishing at infinity is denoted as $C_0(X;\mathbb{R})$. There is an extension of the Stone–Weierstrass theorem to locally compact spaces; see de Branges [dB59].

Theorem 1.1.4. *Suppose* K *is a locally compact topological space and* $\mathcal{A}$ *is a subalgebra of* $C_0(K;\mathbb{R})$ *that separates points such that, for any* $x \in X$, *there is* $f \in \mathcal{A}$ *such that* $f(x) \neq 0$. *Then* $\mathcal{A}$ *is dense in* $C_0(X;\mathbb{R})$.

Let $\mathcal{M}^+(X)$ denote the set of all non-negative measures on X, meaning all non-negative σ-additive set-functions defined on a σ-field of measurable subsets of X. The following is the *Riesz representation theorem*; see Rudin [Rud87, Chapter 2, Theorem 2.14].

Theorem 1.1.5. *Let* X *be a locally compact Hausdorff metric space. Let* f *be a non-negative linear functional defined on the space* $C_c(X)$.

Then there exists a σ*-algebra of measurable sets containing all Borel sets and a unique non-negative measure* $\mu_f \in \mathcal{M}^+(X)$ *such that*

$$\langle f, g\rangle = \int_X g \,\mathrm{d}\mu_f \quad \text{for any } g \in C_c(X).$$

Moreover, the measure μ_f enjoys the following properties:
- *We have $\mu_f[K] < \infty$ for any compact $K \subset X$.*
- *We have*

$$\mu_f[E] = \sup\{\mu_f[K] \mid K \subset E,\ K \text{ compact}\}$$

  for any open set $E \subset X$.
- *We have*

$$\mu_f[V] = \inf\{\mu(E) \mid V \subset E,\ E \text{ open}\}$$

  for any Borel set V.
- *If E is μ_f-measurable, $\mu_f(E) = 0$, and $A \subset E$, then A is μ_f-measurable.*

1.2 Topological spaces

The topological spaces we deal with, besides admitting a σ-field of Borel sets, will satisfy certain *separation* properties. Possibly the weakest assumption in this sense is that a topological space X is completely regular (Tikhonov space), X is Hausdorff, and $C(X)$ separates points from closed sets: for any $x \in X$ and a closed set $F \subset X$ with $x \notin F$, there is $f \in C(X)$ such that $f(x) = 1$, $f|_F = 0$. The topology on a completely regular space is the coarsest topology making all functions from $C(X)$ or $C_b(X)$ continuous. Every subspace of a completely regular space is completely regular. In particular, if Y is completely regular and $X \hookrightarrow Y$ is a continuous injection, then X is completely regular. Any metric space is completely regular. In this book we deal almost exclusively with *topological vector spaces*, where the algebraic operations of addition and multiplication by a scalar are continuous. In particular, any Hausdorff topological vector space is Tikhonov. Topological vector spaces admit a uniform structure. Specifically, any neighborhood $\mathcal{U}(x)$ of a point x can be written as $x + \mathcal{U}$, where $\mathcal{U}$ is a neighborhood of zero. The uniform structure is necessary for a proper definition of some stochastic concepts like convergence in probability.

Most statements in the theory of stochastic PDEs use *Polish spaces*.

Definition 1.2.1. A topological space is *Polish* if the topology on X is separable and completely metrizable.

Later (see Definition 2.1.3) we introduce a larger class of *sub-Polish* spaces. These are, roughly speaking, topological spaces that admit a continuous injection into a Polish space.

The symbol $\mathcal{M}^+_R(X)$ denotes the set of non-negative Radon measures on X, meaning non-negative Borel measures μ such that

$$\mu[E] = \sup\{\mu[K] \mid K \subset E, K \text{ compact}\} \quad \text{for any open set } E \subset X.$$

Proposition 1.2.2. *If X is Polish, then every finite Borel measure is a Radon measure.*

For the proof see, e.g., Bogachev [Bog07].

1.3 Differentiable functions, distributions

The symbol

$$\partial_{y_i} g(y) := \frac{\partial g}{\partial_{y_i}}, \quad y = [y_1, \dots, y_N]$$

stands for the *partial derivative* of a function g defined on an open neighborhood of a point $y \in \mathbb{R}^N$.

The space of functions having k continuous derivatives are denoted C^k. If K is a compact set, then $C^k(K)$ is the space of functions from $C^k(\mathbb{R}^N)$ restricted to K. $C^{k,\nu}(\mathcal{O})$, $\nu \in (0,1)$, is the subspace of $C^k(\mathcal{O})$-functions having their kth derivatives ν-Hölder continuous in $\mathcal{O} \subset \mathbb{R}^N$. $C^{k,1}(\mathcal{O})$ is a subspace of $C^k(\mathcal{O})$ of functions whose kth derivatives are Lipschitz on $\mathcal{O}$. For a bounded domain $\mathcal{O}$, the spaces $C^k(\overline{\mathcal{O}})$ and $C^{k,\nu}(\overline{\mathcal{O}})$, $\nu \in (0,1]$, are Banach spaces with norms

$$\|u\|_{C^k_x} = \max_{|\alpha| \le k} \sup_{x \in \mathcal{O}} |\partial^\alpha u(x)|$$

and

$$\|u\|_{C^{k,\nu}_x} = \|u\|_{C^k_x} + \max_{|\alpha|=k} \sup_{(x,y)\in\mathcal{O}^2, x \ne y} \frac{|\partial^\alpha u(x) - \partial^\alpha u(y)|}{|x-y|^\nu},$$

where $\partial^\alpha u$ stands for the partial derivative $\partial^{\alpha_1}_{x_1} \dots \partial^{\alpha_N}_{x_N} u$ of order $|\alpha| = \sum_{i=1}^N \alpha_i$. The spaces $C^{k,\nu}(\overline{\mathcal{O}}; \mathbb{R}^M)$ are defined in a similar way. However, for notational simplicity the target space $\mathbb{R}^M$ will not be explicitly mentioned. Finally, we set $C^\infty = \bigcap_{k=0}^\infty C^k$.

The symbol $C^k_c(\mathcal{O})$, $k \in \{0, 1, \dots, \infty\}$, denotes the vector space of functions belonging to $C^k(\mathcal{O})$ and having compact support in $\mathcal{O}$. If $\mathcal{O} \subset \mathbb{R}^N$ is an open set, the symbol $\mathcal{D}(\mathcal{O})$ will be used alternatively for the space $C^\infty_c(\mathcal{O})$ endowed with the topology induced by the convergence

$$\varphi_n \to \varphi \quad \text{in } \mathcal{D}(\mathcal{O}),$$

if there is $K \subset \mathcal{O}$, a compact such that $\mathrm{supp}[\varphi_n] \subset K$ for any $k = 0, 1, \dots$ and

$$\varphi_n \to \varphi \quad \text{in } C^k(K). \tag{1.1}$$

The dual space $\mathcal{D}'(\mathcal{O})$ is the space of *distributions* on $\mathcal{O}$. Similarly, we define $\mathcal{D}'(\mathcal{O}; \mathbb{R}^M)$. Continuity of a linear form belonging to $\mathcal{D}'(\mathcal{O})$ is understood with respect to the convergence introduced in (1.1). We also consider the space of periodic distributions $\mathcal{D}'(\mathbb{T}^N)$ defined on the flat torus $\mathbb{T}^N$.

A differential operator ∂^α of order $|\alpha|$ can be identified with a distribution

$$\langle \partial^\alpha v, \varphi \rangle = (-1)^{|\alpha|} \langle v, \partial^\alpha \varphi \rangle = (-1)^{|\alpha|} \int_{\mathcal{O}} v \partial^\alpha \varphi \, dy, \quad \varphi \in \mathcal{D}(\mathcal{O}),$$

where the most right identity makes sense whenever v is a locally integrable function.

1.4 Integrable functions

Let $\mathcal{O}$ be a measurable subset of $\mathbb{R}^N$ and X a separable Banach space with norm $\|\cdot\|_X$. The *Lebesgue space* $L^p(\mathcal{O};X)$ is the space of Bochner measurable functions v ranging in the Banach space X such that the norm

$$\|v\|^p_{L^p_x X} = \int_{\mathcal{O}} \|v(y)\|^p_X \, dy \text{ is finite}, \quad 1 \leq p < \infty.$$

Similarly, $v \in L^\infty(\mathcal{O};X)$ if v is Bochner measurable and

$$\|v\|_{L^\infty_x X} = \operatorname*{ess\,sup}_{y \in \mathcal{O}} \|v(y)\|_X < \infty.$$

The symbol $L^p_{\mathrm{loc}}(\mathcal{O};X)$ denotes the vector space of locally L^p-integrable functions, meaning

$$v \in L^p_{\mathrm{loc}}(\mathcal{O};X) \quad \text{if } v \in L^p(K;X) \text{ for any compact set } K \text{ in } \mathcal{O}.$$

We will omit the target space and write $L^p(\mathcal{O})$ instead of $L^p(\mathcal{O};X)$ whenever no confusion arises.

The dual spaces to the L^p spaces are characterized in the following theorem; see Gajewski et al. [GGZ75, Chapter IV, Theorem 1.14, Remark 1.9], Edwards [Edw94], and Pedregal [Ped97, Chapter 6, Theorem 6.14].

Theorem 1.4.1. (1) *Let $\mathcal{O} \subset \mathbb{R}^N$ be a measurable set, X a Banach space that is reflexive and separable, and $1 \leq p < \infty$. Then any continuous linear form $\xi \in [L^p(\mathcal{O};X)]^*$ admits a unique representation $w_\xi \in L^{p'}(\mathcal{O};X^*)$,*

$$\langle \xi, v \rangle_{L^{p'}(\mathcal{O};X^*);L^p(\mathcal{O};X)} = \int_{\mathcal{O}} \langle w_\xi(y), v(y) \rangle_{X^*;X} \, dy \quad \textit{for all } v \in L^p(\mathcal{O};X),$$

where

$$\frac{1}{p} + \frac{1}{p'} = 1.$$

Moreover, the norm on the dual space is given by

$$\|\xi\|_{[L^p_x X]^*} = \|w_\xi\|_{L^{p'}_x X^*}.$$

Accordingly, the spaces $L^p(\mathcal{O};X)$ are reflexive for $1<p<\infty$ as soon as X is reflexive and separable. Identifying ξ with w_ξ, we obtain the Riesz representation theorem

$$[L^p(\mathcal{O};X)]^* = L^{p'}(\mathcal{O};X^*), \quad \|\xi\|_{[L^p_x X]^*} = \|\xi\|_{L^{p'}_x X^*}, \quad 1\le p<\infty.$$

(2) *If the Banach space X is merely separable, we have*

$$[L^p(\mathcal{O};X)]^* = L^{p'}_{w^*}(\mathcal{O};X^*) \quad \textit{for } 1\le p<\infty,$$

where

$$L^{p'}_{w^*}(\mathcal{O};X^*) := \{\xi : \mathcal{O}\to X^* \mid y\in\mathcal{O}\mapsto \langle \xi(y), v\rangle_{X^*;X} \textit{ measurable } \forall v\in X,$$
$$y\mapsto \|\xi(y)\|_{X^*}\in L^{p'}(\mathcal{O})\}.$$

For L^p-spaces we also report *Hölder's inequality*

$$\|uv\|_{L^r_x}\le \|u\|_{L^p_x}\|v\|_{L^q_x}, \quad \frac{1}{r}=\frac{1}{p}+\frac{1}{q},$$

for any $u\in L^p(\mathcal{O})$, $v\in L^q(\mathcal{O})$, $\mathcal{O}\subset\mathbb{R}^N$, and the *interpolation inequality*

$$\|v\|_{L^r_x}\le \|v\|^{\lambda}_{L^p_x}\|v\|^{(1-\lambda)}_{L^q_x}, \quad \frac{1}{r}=\frac{\lambda}{p}+\frac{1-\lambda}{q}, \quad p<r<q,\ \lambda\in(0,1),$$

for any $v\in L^p\cap L^q(\mathcal{O})$, $\mathcal{O}\subset\mathbb{R}^N$; see Adams [Ada75, Chapter 2].

Finally, we recall the celebrated and frequently used *Gronwall's lemma*; see Carroll [Car13].

Lemma 1.4.2. *Let $a\in L^1(0,T)$, $a\ge 0$, $\beta\in L^1(0,T)$, $b_0\in\mathbb{R}$, and*

$$b(\tau)=b_0+\int_0^\tau \beta(t)\,\mathrm{d}t$$

be given. Let $r\in L^\infty(0,T)$ satisfy

$$r(\tau)\le b(\tau)+\int_0^\tau a(t)r(t)\,\mathrm{d}t \quad \textit{for a.a. } \tau\in[0,T].$$

Then

$$r(\tau)\le b_0\exp\left(\int_0^\tau a(t)\,\mathrm{d}t\right)+\int_0^\tau \beta(t)\exp\left(\int_t^\tau a(s)\,\mathrm{d}s\right)\mathrm{d}t$$

for a.a. $\tau\in[0,T]$.

1.5 Compactness and convergence of integrable functions

Let X be a Banach space, B_X the closed unit ball in X, and B_{X^*} the closed unit ball in the dual space X^*. Then we have:

(1) B_X is weakly compact only if X is reflexive. This is Kakutani's theorem; see Theorem III.6 in Brezis [Bre83].
(2) B_{X^*} is weakly-$*$-compact. This is the Banach–Alaoglu theorem; see Theorem III.15 in Brezis [Bre83].
(3) If X is separable, then B_{X^*} is sequentially weakly-$*$-compact; see Theorem III.25 in Brezis [Bre83].
(4) A non-empty subset of a Banach space X is weakly relatively compact only if it is sequentially weakly relatively compact. This is the Eberlein–Shmuliyan–Grothendieck theorem; see Paragraph 24 in Kothe [KK83].

In view of the above results we get:

- Any bounded sequence in $L^p(\mathcal{O})$, where $1 < p < \infty$ and $\mathcal{O} \subset \mathbb{R}^N$ is a domain, is relatively weakly compact.
- Any bounded sequence in $L^\infty(\mathcal{O})$, where $\mathcal{O} \subset \mathbb{R}^N$ is a domain, is relatively weakly-$*$ compact.

The situation for L^1, which is neither reflexive nor dual of a Banach space, is clarified in the following theorem; see Ekeland–Temam [ET99, Chapter 8, Theorem 1.3] and Pedregal [Ped97, Lemma 6.4].

Theorem 1.5.1. *Let $\mathcal{V} \subset L^1(\mathcal{O})$, where $\mathcal{O} \subset \mathbb{R}^N$ is a bounded measurable set.*

Then the following statements are equivalent:

- *any sequence $(v_n)_{n\in\mathbb{N}} \subset \mathcal{V}$ contains a subsequence weakly converging in $L^1(\mathcal{O})$;*
- *for any $\varepsilon > 0$, there exists $k > 0$ such that*

$$\int_{\{|v|\geq k\}} |v(y)|\,dy \leq \varepsilon \quad \textit{for all } v \in \mathcal{V};$$

- *for any $\varepsilon > 0$, there exists $\delta > 0$ such that, for all $v \in \mathcal{V}$,*

$$\int_M |v(y)|\,dy < \varepsilon,$$

for any measurable set $M \subset \mathcal{O}$, such that

$$|M| < \delta;$$

- *there exists a non-negative function $\Phi \in C([0,\infty))$*

$$\lim_{z\to\infty} \frac{\Phi(z)}{z} = \infty,$$

such that

$$\sup_{v\in\mathcal{V}} \int_{\mathcal{O}} \Phi(|v(y)|)\,dy \leq c.$$

1.6 Sobolev spaces

There is a vast amount of literature devoted to the study of Sobolev spaces. We restrict ourselves to listing some standard results. The reader may consult the monographs by Adams [Ada75], Kufner et al. [KJF77], Maz'ya [Maz13], or Ziemer [Zie89] for more information.

The *Sobolev spaces* $W^{k,p}(\mathcal{O})$, $1 \le p \le \infty$, with k being a positive integer, are the spaces of functions having all distributional derivatives up to order k in $L^p(\mathcal{O})$. The norm in $W^{k,p}(\mathcal{O})$ is defined as

$$\|\mathbf{v}\|_{W_x^{k,p}} = \begin{cases} (\sum_{|\alpha|\le k} \|\partial^\alpha v\|_{L_x^p}^p)^{1/p} & \text{if } 1 \le p < \infty, \\ \max_{|\alpha|\le k}\{\|\partial^\alpha v\|_{L_x^\infty}\} & \text{if } p = \infty, \end{cases}$$

where the symbol ∂^α stands for any partial derivative of order $|\alpha|$.

If $1 \le p < \infty$, then $W^{k,p}(\mathcal{O})$ is separable and the space $C^k(\overline{\mathcal{O}})$ is its dense subspace (if $\mathcal{O}$ has a Lipschitz boundary).

The space $W^{1,\infty}(\mathcal{O})$, where $\mathcal{O}$ is a bounded Lipschitz domain, is isometrically isomorphic to the space $C^{0,1}(\overline{\mathcal{O}})$ of Lipschitz functions on $\overline{\mathcal{O}}$.

The symbol $W_0^{k,p}(\mathcal{O})$ denotes the completion of $C_c^\infty(\mathcal{O})$ with respect to the norm $\|\cdot\|_{W_x^{k,p}}$. In what follows, we identify $W^{0,p}(\mathcal{O}) = W_0^{0,p}(\mathcal{O})$ with $L^p(\mathcal{O})$.

The differentiability of a composition of a Sobolev function with a Lipschitz function is clarified in the following result; see Ziemer [Zie89, Section 2.1].

Lemma 1.6.1. *If $f : \mathbb{R} \to \mathbb{R}$ is a Lipschitz function and $f \circ v \in L^p(\mathcal{O})$ for some $v \in W^{1,p}(\mathcal{O})$, then $f \circ v \in W^{1,p}(\mathcal{O})$ and*

$$\partial_{x_j}[f \circ v](x) = f'(v(x))\partial_{x_j} v(x) \quad \textit{for a.a. } x \in \mathcal{O}.$$

Duals to Sobolev spaces are characterized in the following theorem; see Adams [Ada75, Theorem 3.8] and Maz'ya [Maz13, Section 1.1.14].

Theorem 1.6.2. *Let $\mathcal{O} \subset \mathbb{R}^N$ be a domain and let $1 \le p < \infty$. Then the dual space $[W_0^{k,p}(\mathcal{O})]^*$ is a proper subspace of the space of distributions $\mathcal{D}'(\mathcal{O})$. Moreover, any linear form $f \in [W_0^{k,p}(\mathcal{O})]^*$ admits a representation*

$$\langle f, v\rangle_{[W_0^{k,p}]^*; W_0^{k,p}} = \sum_{|\alpha|\le k} \int_{\mathcal{O}} (-1)^{|\alpha|} w_\alpha \partial^\alpha v \, dx,$$

$$\textit{where } w_\alpha \in L^{p'}(\mathcal{O}), \ \frac{1}{p} + \frac{1}{p'} = 1. \tag{1.2}$$

The norm of f in the dual space is given by

$$\|f\|_{[W_0^{k,p}(Q)]^*} = \begin{cases} \inf\{(\sum_{|\alpha|\le k} \|w_\alpha\|_{L_x^{p'}}^{p'})^{1/p'} \mid w_\alpha \textit{ satisfy } (1.2)\} & \textit{if } 1 < p < \infty, \\ \inf\{\max_{|\alpha|\le k}\{\|w_\alpha\|_{L_x^\infty}\} \mid w_\alpha \textit{ satisfy } (1.2)\} & \textit{if } p = 1. \end{cases}$$

The infimum is attained in both cases.

The dual space of the Sobolev space $W_0^{k,p}(\mathcal{O})$ is denoted as $W^{-k,p'}(\mathcal{O})$. The dual space of the Sobolev space $W^{k,p}(\mathcal{O})$ admits formally the same representation equation (1.2). However, it cannot be identified as a space of distributions on $\mathcal{O}$.

The important result is the Rellich–Kondrachov embedding theorem for Sobolev spaces; see Ziemer [Zie89, Theorem 2.5.1, Remark 2.5.2].

Theorem 1.6.3. *Let $\mathcal{O} \subset \mathbb{R}^N$ be a bounded Lipschitz domain.*

(i) *Then, if $kp < N$ and $p \geq 1$, the space $W^{k,p}(\mathcal{O})$ is continuously embedded in $L^q(\mathcal{O})$ for any*

$$1 \leq q \leq p^* = \frac{Np}{N - kp}.$$

Moreover, the embedding is compact if $k > 0$ and $q < p^*$.

(ii) *If $kp = N$, the space $W^{k,p}(\mathcal{O})$ is compactly embedded in $L^q(\mathcal{O})$ for any $q \in [1, \infty)$.*

(iii) *If $kp > N$, then $W^{k,p}(\mathcal{O})$ is continuously embedded in $C^{k-[N/p]-1,\nu}(\overline{\mathcal{O}})$, where $[\cdot]$ denotes the integer part and*

$$\nu = \begin{cases} [\frac{N}{p}] + 1 - \frac{N}{p} & \text{if } \frac{N}{p} \notin Z, \\ \text{arbitrary positive number in } (0,1) & \text{if } \frac{N}{p} \in Z. \end{cases}$$

Moreover, the embedding is compact if $0 < \nu < [\frac{N}{p}] + 1 - \frac{N}{p}$.

As a straightforward corollary, we get the following dual result.

Theorem 1.6.4. *Let $\mathcal{O} \subset \mathbb{R}^N$ be a bounded domain. Let $k > 0$ and $q < \infty$ satisfy*

$$q > \frac{p^*}{p^* - 1}, \quad \text{where } p^* = \frac{Np}{N - kp} \text{ if } kp < N,$$
$$q > 1 \quad \text{for } kp = N,$$

or

$$q \geq 1 \quad \text{if } kp > N.$$

Then the space $L^q(\mathcal{O})$ is compactly embedded into the space $W^{-k,p'}(\mathcal{O})$, $1/p + 1/p' = 1$.

Remark 1.6.5. We have formulated this section on real-valued functions for the ease of presentation. However, all results extend in a straightforward manner to the case of vectorial functions ranging in $\mathbb{R}^M$ with $M \geq 2$.

1.7 Sobolev spaces of periodic functions

We focus on space periodic functions defined on the flat torus $\mathbb{T}^N$. Although all spaces we shall deal with are real, it is convenient to introduce the complex trigonometric polynomials

$$e_{\mathbf{m}}(x) = \exp(\mathrm{i}\mathbf{m} \cdot \pi x), \quad \mathbf{m} = [m_1, \dots, m_N] \in \mathbb{Z}^N.$$

The space $\mathcal{D}'(\mathbb{T}^N)$ is defined as the space of continuous linear forms on $\mathcal{D}(\mathbb{T}^N) = C_c^\infty(\mathbb{T}^N) = C^\infty(\mathbb{T}^N)$. The vector-valued form $\mathcal{D}'(\mathbb{T}^N; \mathbb{R}^M)$ may be defined analogously. Each distribution $v \in \mathcal{D}'(\mathbb{T}^N)$ can be identified with the infinite sequence of its *Fourier coefficients*, as described by

$$a_{\mathbf{m}}[v] = \frac{1}{(2\pi)^N} \langle v, \overline{e}_{\mathbf{m}} \rangle, \quad \text{formally } v \approx \sum_{\mathbf{m} \in \mathbb{Z}^N} a_{\mathbf{m}}[v] e_{\mathbf{m}},$$

where $\overline{e}_{\mathbf{m}}$ is the complex conjugate.

1.7.1 Hilbertian structure

The *Sobolev spaces* $W^{k,2}(\mathbb{T}^N)$ of periodic functions having derivatives up to the order k in $L^2(\mathbb{T}^N)$ can be characterized as $v \in \mathcal{D}'(\mathbb{T}^N)$ such that

$$\|v\|^2_{W^{k,2}(\mathbb{T}^N)} = \sum_{\mathbf{m} \in \mathbb{Z}^N} (|\mathbf{m}| + 1)^{2k} a^2_{\mathbf{m}}[v] < \infty. \tag{1.3}$$

The definition can be used even for a general exponent $k \in \mathbb{R}$. In particular, we have $(W^{k,2}(\mathbb{T}^N))^* = W^{-k,2}(\mathbb{T}^N)$ for any $k \in \mathbb{R}$. This identification corresponds to the Gelfand triple

$$W^{k,2}(\mathbb{T}^N) \hookrightarrow L^2(\mathbb{T}^N) \approx (L^2(\mathbb{T}^N))^* \hookrightarrow W^{-k,2}(\mathbb{T}^N), \quad k \geq 0,$$

where L^2 has been identified with its dual via Riesz isometry.

The spaces $W^{k,2}$ are separable Hilbert spaces endowed with the scalar product. We have

$$\langle v, w \rangle = \sum_{\mathbf{m} \in \mathbb{Z}^N} (|\mathbf{m}| + 1)^{2k} a_{\mathbf{m}}[v] \overline{a}_{\mathbf{m}}[w].$$

In accordance with Theorem 1.6.3 and Theorem 1.6.4, we have the compact embedding

$$W^{k,2}(\mathbb{T}^N) \overset{c}{\hookrightarrow} C(\mathbb{T}^N) \quad \text{whenever } k > \frac{N}{2},$$

whence we have

$$L^1(\mathbb{T}^N) \overset{c}{\hookrightarrow} W^{-k,2}(\mathbb{T}^N), \quad k > \frac{N}{2}. \tag{1.4}$$

We may also consider time-dependent (periodic) functions defined on the $(N+1)$-dimensional torus

$$\mathbb{T}_L^{N+1} = [-L, L]|_{\{-L,L\}} \times \mathbb{T}^N.$$

We summarize:

- The spaces $W^{k,2}(\mathbb{T}^N)$, $W^{k,2}(\mathbb{T}_L^{N+1})$ are separable Hilbert spaces, in particular Polish spaces.
- If $X \hookrightarrow W^{k,2}(\mathbb{T}^N)$ or $X \hookrightarrow W^{k,2}(\mathbb{T}_L^{N+1})$ with continuous embedding, then X is a completely regular Hausdorff space (Tikhonov space). Moreover, X admits a countable family of continuous functions separating points, namely

$$f_{\mathbf{m}}[v] = a_{\mathbf{m}}[v], \quad \mathbf{m} \in \mathbb{Z}^N.$$

1.7.2 L^p-structure

We start with the presentation of a combination of DeLeeuw's theorem on Fourier multipliers on $\mathbb{T}^N$ (see Stein [Ste70, Chapter 7, Theorem 3.8]) and the Hörmander–Mikhlin theorem (see Stein [Ste70, Chapter 4, Theorem 3]).

Theorem 1.7.1. *Let $\mathbf{M} \in L^\infty(\mathbb{R}^N)$ possess classical derivatives up to order $[N/2]+1$ in $\mathbb{R}^N \setminus \{0\}$ such that*

$$|\partial_\alpha \mathbf{M}(\xi)| \le c_\alpha |\xi|^{-|\alpha|}, \quad |\xi| \neq 0, \ |\alpha| \le [N/2]+1.$$

Then the operator $\mathcal{L}$, since we know

$$\mathcal{L}[v] = \sum_{|\mathbf{m}| \in \mathbb{Z}^N} \mathbf{M}(\mathbf{m}) a_{\mathbf{m}}[v] e_{\mathbf{m}},$$

is bounded on $L^p(\mathbb{T}^N)$, $1 < p < \infty$.

Consider the projection operator

$$\Pi_{\mathbf{M}} : W^{k,2}(\mathbb{T}^N) \to L^2(\mathbb{T}^N) \quad \text{defined as } \Pi_{\mathbf{M}}[v] = \sum_{|m_i| \le M_i, \ i=1,\dots,N} a_{\mathbf{m}}[v] e_{\mathbf{m}}.$$

In accordance with Theorem 1.7.1, $\Pi_{\mathbf{M}}$ is bounded as an operator on $L^p(\mathbb{T}^N)$, $1 < p < \infty$. Moreover (see Weisz [Wei12, Theorem 4.1]), we have

$$\|\Pi_{\mathbf{M}}[v]\|_{L^p_x} \le c_p \|v\|_{L^p_x}, \quad \text{and} \quad \Pi_{\mathbf{M}}[v] \to v \quad \text{in } L^p(\mathbb{T}^N) \text{ as } \min_i\{M_i\} \to \infty. \tag{1.5}$$

1.7.3 Regularization by convolution kernels

Let $\theta_\delta^x \in C^\infty(\mathbb{T}^N)$ be a family of regularizing kernels. More specifically,

$$\theta_\delta^x(x) = \frac{1}{\delta^N}\theta\left(\frac{x}{\delta}\right), \quad \theta \in C_c^\infty((-1,1)^N), \quad \theta(x) = \theta(|x|), \quad \int_{\mathbb{T}^N} \theta(x) = 1. \tag{1.6}$$

For $v \in \mathcal{D}'(\mathbb{T}^N)$, we define its regularization $[v]_{x,\delta}$ as the convolution

$$[v]_{x,\delta}(x) = v * \theta_\delta^x \equiv \langle v, \theta_\delta^x(x - \cdot)\rangle.$$

The following results can be found in Amann [Ama95, Chapter III.4] or Brezis [Bre83, Chapter IV.4]:

- If $v \in L^1(\mathbb{T}^N)$, then we have $[v]_{x,\delta} \in C^\infty(\mathbb{T}^N)$.
- If $v \in L^p(\mathbb{T}^N)$, $1 \le p < \infty$, then

  $$\|[v]_{x,\delta}\|_{L_x^p} \le \|v\|_{L_x^p}$$

  and

  $$[v]_{x,\delta} \to v \quad \text{in } L^p(\mathbb{T}^N) \text{ as } \delta \to 0.$$

- If $v \in L^\infty(\mathbb{T}^N)$, then

  $$\|[v]_{x,\delta}\|_{L_x^\infty} \le \|v\|_{L_x^\infty}.$$

- If $v \in L^1(\mathbb{T}^N)$, then

  $$[v]_{x,\delta}(x) \to v(x) \quad \text{whenever } x \text{ is a Lebesgue point of } v.$$

  In particular,

  $$[v]_{x,\delta} \to v \quad \text{a.e. in } \mathbb{T}^N.$$

We recall that, for $v \in L^1(\mathcal{O};X)$, the Lebesgue points $x \in \mathcal{O}$ are characterized by the property

$$\frac{1}{|B_r(x)|}\int_{B_r(x)} \|v(y) - v(x)\|_X \,\mathrm{d}y \to 0 \quad \text{as } r \to 0,$$

where $B_r(x) \subset \mathcal{O}$ is a ball with radius r, centered at x.

The above concept may be extended to a larger class of generalized functions as long as the operation of convolution with a smooth kernel is well-defined, notably to the space of distributions; see Section 2.2.2.

1.8 Bochner spaces

In this section we present supplementary material for Bochner spaces. They can be seen as particular cases of the vector-valued functions introduced in Sections 1.1 and 1.4, where $\mathcal{O} = (0, T)$. These spaces are of crucial importance for time-dependent PDEs. Sometimes, it will be convenient to consider functions from Bochner spaces (depending on space and time) as space-time distributions in $\mathcal{D}'(\mathbb{R} \times \mathbb{T}^N)$ or even $\mathcal{D}'([-L, L]|_{\{-L,L\}} \times \mathbb{T}^N)$, defined on the space-time torus

$$\mathbb{T}_L^{N+1} = [-L, L]|_{\{-L,L\}} \times \mathbb{T}^N,$$

extending them conveniently outside the interval I. Similarly to (1.3) we have the embedding

$$L^1(0, T; L^1(\mathbb{T}^N)) \overset{c}{\hookrightarrow} W^{-k,2}(\mathbb{T}_L^{N+1}), \quad k > \frac{N+1}{2}, \, L \geq T. \tag{1.7}$$

1.8.1 Time regularity

Let X be a separable Banach space. For $u \in L^1(0, T; X)$ we consider the distribution

$$C_c^\infty((0, T)) \to X, \quad \phi \mapsto \int_0^T u(t)\phi'(t)\,\mathrm{d}t.$$

Let Y be a Banach space with $X \hookrightarrow Y$ continuously. If there is $v \in L^1(0, T; Y)$ such that

$$\int_0^T u(t)\phi'(t)\,\mathrm{d}t = -\int_0^T v(t)\phi(t)\,\mathrm{d}t \quad \text{for all } \phi \in C_c^\infty((0, T)),$$

then we say that v is the weak derivative of u in Y and write $v = \partial_t u$. The space $W^{1,p}(0, T; X)$ consists of those functions from $L^p(0, T; X)$ having weak derivatives in $L^p(0, T; X)$. It is a Banach space with the norm

$$\|u\|_{W^{1,p}(0,T;X)}^p := \|u\|_{L^p(0,T;X)}^p + \|\partial_t u\|_{L^p(0,T;X)}^p.$$

Obviously this can be iterated to define the spaces $W^{k,p}(0, T; X)$, $k \in \mathbb{N}$.

In order to study the time regularity of functions from Bochner spaces, we recall the concept of continuity introduced in Section 1.1.

Definition 1.8.1. Let X be a Banach space with norm $\|\cdot\|_X$, $T > 0$ and $\alpha \in (0, 1]$. Then:
- $C([0, T]; X)$ denotes the set of functions $u : [0, T] \to X$ being continuous with respect to the norm topology, i.e.,

  $$u(t_k) \to u(t_0) \quad \text{in } X,$$

  for any sequence $(t_k)_{k\in\mathbb{N}} \subset [0, T]$ with $t_k \to t_0$.

- $C_w([0,T];X)$ denotes the set of functions $u : [0,T] \to X$ being continuous with respect to the weak topology, i.e.,

$$u(t_k) \rightharpoonup u(t_0) \quad \text{in } X,$$

for any sequence $(t_k)_{k \in \mathbb{N}} \subset [0,T]$ with $t_k \to t_0$. Equivalently, we may say that u belongs to $C_w([0,T];X)$ if the scalar functions $t \mapsto \langle x^*, u(t,\cdot)\rangle$ belong to $C([0,T])$, for any $x^* \in X^*$.
- $C^\alpha([0,T];X)$ denotes the set of functions $u : [0,T] \to X$ being α-Hölder continuous with respect to the norm topology, i.e.,

$$\sup_{t,s \in [0,T]; t \neq s} \frac{\|u(t) - u(s)\|_X}{|t-s|^\alpha} < \infty.$$

Obviously, we have the inclusions

$$C^\alpha([0,T];X) \subset C([0,T];X) \subset C_w([0,T];X),$$

for any $\alpha \in (0,1]$.

We introduce *convergence* in $C_w([0,T];X)$ by stating

$$v_n \to v \text{ in } C_w([0,T];X) \quad \text{if } \sup_{t \in [0,T]} |\langle x^*, v_n - v\rangle_{X^*,X}| \to 0 \; \forall x^* \in X^*.$$

If the space X is separable and reflexive, then the unit ball $B_X \subset X$ is a metrizable compact set and the above convergence generates a metric topology on $C_w([0,T];B_X)$ in the sense specified in Section 1.1.

1.8.2 Compact embeddings

The following theorem shows how to obtain compactness in Bochner spaces. The original version was developed by Aubin and Lions (see Aubin [Aub63], Lions [Lio69, Section 1.5], or the survey paper by Simon [Sim86]).

Theorem 1.8.2. *Let (V,X,Y) be a triple of separable and reflexive Banach spaces such that the embedding $V \hookrightarrow X$ is compact and the embedding $X \hookrightarrow Y$ is continuous. Then the embedding*

$$\{u \in L^p(0,T;V) : \; \partial_t u \in L^p(0,T;Y)\} \hookrightarrow L^p(0,T;X)$$

is compact for $1 < p < \infty$.

In the context of stochastic PDEs we will be confronted with functions having only fractional derivatives in time. We define for $p \in (1,\infty)$ and $\alpha \in (0,1)$ the norm

$$\|u\|^p_{W^{\alpha,p}(0,T;X)} := \|u\|^p_{L^p(0,T;X)} + \int_0^T \int_0^T \frac{\|u(\sigma_1) - u(\sigma_2)\|^p_X}{|\sigma_1 - \sigma_2|^{1+\alpha p}} \,\mathrm{d}\sigma_1 \,\mathrm{d}\sigma_2.$$

The space $W^{\alpha,p}(0,T;X)$ is now defined as the subspace of $L^p(0,T;X)$ consisting of those functions having finite $W^{\alpha,p}(0,T;X)$-norm. It can be shown that this is a complete space and we have $W^{1,p}(0,T;X) \subset W^{\alpha,p}(0,T;X) \subset L^p(0,T;X)$. The following variant of Theorem 1.8.2 holds (see Flandoli–Gątarek [FG95, Theorem 2.1]).

Theorem 1.8.3. *Let (V,X,Y) be a triple of separable and reflexive Banach spaces such that the embedding $V \hookrightarrow X$ is compact and the embedding $X \hookrightarrow Y$ is continuous. Then the embedding*

$$L^p(0,T;V) \cap W^{\alpha,p}(0,T;Y) \hookrightarrow L^p(0,T;X)$$

is compact for $1 < p < \infty$ and $0 < \alpha < 1$.

Using the continuous embedding $C^\alpha([0,T],Y) \hookrightarrow W^{\alpha,p}(0,T;Y)$, we obtain the following.

Corollary 1.8.4. *Let (V,X,Y) be a triple of separable and reflexive Banach spaces such that the embedding $V \hookrightarrow X$ is compact and the embedding $X \hookrightarrow Y$ is continuous. Then the embedding*

$$L^p(0,T;V) \cap C^\alpha([0,T];Y) \hookrightarrow L^p(0,T;X)$$

is compact for $1 < p < \infty$ and $0 < \alpha < 1$.

We will use Corollary 1.8.4 at various occasions in order to obtain compactness for stochastic PDEs. Typically, solutions are Hölder continuous in a negative Sobolev space, so we have $Y = W^{-\ell,2}(\mathbb{T}^N)$ for some $\ell \in \mathbb{N}$. On the other hand, these functions also belong to $L^p(0,T;L^p(\mathbb{T}^N))$ (or $L^p(0,T;W^{1,p}(\mathbb{T}^N))$) for some $p \in (1,\infty)$. This means we have $V = L^p(\mathbb{T}^N)$ (or $V = W^{1,p}(\mathbb{T}^N)$). Corollary 1.8.4 applies with $X = W^{-1,p}(\mathbb{T}^N)$ (or $X = L^p(\mathbb{T}^N)$).

In view of the applications to compressible Navier–Stokes equations, we have to deal with weakly continuous functions. The following result is appropriate to handle this situation.

Theorem 1.8.5. *Let $\alpha \geq 0$, $1 < p < \infty$, and $\ell \in \mathbb{R}$. Then*

$$L^\infty(0,T;L^p(\mathbb{T}^N)) \cap C^\alpha([0,T];W^{\ell,2}(\mathbb{T}^N)) \hookrightarrow C_w([0,T];L^p(\mathbb{T}^N)).$$

If $\alpha > 0$, then the embedding is sequentially compact, meaning any sequence

$$(v_n)_{n\in\mathbb{N}} \quad \textit{bounded in } L^\infty(0,T;L^p(\mathbb{T}^N)) \cap C^\alpha([0,T];W^{\ell,2}(\mathbb{T}^N))$$

contains a subsequence $(v_{n_k})_{k\in\mathbb{N}}$ such that

$$v_{n_k} \to v \quad \textit{in } C_w([0,T];L^p(\mathbb{T}^N)).$$

Proof. First we have to show that

$$\langle x^*, v(t,\cdot)\rangle \in C([0,T]) \quad \text{for any } x^* \in L^{p'}(\mathbb{T}^N),$$

whenever $v \in L^\infty(0,T;L^p(\mathbb{T}^N)) \cap C^\alpha([0,T];W^{\ell,2}(\mathbb{T}^N))$ and p' is the conjugate exponent of p. As the norm in L^p is weakly lower semi-continuous, we deduce $v(t,\cdot) \in B(r)$ for *any* t, where $B(r)$ is a ball in $L^p(\mathbb{T}^N)$ of suitable radius $r > 0$. The collection of trigonometric polynomials $(e_{\mathbf{m}})_{\mathbf{m}\in\mathbb{Z}^N}$ defined in Section 1.7 generates a basis in $W^{\ell,2}(\mathbb{T}^N)$ for any ℓ, and their finite linear combinations are dense in $L^q(\mathbb{T}^N)$ for any $1 \le q < \infty$, in particular, in $L^{p'}(\mathbb{T}^N)$. Consequently,

$$\begin{aligned}
&|\langle x^*, v(t,\cdot)\rangle - \langle x^*, v(s,\cdot)\rangle| \\
&\le \left|\left\langle \sum_{|\mathbf{m}|\le M} \beta_{\mathbf{m}} e_{\mathbf{m}}, v(t,\cdot) - v(s,\cdot)\right\rangle\right| + \left|\left\langle x^* - \sum_{|\mathbf{m}|\le M} \beta_{\mathbf{m}} e_{\mathbf{m}}, v(t,\cdot) - v(s,\cdot)\right\rangle\right| \\
&\le \left|\left\langle \sum_{|\mathbf{m}|\le M} \beta_{\mathbf{m}} e_{\mathbf{m}}, v(t,\cdot) - v(s,\cdot)\right\rangle\right| + r\left\| x^* - \sum_{|\mathbf{m}|\le M} \beta_{\mathbf{m}} e_{\mathbf{m}}\right\|_{L_x^{p'}} \\
&\le c(M,\ell)\|v\|_{C_t^\alpha W_x^{\ell,2}} |t-s|^\alpha + r\left\| x^* - \sum_{|\mathbf{m}|\le M} \beta_{\mathbf{m}} e_{\mathbf{m}}\right\|_{L_x^{p'}},
\end{aligned} \tag{1.8}$$

where the last term can be made small uniformly for all $s,t \in [0,T]$ by taking suitable $\beta_{\mathbf{m}}$ and M large enough.

If $\alpha > 0$ we may apply the abstract Arzelà–Ascoli theorem (Theorem 1.1.1). The ball $B(r)$ is indeed weakly sequentially compact and the desired equi-continuity of the sequence $(v_n)_{n\in\mathbb{N}}$ follows easily from (1.8). □

1.8.3 Regularization by convolution kernels

This section is dedicated to the regularization of time-dependent functions. In order to avoid problems related to progressive measurability (which typically arise in our applications to stochastic PDEs) we regularize functions backwards in time. Consequently, it is convenient to extend them appropriately for $t \le 0$. For $v \in L^1(-1,T;X)$, where X is a Banach space, we consider the time regularization

$$[v]_{t,\delta}(t) = v * \theta_\delta^t(\cdot - \delta) = \int_{-\infty}^{\infty} \theta_\delta^t(t-\delta-s)v(s)\,\mathrm{d}s.$$

Here, the regularizing kernel is a function of t satisfying (1.6) for $N = 1$.

Referring again to Amann [Ama95, Chapter III.4] and Brezis [Bre83, Chapter IV.4], we have:

- If $v \in L^1(-1,T;X)$, then we have $[v]_{t,\delta} \in C^\infty((-1,T);X)$.
- If $v \in L^p(-1,T;X)$, $1 \le p < \infty$, then

  $$\|[v]_{t,\delta}\|_{L_t^p X} \le \|v\|_{L_t^p X}$$

  and

  $$[v]_{t,\delta} \to v \quad \text{in } L^p(0,T;X) \text{ as } \delta \to 0.$$

- If $v \in L^\infty(-1,T;X)$, then

  $$\|[v]_{t,\delta}\|_{L_t^\infty X} \le \|v\|_{L_t^\infty X}.$$

- If $v \in L^1(-1,T;X)$, then

  $$[v]_{t,\delta}(t) \to v(t) \quad \text{in } X \text{ whenever } t \text{ is a Lebesgue point of } v.$$

  In particular,

  $$[v]_{t,\delta} \to v \quad \text{in } X \text{ as } \delta \to 0 \text{ a.e. in } (0,T).$$

2 Elements of stochastic analysis

We introduce the basic stochastic framework used in this book. We present only a selection of the principal concepts and ideas of stochastic analysis, as the reader is expected to be familiar with the basic notions of probability theory. Part of the results presented in this chapter can be found in the literature. The classical and widely used monographs include for instance Karatzas–Shreve [KS91] and Da Prato–Zabczyk [DPZ92] and we invite the reader to consult these textbooks for further details. In addition, we include a number of original results needed for the study of the compressible Navier–Stokes system later on.

To be more precise, in Section 2.2 we introduce the notion of random distributions (see Definition 2.2.1). It is a generalization of stochastic processes which allows one to treat random elements in the weakest possible topology, namely, the weak-$*$ topology of the space of space-time distributions $\mathcal{D}'(I \times \mathbb{T}^N)$, where $I \subset \mathbb{R}$. For the sake of simplicity, the results will be stated only for $I = \mathbb{R}$, with obvious modifications for a general interval I. In the subsequent sections we show how the classical theory of Itô's stochastic integration and its applications to stochastic PDEs can be formulated in the context of random distributions. We believe that this new perspective is interesting in its own right and will prove useful also for researchers working on other models in fluid dynamics or other fields.

2.1 Random variables and stochastic processes

Throughout the book $(\Omega, \mathfrak{F}, \mathbb{P})$ denotes a complete *probability space* with a σ-field $\mathfrak{F}$ and a probability measure $\mathbb{P}$. The probability space $([0,1], \overline{\mathfrak{B}([0,1])}, \mathfrak{L})$, where $\mathfrak{L}$ denotes the Lebesgue measure, is called *standard*. Here, $\overline{\mathfrak{B}}$ denotes the completion $\mathfrak{B}$ and $\mathfrak{L}$ denotes the one-dimenisonal Lebesgues measure. A *filtration* is a non-decreasing family of sub-σ-fields of $\mathfrak{F}$, that is, $\mathfrak{F}_t \subset \mathfrak{F}$ for all $t \geq 0$ and $\mathfrak{F}_s \subset \mathfrak{F}_t$ whenever $s \leq t$. We say that the filtration $(\mathfrak{F}_t)_{t\geq 0}$ satisfies the *usual conditions*, provided it is complete and right-continuous. In other words,

$$\{N \in \mathfrak{F};\ \mathbb{P}(N) = 0\} \subset \mathfrak{F}_0, \quad \mathfrak{F}_t = \mathfrak{F}_{t+} := \bigcap_{s>t} \mathfrak{F}_s \quad \text{for all } t \geq 0.$$

The multiple $(\Omega, \mathfrak{F}, (\mathfrak{F}_t)_{t\geq 0}, \mathbb{P})$ is then called a *stochastic basis* or a *filtered probability space*.

We proceed with basic definitions concerning random variables.

Definition 2.1.1. Let $(X, \mathcal{A})$ be a measurable space. An X-valued *random variable* is a measurable mapping $\mathbf{U} : (\Omega, \mathfrak{F}) \to (X, \mathcal{A})$. We denote by $\sigma(\mathbf{U})$ the smallest σ-field with respect to which $\mathbf{U}$ is measurable. More precisely,

$$\sigma(\mathbf{U}) := \{\{\omega \in \Omega;\ \mathbf{U}(\omega) \in A\};\ A \in \mathcal{A}\}$$

https://doi.org/10.1515/9783110492552-002

and $\sigma(\mathbf{U}) \subset \mathfrak{F}$. In addition, we denote by $\mathcal{L}[\mathbf{U}]$ or also $\mathcal{L}_X[\mathbf{U}]$ the law of $\mathbf{U}$ on X, that is, $\mathcal{L}[\mathbf{U}]$ is the pushforward probability measure on $(X, \mathcal{A})$ given by

$$\mathcal{L}[\mathbf{U}](A) = \mathbb{P}(\mathbf{U} \in A), \quad A \in \mathcal{A}.$$

Definition 2.1.2. Let $(X, \mathcal{A})$ be a measurable space. We say that two X-valued random variables $\mathbf{U}$ and $\mathbf{V}$ are *equal in law*, if $\mathcal{L}[\mathbf{U}]$ and $\mathcal{L}[\mathbf{V}]$ coincide.

We stress that the assumptions on the state space X will vary in the sequel. Most of the notions presented below require a topology on X and therefore we assume that X is a topological space equipped with a Borel σ-field. In addition, it is convenient that the topology on X is completely determined by the family of continuous functions. Specifically, we consider *Tikhonov spaces*, meaning completely regular and Hausdorff. As a matter of fact, we deal almost exclusively with *topological vector spaces*, in particular with the class of locally convex topological vector spaces. These are vector spaces equipped with a topology that renders the vector addition as well as the scalar multiplication continuous and, in addition, the topology is generated by a family of seminorms $(p_\gamma)_{\gamma\in\Gamma}$.

Many concepts in the theory of stochastic processes require a certain *uniformity* of the topology. Simplifications occur in the case of *Polish spaces*, that is, separable spaces that are completely metrizable. This is also the common setting found in the literature. However, the delicate structure of the compressible Navier–Stokes system studied in the main body of this book naturally leads to spaces which are generally not metrizable, such as Banach spaces equipped with weak topology. Hence we will formulate the basic notions on probability theory in a wider generality. In particular, all spaces we shall deal with will admit a countable family of bounded continuous functions that separates points. Given such a family of continuous functions $(g_n)_{n\in\mathbb{N}}$ on X, we define an embedding

$$\jmath : X \to [-1,1]^{\aleph_0}, \quad \jmath(x) = \big(g_n(x)\big)_{n\in\mathbb{N}}.$$

Here, we have tacitly assumed that all functions g_n range in $(-1,1)$. Note that $[-1,1]^{\aleph_0}$ is a compact Polish space. This motivates the following definition.

Definition 2.1.3 (sub-Polish space). Let (X, τ) be a topological space such that there exists a countable family

$$\{g_n : X \to (-1,1);\ n \in \mathbb{N}\}$$

of continuous functions that separate points of X. Then (X, τ) is called a *sub-Polish* space.

The following characterization of equality in law will be frequently used in the sequel.

Lemma 2.1.4. *Let X be a Tikhonov topological space equipped with the Borel σ-field. Let* $\mathbf{U}$ *and* $\mathbf{V}$ *be X-valued random variables. Then $\mathcal{L}[\mathbf{U}] = \mathcal{L}[\mathbf{V}]$, provided*

$$\langle \mathcal{L}[\mathbf{U}], f \rangle = \langle \mathcal{L}[\mathbf{V}], f \rangle,$$

or, equivalently,

$$\mathbb{E}[f(\mathbf{U})] = \mathbb{E}[f(\mathbf{V})]$$

holds true for all $f \in C_b(X)$.

Although several function spaces we use in the book, notably the space of distributions, are not first countable, the convergence results are usually stated in terms of *sequences* rather than nets. The sequential language seems more adequate for describing the asymptotic behavior of stochastic processes and several results are simply only true for sequences. We proceed with various notions on the convergence of random variables. First, we introduce the almost sure convergence which corresponds to the almost everywhere convergence known from measure theory.

Definition 2.1.5. Let X be a topological space equipped with the Borel σ-field and let $\mathbf{U}$ and $\mathbf{U}_n$, $n \in \mathbb{N}$, be X-valued random variables on $(\Omega, \mathfrak{F}, \mathbb{P})$. We say that $\mathbf{U}_n$ converges to $\mathbf{U}$ *almost surely*, provided

$$\mathbb{P}\Big(\omega \in \Omega;\ \lim_{n\to\infty} \mathbf{U}_n(\omega) = \mathbf{U}(\omega)\Big) = 1.$$

In other words, there exists a set of full probability $\Omega^* \subset \Omega$ such that, for every $\omega \in \Omega^*$, the following statement holds: if $\mathcal{U} \subset X$ is an open neighborhood of $\mathbf{U}(\omega)$, then there exists $n_0 \in \mathbb{N}$ such that, for every $n \geq n_0$, we have $\mathbf{U}_n(\omega) \in \mathcal{U}$.

Next, we define the probabilistic analogue of convergence in measure. To this end, we restrict ourselves to the case of topological vector spaces. Recall that, if X is a topological vector space, the topology on X is uniform. This means that any neighborhood $\mathcal{U}(x)$ of a point x takes the form $x + \mathcal{U} = \{x + y;\ y \in \mathcal{U}\}$, where $\mathcal{U}$ is a neighborhood of 0.

Definition 2.1.6. Let X be a topological vector space. Assume that $\mathbf{U}$ and $\mathbf{U}_n$, $n \in \mathbb{N}$, are X-valued random variables on $(\Omega, \mathfrak{F}, \mathbb{P})$. We say that $\mathbf{U}_n$ converges to $\mathbf{U}$ *in probability* if, for every $\mathcal{U} \subset X$ which is an open neighborhood of 0, we have

$$\lim_{n\to\infty} \mathbb{P}(\omega \in \Omega;\ \mathbf{U}_n(\omega) \notin \mathbf{U}(\omega) + \mathcal{U}) = 0. \tag{2.1}$$

Remark 2.1.7. As pointed out, the definition extends easily to *uniform spaces* but we do not need this generality here.

Remark 2.1.8. In locally convex topological vector spaces equipped with a family of semi-norms $(p_\gamma)_{\gamma\in\Gamma}$, condition (2.1) rewrites as follows: for every $\varepsilon > 0$ and $\gamma \in \Gamma$,

$$\lim_{n\to\infty} \mathbb{P}(\omega \in \Omega; p_\gamma(\mathbf{U}_n(\omega) - \mathbf{U}(\omega)) > \varepsilon) = 0.$$

In other words, for all $\varepsilon > 0$ and $\gamma \in \Gamma$, there exists $n_0 \in \mathbb{N}$ such that, for all $n \geq n_0$,

$$\mathbb{P}(\omega \in \Omega; p_\gamma(\mathbf{U}_n(\omega) - \mathbf{U}(\omega)) > \varepsilon) < \varepsilon.$$

Finally, let us introduce convergence in law. Note that here it is not necessary for the random variables to be defined on the same probability space.

Definition 2.1.9. Let X be a Tikhonov topological space equipped with the Borel σ-field and let $\mathbf{U}$ and $\mathbf{U}_n$, $n \in \mathbb{N}$, be X-valued random variables defined on $(\Omega, \mathfrak{F}, \mathbb{P})$ and $(\Omega_n, \mathfrak{F}_n, \mathbb{P}_n)$, $n \in \mathbb{N}$, respectively. We say that $\mathbf{U}_n$ converges to $\mathbf{U}$ *in law*, provided the law $\mathcal{L}[\mathbf{U}_n]$ converges to $\mathcal{L}[\mathbf{U}]$ weakly-$*$ in the sense of probability measures on X. More precisely,

$$\lim_{n\to\infty} \langle \mathcal{L}[\mathbf{U}_n], f\rangle = \langle \mathcal{L}[\mathbf{U}], f\rangle,$$

or, equivalently,

$$\lim_{n\to\infty} \mathbb{E}[f(\mathbf{U}_n)] = \mathbb{E}[f(\mathbf{U})],$$

for every $f \in C_b(X)$.

Note carefully that sequential compactness, sequential continuity and other topological concepts in general do not coincide with compactness, continuity, etc., unless the underlying space is metrizable. Fortunately, we shall almost exclusively deal with families of random variables with uniformly *tight* laws.

Definition 2.1.10. Suppose that X is a topological space and that $\mathcal{A}$ is a σ-field containing the topology. We say that a collection $\mathcal{M}$ of probability measures on $(X, \mathcal{A})$ is *tight* if, for any given $\varepsilon \in (0, 1)$, there exists $K_\varepsilon \in \mathcal{A}$ compact, such that, for all $\mu \in \mathcal{M}$, we have

$$\mu(K_\varepsilon) > 1 - \varepsilon.$$

A sequence $(\mathbf{U}_n)_{n\in\mathbb{N}}$ of random variables in a topological space is uniformly tight if, for any $\varepsilon > 0$, there is a compact set K_ε such that

$$\mathbb{P}(\mathbf{U}_n \notin K_\varepsilon) < \varepsilon \quad \text{uniformly for all } n \in \mathbb{N}.$$

In particular, if X is sub-Polish, then the X-topology on K_ε is metrizable and all topological concepts coincide with their sequential counterparts on K_ε.

We continue our discussion with stochastic processes, that is, families of random variables parametrized by a continuous parameter that represents time.

Definition 2.1.11. Let $(X, \mathcal{A})$ be a measurable space. An X-valued *stochastic process* is a collection of random variables $\mathbf{U} = \{\mathbf{U}(t); t \in [0, \infty)\}$ taking values in $(X, \mathcal{A})$. That is, for all $t \in [0, \infty)$, the mapping $\mathbf{U}(t) : (\Omega, \mathfrak{F}) \to (X, \mathcal{A})$ is measurable.

Definition 2.1.12. An X-valued stochastic process $\mathbf{U} = \{\mathbf{U}(t); t \in [0, \infty)\}$ is called *measurable*, provided the mapping

$$(\Omega \times [0, \infty), \mathfrak{F} \otimes \mathcal{B}([0, \infty))) \to (X, \mathcal{A}), \quad (\omega, t) \mapsto \mathbf{U}(\omega, t),$$

is measurable.

Definition 2.1.13. Let $(\mathfrak{F}_t)_{t\geq 0}$ be a filtration on $(\Omega, \mathfrak{F})$. An X-valued stochastic process $\mathbf{U} = \{\mathbf{U}(t); t \in [0, \infty)\}$ is called $(\mathfrak{F}_t)$-*adapted*, provided that, for every $t \in [0, \infty)$, the random variable $\mathbf{U}(t)$ is $\mathfrak{F}_t$-measurable.

Remark 2.1.14. It follows immediately from Definition 2.1.13 that a stochastic process $\mathbf{U}$ is always adapted to its $\mathbb{P}$-augmented canonical filtration, given by

$$\sigma_t[\mathbf{U}] := \bigcap_{s>t} \sigma(\sigma(\mathbf{U}(r); 0 \leq r \leq s) \cup \{N \in \mathfrak{F}; \mathbb{P}(N) = 0\}), \quad t \geq 0.$$

We proceed with the definition of independence.

Definition 2.1.15. Let $(\Omega, \mathfrak{F}, \mathbb{P})$ be a probability space. We say that two events $A, B \in \mathfrak{F}$ are *independent*, provided their joint probability is equal to the product of their probabilities, that is, $\mathbb{P}(A \cap B) = \mathbb{P}(A)\mathbb{P}(B)$.

We say that two families of events $(A_{\gamma_1})_{\gamma_1 \in \Gamma_1}, (B_{\gamma_2})_{\gamma_2 \in \Gamma_2} \subset \mathfrak{F}$ are *independent*, provided A_{γ_1} and B_{γ_2} are independent for all $\gamma_1 \in \Gamma_1$, and $\gamma_2 \in \Gamma_2$.

We say that two random variables $\mathbf{U}$, $\mathbf{V}$ are *independent*, provided the families of events $\sigma(\mathbf{U})$ and $\sigma(\mathbf{V})$ are independent.

Definition 2.1.16. Let $(\mathfrak{F}_t)_{t\geq 0}$ be a filtration on $(\Omega, \mathfrak{F})$. An X-valued stochastic process $\mathbf{U} = \{\mathbf{U}(t); t \in [0, \infty)\}$ is called $(\mathfrak{F}_t)$-*progressively measurable*, provided that, for every $t \in [0, \infty)$, the mapping

$$(\Omega \times [0, t], \mathfrak{F}_t \otimes \mathcal{B}([0, t])) \to (X, \mathcal{A}), \quad (\omega, s) \mapsto \mathbf{U}(\omega, s)$$

is measurable.

It follows immediately that every $(\mathfrak{F}_t)$-progressively measurable stochastic process is measurable and $(\mathfrak{F}_t)$-adapted. The converse is not always true. Nonetheless, the following classical results hold; see, e.g., Dellacherie–Meyer [DM75, Théorème IV.30].

Proposition 2.1.17. *Let X be a Polish space. If an X-valued stochastic process* $\mathbf{U}$ *is measurable and $(\mathfrak{F}_t)$-adapted, then it has an $(\mathfrak{F}_t)$-progressively measurable modification, that is, there exists an $(\mathfrak{F}_t)$-progressively measurable X-valued stochastic process* $\mathbf{V}$ *such that, for all $t \in [0,\infty)$,*

$$\mathbb{P}\big(\mathbf{U}(t) = \mathbf{V}(t)\big) = 1.$$

For stochastic processes with left- or right-continuous trajectories even more can be proved; cf. Karatzas–Shreve [KS91, Proposition 1.13].

Proposition 2.1.18. *Let X be a Polish space. If an X-valued stochastic process is $(\mathfrak{F}_t)$-adapted and has left- or right-continuous trajectories, then it is $(\mathfrak{F}_t)$-progressively measurable.*

A distinguished class of stochastic processes is given by martingales. To this end, let us introduce the notion of conditional expectation; see, e.g., Da Prato–Zabczyk [DPZ92, Proposition 1.10].

Proposition 2.1.19. *Assume that X is a separable Banach space. Let* $\mathbf{U}$ *be an integrable X-valued random variable defined on $(\Omega, \mathfrak{F}, \mathbb{P})$ and let $\mathfrak{G}$ be a σ-field contained in $\mathfrak{F}$. Then there exists a unique, up to a set of probability zero, integrable X-valued random variable* $\mathbf{Z}$*, measurable with respect to $\mathfrak{G}$, such that*

$$\mathbb{E}[\mathbf{1}_A \mathbf{U}] = \mathbb{E}[\mathbf{1}_A \mathbf{Z}] \quad \text{for any } A \in \mathfrak{G}.$$

Definition 2.1.20. The random variable $\mathbf{Z}$ constructed in Proposition 2.1.19 is called the *conditional expectation* of $\mathbf{U}$ given $\mathfrak{G}$ and will be denoted by $\mathbb{E}[\mathbf{U}|\mathfrak{G}]$.

Definition 2.1.21. Let $\mathbf{U}$ be an X-valued $(\mathfrak{F}_t)$-adapted stochastic process. We say that $\mathbf{U}$ is an $(\mathfrak{F}_t)$-*martingale*, provided $\mathbb{E}[\mathbf{U}(t)|\mathfrak{F}_s] = \mathbf{U}(s)$ $\mathbb{P}$-a.s. for all $0 \le s \le t < \infty$.

Remark 2.1.22. Note that, if $\mathbf{U}$ is an $(\mathfrak{F}_t)$-martingale, it follows from the definition of the conditional expectation that

$$\mathbb{E}[\mathbf{U}(t)] = \mathbb{E}\big[\mathbb{E}[\mathbf{U}(t)|\mathcal{F}_s]\big] = \mathbb{E}[\mathbf{U}(s)] \quad \text{for all } 0 \le s \le t < \infty.$$

Hence the expectation is constant in time.

The following result is a consequence of the so-called Doob–Meyer decomposition for martingales; see Karatzas–Shreve [KS91, Section 1.4] for more details.

Theorem 2.1.23. *Let* $\mathbf{U}$ *be a continuous L^2-integrable real-valued $(\mathfrak{F}_t)$-martingale, that is, $\mathbb{E}|\mathbf{U}(t)|^2 < \infty$ for all $t \in [0,\infty)$. Then there exists a unique stochastic process $\langle\langle \mathbf{U} \rangle\rangle$ such that:*

(1) $\langle\langle \mathbf{U} \rangle\rangle$ *is* $(\mathfrak{F}_t)$*-adapted and has* $\mathbb{P}$*-a.s. non-decreasing trajectories;*
(2) $\langle\langle \mathbf{U} \rangle\rangle(0) = 0$ $\mathbb{P}$*-a.s.;*
(3) $\mathbf{U}^2 - \langle\langle \mathbf{U} \rangle\rangle$ *is a continuous* $(\mathfrak{F}_t)$*-martingale.*

Definition 2.1.24. The stochastic process $\langle\langle \mathbf{U} \rangle\rangle$ constructed in Theorem 2.1.23 is called the *quadratic variation* of $\mathbf{U}$.

Definition 2.1.25. Let $\mathbf{U}, \mathbf{V}$ be stochastic processes satisfying the assumptions of Theorem 2.1.23. The process

$$\langle\langle \mathbf{U}, \mathbf{V} \rangle\rangle := \frac{1}{4}(\langle\langle \mathbf{U} + \mathbf{V} \rangle\rangle - \langle\langle \mathbf{U} - \mathbf{V} \rangle\rangle)$$

is called the *cross variation* of $\mathbf{U}, \mathbf{V}$.

It can be shown that the cross variation is the unique continuous $(\mathfrak{F}_t)$-adapted process starting from 0 such that

$$\mathbf{U}\mathbf{V} - \langle\langle \mathbf{U}, \mathbf{V} \rangle\rangle$$

is a continuous $(\mathfrak{F}_t)$-martingale.

It will be useful to introduce local martingales, a notion that generalizes that of martingales. First, we need the definition of a stopping time associated to a filtration $(\mathfrak{F}_t)_{t\geq 0}$.

Definition 2.1.26. Let $(\Omega, \mathfrak{F})$ be a measurable space equipped with a filtration $(\mathfrak{F}_t)_{t\geq 0}$. A random variable $\tau : \Omega \to [0, \infty]$ is called an $(\mathfrak{F}_t)$-*stopping time*, provided the event $\{\tau \leq t\}$ belongs to the σ-field $\mathfrak{F}_t$ for any $t \in [0, \infty)$.

Definition 2.1.27. Let $\mathbf{U}$ be an X-valued $(\mathfrak{F}_t)$-adapted stochastic process. We say that $\mathbf{U}$ is an $(\mathfrak{F}_t)$-*local martingale*, provided there exists an increasing sequence of stopping times $(\tau_n)_{n\in\mathbb{N}}$, $\tau_n \uparrow \infty$ a.s., such that the stopped process $\mathbf{U}^{\tau_n}(\cdot) := \mathbf{U}(\tau_n \wedge \cdot)$ is an $(\mathfrak{F}_t)$-martingale for all $n \in \mathbb{N}$.

As an important example of a continuous-time stochastic process and a martingale, we recall the definition of a Wiener process.

Definition 2.1.28. An $\mathbb{R}^m$-valued stochastic process W is called an $(\mathfrak{F}_t)$-*Wiener process*, provided:
(1) W is $(\mathfrak{F}_t)$-adapted;
(2) $W(0) = 0$ $\mathbb{P}$-a.s.;
(3) W has continuous trajectories: $t \mapsto W(t)$ is continuous $\mathbb{P}$-a.s.;
(4) W has independent increments: $W(t) - W(s)$ is independent of $\mathfrak{F}_s$ for all $0 \leq s \leq t < \infty$;

(5) W has Gaussian increments: $W(t) - W(s)$ is normally distributed with mean 0 and variance $(t-s)\mathbb{I}$ for all $0 \le s \le t < \infty$.

Remark 2.1.29. In view of Definition 2.1.15, point (4) in the definition of a Wiener process is to be understood as follows: $W(t) - W(s)$ is independent of $\mathfrak{F}_s$, provided the σ-field $\sigma(W(t) - W(s))$ is independent of $\mathfrak{F}_s$.

Remark 2.1.30. To prove the existence of a Wiener process defined on a suitable probability space is a classical problem and there are several approaches available in the literature. One possibility is to write down the finite-dimensional distributions and apply Daniell–Kolmogorov's consistency theorem; see Karatzas–Shreve [KS91, Section 2.2, Theorem 2.2].

Note that trajectories of a Wiener process are a.s. nowhere differentiable. However, it can be shown that they are a.s. Hölder continuous with exponent γ for every $\gamma \in (0, \frac{1}{2})$.

As an infinite-dimensional generalization, let us introduce a cylindrical Wiener process, which will play the role of a driving stochastic process for the Navier–Stokes system studied in the main body of the book.

Definition 2.1.31. Let $\mathfrak{U}$ be a separable Hilbert space with a complete orthonormal system $(e_k)_{k\in\mathbb{N}}$ and let $(W_k)_{k\in\mathbb{N}}$ be a sequence of mutually independent real-valued $(\mathfrak{F}_t)$-Wiener processes. The stochastic process W given by the formal expansion $W(t) = \sum_{k=1}^{\infty} e_k W_k(t)$ is called a *cylindrical* $(\mathfrak{F}_t)$-Wiener process.

Remark 2.1.32. In view of Remark 2.1.30, a cylindrical Wiener process can easily be constructed on some probability space using the formal expansion from Definition 2.1.31.

The reader can easily verify that, if the underlying Hilbert space $\mathfrak{U}$ is finite-dimensional, Definition 2.1.28 and Definition 2.1.31 lead to the same object. In addition, we point out that the expansion of W from Definition 2.1.31 is indeed formal in the sense that the infinite sum does not converge in any reasonable probabilistic sense as a random variable in $\mathfrak{U}$. Nevertheless, it is possible to construct an auxiliary space $\mathfrak{U}_0 \supset \mathfrak{U}$ such that the sum converges in $\mathfrak{U}_0$. More precisely, we define $\mathfrak{U}_0$ as

$$\mathfrak{U}_0 = \left\{ v = \sum_{k=1}^{\infty} \alpha_k e_k;\ \sum_{k=1}^{\infty} \frac{\alpha_k^2}{k^2} < \infty \right\},$$

endowed with the norm

$$\|v\|_{\mathfrak{U}_0}^2 = \sum_{k=1}^{\infty} \frac{\alpha_k^2}{k^2}, \quad v = \sum_{k=1}^{\infty} \alpha_k e_k.$$

The following result holds true and we refer to Da Prato–Zabczyk [DPZ92] for the proof.

Corollary 2.1.33. *Let W be a cylindrical Wiener process. Then trajectories of W are in $C_{\mathrm{loc}}([0,\infty);\mathfrak{U}_0)$ $\mathbb{P}$-a.s. Accordingly, the law of W, denoted by $\mathcal{L}[W]$, is supported on $C_{\mathrm{loc}}([0,\infty);\mathfrak{U}_0)$.*

Let us now recall Lévy's martingale characterization of a Wiener process adapted to our purposes. The proof can be found in Da Prato–Zabczyk [DPZ92, Theorem 4.4].

Theorem 2.1.34. *Let W be a continuous, $\mathfrak{U}_0$-valued stochastic process such that $W(0) = 0$ a.s. Then W is a cylindrical $(\mathfrak{F}_t)$-Wiener process in $\mathfrak{U}$ if and only if, for all $k \in \mathbb{N}$, the process $W_k := \langle W, e_k\rangle$ is a real-valued, square integrable $(\mathfrak{F}_t)$-martingale with quadratic variation $\langle\langle W_k\rangle\rangle(t) = t$, $t \in [0,\infty)$.*

As a consequence of the fact that a real-valued Wiener process is fully determined by its law, we deduce the same for a cylindrical Wiener process.

Lemma 2.1.35. *A cylindrical Wiener process is fully determined by its law. In other words, let $\mathcal{L}[W]$ be the law of a cylindrical Wiener process constructed in Corollary 2.1.33 and let B be a stochastic process in $\mathfrak{U}_0$ defined on some probability space $(\Omega,\mathfrak{F},\mathbb{P})$ and having the law $\mathcal{L}[W]$. Then B is a cylindrical Wiener process with respect to its canonical filtration, namely, there exists a collection of mutually independent real-valued Wiener processes $(B_k)_{k\in\mathbb{N}}$ on $(\Omega,\mathfrak{F},\mathbb{P})$ such that $B = \sum_{k=1}^{\infty} e_k B_k$.*

Proof. We observe that, due to equality of laws, B is a cylindrical Wiener process in $\mathfrak{U}$. Indeed, for all $k \in \mathbb{N}$, we have

$$\mathcal{L}[\langle B, e_k\rangle] = \mathcal{L}[\langle W, e_k\rangle] = \mathcal{L}[W_k].$$

Consequently $B_k := \langle B, e_k\rangle$ are real-valued, mutually independent Wiener processes defined on $(\Omega,\mathfrak{F},\mathbb{P})$ and the expansion $B = \sum_{k=1}^{\infty} e_k B_k$ follows. □

Corollary 2.1.36. *Let W be a cylindrical Wiener process on $(\Omega,\mathfrak{F},\mathbb{P})$ with respect to its canonical filtration $\mathfrak{F}_t := \sigma(W(r); 0 \le r \le t)$, $t \ge 0$. Assume that $(\mathfrak{G}_t)_{t\ge0}$ is a filtration on $(\Omega,\mathfrak{F},\mathbb{P})$ such that $\mathfrak{F}_t \subset \mathfrak{G}_t$ for any $t \ge 0$ and $(\mathfrak{G}_t)_{t\ge0}$ is non-anticipative with respect to W, that is, for any $t \ge 0$, the σ-field $\mathfrak{G}_t$ is independent of $\sigma(W(t+h) - W(t))$ for any $h > 0$. Then W is a cylindrical $(\mathfrak{G}_t)$-Wiener process.*

Proof. In view of the Lévy martingale characterization of the Wiener process, Theorem 2.1.34, we only need to show that W is a martingale relative to $(\mathfrak{G}_t)_{t\ge0}$. Due to the first assumption on $(\mathfrak{G}_t)_{t\ge0}$, W is $(\mathfrak{G}_t)$-adapted. In accordance with Definition 2.1.21

and Proposition 2.1.19, it suffices to note that, since $W(t) - W(s)$ is independent of $\mathfrak{G}_s$ for any $0 \le s \le t < \infty$, we have, for every $A \in \mathfrak{G}_s$,

$$\mathbb{E}[\mathbf{1}_A(W(t) - W(s))] = \mathbb{E}[\mathbf{1}_A]\,\mathbb{E}[W(t) - W(s)] = 0. \qquad \square$$

2.2 Random distributions

In the analysis of stochastic PDEs, the function $\mathbf{U} = \mathbf{U}(\omega, t, x)$ can be seen as a random variable taking values in some path space of functions depending on $t \in I \subset \mathbb{R}$, $x \in \mathbb{T}^N$, or it can be interpreted as a stochastic process $t \mapsto \mathbf{U}(t, \cdot)$ with values in an abstract functional space. If $\mathbf{U}$ is a random variable ranging in the Lebesgue space $L^p(I \times \mathbb{T}^N)$, the instantaneous value $\mathbf{U}(t, \cdot)$ is only defined modulo a set of times of zero measure in I and the interpretation of $\mathbf{U}$ as a stochastic process becomes delicate. To avoid this difficulty, we introduce the space of random distributions and replace distributions by families of suitable regularizations.

The discussion in this section relies on the classical theory of distributions and the reader is referred to the monograph by Duistermaat–Kolk [DK10] for a thorough introduction. Let $\mathcal{D}(I \times \mathbb{T}^N; \mathbb{R}^M)$ be the space of C^∞-functions with compact support ranging in $\mathbb{R}^M$, where $I \subset \mathbb{R}$. In this book, we basically focus on the three canonical cases $I = \mathbb{R}$, $I = (-\infty, T)$, and $I = (0, \infty)$. For the sake of simplicity, the results in this section will be stated for $I = \mathbb{R}$, with obvious modifications in the other cases. The space $\mathcal{D}'(\mathbb{R} \times \mathbb{T}^N; \mathbb{R}^M)$ is the dual to $\mathcal{D}(\mathbb{R} \times \mathbb{T}^N; \mathbb{R}^M)$ endowed with the weak-$*$ topology. The duality between $\mathcal{D}(\mathbb{R}\times\mathbb{T}^N; \mathbb{R}^M)$ and $\mathcal{D}'(\mathbb{R}\times\mathbb{T}^N; \mathbb{R}^M)$ will be denoted by $\langle \cdot, \cdot \rangle$. For notational simplicity we restrict ourselves to $M = 1$ in the sequel and we denote $\mathcal{D}(\mathbb{R} \times \mathbb{T}^N) = \mathcal{D}(\mathbb{R} \times \mathbb{T}^N; \mathbb{R})$ and similarly $\mathcal{D}'(\mathbb{R} \times \mathbb{T}^N) = \mathcal{D}'(\mathbb{R} \times \mathbb{T}^N; \mathbb{R})$. The multi-dimensional analogue of all the results below can be obtained easily from the fact that $\mathcal{D}'(\mathbb{R}\times\mathbb{T}^N; \mathbb{R}^M) = [\mathcal{D}'(\mathbb{R}\times\mathbb{T}^N)]^M$.

Recall that a linear form $\mathbf{U}$ defined on $\mathcal{D}(\mathbb{R} \times \mathbb{T}^N)$ belongs to $\mathcal{D}'(\mathbb{R} \times \mathbb{T}^N)$ if and only if

$$\begin{aligned} \langle \mathbf{U}, \varphi_n \rangle \to 0 \quad & \text{whenever } \operatorname{supp}[\varphi_n] \subset (-L, L) \times \mathbb{T}^N \text{ for some } L > 0, \\ \varphi_n \to 0 \quad & \text{in } C^k([-L, L] \times \mathbb{T}^N) \text{ for any } k \in \mathbb{N}_0. \end{aligned}$$

Note that he latter relation defines topology on $\mathcal{D}(\mathbb{R} \times \mathbb{T}^N)$. Regular functions $\mathbf{U} \in L^1_{\mathrm{loc}}(\mathbb{R} \times \mathbb{T}^N)$ are identified with distributions as

$$\langle \mathbf{U}, \varphi \rangle = \int_{\mathbb{R}\times\mathbb{T}^N} \mathbf{U} \cdot \varphi \,\mathrm{d}t\,\mathrm{d}x.$$

As the space of test functions $\mathcal{D}(\mathbb{R} \times \mathbb{T}^N)$ is separable, there is a countable family $(\varphi_n)_{n\in\mathbb{N}} \subset \mathcal{D}(\mathbb{R} \times \mathbb{T}^N)$ such that the continuous linear functions

$$\mathbf{U} \mapsto \langle \mathbf{U}, \varphi_n \rangle, \quad n \in \mathbb{N}, \quad \text{separate points in } \mathcal{D}'(\mathbb{R} \times \mathbb{T}^N).$$

We define a canonical projection

$$\jmath : \mathcal{D}'(\mathbb{R}\times\mathbb{T}^N) \to [-1,1]^{\aleph_0}, \quad \jmath(\mathbf{U}) = \left(\chi(\langle \mathbf{U}, \varphi_n\rangle)\right)_{n\in\mathbb{N}},$$

where

$$\chi \in C^\infty(\mathbb{R}), \quad -1 < \chi < 1, \qquad \chi(0) = 0, \quad \chi' > 0.$$

Recall that the space $[-1,1]^{\aleph_0}$ endowed with the standard product topology is a compact Polish space.

Definition 2.2.1. Let $(\Omega, \mathfrak{F}, \mathbb{P})$ be a complete probability space. A mapping

$$\mathbf{U} : \Omega \to \mathcal{D}'(\mathbb{R}\times\mathbb{T}^N)$$

is called a *random distribution* if $\langle \mathbf{U}, \varphi\rangle : \Omega \to \mathbb{R}$ is a measurable function for any $\varphi \in \mathcal{D}(\mathbb{R}\times\mathbb{T}^N)$.

The connection between the notion of a random distribution and that of a random variable introduced in Section 2.1 is provided by the following simple result.

Lemma 2.2.2. *The following statements are equivalent:*
- $\mathbf{U}$ *is a random distribution in the sense of Definition* 2.2.1.
- $\mathbf{U} \in \mathcal{D}'(\mathbb{R}\times\mathbb{T}^N)$ *a.s. and all functions* $\langle \mathbf{U}, \varphi_n\rangle$ *are measurable for* $n \in \mathbb{N}$ *(recall that* $(\varphi_n)_{n\in\mathbb{N}}$ *has been introcuded above).*
- $\jmath(\mathbf{U})$ *is a Borel measurable random variable ranging in* $[-1,1]^{\aleph_0}$.
- $\mathbf{U}$ *is a Borel measurable random variable ranging in the topological space* $\mathcal{D}'(\mathbb{R}\times\mathbb{T}^N)$.

2.2.1 Measurability

The following statement shows that measurability of a function $\mathbf{U}$ ranging in a topological vector space X that is continuously embedded in $\mathcal{D}'(\mathbb{R}\times\mathbb{T}^N)$ is determined by its measurability in the sense of distributions as long as its law is tight.

Theorem 2.2.3. *Let* $X \hookrightarrow \mathcal{D}'(\mathbb{R}\times\mathbb{T}^N)$ *be a topological vector space continuously embedded into* $\mathcal{D}'(\mathbb{R}\times\mathbb{T}^N)$. *Let* $\mathbf{U} \in \mathcal{D}'(\mathbb{R}\times\mathbb{T}^N)$ *be a random distribution such that, for any* $\varepsilon > 0$, *there is a compact set* $K_\varepsilon \subset X$ *such that*

$$\mathbb{P}(\mathbf{U} \in K_\varepsilon) > 1 - \varepsilon.$$

Then $\mathbf{U} \in X$ *a.s.,* $\mathbf{U}$ *is a Borel random variable ranging in* X*, and the law of* $\mathbf{U}$ *is a Radon measure on* X.

Remark 2.2.4. Note that $\jmath(\mathbf{U}) : \Omega \to [-1,1]^{\aleph_0}$ is a random variable, that is, the mapping is measurable. As $\jmath(K_\varepsilon)$ are compact in $[-1,1]^{\aleph_0}$, they are closed Borel sets and $\mathbf{U}^{-1}(K_\varepsilon) = \mathbf{U}^{-1}(\jmath^{-1}\jmath(K_\varepsilon))$ are measurable.

Proof of Theorem 2.2.3. To begin, observe that the set $\{\mathbf{U} \notin \bigcup_{\varepsilon>0} K_\varepsilon\}$ is of probability zero (whence measurable) and, consequently, $\mathbf{U} \in X$ a.s.

As X is a topological vector space, in particular Hausdorff completely regular, it suffices to show that $F(\mathbf{U})$ is measurable for any $F \in C_b(X)$. Without loss of generality, we assume $K_{\varepsilon_1} \subseteq K_{\varepsilon_2}$ for $\varepsilon_1 > \varepsilon_2$. By virtue of the Stone–Weierstrass theorem, Theorem 1.1.2, the algebra generated by the family of continuous functions $(\langle \cdot, \varphi_n \rangle)_{n\in\mathbb{N}}$ is dense in $C(K_\varepsilon)$ for any $\varepsilon > 0$. As observed in Remark 2.2.4, the functions $\mathbf{1}_{K_\varepsilon}(\mathbf{U})\langle \mathbf{U}, \varphi_n \rangle$ are measurable; whence $\mathbf{1}_{K_\varepsilon}(\mathbf{U})F(\mathbf{U})$ is measurable for any ε.

Finally, we have $\mathbf{1}_{K_\varepsilon}(\mathbf{U})F(\mathbf{U}) \to F(\mathbf{U})$ a.s.; whence $F(\mathbf{U})$ is measurable. Obviously, the law of $\mathbf{U}$ in X is tight (Radon). □

2.2.2 Regularization

Let $\chi \in C_c^\infty(-2,2)$ be a smooth cut-off function. We have

$$\chi(-z) = \chi(z), \quad z \in \mathbb{R}, \qquad \chi'(z) \le 0 \quad \text{for } z \ge 1, \qquad \chi(z) = 1 \quad \text{for } z \in [-1,1].$$

Any random distribution $\mathbf{U}$ can be regularized by convolution and cut-off. Specifically, we set

$$\begin{aligned} [[\mathbf{U}]]_\delta &= [\chi(\delta\cdot)\mathbf{U}]_{\delta,t,x} = \theta_\delta^t(\cdot - \delta) * [\chi(\delta\cdot)[\theta_\delta^x * \mathbf{U}]], \\ [\mathbf{U}]_{\delta,t,x}(t,x) &:= \langle \mathbf{U}, \theta_\delta^t(t-\delta-\cdot)\theta_\delta^x(x-\cdot)\chi(\delta\cdot)\rangle. \end{aligned} \tag{2.2}$$

Recall that θ_δ^x and θ_δ^t have been introduced in Sections 1.7.3 and 1.8.3. It is easy to check that $[[\mathbf{U}]]_\delta$ takes values in $\mathcal{D}(\mathbb{R} \times \mathbb{T}^N)$. Moreover, if $\mathbf{U}$ is a random distribution, then $\partial_t^k D_x^\alpha [[\mathbf{U}]]_\delta(t,x)$ are measurable for any (t,x), k, and α.

Remark 2.2.5. Note that the regularization is "non-anticipating", meaning that $[[\mathbf{U}]]_\delta(t,x)$ depends only on the action of the distribution $\mathbf{U}$ on functions in $\mathcal{D}((-\infty,t) \times \mathbb{T}^N)$.

Remark 2.2.6. If the space $\mathcal{D}'(I \times \mathbb{T}^N)$, $I = (a,b)$, is considered instead of $\mathcal{D}'(\mathbb{R} \times \mathbb{T}^N)$, then the regularization is well-defined only for $t \in (a + 2\delta, b)$. Of course, this is irrelevant as long as $a = -\infty$.

Lemma 2.2.7. *The mapping* $\mathbf{U} \mapsto [[\mathbf{U}]]_\delta$ *is a bounded continuous linear operator mapping* $\mathcal{D}'(\mathbb{R} \times \mathbb{T}^N)$ *into* $\mathcal{D}(\mathbb{R} \times \mathbb{T}^N)$ *for any* $\delta > 0$.

Proof. As $[[\cdot]]_\delta$ is obviously linear, it suffices to check that it maps bounded sets in $\mathcal{D}'(\mathbb{R}\times\mathbb{T}^N)$ into bounded sets in $\mathcal{D}(\mathbb{R}\times\mathbb{T}^N)$. Recall that $B\subset\mathcal{D}'(\mathbb{R}\times\mathbb{T}^N)$ is bounded in the weak-$*$ topology if and only if

$$\langle \mathbf{U},\varphi\rangle \le c_\varphi \quad \text{for all } \mathbf{U}\in B,\ \varphi\in\mathcal{D}(\mathbb{R}\times\mathbb{T}^N).$$

Since $\operatorname{supp}[[\mathbf{U}]]_\delta$ is contained in the compact set $K_\delta=[-\frac{2}{\delta}-\delta,\frac{2}{\delta}+\delta]\times\mathbb{T}^N$, it suffices to check that, for each $k\ge 0$, there is a constant c_k such that

$$\|[[\mathbf{U}]]_\delta\|_{C^k(K_\delta)}\le c_k \quad \text{whenever } \mathbf{U}\in B. \tag{2.3}$$

Consider the Fréchet space $\mathcal{D}(K_\delta)$ of all functions from $C^\infty(\mathbb{R}\times\mathbb{T}^N)$ with support contained in K_δ. It follows from the Banach–Steinhaus theorem in Fréchet spaces (see Dieudonné [Die70]) that functions $\mathbf{U}\in B$ are equi-continuous in $\mathcal{D}(K_\delta)$, meaning

$$|\langle \mathbf{U},\varphi\rangle|\le c\left(\sum_{k=1}^{\infty}\frac{1}{2^k}\min\{1,\|\varphi\|_{C^k(K_\delta)}\}\right) \quad \text{uniformly for } \mathbf{U}\in B. \tag{2.4}$$

Seeing that $[[\mathbf{U}]]_\delta$ is defined via cut-off and convolution with smooth kernels, we easily observe that (2.4) implies (2.3). □

It follows that, for any random distribution $\mathbf{U}$, its regularization $[[\mathbf{U}]]_\delta$ can be seen as a Borel measurable random variable ranging in the Banach space $C^k_c((-L,L)\times\mathbb{T}^N)$ for a certain $L=L(\delta)$ and $k\ge 0$ arbitrary. Consequently, we obtain the following lemma.

Lemma 2.2.8. *Let $X\subset\mathcal{D}'(\mathbb{R}\times\mathbb{T}^N)$ be a topological vector space such that*

$$C^k_c((-L,L)\times\mathbb{T}^N)\hookrightarrow X \quad \textit{continuously for a certain } k\ge 0 \textit{ and any } L>0.$$

Let $\mathbf{U}$ be a random distribution in $\mathcal{D}'(\mathbb{R}\times\mathbb{T}^N)$. Then $[[\mathbf{U}]]_\delta$ is a Borel measurable random variable in X. Moreover, the σ-field $\sigma_X[[\mathbf{U}]]_\delta$ generated by $[[\mathbf{U}]]_\delta$ in X is included in the σ-field $\sigma[\mathbf{U}]$ generated by $\mathbf{U}$ in $\mathcal{D}'(\mathbb{R}\times\mathbb{T}^N)$. Specifically, let

$$\sigma_X[[\mathbf{U}]]_\delta := \sigma([[\mathbf{U}]]_\delta^{-1}(\mathcal{B}),\ \mathcal{B} \textit{ Borel in } X)$$

and, for a fixed sequence $(\varphi_n)_{n\in\mathbb{N}}$ which is dense in $\mathcal{D}(\mathbb{R}\times\mathbb{T}^N)$, define

$$\sigma[\mathbf{U}] := \sigma\Big(\bigcup_{\varphi_n,M}\{\langle\mathbf{U},\varphi_n\rangle<M\},\ M\in\mathbb{R}\Big).$$

Then

$$\sigma_X[[\mathbf{U}]]_\delta\subseteq\sigma[\mathbf{U}].$$

Corollary 2.2.9. *Let $X \subset \mathcal{D}'(\mathbb{R} \times \mathbb{T}^N)$ be a topological space such that*

$$C_c^k((-L, L) \times \mathbb{T}^N) \hookrightarrow X \quad \textit{continuously for a certain } k \geq 0 \textit{ and any } L > 0.$$

Let $\mathbf{U}$ *be a random distribution in* $\mathcal{D}'(\mathbb{R} \times \mathbb{T}^N)$ *such that* $\mathbf{U} \in X$ *a.s. Suppose that*

$$[[v]]_\delta \to v \textit{ in } X \quad \textit{for any } v \in X. \tag{2.5}$$

Then $\mathbf{U}$ *is a Borel measurable random variable in* X*. Moreover,*

$$\sigma_X[\mathbf{U}] = \sigma\Big(\bigcup_{\delta>0} \sigma_X[[\mathbf{U}]]_\delta\Big) = \sigma[\mathbf{U}].$$

Remark 2.2.10. If X is a Fréchet space equipped with a translation invariant metric d, hypothesis (2.5) can be replaced by

$$\begin{aligned} \sup_{\delta>0} d([[v]]_\delta, 0) \leq c_v, \quad & \text{for any } v \in X, \\ [[v]]_\delta \to v \text{ in } X, \quad & \text{for any } v \text{ belonging to a dense subset of } X. \end{aligned} \tag{2.6}$$

Moreover, observing that $[[v]]_\delta \to v$ in $\mathcal{D}(\mathbb{R} \times \mathbb{T}^N)$ whenever $v \in \mathcal{D}(\mathbb{R} \times \mathbb{T}^N)$, condition (2.6) is automatically satisfied as soon as

$$\sup_{\delta>0} d([[v]]_\delta, 0) \leq c_v \quad \text{for any } v \in X, \ \mathcal{D}(\mathbb{R} \times \mathbb{T}^N) \hookrightarrow X$$

with dense and continuous embedding.

In view of Lemma 2.2.8 and Corollary 2.2.9, we may replace a random distribution $\mathbf{U}$ by the family of regularizations $[[\mathbf{U}]]_\delta$, $\delta > 0$, preserving all the “stochastic” information available. Moreover, we can always work in the largest class $\mathcal{D}'(\mathbb{R} \times \mathbb{T}^N)$ as soon as $\mathbf{U}$ is a random variable in X and $\mathcal{D}(\mathbb{R} \times \mathbb{T}^N)$ is a dense subset of X.

2.2.3 Equality in law

As the next step, we introduce the equality of laws for random distributions in $\mathcal{D}'(\mathbb{R} \times \mathbb{T}^N)$. Recall that the special case of random variables was discussed in Definition 2.1.2 and Lemma 2.1.4.

Definition 2.2.11. We say that two random distributions $\mathbf{U}$, $\tilde{\mathbf{U}}$ coincide in law, $\mathbf{U} \overset{d}{\sim} \tilde{\mathbf{U}}$, if the joint laws

$$\mathcal{L}_{\mathbb{R}^k}[\langle \mathbf{U}, \varphi_1 \rangle, \dots, \langle \mathbf{U}, \varphi_k \rangle] = \mathcal{L}_{\mathbb{R}^k}[\langle \tilde{\mathbf{U}}, \varphi_1 \rangle, \dots, \langle \tilde{\mathbf{U}}, \varphi_k \rangle] \tag{2.7}$$

coincide for any k-tuple of test functions $\varphi_1, \dots, \varphi_k \in \mathcal{D}(\mathbb{R} \times \mathbb{T}^N)$.

In view of Lemma 2.1.4, condition (2.7) can be equivalently rewritten in terms of continuous functions as follows:

$$\mathbb{E}[F(\langle \mathbf{U}, \varphi_1\rangle, \dots, \langle \mathbf{U}, \varphi_k\rangle)] = \mathbb{E}[F(\langle \tilde{\mathbf{U}}, \varphi_1\rangle, \dots, \langle \tilde{\mathbf{U}}, \varphi_k\rangle)], \quad F \in C_b(\mathbb{R}^k). \tag{2.8}$$

Note that the probability space where $\mathbf{U}$ is defined can be different from that of $\tilde{\mathbf{U}}$. We remark that it suffices to require (2.7) only for the functions $(\varphi_n)_{n\in\mathbb{N}}$ forming a dense set in $\mathcal{D}(\mathbb{R}\times\mathbb{T}^N)$.

Theorem 2.2.12. *Let X be a topological vector space and $X \hookrightarrow \mathcal{D}'(\mathbb{R}\times\mathbb{T}^N)$ with continuous embedding. Let $\mathbf{U}$ be a random distribution in $\mathcal{D}'(\mathbb{R}\times\mathbb{T}^N)$ such that, for any $\varepsilon > 0$, there exists a compact $K_\varepsilon \subset X$ such that*

$$\mathbb{P}(\mathbf{U} \in K_\varepsilon) > 1 - \varepsilon.$$

Let $\tilde{\mathbf{U}}$ be another random distribution in $\mathcal{D}'(\mathbb{R}\times\mathbb{T}^N)$ satisfying

$$\mathbf{U} \overset{d}{\sim} \tilde{\mathbf{U}}$$

in the sense of Definition 2.2.11. Then $\tilde{\mathbf{U}} \in X$ a.s., $\tilde{\mathbf{U}}$ is a Borel random variable ranging in X, and

$$\mathcal{L}_X[\mathbf{U}] = \mathcal{L}_X[\tilde{\mathbf{U}}],$$

in particular, the law of $\tilde{\mathbf{U}}$ is a Radon measure on X.

Proof. We know from Theorem 2.2.3 that $\mathbf{U}$ is a random variable ranging in X, the law of which is a Radon measure. Consider the canonical embedding $\jmath : \mathcal{D}' \to [-1,1]^{\aleph_0}$. Since $\mathbf{U} \overset{d}{\sim} \tilde{\mathbf{U}}$, we deduce

$$\mathbb{E}[F(\jmath(\mathbf{U}))] = \mathbb{E}[F(\jmath(\tilde{\mathbf{U}}))],$$

at least for F belonging to the algebra of continuous functions generated by $\langle\cdot, \varphi_n\rangle$ on $[-1,1]^{\aleph_0}$ that separates points. As the space $[-1,1]^{\aleph_0}$ is compact Hausdorff, we may apply the Stone–Weierstrass theorem to conclude

$$\mathcal{L}_{[-1,1]^{\aleph_0}}[\mathbf{U}] = \mathcal{L}_{[-1,1]^{\aleph_0}}[\tilde{\mathbf{U}}].$$

Consequently, as the image of K_ε is compact in $[-1,1]^{\aleph_0}$, we conclude

$$\mathbb{P}(\tilde{\mathbf{U}} \in K_\varepsilon) > 1 - \varepsilon.$$

As $\varepsilon > 0$ was arbitrary, we infer

$$\tilde{\mathbf{U}} \in X \quad a.s.$$

By virtue of Theorem 2.2.3, $\tilde{\mathbf{U}}$ is a Borel measurable random variable in X, whose law is a Radon measure.

It remains to be shown that

$$\mathbb{E}[F(\mathbf{U})] = \mathbb{E}[F(\tilde{\mathbf{U}})] \quad \text{for any } F \in C_b(X). \tag{2.9}$$

Given $F \in C_b(X)$ and a compact set K_ε, we use the Stone–Weierstrass theorem again to find a sequence F_n, where each F_n has the form

$$F_n(v) = G_n(\langle v, \varphi_1\rangle, \dots, \langle v, \varphi_{m_n}\rangle)$$

such that

$$F_n \in C_b(X), \quad \sup_X \|F_n\| \le \sup_X \|F\| \quad \text{for all } n \in \mathbb{N}, \tag{2.10}$$

$$F_n \to F \quad \text{in } C(K_\varepsilon) \text{ as } n \to \infty. \tag{2.11}$$

In accordance with (2.8), we get

$$\begin{aligned}\mathbb{E}[F(\mathbf{U})] &= \mathbb{E}[F(\mathbf{U}) - F_n(\mathbf{U})] + \mathbb{E}[F_n(\mathbf{U})] = \mathbb{E}[F_n(\tilde{\mathbf{U}})] + \mathbb{E}[F(\mathbf{U}) - F_n(\mathbf{U})] \\ &= \mathbb{E}[F(\tilde{\mathbf{U}})] + \mathbb{E}[F(\mathbf{U}) - F_n(\mathbf{U})] + \mathbb{E}[F(\tilde{\mathbf{U}}) - F_n(\tilde{\mathbf{U}})].\end{aligned}$$

Finally,

$$\mathbb{E}[F(\mathbf{U}) - F_n(\mathbf{U})] = \mathbb{E}[1_{K_\varepsilon}(F(\mathbf{U}) - F_n(\mathbf{U}))] + \mathbb{E}[1_{X\setminus K_\varepsilon}(F(\mathbf{U}) - F_n(\mathbf{U}))],$$

where, by virtue of (2.11),

$$\mathbb{E}[1_{K_\varepsilon}(F(\mathbf{U}) - F_n(\mathbf{U}))] \to 0 \quad \text{as } n \to \infty,$$

while, in accordance with (2.10),

$$|\mathbb{E}[1_{X\setminus K_\varepsilon}(F(\mathbf{U}) - F_n(\mathbf{U}))]| < \varepsilon c.$$

Seeing that the term containing $\tilde{\mathbf{U}}$ can be handled in the same way, we obtain (2.9). □

Accordingly, for random variables the law of which is a Radon measure, it is irrelevant which topology we consider to compare their laws. It is therefore convenient to use the weakest form of law equivalence specified in (2.8). Unless otherwise stated, we shall write $\mathbf{U} \overset{d}{\sim} \tilde{\mathbf{U}}$ without specifying the topological space.

2.2.4 Progressive measurability

The notion of progressive measurability for stochastic processes was introduced in Section 2.1, namely, in Definition 2.1.16, and discussed further in the sequel. This is essential in order to keep track of the "arrow of time" which is the key ingredient in the construction of Itô's stochastic integral with respect to a Wiener process, presented in Section 2.3. As we intend to formulate the stochastic integration theory for random distributions, it is necessary to understand what precisely progressive measurability means in this context. In particular, we introduce the notion of adaptedness for random distributions and discuss how it is naturally connected to progressive measurability of stochastic processes.

To this end, let $(\Omega, \mathfrak{F}, (\mathfrak{F}_t)_{t\geq 0}, \mathbb{P})$ be a complete probability space with a complete right-continuous filtration $(\mathfrak{F}_t)_{t\geq 0}$. We consider the σ-field of all progressively measurable sets in $\Omega \times [0,T]$ associated to the filtration $(\mathfrak{F}_t)_{t\geq 0}$. To be more precise, $A \subset \Omega \times [0,T]$ belongs to the progressively measurable σ-field, provided the stochastic process $(\omega, t) \mapsto \mathbf{1}_A(\omega, t)$ is $(\mathfrak{F}_t)$-progressively measurable. We denote by $L^p_{\mathrm{prog}}(\Omega \times [0,T])$ the Lebesgue space of functions that are measurable with respect to the σ-field of $(\mathfrak{F}_t)$-progressively measurable sets in $\Omega \times [0,T]$ and we denote by μ_{prog} the measure $\mathbb{P} \otimes \mathfrak{L}_{[0,T]}$ restricted to the progressively measurable σ-field.

Definition 2.2.13. Let $\mathbf{U}$ be a random distribution in $\mathcal{D}'(\mathbb{R} \times \mathbb{T}^N)$. Then:

- We say that $\mathbf{U}$ is *adapted* to $(\mathfrak{F}_t)_{t\geq 0}$ if $\langle \mathbf{U}, \varphi \rangle$ is $(\mathfrak{F}_t)$-measurable for any $\varphi \in \mathcal{D}((-\infty, t) \times \mathbb{T}^N)$.
- The family of σ-fields $(\sigma_t[\mathbf{U}])_{t\geq 0}$, given as

$$\sigma_t[\mathbf{U}] := \bigcap_{s>t} \sigma\Big(\bigcup_{\varphi \in \mathcal{D}((-\infty,s)\times\mathbb{T}^N)} \{\langle \mathbf{U}, \varphi\rangle < 1\} \cup \{N \in \mathfrak{F}, \mathbb{P}(N) = 0\} \Big),$$

  is called the *history* of $\mathbf{U}$.

Remark 2.2.14. Note that we use the same notation for the history of a random distribution and for the $\mathbb{P}$-augmented canonical filtration of a stochastic process; cf. Remark 2.1.14. This will be further justified below by Lemma 2.2.18.

Remark 2.2.15. The history $(\sigma_t[\mathbf{U}])_{t\geq 0}$ is a complete right-continuous filtration. Note that the value of the distribution $\mathbf{U}$ for $\tau > t$ is irrelevant for determining its history up to the time t. For random variables defined only on the time interval $[0,T]$ it is customary to set $\mathbf{U} = \mathbf{U}_0 = \mathbf{U}(0, \cdot)$ for $t \leq 0$.

Clearly, any random distribution $\mathbf{U}$ is adapted to its history $(\sigma_t[\mathbf{U}])_{t\geq 0}$. Recalling Remark 2.2.5 we easily observe that the regularization $[[\mathbf{U}]]_\delta$ of a random distribution $\mathbf{U}$ in $\mathcal{D}'(\mathbb{R} \times \mathbb{T}^N)$ is $(\mathfrak{F}_t)$-progressively measurable as a stochastic process ranging in $C^k(\mathbb{T}^N)$, $k \geq 0$ arbitrary, whenever the distribution $\mathbf{U}$ is $(\mathfrak{F}_t)$-adapted in the sense of Definition 2.2.13. Moreover, as $[[\mathbf{U}]]_\delta \to \mathbf{U}$ in $\mathcal{D}'(\mathbb{R} \times \mathbb{T}^N)$, we infer

$$\sigma_t[\mathbf{U}] = \sigma\Big(\bigcup_{\delta > 0} \sigma_t[[\mathbf{U}]]_\delta \Big).$$

As the next step, we show that adaptedness is preserved by compositions with Carathéodory functions in the sense of the following definition.

Definition 2.2.16. Let X, Y be topological spaces and $(\Theta, \mathcal{M}, \mu)$ a measure space. We say that $f : \Theta \times X \to Y$ is a *Carathéodory function* if:
(1) $f(\cdot, x)$ is measurable for every $x \in X$;
(2) $f(\theta, \cdot)$ is continuous for almost every $\theta \in \Theta$.

Lemma 2.2.17. *Let* $\mathbf{U}$ *be a random distribution in* $\mathcal{D}'(\mathbb{R}\times\mathbb{T}^N)$ *such that* $\mathbf{U} \in L^1_{\text{loc}}(\mathbb{R}\times\mathbb{T}^N)$ *a.s. Let* $\mathbf{G} : \mathbb{T}^N \times \mathbb{R} \to \mathbb{R}^K$ *be a Carathéodory function and suppose that* $\mathbf{G}(\mathbf{U}) \in L^1_{\text{loc}}(\mathbb{R}\times\mathbb{T}^N)$, *where* $\mathbf{G}(\mathbf{U})(t,x) = \mathbf{G}(x, \mathbf{U}(t,x))$. *Then* $\mathbf{G}(\mathbf{U})$ *is a random distribution in* $\mathcal{D}'(\mathbb{R}\times\mathbb{T}^N)$ *which is adapted to* $(\sigma_t[\mathbf{U}])_{t\geq 0}$ *in the sense of Definition* 2.2.13.

Proof. Consider the regularization $[[\mathbf{U}]]_\delta$ specified through (2.2). Obviously, $[[\mathbf{U}]]_\delta$ is $(\sigma_t[\mathbf{U}])$-adapted (see Remark 2.2.5) and so is the composition $T_M \circ \mathbf{G}([[\mathbf{U}]]_\delta)$, where T_M is the following suitable cut-off function:

$$T_M \in C^\infty(\mathbb{R}^K;\mathbb{R}^K) \cap C_b(\mathbb{R}^K,\mathbb{R}^K), \quad T_M(Z) = Z \quad \text{for } |Z| \leq M, \quad |T_M(Z)| \leq |Z|.$$

Letting $\delta \to 0$ we have

$$T_M \circ \mathbf{G}([[\mathbf{U}]]_\delta) \to T_M \circ \mathbf{G}(\mathbf{U}) \quad \text{in } L^1_{\text{loc}}(\mathbb{R}\times\mathbb{T}^N)$$

and in particular, by the Lebesgue theorem,

$$\int_{\mathbb{R}}\int_{\mathbb{T}^N} T_M \circ \mathbf{G}([[\mathbf{U}]]_\delta) \cdot \varphi \,\mathrm{d}x\,\mathrm{d}t \to \int_{\mathbb{R}}\int_{\mathbb{T}^N} T_M \circ \mathbf{G}(\mathbf{U}) \cdot \varphi \,\mathrm{d}x\,\mathrm{d}t,$$

for any $\varphi \in \mathcal{D}((\infty, t)\times\mathbb{T}^N)$, $t \geq 0$; whence $T_M \circ \mathbf{G}(\mathbf{U})$ is $(\sigma_t[\mathbf{U}])$-adapted.

Letting $M \to \infty$ and using again the Lebesgue theorem, we infer that $\mathbf{G}(\mathbf{U})$ is $(\sigma_t[\mathbf{U}])$-adapted. □

In the abstract definition of the stochastic integral, we use the Sobolev spaces

$$W^{\ell,2}(\mathbb{T}^N) \subset \mathcal{D}'(\mathbb{T}^N)$$

with suitable ℓ not necessarily positive. Note that these are separable Hilbert spaces with an orthogonal basis formed by a countable family $(\phi_n)_{n\in\mathbb{N}}$ of smooth functions (trigonometric polynomials). Accordingly, a $W^{\ell,2}(\mathbb{T}^N)$-valued stochastic process $\mathbf{U}$ is progressively measurable with respect to a filtration $(\mathfrak{F}_t)_{t\geq 0}$ if all scalar functions $\mathbf{U}_\phi := \langle \mathbf{U}, \phi\rangle$ are progressively measurable for any $\phi = \phi_n$, $n \in \mathbb{N}$. Here, the duality product $\langle\cdot,\cdot\rangle$ is understood in $\mathcal{D}'(\mathbb{T}^N)$.

Lemma 2.2.18. *Let* $\mathbf{U}$ *be an* $(\mathfrak{F}_t)$*-adapted random distribution taking values in* $L^2_{\text{loc}}(\mathbb{R}; W^{\ell,2}(\mathbb{T}^N)) \subset \mathcal{D}'(\mathbb{R}\times\mathbb{T}^N)$ *in the sense of Definition* 2.2.1. *Then, for any* $T > 0$, *there exists a stochastic process* $\tilde{\mathbf{U}} \in L^2(0,T; W^{\ell,2}(\mathbb{T}^N))$ *a.s. that is* $(\mathfrak{F}_t)$*-progressively measurable such that*

$$\mathbf{U} = \tilde{\mathbf{U}} \quad \textit{a.a. in } \Omega\times[0,T].$$

Remark 2.2.19. Note that, strictly speaking, the elements of the Lebesgue space $L^2(0,T;W^{\ell,2}(\mathbb{T}^N))$ are classes of equivalence of functions that coincide with a possible exception of a set of zero measure. In this sense, we say $\mathbf{U} = \tilde{\mathbf{U}}$ in $L^2(0,T;W^{\ell,2}(\mathbb{T}^N))$ a.s.

Proof of Lemma 2.2.18. Fix a test function $\phi = \phi_n$. It suffices to show that there exists a progressively measurable real-valued stochastic process $\tilde{\mathbf{U}}_\phi$ such that

$$\mathbf{U}_\phi := \langle \mathbf{U}, \phi \rangle = \tilde{\mathbf{U}}_\phi \quad \text{a.a. in } \Omega \times [0,T].$$

Without loss of generality, we suppose that $\mathbf{U}_\phi$ is uniformly bounded considering a suitable cut-off as the case may be.

Let

$$[\mathbf{U}_\phi]_{t,\delta} = \theta^t_\delta(\cdot - \delta) * \mathbf{U}_\phi$$

be the time regularization of $\mathbf{U}_\phi$. As the distribution $\mathbf{U}$ is adapted, the stochastic process $t \mapsto [\mathbf{U}_\phi]_{t,\delta}(t)$ is continuous and adapted; whence $(\mathfrak{F}_t)$-progressively measurable. Moreover, as $\mathbf{U}_\phi$ is bounded, we have

$$[\mathbf{U}_\phi]_{t,\delta} \in L^p_{\text{prog}}(\Omega \times [0,T]) \quad \text{for any } p \geq 1,$$

where L^p_{prog} is the Lebesgue space of functions that are measurable with respect to the σ-field of all $(\mathfrak{F}_t)$-progressively measurable sets in $\Omega \times [0,T]$.

Letting $\delta \to 0$, we obtain

$$[\mathbf{U}_\phi]_{t,\delta} \to \mathbf{U}_\phi \quad \text{in, say, } L^2_{\text{prog}}(\Omega \times [0,T]).$$

In particular, passing to a subsequence if necessary, we have

$$[\mathbf{U}_\phi]_{t,\delta} \to \mathbf{U}_\phi \quad \mu_{\text{prog}}\text{-a.a.}$$

As the measure μ_{prog} does not need to be complete, the limit is not necessarily prog-measurable. However, changing $[\mathbf{U}_\phi]_{t,\delta}$ on a set of μ_{prog}-measure zero, we get

$$[\mathbf{U}_\phi]_{t,\delta} \to \tilde{\mathbf{U}}_\phi \quad \text{pointwise in } \Omega \times [0,T],$$

where $\tilde{\mathbf{U}}_\phi$ is prog-measurable and whence progressively measurable and $\mathbf{U}_\phi$ coincides with $\tilde{\mathbf{U}}_\phi$ outside of a zero measure set in $\Omega \times [0,T]$. □

2.2.5 Special classes of random distributions

In view of possible applications to stochastic PDEs, the class of random distributions $\mathcal{D}'(\mathbb{R} \times \mathbb{T}^N)$ is too large. As we deal with non-linear compositions, it is convenient if the underlying function space allows multiplication. Accordingly, all random quantities we shall deal with will be distributions of finite order that can be expressed as derivatives of regular functions belonging to the space $L^1_{loc}(\mathbb{R} \times \mathbb{T}^N)$.

Seeing that $W^{\ell,2}(\mathbb{T}^N) \hookrightarrow C(\mathbb{T}^N)$ for $\ell > \frac{N}{2}$, we restrict ourselves to the spaces

$$L^1_{loc}(\mathbb{R}; W^{-\ell,2}(\mathbb{T}^N)) \quad \text{with } \ell > \frac{N}{2}.$$

Dealing with the initial-value problems, the frequently used spaces will be (i) the Hilbert space $L^2(0,T; W^{-\ell,2}(\mathbb{T}^N))$ and (ii) the Banach space $C([0,T]; W^{-\ell,2}(\mathbb{T}^N))$, both Polish spaces.

To accommodate at most exponential growth for $t \to \infty$ of globally defined random variables, we may consider the weighted Sobolev space

$$H^k_\Gamma \equiv L^2_w(\mathbb{R}; W^{k,2}(\mathbb{T}^N)), \quad \|\mathbf{U}\|^2_{H^k_\Gamma} = \int_{\mathbb{R}} \exp(\Gamma|t|)\|\mathbf{U}(t,\cdot)\|^2_{W^{k,2}_x}\,\mathrm{d}t.$$

As

$$\mathcal{D}(\mathbb{R} \times \mathbb{T}^N) \hookrightarrow H^k_\Gamma \hookrightarrow L^2(\mathbb{R} \times \mathbb{T}^N) \quad \text{for } k \geq 0,\ \Gamma \geq 0,$$

we identify the dual spaces as distributions. We have

$$H^k_\Gamma \hookrightarrow \mathcal{D}'(\mathbb{R} \times \mathbb{T}^N) \quad \text{for } k \leq 0,\ \Gamma \leq 0.$$

The space H^k_Γ is a separable Hilbert space. In particular, it is a Polish space. Basically all random distributions we will handle will belong to H^k_Γ for suitable k and Γ. In particular, the space $L^\infty(\mathbb{R}; L^1(\mathbb{T}^N))$ can be identified with a subspace of H^k_Γ with $k < -[\frac{N}{2}]$, $\Gamma < 0$.

Finally, we shall occasionally use the embedding

$$L^1(0,T; L^1(\mathbb{T}^N)) \hookrightarrow W^{-k,2}(\mathbb{T}^{N+1}), \quad k > \frac{N+1}{2}.$$

2.3 Stochastic Itô's integral

As the next step, let us present the definition of a stochastic integral in the Itô sense. To be more precise, we are interested in stochastic integration with respect to the cylindrical Wiener process introduced in Definition 2.1.31. A detailed construction can be found in Da Prato–Zabczyk [DPZ92, Section 4.2–Section 4.4]; see also Prévôt–Röckner [PR07].

Let W be a cylindrical Wiener process on a separable Hilbert space $\mathfrak{U}$. For a stochastic process $\mathbf{G} = \{\mathbf{G}(t); t \in [0,\infty)\}$, taking values in the space of bounded linear operators from $\mathfrak{U}$ to a separable Hilbert space H, we intend to make sense of an integral of the form

$$\int_0^t \mathbf{G}(s)\,\mathrm{d}W(s).$$

In addition, we require that it defines an H-valued martingale; see Definition 2.1.21. We will construct the stochastic integral in such a way that

$$\int_0^t \mathbf{G}(s)\,\mathrm{d}W(s) = \sum_{k=1}^{\infty} \int_0^t \mathbf{G}(s)e_k\,\mathrm{d}W_k(s), \tag{2.12}$$

that is, as a sum of stochastic integrals with respect to real-valued Wiener processes. However, keeping in mind that W does not exist as a stochastic process taking values in $\mathfrak{U}$, it is necessary to impose certain conditions on $\mathbf{G}$ such that the right hand side of (2.12) converges in a suitable sense in H and hence defines an H-valued martingale. It turns out that it is required for the stochastic process $\mathbf{G}$ to take values in the space of Hilbert–Schmidt operators from $\mathfrak{U}$ to H, denoted by $L_2(\mathfrak{U}, H)$. Recall that $L_2(\mathfrak{U}, H)$ contains all bounded linear operators $A \in L(\mathfrak{U}, H)$ such that

$$\|A\|^2_{L_2(\mathfrak{U},H)} := \sum_{k=1}^{\infty} \|Ae_k\|^2_H < \infty.$$

To be more precise, the following result holds true; see Da Prato–Zabczyk [DPZ92, Section 4.2].

Theorem 2.3.1. *Let* $\mathbf{G}$ *be an* $(\mathfrak{F}_t)$*-progressively measurable stochastic process such that*

$$\mathbb{E} \int_0^T \|\mathbf{G}(t)\|^2_{L_2(\mathfrak{U},H)}\,\mathrm{d}t < \infty. \tag{2.13}$$

Then the stochastic Itô integral (2.12) *is a well-defined continuous* H*-valued square integrable* $(\mathfrak{F}_t)$*-martingale.*

Remark 2.3.2. For completeness, let us point out that the stochastic Itô integral can also be constructed under a weaker assumption than (2.13), namely, if

$$\int_0^T \|\mathbf{G}(t)\|^2_{L_2(\mathfrak{U},H)}\,\mathrm{d}t < \infty \quad \mathbb{P}\text{-a.s.}$$

In that case, the stochastic integral is only a local martingale as introduced in Definition 2.1.27; see Da Prato–Zabczyk [DPZ92, Section 4.2] for more details or also Karatzas–Shreve [KS91, Section 3.2] for a finite-dimensional setting.

For the reader's convenience, let us recall the main steps in the construction of the above stochastic Itô integral. First, we introduce a notion of an elementary stochastic process.

Definition 2.3.3. An $L(\mathfrak{U}, H)$-valued $(\mathfrak{F}_t)$-adapted stochastic process Ψ is called *elementary*, provided there exists a sequence $\Psi_0, \dots, \Psi_{k-1}$ of $L(\mathfrak{U}, H)$-valued random variables taking only a finite number of values such that the following holds true: there exists a partition $\{0 = t_0 < t_1 < \cdots < t_k = T\}$ of $[0, T]$ such that Ψ_i is $\mathfrak{F}_{t_i}$-measurable, $i = 0, \dots, k-1$, and

$$\Psi(t) = \sum_{i=0}^{k-1} \Psi_i \mathbf{1}_{(t_i, t_{i+1}]}(t) \quad t \in [0, T].$$

For an elementary $(\mathfrak{F}_t)$-adapted stochastic process Ψ, it is straightforward to define a stochastic integral as follows:

$$\int_0^t \Psi(s)\, \mathrm{d}W(s) := \sum_{i=0}^{k-1} \Psi_i (W(t_{i+1}) - W_{t_i}). \tag{2.14}$$

Note that this is well-defined due to the fact that all Ψ_i are finite-dimensional. The first observation is that the statement of Theorem 2.3.1 holds true for elementary processes.

Proposition 2.3.4. *Let Ψ be an elementary $(\mathfrak{F}_t)$-adapted stochastic process. Then the stochastic integral* (2.14) *defines a continuous H-valued square integrable $(\mathfrak{F}_t)$-martingale and the following holds true:*

$$\mathbb{E}\left\| \int_0^t \Psi(s)\, \mathrm{d}W(s) \right\|_H^2 = \mathbb{E} \int_0^t \|\Psi(s)\|_{L_2(\mathfrak{U},H)}^2 \,\mathrm{d}s \quad t \geq 0, \tag{2.15}$$

$$\mathbb{E} \sup_{t\in[0,T]} \left\| \int_0^t \Psi(s)\, \mathrm{d}W(s) \right\|_H^2 \leq C\, \mathbb{E} \int_0^T \|\Psi(s)\|_{L_2(\mathfrak{U},H)}^2 \,\mathrm{d}s, \tag{2.16}$$

with a constant C independent of Ψ.

The equality (2.15) is the so-called Itô isometry, which remains valid for more general integrands satisfying (2.13).

The final step of the proof of Theorem 2.3.1 relies on the following approximation result. Recall that $L^p_{\mathrm{prog}}(\Omega \times [0, T])$ denotes the Lebesgue space of functions that are measurable with respect to the σ-algebra of $(\mathfrak{F}_t)$-progressively measurable sets in $\Omega \times [0, T]$.

Proposition 2.3.5. *Let $\mathbf{G}$ be an $(\mathfrak{F}_t)$-progressively measurable stochastic process satisfying* (2.13). *Then there exists a sequence of elementary $(\mathfrak{F}_t)$-adapted stochastic processes (Ψ_n) such that*

$$\Psi_n \to \mathbf{G} \quad \text{in } L^2_{\mathrm{prog}}(\Omega \times (0, T); L_2(\mathfrak{U}, H)).$$

As a consequence, in view of (2.16), the stochastic integral

$$\int_0^t \mathbf{G}(s)\, \mathrm{d}W(s)$$

is obtained as a limit in $L^2(\Omega; C([0,T]; H))$ of the corresponding approximate stochastic integrals

$$\int_0^t \Psi_n(s)\,\mathrm{d}W(s).$$

At this stage, it is important to note an immediate consequence of the above construction, namely, the Itô stochastic integral generally cannot be constructed pathwise. More precisely, given a trajectory of the driving Wiener process $t \mapsto W(\omega,t)$ together with a trajectory of the integrand $t \mapsto \mathbf{G}(\omega,t)$, it is not possible to construct a trajectory of the corresponding stochastic integral, unless the trajectories of $\mathbf{G}$ are sufficiently regular. This flaw originates in the low time regularity of the Wiener process: recall that trajectories of W are only Hölder continuous with exponent γ for every $\gamma \in (0,\frac{1}{2})$. Consequently, the stochastic integral cannot be constructed by analytic arguments (i.e., working with a fixed realization ω) unless the irregularity of W is compensated by regularity of $\mathbf{G}$.

To be more precise, if $t \mapsto \mathbf{G}(\omega,t)$ belongs to $C^1([0,T])$, we integrate by parts to obtain

$$\int_0^t \mathbf{G}(s)\,\mathrm{d}W(s) = \mathbf{G}(t)W(t) - \int_0^t \mathbf{G}'(s)W(s)\,\mathrm{d}s,$$

where the right hand side is indeed defined pathwise. This is a particular case of the generalization of the Stieltjes integral introduced by Young in [You36]. Roughly speaking, if $\mathbf{G} \in C^\alpha([0,T])$ and $\mathbf{Z} \in C^\beta([0,T])$, then one can construct the integral of $\mathbf{G}$ against $\mathbf{Z}$ analytically provided $\alpha + \beta > 1$ and this result is sharp. Recall that the regularity of a Wiener process W is strictly smaller than $\frac{1}{2}$. In typical applications considered in this book, the coefficient $\mathbf{G}$ takes the form $\mathbf{G}(\mathbf{U})$ where $\mathbf{U}$ is a solution to a stochastic PDE driven by W. Hence we expect $\mathbf{G}(\mathbf{U})$ to possess the same regularity as W and consequently the Young condition is not satisfied.

Remark 2.3.6. Let us stress that the role of the underlying Hilbert space $\mathfrak{U}$ is purely auxiliary and not important in the current setting. Indeed, by virtue of Parseval's theorem, every separable Hilbert space $\mathfrak{U}$ is isometrically isomorphic to ℓ^2 via

$$\mathfrak{U} \to \ell^2, \quad v = \sum_{k=1}^\infty \langle v, e_k\rangle e_k \mapsto (\langle v, e_k\rangle)_{k\in\mathbb{N}},$$

where $(e_k)_{k\in\mathbb{N}}$ is a complete orthonormal system in $\mathfrak{U}$. Similarly, a cylindrical Wiener process in $\mathfrak{U}$ given by $W = \sum_{k=1}^\infty e_k W_k$ can be identified with the sequence of real-valued Wiener processes $(W_k)_{k\in\mathbb{N}}$. A Hilbert–Schmidt operator $A \in L_2(\mathfrak{U}; H)$ can be identified with the H-valued ℓ^2-summable sequence $(Ae_k)_{k\in\mathbb{N}} \in \ell^2(H)$. Consequently, a stochastic process $\mathbf{G}$ satisfying the assumptions of Theorem 2.3.1 can be identified with $(\mathbf{G}_k)_{k\in\mathbb{N}}$ as an element of $L^2_{\text{prog}}(\Omega\times(0,T);\ell^2(H))$ by $\mathbf{G}_k := \mathbf{G}e_k$ and the corresponding Itô stochastic integral is given by

$$\int_0^t \mathbf{G}(s)\,\mathrm{d}W(s) = \sum_{k=1}^{\infty}\int_0^t \mathbf{G}_k(s)\,\mathrm{d}W_k(s).$$

For simplicity of the presentation, we often do not distinguish between $\mathbf{G}(s)$ as a Hilbert–Schmidt operator on $\mathfrak{U}$ and the sequence $(\mathbf{G}_k(s))_{k\in\mathbb{N}}$ as an element of ℓ^2.

Remark 2.3.7. If W is a real-valued Wiener process, then, as a consequence of Lemma 2.2.18, we observe that the stochastic integral

$$\int_0^t \mathbf{G}(s)\,\mathrm{d}W(s)$$

may be always defined for an integrand $\mathbf{G}\in L^2(0,T;W^{\ell,2}(\mathbb{T}^N))$ a.s. as long as its natural filtration $(\sigma_t[\mathbf{G}])_{t\geq 0}$ is non-anticipative with respect to the associated Wiener process W, or, more precisely, as long as the filtration $(\mathfrak{F}_t)_{t\geq 0}$ generated by the joint history of W and $\mathbf{G}$ satisfies

$$\mathfrak{F}_t := \sigma[\sigma_t[\mathbf{G}]\cup\sigma_t[W]] \quad \text{is independent of } \sigma[W(s)-W(t)] \text{ for any } s>t.$$

The same applies to integrals with a cylindrical Wiener process $W=\sum_{k=1}^{\infty}e_k W_k$ and $\mathbf{G}$ taking values in $L^2(0,T;L_2(\mathfrak{U};W^{\ell,2}(\mathbb{T}^N)))$ a.s. Here, a similar condition must be verified for any $\mathbf{G}_k := \mathbf{G}e_k$ and W_j. In the case $\mathbf{G}_k = \mathbf{G}_k(\mathbf{U})$, the task reduces to checking the same condition for a single filtration

$$\left(\sigma\left[\sigma_t[\mathbf{U}]\cup\bigcup_{k=1}^{\infty}\sigma_t[W_k]\right]\right)_{t\geq 0},$$

in agreement with Lemma 2.2.17.

As the next step, let us recall the so-called Burkholder–Davis–Gundy inequality, which generalizes (2.16).

Proposition 2.3.8 (Burkholder–Davis–Gundy's inequality). *Let H be a separable Hilbert space. Let $p\in(0,\infty)$. There exists a constant $C_p>0$ such that, for every $(\mathfrak{F}_t)$-progressively measurable stochastic process $\mathbf{G}$ satisfying* (2.13), *the following holds:*

$$\mathbb{E}\sup_{t\in[0,T]}\left\|\int_0^t \mathbf{G}(s)\,\mathrm{d}W(s)\right\|_H^p \leq C_p\,\mathbb{E}\left(\int_0^T \|\mathbf{G}(s)\|^2_{L_2(\mathfrak{U},H)}\,\mathrm{d}s\right)^{\frac{p}{2}}. \tag{2.17}$$

Next, we report the following result by Flandoli–Gątarek [FG95, Lemma 2.1], which allows one to show fractional Sobolev regularity in time for a stochastic integral.

Lemma 2.3.9. *Let H be a separable Hilbert space. Let $p \in [2,\infty)$ and $\alpha \in [0,\frac{1}{2})$ be given. Then there exists a constant $c(p,\alpha,T) > 0$ such that, for every progressively measurable process $\mathbf{G} \in L^p(\Omega \times [0,T]; L_2(\mathfrak{U},H))$, we have*

$$\mathbb{E}\left[\left\|\int_0^t \mathbf{G}(s)\,\mathrm{d}W(s)\right\|^p_{W^{\alpha,p}(0,T;H)}\right] \le c(p,\alpha,T)\mathbb{E}\left[\int_0^T \|\mathbf{G}(s)\|^p_{L^2(\mathfrak{U},H)}\,\mathrm{d}s\right]. \tag{2.18}$$

The proof of Lemma 2.3.9 is rather elementary and actually relies on the Burkholder–Davis–Gundy inequality (2.17). Note that, since only the embedding

$$W^{\alpha,p}(0,T;H) \hookrightarrow C^\beta([0,T];H) \quad \text{whenever } \beta \in \left[0, \alpha - \frac{1}{p}\right), \tag{2.19}$$

holds true, the two estimates (2.17) and (2.18) are different and (2.18) does not imply (2.17), not even in the case $p = 2$.

We also record the following result; see Gīhman–Skorokhod [GhS80].

Lemma 2.3.10 (Martingale inequality). *Let H be a separable Hilbert space. Then*

$$\mathbb{P}\left(\int_0^T \left\|\int_0^t \mathbf{G}(s)\,\mathrm{d}W(s)\right\|^2_H \mathrm{d}t > \varepsilon\right) \le T\frac{\delta}{\varepsilon} + \mathbb{P}\left(\int_0^T \|\mathbf{G}(t)\|^2_{L^2(\mathfrak{U},H)}\,\mathrm{d}t > \delta\right) \tag{2.20}$$

for any $\varepsilon > 0$, $\delta > 0$.

The approach to establish Hölder continuity of a stochastic integral then relies on the Kolmogorov continuity theorem. This is a classical result that allows one to show existence of a Hölder continuous modification for stochastic processes. There are several versions with different proofs available in the literature. Let us particularly mention the elegant analytical proof based on the Sobolev embedding, which can be found in Da Prato–Zabczyk [DPZ92, Theorem 3.3].

Theorem 2.3.11 (Kolmogorov continuity theorem). *Let $\mathbf{U}$ be a stochastic process taking values in a separable Banach space X. Assume that there exist constants $K > 0$, $a \ge 1$, $b > 0$ such that, for all $s,t \in [0,T]$,*

$$\mathbb{E}\|\mathbf{U}(t) - \mathbf{U}(s)\|^a_X \le K|t-s|^{1+b}.$$

Then there exists $\mathbf{V}$, a modification of $\mathbf{U}$, which has $\mathbb{P}$-a.s. Hölder continuous trajectories with exponent γ for every $\gamma \in (0,\frac{b}{a})$. In addition, we have

$$\mathbb{E}\|\mathbf{V}\|^a_{C^\gamma_t X} \lesssim K,$$

where the proportional constant does not depend on $\mathbf{V}$.

2.4 Itô's formula

One of the central results in stochastic Itô calculus is the corresponding chain rule called Itô's formula. The proof can be found in Da Prato–Zabczyk [DPZ92, Theorem 4.17].

Let $W = \sum_{k=1}^{\infty} e_k W_k$ be a cylindrical Wiener process in a separable Hilbert space $\mathfrak{U}$. Let $\mathbf{G}$ be an $L_2(\mathfrak{U}; H)$-valued stochastically integrable process, let $\mathbf{g}$ be an H-valued progressively measurable Bochner integrable process, and let $\mathbf{U}(0)$ be an $\mathfrak{F}_0$-measurable H-valued random variable. Then the process

$$\mathbf{U}(t) = \mathbf{U}(0) + \int_0^t \mathbf{g}(s)\,\mathrm{d}s + \int_0^t \mathbf{G}(s)\,\mathrm{d}W(s)$$

is well-defined. Assume that a function $F : H \to \mathbb{R}$ and its derivatives F', F'' are uniformly continuous on bounded subsets of H.

Theorem 2.4.1 (Itô's formula). *Under the above assumptions we have $\mathbb{P}$-a.s., for all $t \in [0, T]$,*

$$\begin{aligned} F(\mathbf{U}(t)) = F(\mathbf{U}(0)) &+ \int_0^t \langle F'(\mathbf{U}(s)), \mathbf{g}(s)\rangle\,\mathrm{d}s + \int_0^t \langle F'(\mathbf{U}(s)), \mathbf{G}(s)\,\mathrm{d}W(s)\rangle \\ &+ \frac{1}{2}\int_0^t \operatorname{Tr}(\mathbf{G}(s)^* F''(\mathbf{U}(s))\mathbf{G}(s))\,\mathrm{d}s \end{aligned} \tag{2.21}$$

where $\operatorname{Tr} A = \sum_{k=1}^{\infty} \langle A e_k, e_k\rangle$ *for A being a bounded linear operator on H.*

Note that, unlike in the deterministic theory, a first order chain rule does not hold for the Itô calculus. Indeed, the above Itô formula contains a term of second order, the so-called Itô correction term. Besides, the classical notation used for the stochastic integral in (2.21) might seem a bit ambiguous at first sight. It is to be interpreted in the following sense:

$$\int_0^t \langle F'(\mathbf{U}(s)), \mathbf{G}(s)\,\mathrm{d}W(s)\rangle = \sum_{k=1}^{\infty} \int_0^t \langle F'(\mathbf{U}(s)), \mathbf{G}(s)e_k\rangle\,\mathrm{d}W_k(s).$$

A special case of this result is Itô's product rule.

Proposition 2.4.2. *Assume that* $\mathbf{U}$, $\mathbf{V}$ *are H-valued stochastic processes defined on the same probability space and satisfying the assumptions of Theorem 2.4.1. Assume that they admit the decompositions*

$$\mathbf{U}(t) = \mathbf{U}(0) + \int_0^t \mathbf{g}(s)\,\mathrm{d}s + \int_0^t \mathbf{G}(s)\,\mathrm{d}W(s),$$

$$\mathbf{V}(t) = \mathbf{V}(0) + \int_0^t \mathbf{h}(s)\,\mathrm{d}s + \int_0^t \mathbf{H}(s)\,\mathrm{d}W(s).$$

Then

$$\begin{aligned}\langle \mathbf{U}(t), \mathbf{V}(t)\rangle &= \langle \mathbf{U}(0), \mathbf{V}(0)\rangle + \int_0^t \langle \mathbf{V}(s), \mathbf{g}(s)\rangle \, \mathrm{d}s + \int_0^t \langle \mathbf{V}(s), \mathbf{G}(s)\, \mathrm{d}W(s)\rangle \\ &\quad + \int_0^t \langle \mathbf{U}(s), \mathbf{h}(s)\rangle \, \mathrm{d}s + \int_0^t \langle \mathbf{U}(s), \mathbf{H}(s)\, \mathrm{d}W(s)\rangle \\ &\quad + \sum_{k=1}^{\infty} \int_0^t \langle \mathbf{G}(s)e_k, \mathbf{H}(s)e_k\rangle \, \mathrm{d}s.\end{aligned}$$

As a first generalization of Theorem 2.4.1, we present an Itô formula for the square of the H-norm obtained for equations in the variational framework. Another generalization is given in Theorem A.4.1.

Let H be a separable Hilbert space with inner product $\langle\cdot,\cdot\rangle$ and H^* its dual. Let V be a Banach space, such that $V \hookrightarrow H$ continuously and densely. Identifying H and H^* via the Riesz isomorphism, we obtain the Gelfand triplet $V \hookrightarrow H \hookrightarrow V^*$.

The following Itô formula can be found in Krylov–Rozovskii [KR79, Theorem I.3.1] or Prévôt–Röckner [PR07, Theorem 4.2.5].

Theorem 2.4.3. *Assume that* $\mathbf{U}$ *is a continuous* V^**-valued stochastic process given by*

$$\mathbf{U}(t) = \mathbf{U}(0) + \int_0^t \mathbf{g}(s)\, \mathrm{d}s + \int_0^t \mathbf{G}(s)\, \mathrm{d}W(s), \quad t \in [0, T],$$

where

$$\mathbf{G} \in L^2(\Omega \times [0,T]; L_2(\mathfrak{U}; H)), \quad \mathbf{g} \in L^2(\Omega \times [0,T]; V^*),$$

are both progressively measurable and $\mathbf{U}(0) \in L^2(\Omega; H)$ *is* $\mathfrak{F}_0$*-measurable.*

If $\mathbf{U} \in L^2(\Omega \times [0,T]; V)$*, then* $\mathbf{U}$ *is an* H*-valued continuous stochastic process,*

$$\mathbb{E} \sup_{t\in[0,T]} \|\mathbf{U}(t)\|_H^2 < \infty,$$

and the following Itô formula holds true $\mathbb{P}$*-a.s.:*

$$\begin{aligned}\|\mathbf{U}(t)\|_H^2 &= \|\mathbf{U}(0)\|_H^2 + 2\int_0^t \langle \mathbf{g}(s), \mathbf{U}(s)\rangle_{V^*,V} \, \mathrm{d}s + 2\int_0^t \langle \mathbf{U}(s), \mathbf{G}(s)\, \mathrm{d}W(s)\rangle \\ &\quad + \int_0^t \|\mathbf{G}(s)\|_{L_2(\mathfrak{U};H)}^2 \, \mathrm{d}s \quad \textit{for any } t \in [0,T].\end{aligned}$$

2.5 Pathwise vs. martingale solutions

From the probabilistic point of view, two concepts of solution are typically considered in the theory of stochastic (partial) differential equations. Namely, pathwise (or strong) solutions and martingale (or weak) solutions. In the former notion, the underlying probability space as well as the driving stochastic process is fixed in advance, whereas in the latter case, these stochastic elements become part of the solution of the problem. We call such solutions martingale solutions due to the connection with the so-called Stroock–Varadhan martingale problem as discussed in Karatzas–Shreve [KS91, Section 5.4]. Clearly, existence of a pathwise solution is stronger and implies existence of a martingale solution.

Consider a stochastic differential equation

$$\mathrm{d}\mathbf{U} = b(\mathbf{U})\,\mathrm{d}t + \sigma(\mathbf{U})\,\mathrm{d}W, \quad \mathbf{U}(0) = \mathbf{U}_0, \tag{2.22}$$

where W is an $\mathbb{R}^d$-valued Wiener process defined on a probability space $(\Omega, \mathfrak{F}, \mathbb{P})$. For simplicity of the presentation we are in finite dimension. So, a solution $\mathbf{U}$ is an $\mathbb{R}^m$-valued stochastic process. The coefficients b, σ are Borel measurable functions

$$b : \mathbb{R}^m \to \mathbb{R}^m, \quad \sigma : \mathbb{R}^m \to \mathbb{R}^{m\times d}.$$

We formulate the following two notions of solution to (2.22).

Definition 2.5.1. Let W be an $\mathbb{R}^d$-valued Wiener process defined on the stochastic basis $(\Omega, \mathfrak{F}, (\mathfrak{F}_t)_{t\geq 0}, \mathbb{P})$ with a complete right-continuous filtration. Let $\mathbf{U}_0$ be a $\mathcal{F}_0$-measurable random variable. An $\mathbb{R}^m$-valued stochastic process X is called a *pathwise solution* to (2.22) with the initial condition $\mathbf{U}_0$, provided:
(1) $\mathbf{U}$ is an $\mathbb{R}^m$-valued $(\mathfrak{F}_t)$-adapted stochastic process with $\mathbb{P}$-a.s. continuous trajectories;
(2) $\mathbf{U}(0) = \mathbf{U}_0$ $\mathbb{P}$-a.s.;
(3) we have

$$\int_0^t |b(\mathbf{U}(s))| + |\sigma(\mathbf{U}(s))|^2\,\mathrm{d}s < \infty \tag{2.23}$$

and

$$\mathbf{U}(t) = \mathbf{U}_0 + \int_0^t b(\mathbf{U}(s))\,\mathrm{d}s + \int_0^t \sigma(\mathbf{U}(s))\,\mathrm{d}W \tag{2.24}$$

$\mathbb{P}$-a.s. for all $t \in [0, \infty)$.

Definition 2.5.2. Let Λ be a Borel probability measure on $\mathbb{R}^m$. A triple

$$((\Omega, \mathfrak{F}, (\mathfrak{F}_t)_{t\geq 0}, \mathbb{P}), \mathbf{U}, W)$$

is called a *martingale solution* to (2.22) with the initial law Λ, provided:
(1) $(\Omega, \mathfrak{F}, (\mathfrak{F}_t)_{t\geq 0}, \mathbb{P})$ is a stochastic basis with a complete right-continuous filtration;
(2) W is an $\mathbb{R}^d$-valued $(\mathfrak{F}_t)$-Wiener process on $(\Omega, \mathfrak{F}, \mathbb{P})$;
(3) $\mathbf{U}$ is an $\mathbb{R}^m$-valued $(\mathfrak{F}_t)$-adapted stochastic process with $\mathbb{P}$-a.s. continuous trajectories;

(4) $\Lambda = \mathcal{L}[\mathbf{U}(0)]$;
(5) (2.23) and (2.24) hold $\mathbb{P}$-a.s. for all $t \in [0, \infty)$.

We observe that, since in the case of martingale solutions the underlying stochastic basis is not known in advance, the initial condition can only be prescribed in the form of an initial law Λ. Furthermore, a martingale solution is not necessarily adapted to the canonical filtration generated by the Wiener process.

The concept of a martingale solution, although weaker than a pathwise solution, is yet extremely useful in both theory and applications. In particular, it allows one to prove existence under much weaker assumptions on the coefficients b, σ. More precisely, let us recall the seminal papers by Itô [Itô46, Itô51] where existence of a unique pathwise solution to (2.22) was proved under the Lipschitz assumption on b and σ. On the other hand, Skorokhod [Sko61, Sko62] showed that there exists a solution if b and σ are only continuous functions of at most linear growth. It was only realized later that two different concepts of a solution are involved. Nowadays, it is well known that this difference is substantial in general: under the assumptions of the Skorokhod theorem pathwise solutions do not necessarily exist; see Barlow [Bar82].

2.5.1 Pathwise uniqueness vs. uniqueness in law

Attached to the two notions of solution are two ways how uniqueness for (2.22) can be understood. More precisely, we address pathwise uniqueness or uniqueness in law.

Definition 2.5.3. We say that *pathwise uniqueness* is a property of (2.22), provided, if $\mathbf{U}$, $\mathbf{V}$ are two solutions to (2.22) defined on the same stochastic basis with the same Wiener process and under the assumption that $\mathbf{U}(0) = \mathbf{V}(0)$ $\mathbb{P}$-a.s., that the processes $\mathbf{U}$ and $\mathbf{V}$ are indistinguishable, i.e.,

$$\mathbb{P}\big(\mathbf{U}(t) = \mathbf{V}(t) \text{ for all } t \geq 0\big) = 1.$$

Definition 2.5.4. We say that *uniqueness in law* is a property of (2.22), provided, if

$$((\Omega, \mathfrak{F}, (\mathfrak{F}_t)_{t\geq 0}, \mathbb{P}), \mathbf{U}, W), \quad ((\tilde{\Omega}, \tilde{\mathfrak{F}}, (\tilde{\mathfrak{F}}_t), \tilde{\mathbb{P}}), \mathbf{V}, \tilde{W})$$

are two solutions to (2.22) with the same initial law, that the two processes $\mathbf{U}$ and $\mathbf{V}$ have the same law.

A classical result of Yamada–Watanabe, as presented for instance in Karatzas–Shreve [KS91, Proposition 3.20], shows that pathwise uniqueness implies uniqueness in law.

2.6 Stochastic compactness method

Several results presented in the main body of this book rely on a stochastic version of the compactness method known from the deterministic setting. Roughly speaking, the problem is the following: one constructs a sequence of approximations and in order to show their convergence it is necessary to establish compactness. In the presence of randomness, which is represented by the probability space $(\Omega, \mathfrak{F}, \mathbb{P})$, one has to be more careful, as we do not assume any topological structure on Ω and therefore the usual tools based on compact embeddings can only be applied to the time and space variable. A classical method to overcome this flaw employs Prokhorov's and Skorokhod's theorems; see Billingsley [Bil99, Theorem 5.1, 5.2] and Dudley [Dud02, Theorem 11.7.2], respectively.

Theorem 2.6.1 (Prokhorov's theorem). *Let X be a Polish space and $\mathcal{M}$ a collection of probability measures on X. Then $\mathcal{M}$ is tight if and only if it is relatively weakly compact.*

Theorem 2.6.2 (Skorokhod's theorem). *Let X be a Polish space and let μ_n, $n \in \mathbb{N}_0$, be probability measures on X such that μ_n converges weakly to μ_0. Then on some probability space there exist X-valued random variables $\mathbf{U}_n$, $n \in \mathbb{N}_0$, such that the law of $\mathbf{U}_n$ is μ_n, $n \in \mathbb{N}_0$, and $\mathbf{U}_n(\omega) \to \mathbf{U}_0(\omega)$ in X a.s.*

Note that Dudley [Dud02, Theorem 11.7.2] only gives the existence of some probability space such that the statement of Theorem 2.6.2 holds true and the corresponding convergence is almost sure. Nevertheless, the so-called Blackwell–Dubbins–Fernique theorem gives the following stronger result; see Blackwell–Dubbins [BD83] and Fernique [Fer88].

Theorem 2.6.3. *Let X be a Polish space. With every Borel probability measure μ on X, one can associate a Borel X-valued random variable ξ_μ defined on the probability space $([0,1], \mathfrak{B}([0,1]), \mathfrak{L})$ such that $\mathcal{L}[\xi_\mu] = \mu$, and, whenever measures μ_n converge weakly-$*$ to μ, one has $\lim_{n\to\infty} \xi_{\mu_n}(\omega) = \xi_\mu(\omega)$ for almost all $\omega \in [0,1]$.*

As an immediate consequence, we obtain a representation for a Wiener process together with an initial condition having a given law.

Corollary 2.6.4. *Let X be a Polish space and let Λ be a Borel probability measure on X. Then there exists a stochastic basis $(\Omega, \mathfrak{F}, (\mathfrak{F}_t)_{t\geq 0}, \mathbb{P})$ with an $\mathfrak{F}_0$-measurable X-valued random variable $\mathbf{U}_0$ having the law Λ and with an $(\mathfrak{F}_t)$-cylindrical Wiener process W.*

Proof. Let Γ denote the law of a cylindrical Wiener process on $\mathfrak{U}$. Recall that, due to Remark 2.1.30, such a probability measure exists and, due to Corollary 2.1.33, it is supported on $C_{\mathrm{loc}}([0,\infty); \mathfrak{U}_0)$. The application of Theorem 2.6.2 to $\mu = \Lambda \otimes \Gamma$ yields a complete probability space $(\Omega, \mathfrak{F}, \mathbb{P})$ together with random variables $\mathbf{U}_0$ and W having the law Λ and Γ, respectively. Hence, according to Lemma 2.1.35, W is a cylindrical Wiener process with respect to its canonical filtration. Moreover, by definition of the joint law μ, it follows that $\mathbf{U}_0$ is independent of W. Therefore, it can be shown from Theorem 2.1.34 that W is a cylindrical Wiener process with respect to the filtration $(\mathfrak{F}_t)_{t\geq 0}$ given by

$$\mathfrak{F}_t := \sigma(\sigma(\mathbf{U}_0) \cup \sigma(W_s; 0 \leq s \leq t)), \quad t \in [0, \infty),$$

which yields the claim. □

The stochastic compactness method proceeds in several steps. First, one shows tightness of laws of the approximate sequence at hand. Second, due to Prokhorov's theorem, there exists a subsequence of these laws which converges weakly. Third, Skorokhod's theorem yields a representation by random variables defined on some probability space and converging a.s. The last step then consists in the identification of the limit as the desired object of interest, for instance a solution to a stochastic PDE.

The identification of the limit typically contains further difficulties, namely in the passage to the limit in the stochastic integral. Indeed, one deals with a sequence of stochastic integrals driven by a sequence of Wiener processes. One possibility is to pass to the limit directly, and such technical convergence results appeared in a number of publications (see, e.g., Bensoussan [Ben95] or Gyöngy–Krylov [GK96]); a detailed proof of the version presented below can be found in Debussche et al. [DGHT11, Lemma 2.1].

Lemma 2.6.5. *Let $(\Omega, \mathfrak{F}, \mathbb{P})$ be a complete probability space and H a separable Hilbert space. For $n \in \mathbb{N}$, let W_n be an $(\mathfrak{F}_t^n)$-cylindrical Wiener process and let $\mathbf{G}_n$ be an $(\mathfrak{F}_t^n)$-progressively measurable stochastic process ranging in $L_2(\mathfrak{U}; H)$. Suppose that*

$$W_n \to W \quad \text{in } C([0,T]; \mathfrak{U}_0) \text{ in probability},$$
$$\mathbf{G}_n \to \mathbf{G} \quad \text{in } L^2(0,T; L_2(\mathfrak{U}; H)) \text{ in probability},$$

where W is a cylindrical Wiener process adapted to a filtration $(\mathfrak{F}_t)_{t\geq 0}$, and $\mathbf{G}$ is $(\mathfrak{F}_t)$-progressively measurable. Then

$$\int_0^{\cdot} \mathbf{G}_n \, \mathrm{d}W_n \to \int_0^{\cdot} \mathbf{G} \, \mathrm{d}W \quad \text{in } L^2(0,T; H) \text{ in probability.}$$

In our applications, the target space X will be taken as $W^{\ell,2}(\mathbb{T}^N)$ with a suitable ℓ. Using Lemma 2.2.18, we rephrase Lemma 2.6.5 as follows.

Lemma 2.6.6. *Let $(\Omega, \mathfrak{F}, \mathbb{P})$ be a complete probability space. For $n \in \mathbb{N}$, let W_n be an $(\mathfrak{F}_t^n)$-cylindrical Wiener process and let $\mathbf{G}_n$ be an $(\mathfrak{F}_t^n)$-progressively measurable stochastic process such that $\mathbf{G}_n \in L^2(0,T; L_2(\mathfrak{U}; W^{\ell,2}(\mathbb{T}^N)))$ a.s. Suppose that*

$$W_n \to W \quad \text{in } C([0,T]; \mathfrak{U}_0) \text{ in probability},$$

$$\mathbf{G}_n \to \mathbf{G} \quad \textit{in } L^2(0,T;L^2(\mathfrak{U};W^{\ell,2}(\mathbb{T}^N))) \textit{ in probability},$$

where $W = \sum_{k=1}^{\infty} e_k W_k$. *Let* $(\mathfrak{F}_t)_{t\geq 0}$ *be the filtration given by*

$$\mathfrak{F}_t = \sigma\left(\bigcup_{k=1}^{\infty} \sigma_t[\mathbf{G}e_k] \cup \sigma_t[W_k]\right).$$

Then after a possible change on a set of zero measure in $\Omega\times(0,T)$, $\mathbf{G}$ *is* $(\mathfrak{F}_t)$*-progressively measurable, and*

$$\int_0^{\cdot} \mathbf{G}_n \,\mathrm{d}W_n \to \int_0^{\cdot} \mathbf{G}\,\mathrm{d}W \quad \textit{in } L^2(0,T;W^{\ell,2}(\mathbb{T}^N)) \textit{ in probability.}$$

Remark 2.6.7. Here, we have tacitly assumed that $\mathbf{G}_{n,k} := \mathbf{G}_n e_k$ and $\mathbf{G}_k := \mathbf{G}e_k$ have been extended for $t<0$ and $t>T$ as random distributions in $\mathcal{D}'(\mathbb{R}\times\mathbb{T}^N)$. While the extension for $t>T$ is basically irrelevant, for $t<0$ we may take $\mathbf{G}_{n,k} = \mathbf{G}_{0,n,k}$ constant in time, $\mathfrak{F}_0^n$-measurable, and such that

$$\mathbf{G}_{0,n} \to \mathbf{G}_0 \quad \text{in } L^2(\mathfrak{U};W^{\ell,2}(\mathbb{T}^N)) \text{ in probability.}$$

Proof of Lemma 2.6.6. In view of Lemma 2.2.18 and Lemma 2.6.5, the proof reduces to showing that the new filtration $(\mathfrak{F}_t)_{t\geq 0}$ is non-anticipative with respect to the limit Wiener process W. However, Lemma 2.6.6 is a fundamental tool used frequently in the existence proof and therefore we give a complete proof for the reader's convenience.

Step 1 [Non-anticipativity of $(\mathfrak{F}_t)_{t\geq 0}$]
To see that the filtration $(\mathfrak{F}_t)_{t\geq 0}$ is non-anticipative with respect to the noise W, it suffices to observe that

$$\begin{aligned}
&\mathbb{E}\big[F(\langle \mathbf{G}_1,\varphi_{1,1}\rangle,\dots,\langle \mathbf{G}_\ell,\varphi_{\ell,m}\rangle, W_1(t_1),\dots,W_m(t_m))\\
&\qquad \times H(W_1(t+s)-W_1(t),\dots,W_k(t+s)-W_k(t))\big]\\
&= \mathbb{E}\big[F(\langle \mathbf{G}_1,\varphi_{1,1}\rangle,\dots,\langle \mathbf{G}_\ell,\varphi_{\ell,m}\rangle, W_1(t_1),\dots W_m(t_m))\big]\\
&\qquad \times \mathbb{E}\big[H(W_1(t+s)-W_1(t),\dots,W_k(t+s)-W_k(t))\big]
\end{aligned}$$

for any $0\leq t_i\leq t$, $s>0$, $\varphi_{j,i}$ supported in $(-\infty,t)$, $i=1,\dots,m$, $j=1,\dots,\ell$, and bounded continuous F and H. Indeed this follows from the fact that the same relation holds for $\mathbf{G}_n$, W_n, $W_n(t+s)-W_n(t)$ and both converge to their respective limits in probability.

Step 2 [Cut-off]
We write the stochastic integrals in the form

$$\begin{aligned}
\int_0^{\tau} \mathbf{G}_n\,\mathrm{d}W_n &= \sum_{k\leq K}\int_0^{\tau} \mathbf{G}_{n,k}\,\mathrm{d}W_{n,k} + \sum_{k>K}\int_0^{\tau} \mathbf{G}_{n,k}\,\mathrm{d}W_{n,k},\\
\int_0^{\tau} \mathbf{G}\,\mathrm{d}W &= \sum_{k\leq K}\int_0^{\tau} \mathbf{G}_k\,\mathrm{d}W_k + \sum_{k>K}\int_0^{\tau} \mathbf{G}_k\,\mathrm{d}W_k.
\end{aligned}$$

Evoking the martingale inequality (2.20), we obtain

$$\begin{aligned}
&\mathbb{P}\left(\int_0^T \left\| \sum_{k>K} \int_0^t \mathbf{G}_{n,k}(s)\,\mathrm{d}W_{n,k}(s) \right\|^2_{W_x^{\ell,2}} \mathrm{d}t > \varepsilon\right) \\
&\quad \le T\varepsilon + \mathbb{P}\left(\int_0^T \sum_{k>K} \|\mathbf{G}_{n,k}\|^2_{W_x^{\ell,2}}\,\mathrm{d}t > \varepsilon^2\right) \\
&\quad \le T\varepsilon + \mathbb{P}\left(\int_0^T \sum_{k>K} \|\mathbf{G}_{n,k} - \mathbf{G}_k\|^2_{W_x^{\ell,2}}\,\mathrm{d}t > \varepsilon^2\right) + \mathbb{P}\left(\int_0^T \sum_{k>K} \|\mathbf{G}_k\|^2_{W_x^{\ell,2}}\,\mathrm{d}t > \varepsilon^2\right)
\end{aligned}$$

for any $\varepsilon > 0$. Seeing that the second term on the right hand side of the above inequality vanishes for $n \to \infty$ and the third one can be made arbitrarily small taking K large enough, we conclude

$$\sum_{k>K} \int_0^\tau \mathbf{G}_{n,k}\,\mathrm{d}W_{n,k} + \sum_{k>K} \int_0^\tau \mathbf{G}_k\,\mathrm{d}W_k \to 0 \quad \text{in } L^2(0,T;W^{\ell,2}(\mathbb{T}^N)) \text{ in probability}$$

for $K \to \infty$, $n \to \infty$. Consequently, the proof reduces to the case of a finite sum in the stochastic integral.

Step 3 [Convergence for a finite sum]

Fixing k we write $\mathbf{G}_n, W_n$ and $\mathbf{G}, W$ instead of $\mathbf{G}_{n,k}, W_{n,k}$ and $\mathbf{G}_k, W_k$, respectively. Using the regularization operator $[[\cdot]]_\delta$ introduced in Section 2.2.2, we write

$$\begin{aligned}
&\left\| \int_0^\tau \mathbf{G}_n\,\mathrm{d}W_n - \int_0^\tau \mathbf{G}\,\mathrm{d}W \right\|_{W_x^{\ell,2}} \\
&\quad \le \left\| \int_0^\tau (\mathbf{G}_n - [[\mathbf{G}_n]]_\delta)\,\mathrm{d}W_n \right\|_{W_x^{\ell,2}} + \left\| \int_0^\tau ([[\mathbf{G}]]_\delta - \mathbf{G})\,\mathrm{d}W \right\|_{W_x^{\ell,2}} \\
&\qquad + \left\| \int_0^\tau [[\mathbf{G}_n]]_\delta\,\mathrm{d}W_n - \int_0^\tau [[\mathbf{G}]]_\delta\,\mathrm{d}W \right\|_{W_x^{\ell,2}}.
\end{aligned}$$

The last term is easy to handle as we have

$$\begin{aligned}
\int_0^\tau [[\mathbf{G}_n]]_\delta\,\mathrm{d}W_n - \int_0^\tau [[\mathbf{G}]]_\delta\,\mathrm{d}W &= [[\mathbf{G}_n]]_\delta W_n(\tau) - [[\mathbf{G}]]_\delta W(\tau) \\
&\quad - \int_0^\tau (\partial_t [[\mathbf{G}_n]]_\delta W_n - \partial_t [[\mathbf{G}]]_\delta W)\,\mathrm{d}t.
\end{aligned}$$

Thus the desired convergence follows as long as $\delta > 0$.

Consequently, it suffices to show

$$\int_0^T \left\| \int_0^\tau (\mathbf{G}_n - [[\mathbf{G}_n]]_\delta)\,\mathrm{d}W_n \right\|^2_{W_x^{\ell,2}} \mathrm{d}\tau \to 0 \quad \text{in probability as } \delta \to 0 \qquad (2.25)$$

uniformly in n, and

$$\int_0^T \left\| \int_0^\tau (\mathbf{G} - [[\mathbf{G}]]_\delta)\,\mathrm{d}W \right\|_{W_x^{\ell,2}}^2 \mathrm{d}\tau \to 0 \quad \text{in probability as } \delta \to 0. \tag{2.26}$$

Observe that (2.26) follows from the fact that

$$[[\mathbf{G}]]_\delta \to \mathbf{G} \quad \text{in } L^2(0,T;W^{\ell,2}(\mathbb{T}^N)) \text{ a.s.}$$

Finally, we use again the martingale inequality (2.20) to deduce

$$\begin{aligned}
&\mathbb{P}\left(\int_0^T \left\| \int_0^\tau (\mathbf{G}_n - [[\mathbf{G}_n]]_\delta)\,\mathrm{d}W_n \right\|_{W_x^{\ell,2}}^2 \mathrm{d}t > \varepsilon \right) \\
&\lesssim \varepsilon + \mathbb{P}\left(\int_0^T \|\mathbf{G}_n - [[\mathbf{G}_n]]_\delta\|_{W_x^{\ell,2}}^2 \,\mathrm{d}t \geq \varepsilon^2 \right) \\
&\lesssim \varepsilon + \mathbb{P}\left(\int_0^T \|\mathbf{G}_n - \mathbf{G}\|_{W_x^{\ell,2}}^2 \,\mathrm{d}t \geq \varepsilon^2 \right) + \mathbb{P}\left(\int_0^T \|[[\mathbf{G}_n]]_\delta - [[\mathbf{G}]]_\delta\|_{W_x^{\ell,2}}^2 \,\mathrm{d}t \geq \varepsilon^2 \right) \\
&\quad + \mathbb{P}\left(\int_0^T \|[[\mathbf{G}]]_\delta - \mathbf{G}\|_{W_x^{\ell,2}}^2 \,\mathrm{d}t \geq \varepsilon^2 \right) \\
&\lesssim \varepsilon + \mathbb{P}\left(\int_0^T \|\mathbf{G}_n - \mathbf{G}\|_{W_x^{\ell,2}}^2 \,\mathrm{d}t \geq \varepsilon^2 \right) + \mathbb{P}\left(\int_0^T \|[[\mathbf{G}]]_\delta - \mathbf{G}\|_{W_x^{\ell,2}}^2 \,\mathrm{d}t \geq \varepsilon^2 \right).
\end{aligned}$$

Thus we conclude by using (2.26) and the hypotheses imposed on $(\mathbf{G}_n)_{n\in\mathbb{N}}$. □

For completeness, let us also mention that there are other ways to identify the limit stochastic integral. Namely, one can show that the limit process is a martingale, identify its quadratic variation, and apply an integral representation theorem for martingales, if available (see Da Prato–Zabczyk [DPZ92]). Another approach follows a rather general and elementary method introduced in Brzeźniak–Ondreját [BO07] and has already been generalized to different settings. The keystone is to identify not only the quadratic variation of the corresponding martingale but also its cross variation with the limit Wiener process obtained through compactness. This permits to conclude directly without the use of any further difficult results. The following result summarizes the simple observation made by Brzeźniak–Ondreját [BO07]; for a proof of a more general version, see [Hof13, Proposition A.1].

Lemma 2.6.8. *Let $(\Omega, \mathfrak{F}, (\mathfrak{F}_t)_{t\geq 0}, \mathbb{P})$ be a stochastic basis and W an $(\mathfrak{F}_t)$-cylindrical Wiener process in $\mathfrak{U}$ given by $W = \sum_{k=1}^\infty e_k W_k$. Let $\mathbf{G}$ be an $(\mathfrak{F}_t)$-progressively measurable $L_2(\mathfrak{U};H)$-valued stochastically integrable stochastic process, i.e., satisfying (2.13). If $\mathbf{M}$ is an H-valued square integrable continuous $(\mathfrak{F}_t)$-martingale such that, for all $\varphi \in H$, the processes*

$$\langle \mathbf{M}(\cdot), \varphi\rangle^2 - \sum_{k=1}^\infty \int_0^\cdot \langle \mathbf{G}(s)e_k, \varphi\rangle^2 \,\mathrm{d}s, \quad \langle \mathbf{M}(\cdot), \varphi\rangle W_k(\cdot) - \int_0^\cdot \langle \mathbf{G}(s)e_k, \varphi\rangle \,\mathrm{d}s$$

are $(\mathfrak{F}_t)$*-martingales, then* $\mathbb{P}$*-a.s.*

$$\mathbf{M}(\cdot) = \int_0^{\cdot} \mathbf{G}(s)\,\mathrm{d}W(s) \quad \textit{for all } t \in [0,T].$$

2.7 Jakubowski–Skorokhod representation theorem

Let us stress that both Prokhorov's theorem and Skorokhod's theorem above are restricted to Polish spaces. Nevertheless, the specific structure of the compressible Navier–Stokes equations naturally leads to weak convergences, that is, to convergence in non-metrizable spaces. Jakubowski [Jak97] gave a suitable generalization of the Prokhorov and Skorokhod theorems that holds true in the class of sub-Polish spaces introduced in Definition 2.1.3. These are topological spaces that are not necessarily metrizable but retain several important properties of Polish spaces. Let us recall that a topological space is called sub-Polish if there exists a countable family

$$\{g_n : X \to (-1,1);\ n \in \mathbb{N}\}$$

of continuous functions that separate points of X; the reader is referred to Brzeźniak et al. [BOS16, Section 3] for further discussion.

It can be seen that for instance Polish spaces, separable Banach spaces equipped with the weak topology, and spaces of weakly continuous functions with values in a separable Banach space belong to this class. In fact, all the spaces considered in this book are continuously embedded in the space of distributions $\mathcal{D}'$ so they are sub-Polish.

Among the properties of sub-Polish spaces needed in this book is the following result; see [Jak97, Theorem 2].

Theorem 2.7.1 (Jakubowski–Skorokhod representation theorem). *Let* (X,τ) *be a sub-Polish space and let* $\mathcal{S}$ *be the* σ*-field generated by* $\{f_n;\ n \in \mathbb{N}\}$*. If* $(\mu_n)_{n\in\mathbb{N}}$ *is a tight sequence of probability measures on* $(X,\mathcal{S})$*, then there exists a subsequence* (n_k) *and* X*-valued Borel measurable random variables* $(\mathbf{U}_k)_{k\in\mathbb{N}}$ *and* $\mathbf{U}$ *defined on the standard probability space* $([0,1],\overline{\mathfrak{B}([0,1])},\mathfrak{L})$*, such that* μ_{n_k} *is the law of* $\mathbf{U}_k$ *and* $\mathbf{U}_k(\omega)$ *converges to* $\mathbf{U}(\omega)$ *in* X *for every* $\omega \in [0,1]$*. Moreover, the law of* $\mathbf{U}$ *is a Radon measure.*

We point out that, basically in all situations considered in this book, the use of Jakubowski's extension of the Skorokhod theorem can be avoided. A typical situation is a family of random variables $(\mathbf{U}_n)_{n\in\mathbb{N}}$ ranging in a dual X^* of a separable Banach space X such that $\|\mathbf{U}_n\|_{X^*}$ is tight on $[0,\infty)$. Our goal is to show that there exist random variables $\tilde{\mathbf{U}}_n$, $n \in \mathbb{N}$, and $\mathbf{U}_0$ defined on the standard probability space $([0,1];\overline{\mathfrak{B}([0,1])},\mathfrak{L})$ and satisfying

$$\mathbf{U}_n \overset{d}{\sim}_{X^*} \tilde{\mathbf{U}}_n, \quad \tilde{\mathbf{U}}_{n_k} \overset{*}{\rightharpoonup} \tilde{\mathbf{U}}_0 \quad \text{in } X^*\ \mathfrak{L}\text{-a.e.}$$

Suppose that there is a separable Hilbert space Y,

$$Y \overset{c}{\hookrightarrow} X, \quad Y \text{ dense in } X. \tag{2.27}$$

Consequently, the norm in X^* is given by

$$\|Z\|_{X^*} = \sup_{h_m \in Y,\ \|h_m\|_X \leq 1} \langle Z, h_m \rangle \quad \text{for a countable family } (h_m)_{m \in \mathbb{N}} \subset Y.$$

Let $[\mathbf{U}_n, \|\mathbf{U}_n\|_{X^*}]$ be an extended sequence of random variables taking values in $[Y^*, \mathbb{R}]$. In view of (2.27) and tightness of $\|\mathbf{U}_n\|_{X^*}$, the sequence

$$(\mathbf{U}_n, \|\mathbf{U}_n\|_{X^*})_{n \in \mathbb{N}} \quad \text{is tight in the Polish space } Y^* \times \mathbb{R}.$$

Applying the standard Skorokhod theorem (see Theorem 2.6.2) we obtain a subsequence $(\mathbf{U}_{n_k})_{k \in \mathbb{N}}$, together with a sequence of $Y^* \times \mathbb{R}$-valued random variables $(\tilde{\mathbf{U}}_{n_k}, \tilde{N}_{n_k})_{k \in \mathbb{N}}$ defined on the standard probability space $([0,1], \overline{\mathfrak{B}([0,1])}, \mathfrak{L})$, such that

$$\begin{aligned}
&[\mathbf{U}_{n_k}, \|\mathbf{U}_{n_k}\|_{X^*}] \overset{d}{\sim}_{Y^* \times \mathbb{R}} [\tilde{\mathbf{U}}_{n_k}, \tilde{N}_{n_k}], \\
&\mathbf{U}_{n_k} \overset{d}{\sim}_{Y^*} \tilde{\mathbf{U}}_{n_k}, \quad \tilde{\mathbf{U}}_{n_k} \to \tilde{\mathbf{U}}_0 \quad \text{in } Y^* \ \mathfrak{L}\text{-a.e.}, \\
&\|\mathbf{U}_{n_k}\|_{X^*} \overset{d}{\sim}_{\mathbb{R}} \tilde{N}_k, \quad \tilde{N}_k \to \tilde{N} \ \mathfrak{L}\text{-a.e.}
\end{aligned} \tag{2.28}$$

As Y is densely embedded in X, we get

$$\mathbf{U}_{n_k} \overset{d}{\sim}_{X^*} \tilde{\mathbf{U}}_{n_k}.$$

Finally, the norm $\|Z\|_{X^*}$ is a supremum of $Z \mapsto \langle Z; h_m \rangle$, $h_m \in Y$; whence a supremum of continuous functions on Y^*. Consequently, in accordance with (2.28),

$$\mathcal{L}_{\mathbb{R}}[\|\tilde{\mathbf{U}}_{n_k}\|_{X^*} - \tilde{N}_{n_k}] = \mathcal{L}_{\mathbb{R}}[\|\mathbf{U}_{n_k}\|_{X^*} - \|\mathbf{U}_{n_k}\|_{X^*}] = \delta_0;$$

whence

$$\|\tilde{\mathbf{U}}_{n_k}\|_{X^*} = \tilde{N}_{n_k} \quad \mathfrak{L}\text{-a.e.}$$

Thus it follows from (2.28) that $\|\tilde{\mathbf{U}}_{n_k}\|_{X^*}$ is bounded $\mathfrak{L}$-a.e. which implies the desired result

$$\tilde{\mathbf{U}}_{n_k} \overset{*}{\rightharpoonup} \tilde{\mathbf{U}}_0 \text{ in } X^* \quad \mathfrak{L}\text{-a.e.}$$

Note that in applications one usually has $X = L^p(\mathcal{O})$, for some $1 \leq p < \infty$ and $Y = W_0^{k,2}(\mathcal{O})$ with sufficiently large k.

2.8 Random distributions in L^p and Young measures

Very often, we will have to handle the following situation. Let $Q = [0, T] \times \mathbb{T}^N$. Given a sequence of random distributions $(\mathbf{U}_n)_{n\in\mathbb{N}}$ satisfying the estimate

$$\mathbb{E}[\|\mathbf{U}_n\|^p_{L^p_{t,x}}] < \infty \quad \text{for a certain } p \in (1, \infty).$$

Obviously, the set of laws of $(\mathbf{U}_n)_{n\in\mathbb{N}}$ is tight on $L^p(Q)$. In analogy with the deterministic theory, we wish to conclude that, up to a subsequence,

$$G(\mathbf{U}_n) \rightharpoonup \overline{G(\mathbf{U})} \quad \text{in } L^{p/q}(Q) \text{ a.s.}$$
$$\text{for any } G \in C(\mathbb{R}^M; \mathbb{R}), \quad |G(Z)| \le c(1 + |Z|^q) \quad 1 \le q < p,$$

and for certain limit function $\overline{G(\mathbf{U})}$. This is indeed the case if we change appropriately the probability space as in Skorokhod's theorem, Theorem 2.6.2.

Theorem 2.8.1. *Let $(\mathbf{U}_n)_{n\in\mathbb{N}}$ be a sequence of random distributions in $\mathcal{D}'(\mathbb{R}\times\mathbb{T}^N)$ such that*

$$\mathbb{E}[\|\mathbf{U}_n\|^p_{L^p_{t,x}}] < \infty \quad \text{for a certain } p \in (1, \infty). \tag{2.29}$$

Then there exists a new sequence $(\tilde{\mathbf{U}}_n)_{n\in\mathbb{N}}$ defined on the standard probability space $([0,1], \overline{\mathfrak{B}[0,1]}, \mathfrak{L})$, such that

$$\mathbf{U}_n \overset{d}{\sim} \tilde{\mathbf{U}}_n$$

and

$$G(\tilde{\mathbf{U}}_n) \rightharpoonup \overline{G(\tilde{\mathbf{U}})} \quad \text{in } L^{p/q}(Q)\ \mathfrak{L}\text{-a.s.}$$
$$\text{for any } G \in C(\mathbb{R}^M; \mathbb{R}), \quad |G(Z)| \le c(1 + |Z|^q), \quad 1 \le q < p. \tag{2.30}$$

Proof. The proof leans on a suitable application of Skorokhod's theorem, Theorem 2.6.2. Let $(G_m)_{m\in\mathbb{N}}$ be a family of functions in $C_b(\mathbb{R}^M)$ that are dense in the space of continuous functions satisfying the growth restriction (2.30). Specifically, for each $G \in C(\mathbb{R}^M; \mathbb{R})$ satisfying

$$|G(Z)| \le c(1 + |Z|^q), \quad 1 \le q < p,$$

there is $(G_n)_{n\in\mathbb{N}} \subset (G_m)_{m\in\mathbb{N}}$ such that

$$|G_n| \le |G|, \quad G_n \to G \text{ in } C_{\text{loc}}(\mathbb{R}^M; \mathbb{R}).$$

Extending $\mathbf{U}_n$ to be zero outside Q and rescaling the time variable as the case may be, we may assume that $Q = \mathbb{T}^{N+1}$ – the time-space torus. We define a Hilbert space $\mathcal{X}$,

$$\mathcal{X} = W^{-k,2}(Q) \times (W^{-k,2}(Q))^{\aleph_0} \times \mathbb{R}, \quad k > \frac{N}{2} + 1,$$

endowed with the scalar product

$$\langle [\mathbf{U}_1, (G^1_m)_{m\in\mathbb{N}}, r^1]; [\mathbf{U}_2, (G^2_m)_{m\in\mathbb{N}}, r^2] \rangle$$
$$= \langle \mathbf{U}_1, \mathbf{U}_2 \rangle_{W^{-k,2}_x} + \sum_{m=1}^{\infty} \frac{1}{2^m} \langle G^1_m; G^2_m \rangle_{W^{-k,2}_x} + r^1 r^2.$$

Now, it suffices to apply the Skorokhod representation theorem to

$$[\mathbf{U}_n, (G_m(\mathbf{U}_n))_{m\in\mathbb{N}}, \|\mathbf{U}_n\|_{L^p_{t,x}}]_{n\in\mathbb{N}}.$$

Indeed, convergence in $W^{-k,2}$ and boundedness in L^p, $p > 1$, imply weak convergence in L^q, $q < p$. Accordingly, we get (2.30) for any G_m, and finally, by density, for any G satisfying the growth conditions in (2.30). □

In view of Theorem 2.8.1, a random distribution $\mathbf{U}$ ranging in the Lebesgue spaces L^p endowed with the weak topology should be regarded as a pair $[\mathbf{U}, \|\mathbf{U}\|_{L^p_{t,x}}]$ ranging in the Polish space $W^{-k,2} \times \mathbb{R}$.

In view of future applications to a large class of compositions, we evoke the fundamental theorem in the theory of Young measures; see Ball [Bal89] and Pedregal [Ped97, Chapter 6, Theorem 6.2]. Recall that the notion of a Carathéodory function was introduced in Definition 2.2.16.

Theorem 2.8.2. *Let $(\mathbf{v}_n)_{n\in\mathbb{N}}$, $\mathbf{v}_n : \mathcal{O} \subset \mathbb{R}^N \to \mathbb{R}^M$ be a weakly convergent sequence of functions in $L^1(\mathcal{O}; \mathbb{R}^M)$, where $\mathcal{O}$ is a domain in $\mathbb{R}^N$. Then there exist a subsequence (not relabeled) and a parametrized family $\{\nu_x\}_{x\in\mathcal{O}}$ of probability measures on $\mathbb{R}^M$ depending measurably on $x \in \mathcal{O}$ with the following property: for any Carathéodory function $G = G(x, Z)$, $x \in \mathcal{O}$, $Z \in \mathbb{R}^M$, such that*

$$G(\cdot, \mathbf{v}_n) \to \overline{G} \quad \text{weakly in } L^1(\mathcal{O}),$$

we have

$$\overline{G}(x) = \int_{\mathbb{R}^M} G(x, z)\, \mathrm{d}\nu_x(z) \quad \text{for a.a. } x \in \mathcal{O}.$$

Combining Theorem 2.8.1 and Theorem 2.8.2 we immediately obtain the following.

Corollary 2.8.3. *Under the hypotheses of Theorem* 2.8.1, *the validity of* (2.30) *may be extended to any Carathéodory function $G = G(x, Z)$ satisfying the growth restrictions in* (2.30) *uniformly in x.*

The parametrized measure in Theorem 2.8.2 is called the Young measure associated to the sequence $(\mathbf{v}_n)_{n\in\mathbb{N}}$. The concept of Young measures was introduced by Young [You69] as a technical tool for describing composite limits of non-linear functions with weakly convergent sequences; for further reading we refer the reader also to Castaing et al. [CdFV04] for a thorough exposition. There is another way of proving Theorem 2.8.1 that relies on compactness of the corresponding Young measures.

In what follows, we denote by $\mathcal{P}(\mathbb{R})$ the set of probability measures on $\mathbb{R}$.

Definition 2.8.4 (Young measure). Let (X,λ) be a σ-finite measure space. A mapping $\nu : X \to \mathcal{P}(\mathbb{R})$ is called a *Young measure*, provided it is weak-$*$-measurable, that is, for all $\phi \in C_b(\mathbb{R})$, the mapping $z \mapsto \nu_z(\phi)$ from X to $\mathbb{R}$ is measurable. A Young measure ν is said to vanish at infinity if

$$\int_X \int_{\mathbb{R}} |\xi| \, \mathrm{d}\nu_z(\xi) \, \mathrm{d}\lambda(z) < \infty.$$

Proposition 2.8.5 (Compactness of Young measures). *Let (X,λ) be a σ-finite measure space such that $L^1(X)$ is separable. Let $(\nu^n)_{n\in\mathbb{N}}$ be a sequence of Young measures on X such that, for some $p \in [1,\infty)$,*

$$\sup_{n\in\mathbb{N}} \int_X \int_{\mathbb{R}} |\xi|^p \, \mathrm{d}\nu_z^n(\xi) \, \mathrm{d}\lambda(z) < \infty. \tag{2.31}$$

Then there exists a Young measure ν on X and a subsequence still denoted by $(\nu^n)_{n\in\mathbb{N}}$ such that, for all $\psi \in L^1(X)$ and all $\phi \in C_b(\mathbb{R})$,

$$\lim_{n\to\infty} \int_X \psi(z) \int_{\mathbb{R}} \phi(\xi) \, \mathrm{d}\nu_z^n(\xi) \, \mathrm{d}\lambda(z) = \int_X \psi(z) \int_{\mathbb{R}} \phi(\xi) \, \mathrm{d}\nu_z(\xi) \, \mathrm{d}\lambda(z). \tag{2.32}$$

Various results of this form can be found in the literature; a proof for the case of (X,λ) being a finite measure space can be found in Debussche–Vovelle [DV10, Theorem 5, Corollary 6]. However, one can actually observe that this additional assumption is not used in the proof and therefore the same proof applies to our setting of (X,λ) being a σ-finite measure space. For a discussion of similar issues in the case of $X = \mathbb{R}^N$ we refer the reader to Málek et al. [MNRR96, Theorem 4.2.1].

Note that the separability of $L^1(X)$ follows once the corresponding σ-algebra on X is countably generated; see, e.g., Cohn [Coh13, Proposition 3.4.5].

We observe that a Young measure is actually an element of the unit sphere of $L^\infty_{w^*}(X;\mathcal{M}_b(\mathbb{R}))$, the space of weakly-$*$-measurable and bounded mappings from X to $\mathcal{M}_b(\mathbb{R})$. Consequently, any sequence of Young measures $(\nu^n)_{n\in\mathbb{N}}$ is bounded in $L^\infty_{w^*}(X;\mathcal{M}_b(\mathbb{R}))$ and, due to Banach–Alaoglu's theorem, there exists a subsequence converging weakly-$*$ to some limit $\nu \in L^\infty_{w^*}(X;\mathcal{M}_b(\mathbb{R}))$. It should be noted, however, that the limit is generally not a Young measure. Indeed, setting $\nu_z^n(\cdot) := \delta_n(\cdot)$, we see that mass can be escaping at infinity and therefore only $\operatorname{esssup}_{z\in X} \|\nu_z\|_{\mathcal{M}_b} \le 1$ holds true in general. This is a consequence of a more general fact, namely that the unit sphere is not closed under weak-$*$-convergence. The sufficient condition that guarantees that also the limit ν is a Young measure, that is, a parametrized probability measure, is (2.31), which yields the necessary tightness.

Let $(L^\infty_{w^*}(X;\mathcal{P}(\mathbb{R})), w^*)$ denote the space of Young measures equipped with the topology given by (2.32). To summarize our discussion, this topology is not the subspace topology of the weak-$*$-topology on $L^\infty_{w^*}(X;\mathcal{M}_b(\mathbb{R}))$. The topology of Young measures is finer – it has more open sets – and consequently there are less compact sets. Thus, any continuous map $f : (L^\infty_{w^*}(X;\mathcal{M}_b(\mathbb{R})), w^*) \to (-1,1)$ is also continuous from $(L^\infty_{w^*}(X;\mathcal{P}(\mathbb{R})), w^*)$ to $(-1,1)$. Recall that $C_0(\mathbb{R})$, the space of continuous functions vanishing at infinity is separable and it is the predual of $\mathcal{M}_b(\mathbb{R})$. Thus $L^\infty_{w^*}(X;\mathcal{M}_b(\mathbb{R}))$ is the dual of the separable Banach space $L^1(X;C_0(\mathbb{R}))$ (see Edwards [Edw94, Section 8.18]) and hence it is sub-Polish. We deduce that $(L^\infty_{w^*}(X;\mathcal{P}(\mathbb{R})), w^*)$ is also a sub-Polish space with the same separating sequence of continuous functions.

As a consequence of Proposition 2.8.5 we deduce the following.

Corollary 2.8.6. *Let (X,λ) be a σ-finite measure space such that $L^1(X)$ is separable. Let $R \in \mathbb{N}$ and $p \in [1,\infty)$. Then the set*

$$B_{R,p} := \left\{ \nu \in L^\infty_{w^*}(X;\mathcal{P}(\mathbb{R}));\ \int_X \int_{\mathbb{R}} |\xi|^p \,\mathrm{d}\nu_z(\xi)\,\mathrm{d}\lambda(z) \le R \right\}$$

is relatively compact in $(L^\infty_{w^*}(X;\mathcal{P}(\mathbb{R})), w^*)$.

Let $\mathbf{U}$ be a random distribution whose law is tight on $L^p(\mathbb{R}\times\mathbb{T}^N)$ for some $p \in (1,\infty)$. Then one may define a Young measure on $\Omega\times\mathbb{R}\times\mathbb{T}^N$ associated to $\mathbf{U}$ by

$$\nu_{\omega,t,x}(\cdot) := \delta_{\mathbf{U}(\omega,t,x)}(\cdot).$$

In addition, ν can be regarded as a random variable taking values in the space of Young measures on $\mathbb{R}\times\mathbb{T}^N$, i.e., $L^\infty_{w^*}(\mathbb{R}\times\mathbb{T}^N;\mathcal{P}(\mathbb{R}))$. This leads to an alternative proof of Theorem 2.8.1.

Another proof of Theorem 2.8.1. For $n \in \mathbb{N}$, let ν_n denote the Young measure associated to $\mathbf{U}_n$, regarded as a random variable taking values in $(L^\infty_{w^*}(\mathbb{R}\times\mathbb{T}^N;\mathcal{P}(\mathbb{R})), w^*)$. Denote by $\mathcal{L}[\nu_n]$ its law. Let $R \in \mathbb{N}$ and let $B_{R,p}$ be the relatively compact set defined in Corollary 2.8.6. As a consequence of (2.29), we obtain

$$\mathcal{L}[\nu_n](B^c_{R,p}) = \mathbb{P}(\nu_n \in B^c_{R,p}) = \mathbb{P}\left(\int_{\mathbb{R}}\int_{\mathbb{T}^N} |\mathbf{U}_n|^p \,\mathrm{d}x\,\mathrm{d}t > R\right) \le \frac{1}{R}\mathbb{E}\|\mathbf{U}_n\|^p_{L^p_{t,x}} \le \frac{C}{R}.$$

Thus, the family of laws $\{\mathcal{L}[\nu_n];\ n \in \mathbb{N}\}$ is tight on $(L^\infty_{w^*}(\mathbb{R}\times\mathbb{T}^N, \mathcal{P}(\mathbb{R})), w^*)$. Next, we observe that the set

$$B_R := \{\mathbf{U} \in L^p(\mathbb{R}\times\mathbb{T}^N);\ \|\mathbf{U}\|_{L^p_{t,x}} \le R\}$$

is relatively compact with respect to the weak topology on $L^p(\mathbb{R}\times\mathbb{T}^N)$. If $\mathcal{L}[\mathbf{U}_n]$ denotes the law of $\mathbf{U}_n$ on $(L^p(\mathbb{R}\times\mathbb{T}^N), w)$, we obtain

$$\mathcal{L}[\mathbf{U}_n](B^c_R) \le \frac{1}{R^p}\mathbb{E}\|\mathbf{U}_n\|^p_{L^p_{t,x}} \le \frac{C}{R^p}.$$

This implies the tightness of $\{\mathcal{L}[\mathbf{U}_n]; n \in \mathbb{N}\}$ on $(L^p(\mathbb{R} \times \mathbb{T}^N), w)$.

Consequently, we deduce that the family of joint laws $\{\mathcal{L}[\mathbf{U}_n, \nu_n]; n \in \mathbb{N}\}$ is tight. Hence we may apply Jakubowski–Skorokhod's theorem, Theorem 2.7.1, to obtain a complete probability space $(\tilde{\Omega}, \tilde{\mathfrak{F}}, \tilde{\mathbb{P}})$ with Borel random variables $[\tilde{\mathbf{U}}_n, \tilde{\nu}_n]$ having the law $\mathcal{L}[\mathbf{U}_n, \nu_n]$ and converging $\tilde{\mathbb{P}}$-a.s. to some Borel random variable $[\tilde{\mathbf{U}}, \tilde{\nu}]$ taking values in $(L^p(\mathbb{R} \times \mathbb{T}^N), w) \times (L^\infty_{w^*}(\mathbb{R} \times \mathbb{T}^N; \mathcal{P}(\mathbb{R})), w^*)$.

As the next step, we observe that, as a consequence of equality of laws, we have

$$\begin{aligned}
1 = \mathbb{P}\Big(&\int_{\mathbb{R}} \int_{\mathbb{T}^N} \psi(t,x) \int_{\mathbb{R}} \phi(\xi)\, \mathrm{d}\nu_{n,t,x}(\xi)\, \mathrm{d}x\, \mathrm{d}t = \int_{\mathbb{R}} \int_{\mathbb{T}^N} \psi(t,x)\phi(\mathbf{U}_n)\, \mathrm{d}x\, \mathrm{d}t \\
&\text{for all } \psi \in L^1(\mathbb{R} \times \mathbb{T}^N),\ \phi \in C_b(\mathbb{R})\Big) \\
= \tilde{\mathbb{P}}\Big(&\int_{\mathbb{R}} \int_{\mathbb{T}^N} \psi(t,x) \int_{\mathbb{R}} \phi(\xi)\, \mathrm{d}\tilde{\nu}_{n,t,x}(\xi)\, \mathrm{d}x\, \mathrm{d}t = \int_{\mathbb{R}} \int_{\mathbb{T}^N} \psi(t,x)\phi(\tilde{\mathbf{U}}_n)\, \mathrm{d}x\, \mathrm{d}t \\
&\text{for all } \psi \in L^1(\mathbb{R} \times \mathbb{T}^N),\ \phi \in C_b(\mathbb{R})\Big).
\end{aligned}$$

Note that this is valid due to tightness even though the space $C_b(\mathbb{R})$ is not separable. We deduce that $\tilde{\nu}_{n,\omega,t,x}(\cdot) = \delta_{\tilde{\mathbf{U}}_n(\omega,t,x)}(\cdot)$.

Finally, for a fixed ω from a set of full probability, we may apply the fundamental theorem of Young measures (Theorem 2.8.2; see also Málek et al. [MNRR96, Theorem 4.2.1, Corollary 4.2.10]) and (2.30) follows with

$$\overline{G(\tilde{\mathbf{U}})}(\omega, t, x) = \int_{\mathbb{R}} G(\xi)\, \mathrm{d}\tilde{\nu}_{\omega,t,x}(\xi). \qquad \square$$

2.9 Stochastic partial differential equations

Formally, all stochastic PDEs considered in this book can be written in the form

$$\mathrm{d}D(\mathbf{U}) + \operatorname{div} \mathbf{F}(\mathbf{U})\, \mathrm{d}t = \mathbf{G}(\mathbf{U})\mathrm{d}W, \quad D(\mathbf{U}_0) = D_0, \tag{2.33}$$

or

$$D(\mathbf{U})(\tau) - D_0 + \int_0^\tau \operatorname{div} \mathbf{F}(\mathbf{U})\, \mathrm{d}t = \int_0^\tau \mathbf{G}(\mathbf{U})\, \mathrm{d}W.$$

Using x-dependent test functions $\varphi \in C^\infty(\mathbb{T}^N)$, this can be relaxed to

$$\langle D(\mathbf{U})(\tau), \varphi\rangle - \langle D_0, \varphi\rangle - \int_0^\tau \langle \mathbf{F}(\mathbf{U}), \nabla\varphi\rangle\, \mathrm{d}t = \int_0^\tau \langle \mathbf{G}(\mathbf{U}), \varphi\rangle\, \mathrm{d}W.$$

Finally, applying formally Itô's formula, Theorem 2.4.1, we get

$$\int_0^T [\partial_t \psi \langle D(\mathbf{U}), \varphi\rangle + \psi\langle \mathbf{F}(\mathbf{U}), \nabla\varphi\rangle]\, \mathrm{d}t + \int_0^T \psi\langle \mathbf{G}(\mathbf{U}), \varphi\rangle\, \mathrm{d}W + \psi(0)\langle D_0, \varphi\rangle = 0, \tag{2.34}$$

for any $\varphi \in C^\infty(\mathbb{T}^N)$, $\psi \in C_c^\infty([0,T))$. The family of integral identities (2.34) can be interpreted as a weak formulation of (2.33). Note that

$$\int_0^T \psi\langle \mathbf{G}(\mathbf{U}), \varphi\rangle \, \mathrm{d}W = \sum_{k=1}^{\infty} \int_0^T \psi\langle \mathbf{G}_k(\mathbf{U}), \varphi\rangle \, \mathrm{d}W_k,$$

where $W = (W_k)_{k\in\mathbb{N}}$ is a cylindrical Wiener process and the coefficient $\mathbf{G}$ is identified with $(\mathbf{G}_k)_{k\in\mathbb{N}}$; see Section 2.3 for more details.

The operators D, $\mathbf{F}$, and $\mathbf{G}_k$, $k \in \mathbb{N}$, will be non-linear superposition operators given by *Carathéodory functions* in the sense of Definition 2.2.16. More precisely,

$$D = D(x,\mathbf{U}), \quad \mathbf{F} = \mathbf{F}(x,\mathbf{U}), \quad \mathbf{G}_k = \mathbf{G}_k(x,\mathbf{U}), \quad x \in \mathbb{T}^N, \quad \mathbf{U} \in \mathbb{R}^M,$$

where $D(\cdot,\mathbf{U})$ is measurable for any $\mathbf{U}$, $D(x,\cdot)$ is continuous for almost every $x \in \mathbb{T}^N$, and similarly for $\mathbf{F}$ and $\mathbf{G}_k$, $k \in \mathbb{N}$. Consequently, the most general assumption on regularity of $\mathbf{U}$ requires at least $\mathbf{U} \in L^1((0,T) \times \mathbb{T}^N)$ $\mathbb{P}$-a.s. Extending $\mathbf{U}$ suitably outside $(0,T)$ we may therefore assume that $\mathbb{P}$-a.s.

$$\mathbf{U} \in L^1_{\mathrm{loc}}(\mathbb{R}; L^1(\mathbb{T}^N)) \subset \mathcal{D}'(\mathbb{R} \times \mathbb{T}^N). \tag{2.35}$$

In addition, in order to give sense to (2.34), we suppose

$$D(\mathbf{U}), \mathbf{F}(\mathbf{U}), \mathbf{G}_k(\mathbf{U}) \in L^1_{\mathrm{loc}}(\mathbb{R}; L^1(\mathbb{T}^N)) \quad \mathbb{P}\text{-a.s.} \tag{2.36}$$

In view of Section 2.3 and in particular Remark 2.3.7, the stochastic integral in (2.34) is well-defined, provided, at least,

$$\int_0^T \sum_{k=1}^{\infty} |\langle \mathbf{G}_k(\mathbf{U}), \varphi\rangle|^2 \, \mathrm{d}t < \infty \quad \mathbb{P}\text{-a.s.}, \tag{2.37}$$

meaning

$$(\langle \mathbf{G}_k(\mathbf{U}), \varphi\rangle)_{k\in\mathbb{N}} \in L^2(0,T; L_2(\mathfrak{U}; \mathbb{R})) \quad \mathbb{P}\text{-a.s. for any } \varphi \in C^\infty(\mathbb{T}^N).$$

Finally, the random distribution $\mathbf{U}$ must be adapted to the filtration $(\mathfrak{F}_t)_{t\geq 0}$ associated to the Wiener process. Note that, if (2.37) is satisfied for every $\varphi \in W^{\ell,2}(\mathbb{T}^N)$, the stochastic integral can be interpreted as a local martingale taking values in the Hilbert space $W^{-\ell,2}(\mathbb{T}^N)$.

It follows from (2.34) that the function $t \mapsto \langle D(\mathbf{U}), \varphi\rangle$ is continuous in $[0,T]$ $\mathbb{P}$-a.s. and

$$\langle D(\mathbf{U})(0), \varphi\rangle = \langle D_0, \varphi\rangle.$$

We shall suppose that $D_0 = D(\mathbf{U}_0)$ for a certain $\mathbf{U}_0 \in L^1(\mathbb{T}^N)$ $\mathbb{P}$-a.s. Accordingly, we extend $\mathbf{U}(t)$ by $\mathbf{U}_0$ for all $t \leq 0$.

We have the following result about equivalence in law for problem (2.34).

Theorem 2.9.1. *Let $\mathbf{U} \in L^1_{\mathrm{loc}}(\mathbb{R}; L^1(\mathbb{T}^N))$ $\mathbb{P}$-a.s., $\mathbf{U}(t) = \mathbf{U}_0$ for $t \le 0$, be a random distribution belonging to the regularity class (2.35)–(2.37). Let $W = (W_k)_{k\in\mathbb{N}}$ be a cylindrical Wiener process. Suppose that the filtration*

$$\mathfrak{F}_t = \sigma\left(\sigma_t[\mathbf{U}] \cup \bigcup_{k=1}^{\infty} \sigma_t[W_k] \right), \quad t \ge 0,$$

is non-anticipative with respect to W. Let $\tilde{\mathbf{U}}$ be another random distribution and $\tilde{W}$ another stochastic process such that their joint laws coincide, namely

$$\mathcal{L}[\mathbf{U}, W] = \mathcal{L}[\tilde{\mathbf{U}}, \tilde{W}] \quad or \quad [\mathbf{U}, W] \overset{d}{\sim} [\tilde{\mathbf{U}}, \tilde{W}].$$

Then $\tilde{W}$ is a cylindrical Wiener process, the filtration

$$\tilde{\mathfrak{F}}_t = \sigma\left(\sigma_t[\tilde{\mathbf{U}}] \cup \bigcup_{k=1}^{\infty} \sigma_t[\tilde{W}_k] \right), \quad t \ge 0,$$

is non-anticipative with respect to $\tilde{W}$, and

$$\begin{aligned} \mathcal{L}_{\mathbb{R}}\bigg[\int_0^T & [\partial_t \psi \langle D(\mathbf{U}), \varphi\rangle + \psi \langle \mathbf{F}(\mathbf{U}), \nabla\varphi\rangle]\,\mathrm{d}t \\ & + \int_0^T \psi \langle \mathbf{G}(\mathbf{U}), \varphi\rangle \,\mathrm{d}W + \psi(0)\langle D(\mathbf{U}_0), \varphi\rangle \bigg] \\ = \mathcal{L}_{\mathbb{R}}\bigg[\int_0^T & [\partial_t \psi \langle D(\tilde{\mathbf{U}}), \varphi\rangle + \psi \langle \mathbf{F}(\tilde{\mathbf{U}}), \nabla\varphi\rangle]\,\mathrm{d}t \\ & + \int_0^T \psi \langle \mathbf{G}(\tilde{\mathbf{U}}); \varphi\rangle \,\mathrm{d}\tilde{W} + \psi(0)\langle D(\tilde{\mathbf{U}}_0), \varphi\rangle \bigg] \end{aligned} \tag{2.38}$$

for any deterministic $\psi \in C_c^\infty([0,T))$, $\phi \in C^\infty(\mathbb{T}^N)$.

Remark 2.9.2. The meaning of $[\mathbf{U}, W] \overset{d}{\sim} [\tilde{\mathbf{U}}, \tilde{W}]$ is that the vector-valued distributions $[\mathbf{U}, W_1, \dots, W_M]$ and $[\tilde{\mathbf{U}}, \tilde{W}_1, \dots, \tilde{W}_M]$ satisfy

$$[\mathbf{U}, W_1, \dots, W_M] \overset{d}{\sim} [\tilde{\mathbf{U}}, \tilde{W}_1, \dots, \tilde{W}_M] \quad \text{for any } M \in \mathbb{N},$$

in the sense of Definition 2.2.11.

Proof of Theorem 2.9.1. First of all, we observe that, according to Lemma 2.1.35, $\tilde{W}$ is a cylindrical Wiener process with respect to its canonical filtration $(\sigma_t[\tilde{W}])_{t\ge0}$.

Fix the test functions φ and ψ in (2.38). The proof will be carried out by performing several regularizations. Note that $\mathbf{U}$ is defined as $\mathbf{U}_0$ for $t < 0$ and the value of $\mathbf{U}$ for $t > T$ is irrelevant as ψ is compactly supported in $[0, T)$. Then:

- We regularize $\mathbf{U}$, $\tilde{\mathbf{U}}$ by means of the convolution kernels. We have

$$[\mathbf{U}]_{t,x,\delta} = \theta^t_\delta(\cdot - \delta) * [\theta^t_x * \mathbf{U}], \quad [\tilde{\mathbf{U}}]_{t,x,\delta} = \theta^t_\delta(\cdot - \delta) * [\theta^x_\delta * \tilde{\mathbf{U}}], \quad \delta > 0.$$

Note that all derivatives of $[\mathbf{U}]_{t,x,\delta}$, $[\tilde{\mathbf{U}}]_{t,x,\delta}$ are continuous in $(-\infty, T]$ and their laws coincide pointwise. Moreover,

$$[\mathbf{U}]_{t,x,\delta} = [\mathbf{U}_0]_{x,\delta}, \quad [\tilde{\mathbf{U}}]_{t,x,\delta} = [\tilde{\mathbf{U}}_0]_{x,\delta} \quad \text{for } t \le 0.$$

Finally, observe that $[\tilde{\mathbf{U}}]_{t,x,\delta}$ is adapted to the filtration

$$\tilde{\mathfrak{F}}_t = \sigma\left[\sigma_t[\tilde{\mathbf{U}}] \cup \bigcup_{k=1}^{\infty} \sigma_t[\tilde{W}_k]\right], \quad t \ge 0,$$

the latter one being non-anticipative with respect to the noise $\tilde{W}$.
- We replace the non-linearities D, $\mathbf{F}$, and $\mathbf{G}_k$ by their cut-offs,

$$D_M = T_M \circ D, \quad \mathbf{F}_M = T_M \circ \mathbf{F}, \quad \mathbf{G}_{k,M} = T_M \circ \mathbf{G}_k,$$
$$\text{where } T_M \in C^\infty(\mathbb{R}),\ 0 \le T_M \le 2M, \quad T_M(Z) = Z \text{ for } |Z| \le M,\ T_M' > 0.$$

- We take $\Delta_J = \frac{T}{J}$ and set

$$t_0 = 0, \quad t_{j+1} = t_j + \Delta_J, \quad j = 0, \ldots, J-1.$$

Step 1: Obviously,

$$\begin{aligned}
&\mathcal{L}_{\mathbb{R}}\Bigg[\int_0^T [\partial_t \psi \langle D_M([\mathbf{U}]_{t,x,\delta}), \varphi\rangle + \psi\langle \mathbf{F}_M([\mathbf{U}]_{t,x,\delta}), \nabla\varphi\rangle]\,\mathrm{d}t \\
&\quad + \sum_{k=1}^{K}\left(\sum_{j=1}^{J-1} \psi(t_j)\langle \mathbf{G}_{k,M}([\mathbf{U}]_{t,x,\delta})(t_j), \varphi\rangle (W_k(t_{j+1}) - W_k(t_j))\right) + \psi(0)\langle D_M([\mathbf{U}_0]_{x,\delta}), \varphi\rangle\Bigg] \\
&= \mathcal{L}_{\mathbb{R}}\Bigg[\int_0^T [\partial_t \psi \langle D_M([\tilde{\mathbf{U}}]_{t,x,\delta}), \varphi\rangle + \psi\langle \mathbf{F}_M([\tilde{\mathbf{U}}]_{t,x,\delta}), \nabla\varphi\rangle]\,\mathrm{d}t \\
&\quad + \sum_{k=1}^{K}\left(\sum_{j=1}^{J-1} \psi(t_j)\langle \mathbf{G}_{k,M}([\tilde{\mathbf{U}}]_{t,x,\delta})(t_j), \varphi\rangle (\tilde{W}_k(t_{j+1}) - \tilde{W}_k(t_j))\right) + \psi(0)\langle D_M([\tilde{\mathbf{U}}_0]_{x,\delta}), \varphi\rangle\Bigg].
\end{aligned}$$

Letting $J \to \infty$, we obtain

$$\begin{aligned}
&\mathcal{L}_{\mathbb{R}}\Bigg[\int_0^T [\partial_t \psi \langle D_M([\mathbf{U}]_{t,x,\delta}), \varphi\rangle + \psi\langle \mathbf{F}_M([\mathbf{U}]_{t,x,\delta}), \nabla\varphi\rangle]\,\mathrm{d}t \\
&\quad + \sum_{k=1}^{K}\int_0^T \psi\langle \mathbf{G}_{k,M}([\mathbf{U}]_{t,x,\delta}), \varphi\rangle\,\mathrm{d}W_k + \psi(0)\langle D_M([\mathbf{U}_0]_{x,\delta}), \varphi\rangle\Bigg] \\
&= \mathcal{L}_{\mathbb{R}}\Bigg[\int_0^T [\partial_t \psi \langle D_M([\tilde{\mathbf{U}}]_{t,x,\delta}), \varphi\rangle + \psi\langle \mathbf{F}_M([\tilde{\mathbf{U}}]_{t,x,\delta}), \nabla\varphi\rangle]\,\mathrm{d}t
\end{aligned}$$

$$+\sum_{k=1}^{K}\int_0^T \psi\langle \mathbf{G}_{k,M}([\tilde{\mathbf{U}}]_{t,x,\delta}),\varphi\rangle \mathrm{d}\tilde{W}_k + \psi(0)\langle D_M([\tilde{\mathbf{U}}_0]_{x,\delta}),\varphi\rangle\Bigg], \tag{2.39}$$

as a consequence of the construction of the stochastic integral in Section 2.3. Note that the functions $[\tilde{\mathbf{U}}]_{t,x,\delta}$ are smooth. In particular, the Riemann sums in the stochastic integral converge in probability and unconditionally to their limits.

Step 2: The next step is to let $\delta \to 0$ in (2.39). As $\mathbf{U} \in L^1_{\text{loc}}(\mathbb{R}; L^1(\mathbb{T}^N))$ a.s., we have

$$[\mathbf{U}]_{t,x,\delta} \to \mathbf{U} \quad \text{in } L^1_{\text{loc}}(\mathbb{R}; L^1(\mathbb{T}^N)) \text{ as } \delta \to 0 \ \mathbb{P}\text{-a.s.}$$

In particular, $\mathbf{U}$ is a random variable in $L^1_{\text{loc}}(\mathbb{R}; L^1(\mathbb{T}^N))$. By virtue of Theorem 2.2.12, $\tilde{\mathbf{U}}$ is a random variable in $L^1(-L, L; L^1(\mathbb{T}^N))$ for any $L > 0$. Consequently, we have

$$[\tilde{\mathbf{U}}]_{t,x,\delta} \to \tilde{\mathbf{U}} \quad \text{in } L^1_{\text{loc}}(\mathbb{R}; L^1(\mathbb{T}^N)) \text{ as } \delta \to 0 \text{ a.s.}$$

Using Lemma 2.6.6, we apply the limit $\delta \to 0$ in (2.39) to obtain

$$\begin{aligned}
&\mathcal{L}_{\mathbb{R}}\Bigg[\int_0^T [\partial_t\psi\langle D_M(\mathbf{U}),\varphi\rangle + \psi\langle \mathbf{F}_M(\mathbf{U}),\nabla\varphi\rangle]\,\mathrm{d}t \\
&\quad + \sum_{k=1}^{K}\int_0^T \psi\langle \mathbf{G}_{k,M}(\mathbf{U}),\varphi\rangle\,\mathrm{d}W_k + \psi(0)\langle D_M(\mathbf{U}_0),\varphi\rangle\Bigg] \\
&= \mathcal{L}_{\mathbb{R}}\Bigg[\int_0^T [\partial_t\psi\langle D_M(\tilde{\mathbf{U}}),\varphi\rangle + \psi\langle \mathbf{F}_M(\tilde{\mathbf{U}}),\nabla\varphi\rangle]\,\mathrm{d}t \\
&\quad + \sum_{k=1}^{K}\int_0^T \psi\langle \mathbf{G}_{k,M}(\tilde{\mathbf{U}}),\varphi\rangle\,\mathrm{d}\tilde{W}_k + \psi(0)\langle D_M(\tilde{\mathbf{U}}_0),\varphi\rangle\Bigg].
\end{aligned} \tag{2.40}$$

Step 3: Finally, making use of Lemma 2.6.6 again, we successively carry out the limits $M \to \infty$ and then $K \to \infty$ in (2.40) to obtain the desired conclusion (2.38). □

Let us conclude this section with the following simple observation.

Lemma 2.9.3. *Let $\mathbf{U}_n$, $n \in \mathbb{N}$, and $\mathbf{U}$ be random distributions on $(\Omega, \mathfrak{F}, \mathbb{P})$ and W_n, $n \in \mathbb{N}$, and W cylindrical Wiener processes defined on the same probability space. Assume that the filtration*

$$\sigma(\sigma_t[\mathbf{U}_n] \cup \sigma_t[W_n]), \quad t \geq 0, \tag{2.41}$$

is non-anticipative with respect to W_n for every $n \in \mathbb{N}$. If

$$\begin{aligned}
&\langle \mathbf{U}_n, \varphi\rangle \to \langle \mathbf{U}, \varphi\rangle \quad \text{in probability for any } \varphi \in \mathcal{D}(\mathbb{R}\times\mathbb{T}^N), \\
&W_n \to W \quad \text{in } C([0,T]; \mathfrak{U}_0) \text{ in probability,}
\end{aligned}$$

then the filtration

$$\sigma(\sigma_t[\mathbf{U}] \cup \sigma_t[W]), \quad t \geq 0,$$

is non-anticipative with respect to W.

Proof. Since the filtration (2.41) is non-anticipative with respect to W_n, we deduce

$$\begin{aligned}
&\mathbb{E}\big[F(\langle \mathbf{U}_n, \varphi_1\rangle, \dots, \langle \mathbf{U}_n, \varphi_m\rangle, W_{n,1}(t_1), \dots, W_{n,\ell}(t_\ell))\\
&\quad \times H(W_{n,1}(t+s) - W_{n,1}(t), \dots, W_{n,k}(t+s) - W_{n,k}(t))\big]\\
&= \mathbb{E}\big[F(\langle \mathbf{U}_n, \varphi_1\rangle, \dots, \langle \mathbf{U}_n, \varphi_m\rangle, W_{n,1}(t_1), \dots, W_{n,\ell}(t_\ell))\big]\\
&\quad \times \mathbb{E}\big[H(W_{n,1}(t+s) - W_{n,1}(t), \dots, W_{n,k}(t+s) - W_{n,k}(t))\big]
\end{aligned}$$

for any $0 \le t_i \le t$, $i = 1, \dots, \ell$, $s > 0$, ϕ_j supported in $(-\infty, t)$, $j = 1, \dots, m$, and bounded continuous F and H. Due to the assumptions we may pass to the limit to obtain the corresponding statement for the pair $[\mathbf{U}, W]$. □

2.10 Gyöngy–Krylov lemma

The application of Skorokhod's theorem within the stochastic compactness method discussed in Section 2.6 inevitably leads to the probabilistically weak notion of solution as introduced in Section 2.5. To be more precise, the probability space as well as the driving Wiener process cannot be given in advance and become part of the solution. However, in certain situations, it is possible to establish compactness on the original probability space and hence construct solutions on every given probability space, that is, existence of a pathwise solution can be proved.

Due to a classical Yamada–Watanabe type argument (see, e.g., Karatzas–Shreve [KS91, Section 5.3.D]), existence of a martingale solution together with pathwise uniqueness implies existence of a pathwise solution. For problems where existence of a martingale solution follows from the stochastic compactness method, one can give a rather straightforward proof of this result based on the following characterization of convergence in probability observed in Gyöngy–Krylov [GK96, Lemma 1.1].

We recall that convergence in probability was defined in Definition 2.1.6 in the setting of topological vector spaces and also discussed in Remark 2.1.8 in the setting of locally convex topological vector spaces.

Lemma 2.10.1. *Let X be a Polish space equipped with the Borel σ-algebra. A sequence of X-valued random variables $(\mathbf{U}_n)_{n\in\mathbb{N}}$ converges in probability if and only if for every sequence of joint laws of $(\mathbf{U}_{n_k}, \mathbf{U}_{m_k})_{k\in\mathbb{N}}$ there exists a further subsequence which converges weakly to a probability measure μ such that*

$$\mu\big((x, y) \in X \times X;\ x = y\big) = 1.$$

Before extending this result to the general setting of sub-Polish spaces, let us study the case of separable Hilbert spaces equipped with weak topology.

Lemma 2.10.2. *Let H be a separable Hilbert space with a countable dense set $(h_m)_{m\in\mathbb{N}}$. Suppose that $(\mathbf{U}_n)_{n\in\mathbb{N}}$ is a sequence of H-valued random variables such that*

$$\mathbf{U}_n \rightharpoonup \mathbf{U} \quad \text{in } L^2(\Omega; H) \tag{2.42}$$

and

$$\langle \mathbf{U}_n, h_m \rangle \to \langle \mathbf{U}, h_m \rangle \quad \text{in probability for any } m \in \mathbb{N}. \tag{2.43}$$

Then

$$\mathbf{U}_n \to \mathbf{U} \quad \text{in } (H, w) \text{ in probability.} \tag{2.44}$$

Proof of Lemma 2.10.2. In agreement with Definition 2.1.6 and Remark 2.1.8, statement (2.44) is equivalent to showing that, for any $h \in H$ and $\varepsilon > 0$, there exists $n_0 = n_0(h, \varepsilon)$ such that

$$\mathbb{P}(|\langle h, \mathbf{U}_n - \mathbf{U}\rangle| < 1) > 1 - \varepsilon \quad \text{for all } n \geq n_0. \tag{2.45}$$

To see (2.45), first observe that (2.42) implies

$$\mathbb{E}[\|\mathbf{U}_n\|_H^2],\ \mathbb{E}[\|\mathbf{U}\|_H^2] \leq c \quad \text{uniformly for all } n \in \mathbb{N}. \tag{2.46}$$

In particular, given $\varepsilon > 0$, there exists $M = M(\varepsilon)$ such that

$$\mathbb{P}(\{\|\mathbf{U}_n\|_H \geq M\} \cup \{\|\mathbf{U}\|_H \geq M\}) < \frac{\varepsilon}{2} \quad \text{for all } n \in \mathbb{N}.$$

Next, we choose m such that

$$M\|h_m - h\|_H \leq \frac{1}{4}.$$

Consequently, we have

$$\begin{aligned} |\langle h, \mathbf{U}_n - \mathbf{U}\rangle| &\leq |\langle h - h_m, \mathbf{U}_n - \mathbf{U}\rangle| + |\langle h_m, \mathbf{U}_n - \mathbf{U}\rangle| \\ &\leq \frac{1}{2} + |\langle h_m, \mathbf{U}_n - \mathbf{U}\rangle| \end{aligned} \tag{2.47}$$

for any $\omega \in \Omega$ such that $\|\mathbf{U}_n(\omega)\|_H + \|\mathbf{U}(\omega)\|_H \leq 2M$. Finally, as

$$|\langle h_m, \mathbf{U}_n - \mathbf{U}\rangle| \to 0 \quad \text{as } n \to \infty \text{ in probability,}$$

there exists n_0 such that

$$\mathbb{P}\left(|\langle h_m, \mathbf{U}_n - \mathbf{U}\rangle| \geq \frac{1}{2}\right) < \frac{\varepsilon}{2} \quad \text{for all } n \geq n_0. \tag{2.48}$$

Combining (2.46)–(2.48) we obtain the desired conclusion (2.45). □

In applications, typically $H = L^2(\mathcal{O})$, $\mathcal{O} \subset \mathbb{R}^N$ a bounded domain, and h_m smooth functions, say $h_m \in W_0^{1,2}(\mathcal{O})$. Hypothesis (2.42) will follow (up to a suitable subsequence) if

$$\mathbb{E}[\|\mathbf{U}_n\|_{L^2_x}^2] \leq c, \tag{2.49}$$

$$\mathbf{U}_n \to \mathbf{U} \quad \text{in } W^{-1,2}(\mathcal{O}) \text{ in probability.} \tag{2.50}$$

Condition (2.49) usually follows from suitable *a priori* bounds, while (2.50) can be deduced by means of the standard Gyöngy–Krylov theorem. Note that $L^2(\mathcal{O}) \overset{c}{\hookrightarrow} W^{-1,2}(\mathcal{O})$ and $W^{-1,2}(\mathcal{O})$ is a Polish space.

In a similar way, we show the following general result.

Theorem 2.10.3. *Let X be a sub-Polish topological vector space with a family $(f_n)_{n\in\mathbb{N}}$ of continuous functions that separate points. Let $(\mathbf{U}_n)_{n\in\mathbb{N}}$ be a sequence of random variables that is tight in X. Suppose that every subsequence $(\mathbf{U}_n, \mathbf{U}_m)_{n,m\in\mathbb{N}}$ admits a subsequence such that its joint laws satisfy*

$$\mathcal{L}_{X\times X}[\mathbf{U}_{n_k}, \mathbf{U}_{m_k}] \overset{*}{\rightharpoonup} \mu,$$

where μ is a probability measure on $X \times X$ supported by the diagonal $\{[x,x];\ x \in X\}$. Then there exists a random variable $\mathbf{U}$ ranging in X such that

$$\mathbf{U}_{n_k} \to \mathbf{U} \quad \text{in } X \text{ in probability,}$$

at least for a suitable subsequence. In addition, $\mathbf{U}$ is tight in X, meaning its law is a Radon measure on X.

Remark 2.10.4. The assumption that X is a topological vector space is only to make sense of convergence in probability. As already pointed out, any *uniform* topology on X would do so.

Proof of Theorem 2.10.3. We define an injection of X into the space $\mathbb{R}^{\aleph_0}$ (recall that $\aleph_0$ denotes the cardinality of the set of integers) by setting

$$\mathbf{V} \in X \mapsto (f_n(\mathbf{V}))_{n\in\mathbb{N}}.$$

As f_n are continuous and separating points, this is a continuous injection. In what follows, we identify points in X with their images in $\mathbb{R}^{\aleph_0}$.

Since the family of laws $(\mathcal{L}[\mathbf{U}_n])_{n\in\mathbb{N}}$ is tight on X by assumption, there exists a family of compact sets K_M, $M \in \mathbb{N}$, such that

$$\mathbb{P}(\mathbf{U}_n \in K_M) \geq 1 - \frac{1}{M} \quad \text{for all } n \in \mathbb{N}. \tag{2.51}$$

As $\mathbb{R}^{\aleph_0}$ is a Polish space, we may use the standard version of the Gyöngy–Krylov lemma to obtain a subsequence such that

$$\mathbf{U}_{n_k} \to \mathbf{U} \quad \mathbb{P}\text{-a.s. in } \mathbb{R}^{\aleph_0} \tag{2.52}$$

for a certain $\mathbb{R}^{\aleph_0}$-valued random variable $\mathbf{U}$. Specifically, we have $\mathbb{P}$-a.s.

$$d[\mathbf{U}_{n_k}, \mathbf{U}] \to 0, \quad d[\mathbf{U}, \mathbf{V}] = \sum_{n=1}^{\infty} \frac{1}{2^n} \min\{|f_n(\mathbf{U}) - f_n(\mathbf{V})|, 1\}. \tag{2.53}$$

The next observation is that $\mathbf{U} \in X$ $\mathbb{P}$-a.s. To see this, we note that the sets K_M are compact in $\mathbb{R}^{\aleph_0}$. In particular, the functions

$$\mathbf{V} \mapsto \mathrm{dist}[\mathbf{V}, K_M] \text{ are continuous in } \mathbb{R}^{\aleph_0}.$$

Thus, relations (2.51) and (2.52) imply that

$$\mathbb{P}(\mathbf{U} \in K_M) \geq 1 - \frac{1}{M} \quad \text{for any } M \text{ yielding } \mathbf{U} \in \bigcup_M K_M \subset X \ \mathbb{P}\text{-a.s.}$$

Finally, we show that

$$\mathbf{U}_{n_k} \to \mathbf{U} \quad \text{in } X \text{ in probability.}$$

To this end, we observe that the mapping d introduced in (2.53) defines a metric equivalent to the original topology on each compact K_M. Indeed, by the Stone–Weierstrass theorem, the subalgebra generated by the functions $(f_n)_{n \in \mathbb{N}}$ is dense in $C(K_M)$; whence the original topology is metrizable on K_M by the metric d.

Our goal is to show that, for any neighborhood $\mathcal{U}(\mathbf{U})$ of $\mathbf{U}$ in X and any $\varepsilon > 0$, there is k_0 such that

$$\mathbb{P}(\mathbf{U}_{n_k} \in \mathcal{U}(\mathbf{U})) > 1 - \varepsilon \quad \text{for all } k \geq k_0. \tag{2.54}$$

As the underlying space is a topological vector space, (2.54) makes sense. Specifically,

$$\mathbf{U}_{n_k} \in \mathcal{U}(\mathbf{U}) \quad \text{means } \mathbf{U}_{n_k} - \mathbf{U} \in \mathcal{U}, \ \mathcal{U} \text{ a neighborhood of } 0.$$

To see (2.54), we fix $M > 0$ such that

$$\mathbb{P}(\{\mathbf{U}_{n_k} \notin K_M\} \cup \{\mathbf{U} \notin K_M\}) < \frac{\varepsilon}{2}. \tag{2.55}$$

As the metric d yields the X topology on K_M, there exists $\delta > 0$ such that

$$\mathbf{V} \in \mathcal{U}(\mathbf{U}) \quad \text{provided } d[\mathbf{V}, \mathbf{U}] < \delta, \quad \mathbf{V}, \mathbf{U} \in K_M. \tag{2.56}$$

Finally, in view of (2.53), we find k_0 such that

$$\mathbb{P}(d[\mathbf{U}_{n_k}, \mathbf{U}] > \delta) < \frac{\varepsilon}{2} \quad \text{whenever } k \geq k_0. \tag{2.57}$$

Combining (2.55)–(2.57) we obtain (2.54). □

2.11 Stationarity

As the next step, we present several original results concerning stationarity of stochastic processes or more generally random variables ranging in certain spaces of space-time distributions. To begin with, let us recall the standard definition of stationarity for stochastic processes wirh values in a topological space X.

Definition 2.11.1. Let $\mathbf{U} = \{\mathbf{U}(t); t \in [0,\infty)\}$ be an X-valued measurable stochastic process. We say that $\mathbf{U}$ is *stationary*, provided the joint laws

$$\mathcal{L}(\mathbf{U}(t_1+\tau),\dots,\mathbf{U}(t_n+\tau)),\quad \mathcal{L}(\mathbf{U}(t_1),\dots,\mathbf{U}(t_n))$$

coincide on X^n for all $\tau \geq 0$, for all $t_1,\dots,t_n \in [0,\infty)$.

The above defined notion of stationarity is not well suited for the purposes of this book. Indeed, it will be seen later that not all the objects needed for the description of a compressible fluid flow can be understood as stochastic processes in the classical sense. To overcome this flaw, we introduce a weaker notion of stationarity which applies to random distributions in $\mathcal{D}'(\mathbb{R}\times\mathbb{T}^N)$ as introduced in Section 2.2. This is motivated by the notion of stationarity considered by Itô–Nisio [IN64].

Definition 2.11.2. Let $\mathbf{U}$ be a random distribution in $\mathcal{D}'(\mathbb{R}\times\mathbb{T}^N)$ in the sense of Definition 2.2.1. Let $\mathcal{S}_\tau$ be the time shift on the space of trajectories given by $\mathcal{S}_\tau\varphi(t) = \varphi(t+\tau)$. We say that $\mathbf{U}$ is *stationary*, provided the laws

$$\mathcal{L}(\langle\mathbf{U},\mathcal{S}_{-\tau}\varphi_1\rangle,\dots,\langle\mathbf{U},\mathcal{S}_{-\tau}\varphi_n\rangle),\quad \mathcal{L}(\langle\mathbf{U},\varphi_1\rangle,\dots,\langle\mathbf{U},\varphi_n\rangle)$$

coincide on $\mathbb{R}^n$ for all $\tau \geq 0$ and all $\varphi_1,\dots,\varphi_n \in \mathcal{D}(\mathbb{R}\times\mathbb{T}^N)$.

Alternatively, we define a time shift of a distribution $\mathbf{U}$ by

$$\langle\mathcal{S}_\tau\mathbf{U},\varphi\rangle := \langle\mathbf{U},\mathcal{S}_{-\tau}\varphi\rangle \quad \text{for all } \varphi \in \mathcal{D}(\mathbb{R}\times\mathbb{T}^N).$$

Then a random distribution is stationary, provided $\mathcal{S}_\tau\mathbf{U} \overset{d}{\sim} \mathbf{U}$ for all $\tau \geq 0$ in the sense of Definition 2.2.11.

Remark 2.11.3. Using the notation of Section 2.2 and in particular (2.2), we observe that the regularization $[[\mathbf{U}]]_\delta$ of a stationary random distribution $\mathbf{U}$ in $\mathcal{D}'(\mathbb{R}\times\mathbb{T}^N)$ is a stationary stochastic process with continuous trajectories.

It follows immediately that the notion of stationarity for random distributions is stable under weak convergence.

Lemma 2.11.4. *Let* $(\mathbf{U}_n)_{n\in\mathbb{N}}$ *be a sequence of stationary random distributions such that*

$$\mathbf{U}_n \to \mathbf{U} \quad \textit{in } \mathcal{D}'(\mathbb{R}\times\mathbb{T}^N)\ \mathbb{P}\textit{-a.s.}$$

Then $\mathbf{U}$ *is stationary.*

Proof. For all $\phi \in \mathcal{D}(\mathbb{R}\times\mathbb{T}^N)$, the mapping $\langle \mathbf{U}, \phi\rangle$ is a limit of measurable mappings. It is also measurable and, consequently, $\mathbf{U}$ is weakly measurable. The weak stationarity of $\mathbf{U}$ then follows from the dominated convergence theorem. □

Very often, the objects of interest are random variables with values in some Polish space. It is then possible to deduce a stronger notion of stationarity, somewhat closer to the classical concept from Definition 2.11.1. To this end, we say that $Y \subset \mathcal{D}'(\mathbb{R}\times\mathbb{T}^N)$ is positive shift invariant if $\mathcal{S}_\tau\mathbf{U}$ belongs to Y whenever $\mathbf{U} \in Y$.

Lemma 2.11.5. *Let* Y *be a topological vector space which is positive shift invariant and continuously embedded in* $\mathcal{D}'(\mathbb{R}\times\mathbb{T}^N)$. *Assume that* $\mathbf{U}$ *is a random distribution in* $\mathcal{D}'(\mathbb{R}\times\mathbb{T}^N)$ *such that its law is tight on* Y. *If* $\mathbf{U}$ *is stationary, then*

$$\mathcal{L}_Y[\mathcal{S}_\tau\mathbf{U}] = \mathcal{L}_Y[\mathbf{U}] \quad \textit{for all } \tau \geq 0.$$

Proof. The claim is a direct consequence of Theorem 2.2.12. □

Corollary 2.11.6. *Let* Y *be a Polish space which is positive shift invariant and continuously embedded in* $\mathcal{D}'(\mathbb{R}\times\mathbb{T}^N)$. *Assume that* $\mathbf{U}$ *is a random distribution in* $\mathcal{D}'(\mathbb{R}\times\mathbb{T}^N)$ *such that* $\mathbf{U} \in Y$ $\mathbb{P}$*-a.s. If* $\mathbf{U}$ *is stationary, then*

$$\mathcal{L}_Y[\mathcal{S}_\tau\mathbf{U}] = \mathcal{L}_Y[\mathbf{U}] \quad \textit{for all } \tau \geq 0.$$

Proof. First, we note that the statement of Theorem 2.2.12 as well as Lemma 2.11.5 remains valid for Polish spaces. Since every probability measure on a Polish space equipped with the Borel σ-field is Radon, the conclusion then follows from Lemma 2.11.5. □

Next, we show that, for the case of stochastic processes with continuous trajectories, the two definitions of stationarity are equivalent. Note that, if a separable Banach space X is continuously embedded into $\mathcal{D}'(\mathbb{T}^N)$, every X-valued measurable stochastic process with locally L^1-integrable trajectories defines a space-time distribution given as follows:

$$\langle \mathbf{U}, \psi\varphi\rangle = \int_{\mathbb{R}} \langle \mathbf{U}(t), \varphi\rangle \psi(t)\,\mathrm{d}t, \quad \psi \in C_c^\infty(\mathbb{R}),\ \varphi \in C^\infty(\mathbb{T}^N),$$

where we set $\mathbf{U}(t) = \mathbf{U}(0)$ for $t \leq 0$. Consequently, we find the following link between the two notions of stationarity.

Lemma 2.11.7. *Let X be a separable Banach space continuously embedded into $\mathcal{D}'(\mathbb{T}^N)$. An X-valued stochastic process with continuous trajectories is stationary if and only if it defines a random distribution which is stationary in the sense of Definition* 2.11.2.

Proof. Let $\mathbf{U}$ be an X-valued stochastic process which is stationary in the sense of Definition 2.11.2 (with the usual extension $\mathbf{U}(t) = \mathbf{U}(0)$ for $t \leq 0$). According to Corollary 2.11.6, setting $Y = C_{\mathrm{loc}}(\mathbb{R}; X)$ yields

$$\mathcal{L}_Y[\mathcal{S}_\tau \mathbf{U}] = \mathcal{L}_Y[\mathbf{U}] \quad \text{for all } \tau \geq 0.$$

Since the point evaluation $\mathbf{V} \mapsto \mathbf{V}(t)$ is a continuous function from Y to X, we deduce that, for every $F \in C_b(X^n)$ and all $t_1, \dots, t_n \in [0, \infty)$,

$$\mathbf{V} \mapsto (\mathbf{V}(t_1), \dots, \mathbf{V}(t_n)) \mapsto F(\mathbf{V}(t_1), \dots, \mathbf{V}(t_n)) \in C_b(Y).$$

Hence

$$\mathbb{E}[F(\mathbf{U}(t_1 + \tau), \dots, \mathbf{U}(t_n + \tau))] = \mathbb{E}[F(\mathbf{U}(t_1), \dots, \mathbf{U}(t_n))]$$

and $\mathbf{U}$ is stationary in the classical sense.

To show the converse implication, let us fix an equi-distant partition $0 = t_1 < \cdots < t_n < \cdots < \infty$ with mesh size $\Delta t = \frac{1}{m}$ for some $m \in \mathbb{N}$. We have

$$\begin{aligned}
&\mathbb{E}[F(\langle \mathbf{U}, \varphi_1 \rangle, \dots, \langle \mathbf{U}, \varphi_j \rangle)] \\
&= \lim_m \mathbb{E}\left[F\left(\frac{1}{m} \sum_n \int_{\mathbb{T}^N} \mathbf{U}(t_n) \varphi_1(t_n)\, \mathrm{d}x, \dots, \frac{1}{m} \sum_n \int_{\mathbb{T}^N} \mathbf{U}(t_n) \varphi_j(t_n)\, \mathrm{d}x\right)\right]
\end{aligned}$$

and, similarly,

$$\begin{aligned}
&\mathbb{E}[F(\langle \mathcal{S}_{-\tau} \mathbf{U}, \varphi_1 \rangle, \dots, \langle \mathcal{S}_{-\tau} \mathbf{U}, \varphi_j \rangle)] \\
&= \lim_m \mathbb{E}\left[F\left(\frac{1}{m} \sum_n \int_{\mathbb{T}^N} \mathbf{U}(t_n + \tau) \varphi_1(t_n)\, \mathrm{d}x, \dots, \frac{1}{m} \sum_n \int_{\mathbb{T}^N} \mathbf{U}(t_n + \tau) \varphi_j(t_n)\, \mathrm{d}x\right)\right],
\end{aligned}$$

for any $\varphi_1, \dots, \varphi_j \in \mathcal{D}(\mathbb{R} \times \mathbb{T}^N)$. As the process $\mathbf{U}$ is stationary in the classical sense, the expressions on the right hand sides of the above limits coincide for any fixed m and the desired conclusion follows. □

The next result proves that weak continuity together with a uniform bound suffices for the equivalence of the two notions of stationarity.

Corollary 2.11.8. *Let X be a separable Banach space continuously embedded into $\mathcal{D}'(\mathbb{T}^N)$ such that, for all $\mathbf{V} \in X$, $[\mathbf{V}]_{x,\delta} \to \mathbf{V}$ in X. Let $\mathbf{U}$ be an X-valued stochastic process with weakly continuous trajectories.*

Then $\mathbf{U}$ is stationary if and only if it is stationary in the sense of Definition 2.11.2.

Proof. As any weakly convergent sequence is bounded in the norm, we have

$$\sup_{t\in[0,T]} \|\mathbf{U}\|_X < \infty \quad \mathbb{P}\text{-a.s.} \tag{2.58}$$

As before, we define $\mathbf{U}(t) = \mathbf{U}(0)$ for $t \le 0$. Since $\mathbf{U}$ has weakly continuous trajectories in X and satisfies (2.58), the process $[\mathbf{U}]_{x,\delta}$ has strongly continuous trajectories in X. Hence the equivalence of the two notions of stationarity from Lemma 2.11.7 holds.

Now, let $\mathbf{U}$ be stationary. Then Corollary 2.11.6 implies that, for every $f \in C_b(L^1_{\mathrm{loc}}(\mathbb{R}; X))$, we have

$$\mathbb{E}[f(\mathcal{S}_\tau \mathbf{U})] = \mathbb{E}[f(\mathbf{U})].$$

Since $\mathbf{U} \mapsto f([\mathbf{U}]_{x,\delta})$ also belongs to $C_b(L^1_{\mathrm{loc}}(\mathbb{R}; X))$, we deduce

$$\mathbb{E}[f([\mathbf{U}]_{x,\delta})] = \mathbb{E}[f([\mathcal{S}_\tau \mathbf{U}]_{x,\delta})] = \mathbb{E}[f(\mathcal{S}_\tau [\mathbf{U}]_{x,\delta})].$$

So $[\mathbf{U}]_{x,\delta}$ is stationary in the sense of Definition 2.11.2 and, due to Lemma 2.11.7, it is stationary in the sense of Definition 2.11.1. In addition, due to the assumptions we have, $[\mathbf{U}]_{x,\delta}(t) \to \mathbf{U}(t)$ strongly in X for every $t \in \mathbb{R}$. Therefore, if $g \in C_b(X^n)$, we may use dominated convergence in order to pass to the limit in expressions of the form

$$\mathbb{E}[g([\mathbf{U}]_{x,\delta}(t_1), \dots, [\mathbf{U}]_{x,\delta}(t_n))] = \mathbb{E}[g([\mathbf{U}]_{x,\delta}(t_1 + \tau), \dots, [\mathbf{U}]_{x,\delta}(t_n + \tau))].$$

Stationarity of $\mathbf{U}$ in the sense of Definition 2.11.1 follows.

To show the converse implication, assume that $\mathbf{U}$ is stationary in the sense of Definition 2.11.1. By the same argument as above, it follows that $[\mathbf{U}]_{x,\delta}$ is stationary in the sense of Definition 2.11.1 and hence stationary due to Lemma 2.11.7. In other words, Corollary 2.11.6 gives, for every $f \in C_b(L^1_{\mathrm{loc}}(\mathbb{R}; X))$,

$$\mathbb{E}[f([\mathbf{U}]_{x,\delta})] = \mathbb{E}[f(\mathcal{S}_\tau [\mathbf{U}]_{x,\delta})].$$

In accordance with (2.58) we obtain $[\mathbf{U}]_{x,\delta} \to \mathbf{U}$ in $L^1_{\mathrm{loc}}(\mathbb{R}; X)$ and the dominated convergence theorem yields the claim, exactly as in Lemma 2.11.7. □

Let us conclude with a simple observation that stationarity is preserved under composition with measurable functions.

Corollary 2.11.9. *Let $\mathbf{U}$ be an X-valued stationary stochastic process. Then, for every Borel measurable function $F : X \to \mathbb{R}$, the $\mathbb{R}$-valued stochastic process $F(\mathbf{U})$ is stationary.*

Proof. The proof follows immediately from the corresponding equality of joint laws of $(\mathbf{U}(t_1), \dots, \mathbf{U}(t_n))$ and $(\mathbf{U}(t_1 + \tau), \dots, \mathbf{U}(t_n + \tau))$. □

Corollary 2.11.10. *Let X be a separable Banach space continuously embedded into $\mathcal{D}'(\mathbb{T}^N)$. Assume that $\mathbf{U}$ is a random variable taking values in $L^1_{\mathrm{loc}}(\mathbb{R};X)$. If $\mathbf{U}$ is stationary, then, for every Borel measurable function $F : X \to \mathbb{R}$ and a.e. $s,t \in \mathbb{R}$, the laws of $F(\mathbf{U}(s))$ and $F(\mathbf{U}(t))$ coincide on X.*

Proof. First, we notice that the mapping $\mathbf{U} \mapsto \mathbf{U}(t) \mapsto F(\mathbf{U}(t))$ is measurable from $L^1_{\mathrm{loc}}(\mathbb{R};X)$ to $\mathbb{R}$ for a.e. $t \in \mathbb{R}$. For the same reasons, the mapping $\mathcal{S}_{s-t} : \mathbf{U} \mapsto \mathbf{U}(s) \mapsto F(\mathbf{U}(s))$ is measurable from $L^1_{\mathrm{loc}}(\mathbb{R};X)$ to $\mathbb{R}$ for a.e. $s,t \in \mathbb{R}$. Hence the claim follows from the equality of laws of $\mathbf{U}$ and $\mathcal{S}_{s-t}\mathbf{U}$ on $L^1_{\mathrm{loc}}(\mathbb{R};X)$. □

Remark 2.11.11. Note that, in view of Corollary 2.11.10, weak stationarity implies the following almost everywhere version of Definition 2.11.1: under the assumptions of Corollary 2.11.10, the joint laws

$$\mathcal{L}(\mathbf{U}(t_1+\tau),\dots,\mathbf{U}(t_n+\tau)), \quad \mathcal{L}(\mathbf{U}(t_1),\dots,\mathbf{U}(t_n))$$

coincide on X^n for a.e. $\tau \geq 0$, for a.e. $t_1,\dots,t_n \in [0,\infty)$.

2.12 Krylov–Bogoliubov method

In this section, we recall the so-called Krylov–Bogoliubov method, which is commonly used in order to construct invariant measures for dynamical systems generated by Markovian transition semi-groups on Polish spaces. However, for simplicity of presentation we explain these concepts with the example of a Markovian system generated by a stochastic differential equation in a Hilbert space H. A more detailed exposition can be found in Da Prato–Zabczyk [DPZ96]. An application of this method in the context of the compressible Navier–Stokes system is presented in Chapter 7.

For simplicity, we consider a stochastic differential equation in a separable Hilbert space H. We have

$$\mathrm{d}\mathbf{U} = b(\mathbf{U})\,\mathrm{d}t + \sigma(\mathbf{U})\,\mathrm{d}W, \quad \mathbf{U}(0) = x \in H, \tag{2.59}$$

where W is a cylindrical Wiener process on a separable Hilbert space $\mathfrak{U}$ relative to the filtration $(\mathfrak{F}_t)_{t\geq 0}$. Assume that the coefficients are Lipschitz continuous in the following sense:

$$\|b(\mathbf{V}_1) - b(\mathbf{V}_2)\|_H + \|\sigma(\mathbf{V}_1) - \sigma(\mathbf{V}_2)\|_{L_2(\mathfrak{U};H)} \leq C\|\mathbf{V}_1 - \mathbf{V}_2\|_H, \quad \text{for any } \mathbf{V}_1, \mathbf{V}_2 \in H.$$

Then it can be shown, for instance by means of a Picard iteration and the Banach fixed point theorem, that there exists a unique pathwise solution to (2.59). Let $\mathbf{U}^x_t$ denote the solution at time t starting from x at time 0. The dependence of $\mathbf{U}_t^x$ on x is measurable and hence we can define the operators $P_t : B_b(H) \to B_b(H)$ [1] by

$$(P_t\varphi)(x) = \mathbb{E}[\varphi(\mathbf{U}_t^x)]. \tag{2.60}$$

Let $\mathbf{U}_{t_0,t}^\eta$ denote the unique solution to (2.59) starting at time t_0 from an $\mathfrak{F}_{t_0}$-measurable initial condition η. Under the above Lipschitz assumption, the continuous dependence on the initial condition holds true for (2.59). More precisely, for all $T > 0$ and $p \in [2, \infty)$, we have

$$\mathbb{E}\|\mathbf{U}_{s,t}^\eta - \mathbf{U}_{s,t}^\theta\|_H^p \le C_{p,T}\mathbb{E}\|\eta - \theta\|_H^p \quad \text{for all } 0 \le s \le t \le T.$$

As an immediate consequence, we deduce that $P_t : C_b(H) \to C_b(H)$ for all $t \ge 0$. This is the so-called Feller property. Furthermore, we have the following result.

Lemma 2.12.1. *The equation* (2.59) *defines a Markov process in the sense that*[2]

$$\mathbb{E}[\varphi(\mathbf{U}_{t+s}^x)|\mathfrak{F}_t] = (P_s\varphi)(\mathbf{U}_t^x) \tag{2.61}$$

holds true for all $\varphi \in C_b(H)$, $x \in H$ and $t, s > 0$.

Proof. We have to prove that, for all $\mathfrak{F}_t$-measurable random variables Z, we have

$$\mathbb{E}[\varphi(\mathbf{U}_{t+s}^x)Z] = \mathbb{E}[(P_s\varphi)(\mathbf{U}_t^x)Z].$$

By uniqueness we have

$$\mathbf{U}_{t+s}^x = \mathbf{U}_{t,t+s}^{\mathbf{U}_t^x} \quad \mathbb{P}\text{-a.s.}$$

In view of Proposition 2.1.19, it is sufficient to show that

$$\mathbb{E}[\varphi(\mathbf{U}_{t,t+s}^\eta)Z] = \mathbb{E}[(P_s\varphi)(\eta)Z]$$

for every $\mathfrak{F}_t$-measurable random variable η. By approximation, it suffices to prove it for random variables $\eta = \sum_{i=1}^k \eta^i \mathbf{1}_{A^i}$ with $\eta^i \in H$ and $A^i \in \mathfrak{F}_t$. Indeed, this is a consequence of the dominated convergence theorem and the fact that $\eta_n \to \eta$ in H implies $P_t\varphi(\eta_n) \to P_t\varphi(\eta)$ in $\mathbb{R}$ $\mathbb{P}$-a.s. Moreover, it suffices to show the claim for every deterministic η. Now, the random variable $\mathbf{U}_{t,t+s}^\eta$ depends only on the increments of the Wiener process between t and $t+s$, hence it is independent of $\mathfrak{F}_t$ due to Definition 2.1.28. Therefore

$$\mathbb{E}[\varphi(\mathbf{U}_{t,t+s}^\eta)Z] = \mathbb{E}[\varphi(\mathbf{U}_{t,t+s}^\eta)]\mathbb{E}[Z].$$

1 $B_b(H)$ denotes the set of bounded Borel functions from H to $\mathbb{R}$.

2 Recall that the notion of conditional expectation was introduced in Definition 2.1.20.

Since $\mathbf{U}^{\eta}_{t,t+s}$ has the same law as $\mathbf{U}^{\eta}_{s}$ by uniqueness, we have

$$\mathbb{E}[\varphi(\mathbf{U}^{\eta}_{t,t+s})Z] = \mathbb{E}[\varphi(\mathbf{U}^{\eta}_{s})]\mathbb{E}[Z] = P_s\varphi(\eta)\mathbb{E}[Z] = \mathbb{E}[P_s\varphi(\eta)Z]$$

and the proof is complete. □

As the next step, we observe that taking expectation on the left hand side of (2.61) gives

$$\mathbb{E}[\mathbb{E}[\varphi(\mathbf{U}^{x}_{t+s})|\mathfrak{F}_t]] = \mathbb{E}[\varphi(\mathbf{U}^{x}_{t+s})] = (P_{t+s}\varphi)(x),$$

whereas taking expectation on the right hand side of (2.61) yields

$$\mathbb{E}[(P_s\varphi)(\mathbf{U}^{x}_{t})] = (P_t(P_s\varphi))(x).$$

Thus the semi-group property $P_{t+s} = P_t \circ P_s$ follows and the family $(P_t)_{t\geq 0}$ is called the transition semi-group associated to (2.59). We say that equation (2.59) defines a Feller–Markov process.

Let us now denote by $\mu_{t,x}$ the law of $\mathbf{U}^{x}_{t}$. We have

$$P_t\varphi(x) = \mathbb{E}[\varphi(\mathbf{U}^{x}_{t})] = \int_H \varphi(y)\mu_{t,x}(\mathrm{d}y)$$

and, denoting by $\langle\cdot,\cdot\rangle$ the duality product between continuous bounded functions and probability measures, we obtain

$$P_t\varphi(x) = \langle\varphi, \mu_{t,x}\rangle = \langle P_t\varphi, \delta_x\rangle.$$

Thus it follows that $\mu_{t,x} = P^*_t\delta_x$. More generally, if we consider a solution to (2.59) with the initial condition $\mathbf{U}_0$ having the initial law μ, we have $\mu_{t,\mathbf{U}_0} = P^*_t\mu$.

Definition 2.12.2. We say that a probability measure μ on H is an invariant measure with respect to the transition semi-group $(P_t)_{t\geq 0}$, provided

$$P^*_t\mu = \mu \quad \text{for all } t \geq 0.$$

Then, if a solution has the law μ at time s, it is the case for all $t \geq s$. In fact, for such a solution it can be shown by the Markov property that, for all $(t_1, \dots, t_n)$ and $\tau > 0$, the laws of $(\mathbf{U}_{t_1+\tau}, \dots, \mathbf{U}_{t_n+\tau})$ and $(\mathbf{U}_{t_1}, \dots, \mathbf{U}_{t_n})$ coincide. Therefore, the solution is a stationary stochastic process in the sense of Definition 2.11.1.

Analogously, one can define transition semi-groups with the above properties on general Polish spaces. As the next step, we state the Krylov–Bogoliubov theorem. The result is taken from Da Prato–Zabczyk [DPZ96, Theorem 3.1.1] and will be applied in our construction of stationary solutions in Chapter 7, namely in Section 7.1.

Theorem 2.12.3. *Let X be a Polish space and let $(P_t)_{t\geq 0}$ be a Markovian Feller transition semi-group on $C_b(X)$. For a random variable $\mathbf{U}_0$ having the law ν, denote $\mu_{t,\mathbf{U}_0} = P_t^*\nu$. Assume that there exists a sequence $T_n \uparrow \infty$ and a probability measure μ on X such that*

$$\frac{1}{T_n}\int_0^{T_n} \mu_{s,\mathbf{U}_0}\,\mathrm{d}s \quad \text{converges weakly to } \mu \text{ as } n \to \infty.$$

Then μ is an invariant for $(P_t)_{t\geq 0}$, namely, $P_t^*\mu = \mu$ for all $t \geq 0$.

Corollary 2.12.4. *If, for some random variable $\mathbf{U}_0$ and some sequence $T_n \uparrow \infty$, the set*

$$\left\{\frac{1}{T_n}\int_0^{T_n} \mu_{s,\mathbf{U}_0}\,\mathrm{d}s;\; n \in \mathbb{N}\right\}$$

is tight on X, then there exists an invariant measure for $(P_t)_{t\geq 0}$.

Remark 2.12.5. Consider again the transition semi-group generated by (2.59) and let $x \in H$. Then the tightness of

$$\left\{\frac{1}{T_n}\int_0^{T_n} \mu_{s,x}\,\mathrm{d}s;\; n \in \mathbb{N}\right\} \tag{2.62}$$

implies in particular that, for all $\varepsilon > 0$, there exists $R > 0$ such that, for all $T \geq 1$,

$$\frac{1}{T}\int_0^T \mathbb{P}(\|\mathbf{U}_t^x\|_H \geq R)\,\mathrm{d}t < \varepsilon. \tag{2.63}$$

Indeed, this is a consequence of the fact that a compact set is in particular bounded. Note that (2.63) implies the tightness of (2.62) and hence existence of an invariant measure, provided $\dim H < \infty$. However, if $\dim H = \infty$, then this is no longer true.

Part II: **Existence theory**

3 Modeling fluid motion subject to random effects

Continuum fluid mechanics relies on a system of partial differential equations which has been derived from basic physical principles under the assumption that all quantities – *fields* – are smooth. On the other hand, however, and in spite of great effort of generations of excellent mathematicians, many fundamental problems related to dynamics of fluids remain largely open. Solvability of the Navier–Stokes system describing the motion of an incompressible viscous fluid is one of the Millennium Prize Problems proposed by the Clay Institute. Apparently, the problem becomes even more involved when general compressible fluids are considered. In contrast with these theoretical difficulties, the Navier–Stokes system became a well-established model serving as a reliable basis of investigation both for theoretical issues and in engineering. As the modeled fluids often exhibit very complicated and chaotic behavior, commonly denoted as turbulent phenomena, the relation of the Navier–Stokes system to the phenomena of hydrodynamic turbulence is regarded as one of the most fascinating open problems in physics.

Turbulence is frequently associated with an intrinsic presence of randomness and also the experimental studies of turbulence lead rather to a statistical approach than to a deterministic one. Therefore several ways of encoding randomness into the governing system of PDEs have been proposed. Let us particularly mention the seminal work of Foias [Foi72a, Foi72b], where a notion of statistical solution to the Navier–Stokes system was introduced, that is, a solution with a given initial probability distribution. Another approach is based on the ideas introduced by Mikulevicius–Rozovskii [MR04]: they start with the Lagrangian formalism and split the velocity field into a sum of slow oscillating (deterministic) and fast oscillating (turbulent, stochastic) components. The subsequent derivation of the Navier–Stokes system in the Eulerian coordinates then leads to a multiplicative noise depending not only on the velocity but also on the velocity gradient. More references and further discussion may be found in the review paper by Weinan [Wei01].

Apart from that, the addition of stochastic terms to the basic governing equations is often used to account for other numerical, empirical or physical uncertainties. Another example of how randomness can influence a fluid body comes with the modeling of earthquakes, which also exhibit a certain random character. To summarize, in view of important applications ranging from climatology to turbulence theory, there is an essential need to develop the mathematical foundation of stochastic PDEs of fluid flow.

Nowadays there exists an abundant amount of literature concerning the dynamics of *incompressible* fluids driven by stochastic forcing. The first results are found in the pioneering work by Bensoussan–Temam [BT73]. We refer to the lecture notes by Debussche [Deb13], Flandoli [Fla08], and the monograph by Kuksin–Shyrikian [KS12] as well as the references cited therein for a recent overview. However, definitely much

https://doi.org/10.1515/9783110492552-003

less is known if compressibility of the fluid is taken into account. Fundamental questions of well-posedness and even mere existence of solutions to problems dealing with stochastic perturbations of compressible fluids are, to the best of our knowledge, largely open, with only a few rigorous results available.

Our aim is to systematically develop a consistent mathematical theory of compressible fluid flows driven by random initial data and stochastic external forces in the context of classical *continuum fluid mechanics*. For the sake of simplicity, we focus only on the purely mechanical aspects, neglecting the thermal effects. Accordingly, the state of a fluid at a given time $t \in (0,T)$ and a spatial position $x \in \mathbb{R}^N$, $N = 1,2,3$, is determined by two fundamental state variables: the mass density $\varrho = \varrho(t,x)$ and the bulk velocity $\mathbf{u} = \mathbf{u}(t,x)$. In view of problems involving random phenomena, it is more convenient to consider $t \mapsto \varrho(t,\cdot)$ together with the momentum $t \mapsto \varrho\mathbf{u}(t,\cdot)$, both as stochastic processes depending on t and ranging in suitable function spaces specified below. Thus the initial state of the fluid will be often described in terms of ϱ and $\varrho\mathbf{u}$ rather than ϱ and $\mathbf{u}$.

Although the boundary conditions in the real world applications may be quite complicated and of substantial influence on the fluid motion, our goal is to focus on the effect of stochastic perturbations imposed through stochastic volume forces. Accordingly, we consider the periodic boundary conditions, where the physical domain may be identified with the flat torus

$$\mathbb{T}^N = \left([-1,1]|_{\{-1,1\}}\right)^N.$$

The fact that we have chosen the same period, namely 2, in all directions plays no role in the subsequent analysis and may be relaxed. We also claim that basically all results proved in the space-periodic setting may be reproduced for the commonly used *no-slip* boundary conditions

$$\mathbf{u}|_{\partial\mathcal{O}} = 0,$$

with $\mathcal{O} \subset \mathbb{R}^N$ being the physical domain occupied by the fluid. Unbounded domains could be considered as well.

3.1 Field equations

The basic *field equations* of continuum fluid mechanics, written in the Eulerian reference system attached to the physical domain, govern the time evolution of density ϱ and velocity $\mathbf{u}$ of a compressible viscous fluid. More precisely, they read

$$\mathrm{d}\varrho + \mathrm{div}\,(\varrho\mathbf{u})\,\mathrm{d}t = 0, \tag{3.1}$$

$$\mathrm{d}(\varrho\mathbf{u}) + \mathrm{div}\,(\varrho\mathbf{u}\otimes\mathbf{u})\,\mathrm{d}t = \mathrm{div}\,\mathbb{T}\,\mathrm{d}t + \mathbb{G}(\varrho,\varrho\mathbf{u})\,\mathrm{d}W. \tag{3.2}$$

Equation (3.1) – *the equation of continuity* – represents a mathematical formulation of the physical principle of mass conservation. Equation (3.2) – *the momentum equation* – reflects Newton's second law of momentum conservation. Here $\mathbb{T}$ is the Cauchy stress and random effects are incorporated in the forcing term $\mathbb{G}(\varrho,\varrho\mathbf{u})\,\mathrm{d}W$. The driving stochastic process W is a cylindrical Wiener process as introduced in Definition 2.1.31 and the stochastic integral is understood in the Itô sense presented in Section 2.3. The precise assumptions on the coefficient $\mathbb{G}$ as well as further details are given in Section 3.2.2.

3.1.1 Constitutive relations – Navier–Stokes system

Fluids in continuum mechanics are characterized by Stokes' law as follows:

$$\mathbb{T} = \underbrace{\mathbb{S}}_{\text{viscous stress}} - \underbrace{p(\varrho)}_{\text{pressure}}\,\mathbb{I}. \tag{3.3}$$

Here the symbol p denotes the pressure – a quantity that is constitutively related to the density ϱ and the temperature θ of the fluid. As thermal effects are neglected in this book, the pressure $p = p(\varrho)$ will depend only on ϱ. Such a relation can be determined by considering purely material properties of a given fluid – the barotropic pressure-density state relation, or by considering the isentropic/isothermal approximation, where the entropy/temperature of the fluid are assumed to be constant. Under such circumstances, the pressure-density state equation is given by

$$p(\varrho) = a\varrho^{\gamma}, \quad a > 0, \tag{3.4}$$

where $\gamma > 1$ is the adiabatic exponent in the isentropic case, and $\gamma = 1$ corresponds to the isothermal case for the perfect gas $p = \varrho\theta$, θ constant. More specifically, the physically relevant values of γ belong to the range $1 \le \gamma \le \frac{5}{3}$, where the upper bound $\gamma = \frac{5}{3}$ characterizes the adiabatic coefficient of the monoatomic gas; see, e.g., Becker [Bec66]. If the system of field equations is rewritten in the dimensionless form, the parameter a can be interpreted as the squared reciprocal of the Mach number, that is, the ratio of flow velocity and speed of sound. For the sake of simplicity, we shall always consider p in the form (3.4), although a major part of the results apply also to the more general case

$$p \in C^1((0,\infty)) \cap C([0,\infty)), \quad p'(\varrho) \ge 0 \text{ for } \varrho > 0, \quad p'(\varrho) \approx \varrho^{\gamma-1} \text{ for } \varrho \to \infty. \tag{3.5}$$

In addition, the adiabatic exponent will be often subjected to further technical restriction $\gamma > \frac{N}{2}$ if $N = 2,3$ and $\gamma \ge 1$ if $N = 1$. This may be seen as a drawback of the present theory related to the lack of suitable *a priori* bounds in the multi-dimensional case. Note that the critical case $\gamma = 1$, $N = 2$ has been solved in the deterministic setting quite recently by Plotnikov–Weigant [PW15].

We focus on Newtonian fluids, for which the viscous stress is a linear function of the velocity gradient. More specifically,

$$\mathbb{S} = \mathbb{S}(\nabla \mathbf{u}) = \mu\left(\nabla \mathbf{u} + \nabla^t \mathbf{u} - \frac{2}{3} \mathrm{div}\, \mathbf{u} \mathbb{I} \right) + \lambda\, \mathrm{div}\, \mathbf{u} \mathbb{I}, \tag{3.6}$$

where $\mu > 0$, $\lambda \geq 0$ are constant viscosity coefficients. It can be shown that (3.6) characterizes any linearly viscous isotropic fluid, meaning a fluid for which its material properties are invariant with respect to the translation and rotation of the reference frame; see, e.g., Chorin–Marsden [CM90].

As long as the viscosity coefficients are constant, we may rewrite (3.1)–(3.6) in the form

$$\mathrm{d}\varrho + \mathrm{div}\,(\varrho \mathbf{u})\, \mathrm{d}t = 0, \tag{3.7}$$

$$\mathrm{d}(\varrho \mathbf{u}) + \mathrm{div}\,(\varrho \mathbf{u} \otimes \mathbf{u})\, \mathrm{d}t + \nabla p(\varrho)\, \mathrm{d}t = [\mu \Delta \mathbf{u} + \eta \nabla \mathrm{div}\, \mathbf{u}]\, \mathrm{d}t + \mathbb{G}(\varrho, \varrho \mathbf{u})\, \mathrm{d}W, \tag{3.8}$$

where $\eta = \frac{\mu}{3} + \lambda > 0$. Equations (3.7)–(3.8) are commonly known as barotropic *Navier–Stokes system*. For notational simplicity, unless specified otherwise, we restrict ourselves to the most difficult and physically relevant case of three space dimensions. However, our results extend to the one- and two-dimensional cases as well.

3.2 Random phenomena

Let us now set up the precise assumptions on the random effects included in the system (3.1)–(3.6). As already mentioned above, we consider perturbations encoded in problem (3.1)–(3.6) in two ways: by random initial data and by a stochastic forcing term $\mathbb{G}(\varrho, \varrho \mathbf{u})\, \mathrm{d}W$ in the momentum equation. Throughout the book, $(\Omega, \mathfrak{F}, (\mathfrak{F}_t)_{t \geq 0}, \mathbb{P})$ denotes a stochastic basis with a complete, right-continuous filtration.

3.2.1 Initial data

The original state of the system is given through the initial conditions

$$\varrho(0) = \varrho_0, \quad \varrho \mathbf{u}(0) = \mathbf{q}_0. \tag{3.9}$$

Both ϱ_0 and $\mathbf{q}_0$ are generally random variables satisfying the physically relevant conditions

$$\varrho_0 \geq 0, \quad \mathbf{q}_0 = 0 \quad \text{a.e. on the set } \{\varrho_0 = 0\}\ \mathbb{P}\text{-a.s.} \tag{3.10}$$

This is apparently a dubious definition unless we specify the function spaces in which $(\varrho_0, \mathbf{q}_0)$ live. To motivate our choice of the phase space, consider the total energy balance associated to system (3.7)–(3.8):

$$\begin{aligned} &\mathrm{d}\int_{\mathbb{T}^N}\left[\frac{1}{2}\varrho|\mathbf{u}|^2+P(\varrho)\right]\mathrm{d}x+\int_{\mathbb{T}^N}\mathbb{S}(\nabla\mathbf{u}):\nabla\mathbf{u}\,\mathrm{d}x\,\mathrm{d}t \\ &=\frac{1}{2}\sum_{k=1}^{\infty}\int_{\mathbb{T}^N}\varrho^{-1}|\mathbf{G}_k(\varrho,\varrho\mathbf{u})|^2\,\mathrm{d}x\,\mathrm{d}t+\int_{\mathbb{T}^N}\mathbb{G}(\varrho,\varrho\mathbf{u})\cdot\mathbf{u}\,\mathrm{d}x\,\mathrm{d}W, \end{aligned} \tag{3.11}$$

where

$$P(\varrho)=\varrho\int_1^{\varrho}\frac{p(z)}{z^2}\,\mathrm{d}z \tag{3.12}$$

is called *pressure potential* and the coefficient $\mathbb{G}$ taking values in the space of Hilbert–Schmidt operators is identified with the sequence $(\mathbf{G}_k)_{k\in\mathbb{N}}$. The precise meaning of the terms on the right hand side of (3.11) will be given in Section 3.2.2. Relation (3.11) can be derived from (3.7)–(3.8) by applying Itô's product rule, Theorem 2.4.2, to the functional

$$[\varrho,\mathbf{q}]\mapsto\frac{1}{2}\int_{\mathbb{T}^N}\frac{|\mathbf{q}|^2}{\varrho}\,\mathrm{d}x,$$

where $\mathbf{q}$ corresponds to $\varrho\mathbf{u}$. The quantity

$$\int_{\mathbb{T}^N}\left[\frac{1}{2}\varrho|\mathbf{u}|^2+P(\varrho)\right]\mathrm{d}x$$

is the total energy of the fluid. In particular, if the pressure is given by the isentropic relation (3.4), all *finite energy* solutions $[\varrho,\mathbf{q}=\varrho\mathbf{u}]$ and, accordingly, the initial data $[\varrho_0,\mathbf{q}_0]$ will belong to the space $L^\gamma(\mathbb{T}^N)\times L^{\frac{2\gamma}{\gamma+1}}(\mathbb{T}^N)$, where γ is the adiabatic exponent from (3.4). In a more general setting, the data can be interpreted in the space of random distributions introduced in Section 2.2, where measurability is enforced through the infinite family of test functions. More specifically, ϱ_0, $\mathbf{q}_0$ are $\mathbb{P}$-measurable if

$$\int_{\mathbb{T}^N}\varrho_0\varphi\,\mathrm{d}x,\ \int_{\mathbb{T}^N}\mathbf{q}_0\cdot\boldsymbol{\varphi}\,\mathrm{d}x\quad\text{are }\mathbb{P}\text{-measurable}$$

for all $\varphi\in C^\infty(\mathbb{T}^N)$, $\boldsymbol{\varphi}\in C^\infty(\mathbb{T}^N)$, respectively.

Finally, in order to guarantee solvability of the Cauchy problem, the initial condition must be independent of the future of the driving Wiener process W. More precisely, as discussed in Section 2.3 and in particular in Remark 2.3.7, the random distributions ϱ, $\varrho\mathbf{u}$ must be adapted to $(\mathfrak{F}_t)_{t\geq0}$ (in the sense of Definition 2.2.13) for the stochastic integral to be well-defined. In particular, the initial conditions ϱ_0, $\mathbf{q}_0$ must be $\mathfrak{F}_0$-measurable. In the case of martingale solutions, that is, solutions that are weak in the probabilistic sense, the initial conditions (3.9) as well as the compatibility conditions (3.10) will be enforced through the associated *initial law* defined on a suitable function space (cf. Section 2.5).

3.2.2 Driving force

The stochastic process W is a cylindrical $(\mathfrak{F}_t)$-Wiener process in a separable Hilbert space $\mathfrak{U}$ as introduced in Definition 2.1.31. It is formally given by the expansion $W(t) = \sum_{k=1}^{\infty} e_k W_k(t)$, where $(W_k)_{k\in\mathbb{N}}$ is a sequence of mutually independent real-valued Wiener processes relative to $(\mathfrak{F}_t)_{t\geq 0}$ and $(e_k)_{k\in\mathbb{N}}$ is a complete orthonormal system in $\mathfrak{U}$. Accordingly, the diffusion coefficient $\mathbb{G}$ is defined as a superposition operator $\mathbb{G}(\varrho,\mathbf{q}) : \mathfrak{U} \to L^1(\mathbb{T}^N)$,

$$\mathbb{G}(\varrho,\mathbf{q})e_k = \mathbf{G}_k(\cdot,\varrho(\cdot),\mathbf{q}(\cdot)).$$

The coefficients $\mathbf{G}_k = \mathbf{G}_k(x,\varrho,\mathbf{q}) : \mathbb{T}^N \times [0,\infty) \times \mathbb{R}^N \to \mathbb{R}^N$ are C^1-functions such that there exist constants $(g_k)_{k\in\mathbb{N}} \subset [0,\infty)$ with $\sum_{k=1}^{\infty} g_k^2 < \infty$ and uniformly in $x \in \mathbb{T}^3$

$$|\mathbf{G}_k(x,\varrho,\mathbf{q})| \leq g_k(\varrho + |\mathbf{q}|), \tag{3.13}$$

$$|\nabla_{\varrho,\mathbf{q}}\mathbf{G}_k(x,\varrho,\mathbf{q})| \leq g_k. \tag{3.14}$$

Note that under these circumstances, the integrals on the right hand side of the energy balance (3.11) are well-defined. The specific form of the forcing term as well as a proper choice of the stochastic integration (Itô vs. Stratonovich) and its physical interpretation might be subject to discussion. Our setting includes, in particular, the case

$$\mathbb{G}(\varrho,\varrho\mathbf{u})\,\mathrm{d}W = \mathbb{G}_1(\varrho)\,\mathrm{d}W^1 + \mathbb{G}_2(\varrho\mathbf{u})\,\mathrm{d}W^2,$$

with two independent cylindrical Wiener processes W^1 and W^2 and suitable growth assumptions of $\mathbb{G}_1$ and $\mathbb{G}_2$. This is the main example we have in mind. Here, the first term describes some external force, whereas the second one may be interpreted as a friction force of Brinkman's type; see, e.g., Angot et al. [ABF99]. Of course, much more general forcing can be included. The interested reader may consult the relevant discussion by Mikulevicius–Rozovskii [MR04, MR05].

Note that, according to (3.13), if $\mathbf{q} = \varrho\mathbf{u}$, the coefficients in the noise can be written as $\mathbf{G}_k(x,\varrho,\varrho\mathbf{u}) = \varrho\,\mathbf{F}_k(x,\varrho,\mathbf{u})$ for suitable functions $\mathbf{F}_k$. In other words, the noise vanishes on the vacuum regions, i.e., when the density is zero. Recall that this is consistent with the deterministic theory, where an external force is always considered in the form $\varrho\mathbf{f}\,\mathrm{d}t$. This is essential in order to derive the corresponding energy estimate. In this sense, the noise of the form $\mathbf{G}_k(x,\varrho,\varrho\mathbf{u}) = \varrho\,\mathbf{F}_k(x)$ plays the role of the so-called additive noise for the system (3.1)–(3.6), i.e., a noise which does not depend on the solution. After applying a suitable transformation, the system can be solved pathwise, and existence of a weak solution can be established using deterministic arguments; see [FMN13]. Nevertheless, the obvious drawback of this method is that the constructed solutions do not necessarily satisfy an energy inequality and are not progressively measurable (hence, the stochastic integral is not well-defined).

For the sake of simplicity, we focus only on the effect of stochastic driving forces. A *deterministic* perturbation of the form $\varrho\mathbb{H}(\varrho,\varrho\mathbf{u})$ could be added to the system without any extra difficulty as long as $\mathbb{H}$ satisfies suitable growth conditions. In particular, the Itô integration may be replaced by Stratonovich and *vice versa*.

The space $L^1(\mathbb{T}^N)$ is the natural range for the operator $\mathbb{G}(\varrho,\varrho\mathbf{u})$, at least if finite energy solutions are considered. In general, due to limited *a priori* estimates provided almost exclusively by the energy balance (3.11), it is not possible to consider $\mathbb{G}(\varrho,\varrho\mathbf{v})$ as a mapping with values in a space with higher integrability. Because of this fact, certain difficulties arise concerning a proper definition of the stochastic integral in (3.2). Recall that in Section 2.3 we introduced the stochastic Itô integration theory in Hilbert spaces. Apart from that, the theory is well established within the class of 2-smooth Banach spaces or the so-called UMD Banach spaces; see, e.g., Brzeźniak [Brz95], van Neerven et al. [vNVW07], and Ondreját [Ond04]. Nevertheless, the space $L^1(\mathbb{T}^N)$ belongs to neither of those classes. Since we expect the momentum equation (3.2) to be satisfied only in the sense of distributions anyway, we make use of the embedding $L^1(\mathbb{T}^N) \hookrightarrow W^{-b,2}(\mathbb{T}^N)$, which is true, provided $b > \frac{N}{2}$. Hence we shall understand the stochastic integral, in the sense presented in Section 2.3, as a stochastic process in the Hilbert space $W^{-b,2}(\mathbb{T}^N)$. To be more precise, it is easy to check that, under the above assumptions on ϱ and $\mathbf{u}$, the operator $\mathbb{G}(\varrho,\varrho\mathbf{u})$ belongs to $L_2(\mathfrak{U}; W^{-b,2}(\mathbb{T}^N))$, the space of Hilbert–Schmidt operators from $\mathfrak{U}$ to $W^{-b,2}(\mathbb{T}^N)$. Indeed, due to (3.13) we have

$$
\begin{aligned}
\|\mathbb{G}(\varrho,\varrho\mathbf{u})\|^2_{L_2(\mathfrak{U};W_x^{-b,2})} &= \sum_{k=1}^{\infty} \|\mathbf{G}_k(\varrho,\varrho\mathbf{u})\|^2_{W_x^{-b,2}} \le C \sum_{k=1}^{\infty} \|\mathbf{G}_k(\varrho,\varrho\mathbf{u})\|^2_{L^1_x} \\
&\lesssim \int_{\mathbb{T}^N} \left(\sum_{k=1}^{\infty} \varrho^{-1} |\mathbf{G}_k(x,\varrho,\varrho\mathbf{u})|^2 \right) \mathrm{d}x \\
&\lesssim \int_{\mathbb{T}^N} (\varrho + \varrho|\mathbf{u}|^2)\, \mathrm{d}x.
\end{aligned} \tag{3.15}
$$

Consequently, we see that the right hand side of (3.15) will be finite whenever we deal with finite energy solutions. More precisely, if[1]

$$
\begin{aligned}
\varrho &\in L^{\gamma}_{\mathrm{prog}}(\Omega \times (0,T); L^{\gamma}(\mathbb{T}^N)), \quad \gamma \ge 1, \\
\sqrt{\varrho}\mathbf{u} &\in L^2_{\mathrm{prog}}(\Omega \times (0,T); L^2(\mathbb{T}^N)),
\end{aligned} \tag{3.16}
$$

then the stochastic integral in (3.2) is a well-defined $(\mathfrak{F}_t)$-martingale taking values in $W^{-b,2}(\mathbb{T}^N)$. Accordingly, the stochastic driving force is represented by the stochastic differential of the form

$$
\mathbb{G}(\varrho,\varrho\mathbf{u})\,\mathrm{d}W = \sum_{k=1}^{\infty} \mathbf{G}_k(x,\varrho,\varrho\mathbf{u})\,\mathrm{d}W_k.
$$

1 Recall that L^p_{prog} denotes the space of progressively measurable L^p-integrable functions; see Section 2.2.4.

Note that, in accordance with Remark 2.3.2, the integrability in (3.16) can be weakened to

$$\varrho \in L^\gamma(0,T;L^\gamma(\mathbb{T}^N)) \quad \mathbb{P}\text{-a.s.}, \quad \sqrt{\varrho}\mathbf{u} \in L^2(0,T;L^2(\mathbb{T}^N)) \quad \mathbb{P}\text{-a.s.}$$

In this case, the stochastic integral is generally only a local martingale. Moreover, in accordance with Lemma 2.2.18, the progressive measurability of the integrand (or, more precisely, existence of a progressively measurable representative) with respect to $(\mathfrak{F}_t)_{t\geq 0}$ follows once we have shown that the random distributions ϱ, $\mathbf{v}$ are adapted to $(\mathfrak{F}_t)_{t\geq 0}$ in the sense of Definition 2.2.13. Finally, note that the continuity equation (3.1) implies that the total mass (the mean value of the density)

$$\int_{\mathbb{T}^N} \varrho(t,x)\,\mathrm{d}x = \left(\varrho(t)\right)_{\mathbb{T}^N}$$

is constant in time (but in general may depend on ω).

3.3 Strong pathwise solutions

Although most of the results discussed in the monograph are based on the concept of weak solutions, it is natural to examine the system (3.1)–(3.6) first in the framework of strong solutions both in the PDE and the probabilistic sense. Such solutions are sufficiently regular in the space variable and consequently (3.1)–(3.6) is satisfied pointwise, whereas the stochastic integral is to be defined for any fixed x. In addition, these solutions are defined on a given probability space with a given cylindrical Wiener process. Similarly to the deterministic case, however, existence of such solutions globally in time is currently out of reach. Instead we consider *local* solutions defined on a maximal time interval bounded above by a positive stopping time that may depend on the size of the initial data.

The deterministic approach to the local existence problem for the compressible Navier–Stokes system is usually based on energy estimates. These are derived first for the unknown functions ϱ, $\mathbf{u}$ and then, repeatedly, for their time derivatives up to a sufficient order to guarantee the required smoothness. The nowadays probably optimal result in this direction has been achieved by Cho et al. [CCK04]. However, for obvious reasons related to the irregularity of sample paths of the Brownian motion, this technique is not suitable in the stochastic setting. Instead, the required space regularity must be obtained by differentiating the equations only with respect to the space variables – a typical approach applicable to purely hyperbolic systems without viscosity. The related references include works on the incompressible stochastic Navier–Stokes system by Bensoussan–Frehse [BF00] and Brzeźniak–Peszat [BP99], the incompressible stochastic Euler equations by Glatt-Holtz and Vicol [GHV14], and also quasilinear hyperbolic systems by Kim [Kim11].

Given a time interval $(0, T)$, we introduce the notion of *local strong pathwise solutions* which exist up to a suitable stopping time τ that may be strictly less than T; see Definition 3.3.1. Next, we consider *maximal strong pathwise solutions*, which live on a maximal random time interval determined by a possible blow-up of the $W^{2,\infty}$-norm of the velocity $\mathbf{u}$; see Definition 3.3.2. Although the natural functional framework here is given by spaces of continuously differentiable functions, the strong solutions constructed in this monograph live in the energy spaces $W^{s,2}$, where s is chosen large enough. These are Hilbert spaces induced by the energy method used in the existence proof in Chapter 5. Note that $W^{s,2}(\mathbb{T}^N) \hookrightarrow C^k(\mathbb{T}^N)$ as soon as $s > \frac{N}{2} + k$.

Definition 3.3.1 (Local strong pathwise solution). Let $(\Omega, \mathfrak{F}, (\mathfrak{F}_t)_{t\geq 0}, \mathbb{P})$ be a stochastic basis with a complete right-continuous filtration and let W be an $(\mathfrak{F}_t)$-cylindrical Wiener process. Let $(\varrho_0, \mathbf{u}_0)$ be an $\mathfrak{F}_0$-measurable random variable in the space $W^{s,2}(\mathbb{T}^N) \times W^{s,2}(\mathbb{T}^N)$ for some $s > \frac{N}{2} + 2$. A triplet $(\varrho, \mathbf{u}, \mathfrak{t})$ is called a *local strong pathwise solution* to system (3.1)–(3.6), provided:

(1) $\mathfrak{t}$ is an $\mathbb{P}$-a.s. strictly positive $(\mathfrak{F}_t)$-stopping time;

(2) the density ϱ is a $W^{s,2}(\mathbb{T}^N)$-valued $(\mathfrak{F}_t)$-progressively measurable stochastic process such that

$$\varrho(\cdot \wedge \mathfrak{t}) > 0, \quad \varrho(\cdot \wedge \mathfrak{t}) \in C([0, T]; W^{s,2}(\mathbb{T}^N)) \quad \mathbb{P}\text{-a.s.};$$

(3) the velocity $\mathbf{u}$ is a $W^{s,2}(\mathbb{T}^N)$-valued $(\mathfrak{F}_t)$-progressively measurable stochastic process such that

$$\mathbf{u}(\cdot \wedge \mathfrak{t}) \in C([0, T]; W^{s,2}(\mathbb{T}^N)) \cap L^2(0, T; W^{s+1,2}(\mathbb{T}^N)) \quad \mathbb{P}\text{-a.s.};$$

(4) the equation of continuity

$$\varrho(t \wedge \mathfrak{t}) = \varrho_0 - \int_0^{t\wedge\mathfrak{t}} \operatorname{div}(\varrho\mathbf{u})\,\mathrm{d}s$$

holds for all $t \in [0, T]$ $\mathbb{P}$-a.s.;

(5) the momentum equation

$$\begin{aligned}(\varrho\mathbf{u})(t \wedge \mathfrak{t}) = \varrho_0\mathbf{u}_0 &- \int_0^{t\wedge\mathfrak{t}} \operatorname{div}(\varrho\mathbf{u} \otimes \mathbf{u})\,\mathrm{d}s \\ &+ \int_0^{t\wedge\mathfrak{t}} \operatorname{div}\mathbb{S}(\nabla\mathbf{u})\,\mathrm{d}s - \int_0^{t\wedge\mathfrak{t}} \nabla p(\varrho)\,\mathrm{d}s + \int_0^{t\wedge\mathfrak{t}} \mathbb{G}(\varrho, \varrho\mathbf{u})\,\mathrm{d}W\end{aligned}$$

holds for all $t \in [0, T]$ $\mathbb{P}$-a.s.

In the above definition, we have assumed that s is large enough in order to provide sufficient regularity for the strong solutions. Classical solutions require two spatial derivatives of $\mathbf{u}$ to be continuous $\mathbb{P}$-a.s. This motivates the following definition.

Definition 3.3.2 (Maximal strong pathwise solution). Fix a stochastic basis with a cylindrical Wiener process and an initial condition exactly as in Definition 3.3.1. A quadruplet

$$(\varrho, \mathbf{u}, (\mathfrak{t}_R)_{R\in\mathbb{N}}, \mathfrak{t})$$

is called a *maximal strong pathwise solution* to system (3.1)–(3.6), provided:
(1) $\mathfrak{t}$ is a $\mathbb{P}$-a.s. strictly positive $(\mathfrak{F}_t)$-stopping time;
(2) $(\mathfrak{t}_R)_{R\in\mathbb{N}}$ is an increasing sequence of $(\mathfrak{F}_t)$-stopping times such that $\mathfrak{t}_R < \mathfrak{t}$ on the set $[\mathfrak{t} < T]$, $\lim_{R\to\infty} \mathfrak{t}_R = \mathfrak{t}$ a.s., and

$$\sup_{t\in[0,\mathfrak{t}_R]} \|\mathbf{u}(t)\|_{W^{2,\infty}_x} \geq R \quad \text{on } [\mathfrak{t} < T]; \tag{3.17}$$

(3) each triplet $(\varrho, \mathbf{u}, \mathfrak{t}_R)$, $R \in \mathbb{N}$, is a local strong pathwise solution in the sense of Definition 3.3.1.

The stopping times $\mathfrak{t}_R$ in Definition 3.3.2 announce the stopping time $\mathfrak{t}$, which is therefore predictable. It denotes the maximal life span of the solution, which is determined by the time of explosion of the $W^{2,\infty}$-norm of the velocity field. Indeed, from (3.17) we deduce

$$\sup_{t\in[0,\mathfrak{t})} \|\mathbf{u}(t)\|_{W^{2,\infty}_x} = \infty \quad \text{on } [\mathfrak{t} < T].$$

Note that the announcing sequence $(\mathfrak{t}_R)_{R\in\mathbb{N}}$ is not unique. Therefore, uniqueness for maximal strong solutions is understood in the sense that only the solution $(\varrho, \mathbf{u})$ and its blow-up time $\mathfrak{t}$ are unique.

Remark 3.3.3. Relation (3.17) should not be confused with a *blow-up* criterion. Boundedness of $\|\mathbf{u}(t)\|_{W^{2,\infty}_x}$ may follow from boundedness of a lower order norm. In the deterministic case, $\|\mathbf{u}(t)\|_{W^{2,\infty}_x}$ can be replaced by the much weaker norm $\|\varrho\|_{L^\infty_x}$; see Sun et al. [SWZ11]. Analogous results for the stochastically driven system are not known.

Existence of a unique maximal strong pathwise solution is presented in Chapter 5. Our approach relies on rewriting the problem as a symmetric hyperbolic system augmented by partial diffusion. The latter is solved with a suitable approximation procedure using the stochastic compactness method and the Yamada–Watanabe type argument based on the Gyöngy–Krylov characterization of convergence in probability, presented in Section 2.10.

3.4 Dissipative martingale solutions

A rigorous treatment of almost all real world problems requires solutions defined globally in time and for the initial data of arbitrary size. As the strong solutions introduced in the previous section are not known to enjoy these properties, the concept of dissipative martingale solutions developed in this monograph may be seen as a suitable physically grounded alternative. Here, the literature concerning the deterministic counterpart of (3.1)–(3.6) is rather extensive. The existence of weak solutions has been settled in the space dimension one (see, e.g., Antontsev et al. [AKM83] and Serre [Ser86]), but the truly multi-dimensional case seems much more complicated. The obvious mathematical difficulties, in particular the lack of control of possible density oscillations, led to the development of new concepts such as that of renormalized solutions. They have been introduced by DiPerna–Lions [DL89] and subsequently adapted by many authors in rather different contexts. The first positive existence result in two and three space dimensions for general initial data was given by Lions [Lio98] and later extended in [FNP01]; see also the monograph in [Fei04] for further references. At this point, it is worth mentioning the existence theorem by Vaigant–Kazhikhov [VK95] providing even strong solutions in the two-dimensional setting. It is based on the hypothesis of density-dependent shear and bulk viscosities coupled in a physically unrealistic way. A similar approach based on density-dependent viscosity coefficients has been developed by Bresch–Desjardins [BD07] and Bresch et al. [BDGV07].

The first existence results in the stochastic setting were based on a suitable transformation formula that allows one to reduce the problem to a system of PDEs with random coefficients, where the stochastic integral no longer appears and deterministic methods are applicable; see Tornatore–Yashima [TFY97] for the one-dimensional case and Tornatore [Tor00] for a rather special periodic two-dimensional case. Finally, the three-dimensional case was dealt with in [FMN13]. The first "truly" stochastic existence result for the compressible Navier–Stokes system perturbed by a general nonlinear multiplicative noise was obtained in [BH16]. The existence of the so-called finite energy weak martingale solutions in three space dimensions with periodic boundary conditions was established. Extensions of this result to the case of zero Dirichlet boundary conditions were given by Smith [Smi15] and Wang–Wang [WW15].

Our approach to (3.1)–(3.6) is based on the concept of *dissipative martingale solution* introduced in [BFH17]. These are, roughly speaking, weak martingale solutions satisfying in addition a variant of the energy balance. The idea to include some form of the energy/entropy balance as an integral part of a weak formulation of conservation laws goes back to Dafermos [Daf79]. Germain [Ger11] introduced a similar concept in the context of the deterministic compressible Navier–Stokes system for a class of "regular" weak solutions. Finally, the theory to weak solutions was extended in [FJN12], where it proved to be an important and rather versatile tool with various applications.

From the probabilistic point of view, dissipative martingale solutions to (3.1)–(3.6) are weak solutions in the sense that neither the underlying probability space nor the driving Wiener process can be specified in advance and these stochastic elements become part of the solution; see Section 2.5 for more details. As discussed in Section 2.10, this is intimately related to the lack of uniqueness. For the incompressible counterpart, the existence of global in time classical solutions can be established at least in the case of two space dimensions. However, global smooth solutions to compressible fluid systems are not known to exist unless the motion is drastically simplified and restricted to the one-dimensional geometry. As noted above, the only rather unphysical exception was studied in the deterministic case by Vaĭgant–Kazhikhov [VK95] and in the stochastic version by Tornatore [Tor00]. In hand with this issue goes the manner in which the initial condition is posed: one is given a probability measure which plays the role of initial law of the system (3.1)–(3.6).

From the PDE point of view, dissipative martingale solutions are also weak, that is, (3.1)–(3.6) are satisfied in the sense of distributions. In addition, the continuity equation (3.1) is required to hold in the renormalized sense and an energy inequality holds true. The concept of renormalized solution was introduced by DiPerna–Lions [DL89] in the context of linear transport equations. It is an essential tool in the study of the compressible Navier–Stokes system providing the necessary information on possible density oscillations. The energy inequality has to be understood as an integral part of the definition of a dissipative martingale solution. It encodes a certain piece of information concerning stability which would be otherwise lost in the construction process of conventional weak martingale solutions. Its important role is demonstrated by the fact that it allows one to prove a weak–strong uniqueness principle as well as an inviscid-incompressible limit result, both discussed in this monograph.

3.4.1 Weak formulation

There is an additional difficulty when dealing with weak solutions in the context of stochastic processes. The best available *a priori* bounds are provided solely by the energy balance (3.11). Specifically, we know only that $\mathbb{P}$-a.s.

$$\varrho \in L^\infty(0,T;L^\gamma(\mathbb{T}^N)), \quad \varrho\mathbf{u} \in L^\infty(0,T;L^{\frac{2\gamma}{\gamma+1}}(\mathbb{T}^N)), \quad \mathbf{u} \in L^2(0,T;W^{1,2}(\mathbb{T}^N)).$$

In particular, these quantities are *a priori* determined only for a.a. $t \in (0,T)$ and as such cannot be interpreted as standard stochastic processes. Although one finds a partial remedy by using the fact that ϱ, $\varrho\mathbf{u}$ solve the evolutionary equations (3.7)–(3.8), whence $\mathbb{P}$-a.s.

$$\varrho \in C_w(0,T;L^\gamma(\mathbb{T}^N)), \quad \varrho\mathbf{u} \in C_w(0,T;L^{\frac{2\gamma}{\gamma+1}}(\mathbb{T}^N)),$$

where the velocity $\mathbf{u}$ remains undetermined on the hypothetical vacuum zones where ϱ vanishes. It seems therefore more convenient to interpret these fields as random distributions as introduced in Section 2.2. Specifically, we may use the space $\mathcal{D}'((-\infty,T)\times\mathbb{T}^N)$ extending ϱ, $\mathbf{u}$ (and $\varrho\mathbf{u}$) by their initial values:

$$\varrho(t,\cdot) = \varrho_0, \quad \mathbf{u}(t,\cdot) = \mathbf{u}_0, \quad \varrho_0\mathbf{u}_0 = \mathbf{q}_0 \quad \text{for } t \le 0. \tag{3.18}$$

As system (3.7)–(3.8) contains superpositions of these quantities with non-linear functions, we shall always assume that $\mathbb{P}$-a.s.

$$\varrho, \mathbf{u} \in L^1_{\text{loc}}(-\infty, T; L^1(\mathbb{T}^N)).$$

In particular, $\varrho_0, \mathbf{u}_0 \in L^1(\mathbb{T}^N)$ and (3.18) makes sense. Moreover, in view of Lemma 2.2.8 and Corollary 2.2.9, it is equivalent to consider ϱ, $\mathbf{u}$ and their initial values ϱ_0, $\mathbf{u}_0$ either as random distributions in $L^1(\mathbb{T}^N)$ or $\mathcal{D}'(\mathbb{T}^N)$.

This motivates the following definition of dissipative martingale solutions.

Definition 3.4.1 (Dissipative martingale solution). Let $\Lambda = \Lambda(\varrho, \mathbf{q})$ be a Borel probability measure on $L^1(\mathbb{T}^N) \times L^1(\mathbb{T}^N)$ such that

$$\Lambda\{\varrho \geq 0\} = 1, \qquad \int_{L^1_x \times L^1_x} \left| \int_{\mathbb{T}^N} \left[\frac{|\mathbf{q}|^2}{\varrho} + P(\varrho) \right] \mathrm{d}x \right|^r \mathrm{d}\Lambda(\varrho, \mathbf{q}) < \infty, \tag{3.19}$$

where $P(\varrho)$ is the pressure potential introduced in (3.12) and $r \geq 1$ will be determined below.

The quantity $((\Omega, \mathfrak{F}, (\mathfrak{F}_t)_{t\geq 0}, \mathbb{P}), \varrho, \mathbf{u}, W)$ is called a *dissipative martingale solution* to (3.1)–(3.6) with the initial law Λ if:

(1) $(\Omega, \mathfrak{F}, (\mathfrak{F}_t)_{t\geq 0}, \mathbb{P})$ is a stochastic basis with a complete right-continuous filtration;

(2) W is a cylindrical $(\mathfrak{F}_t)$-Wiener process;

(3) the density ϱ and the velocity $\mathbf{u}$ are random distributions adapted to $(\mathfrak{F}_t)_{t\geq 0}$, $\varrho \geq 0$ $\mathbb{P}$-a.s.;

(4) there exists an $\mathfrak{F}_0$-measurable random variable $[\varrho_0, \mathbf{u}_0]$ such that $\Lambda = \mathcal{L}[\varrho_0, \varrho_0 \mathbf{u}_0]$;

(5) the equation of continuity

$$-\int_0^T \partial_t \phi \int_{\mathbb{T}^N} \varrho \psi \,\mathrm{d}x\,\mathrm{d}t = \phi(0) \int_{\mathbb{T}^N} \varrho_0 \psi \,\mathrm{d}x + \int_0^T \phi \int_{\mathbb{T}^N} \varrho \mathbf{u} \cdot \nabla \psi \,\mathrm{d}x\,\mathrm{d}t \tag{3.20}$$

holds for all $\phi \in C_c^\infty([0,T))$ and all $\psi \in C^\infty(\mathbb{T}^N)$ $\mathbb{P}$-a.s.;

(6) the momentum equation

$$\begin{aligned}
&-\int_0^T \partial_t \phi \int_{\mathbb{T}^N} \varrho \mathbf{u} \cdot \boldsymbol{\varphi} \,\mathrm{d}x\,\mathrm{d}t - \phi(0) \int_{\mathbb{T}^N} \varrho_0 \mathbf{u}_0 \cdot \boldsymbol{\varphi} \,\mathrm{d}x \\
&= \int_0^T \phi \int_{\mathbb{T}^N} [\varrho \mathbf{u} \otimes \mathbf{u} : \nabla \boldsymbol{\varphi} + p(\varrho) \operatorname{div} \boldsymbol{\varphi}] \,\mathrm{d}x\,\mathrm{d}t - \int_0^T \phi \int_{\mathbb{T}^N} \mathbb{S}(\nabla \mathbf{u}) : \nabla \boldsymbol{\varphi} \,\mathrm{d}x\,\mathrm{d}t \\
&\quad + \sum_{k=1}^\infty \int_0^T \phi \int_{\mathbb{T}^N} \mathbf{G}_k(\varrho, \varrho \mathbf{u}) \cdot \boldsymbol{\varphi} \,\mathrm{d}x\,\mathrm{d}W_k
\end{aligned} \tag{3.21}$$

holds for all $\phi \in C_c^\infty([0,T))$ and all $\boldsymbol{\varphi} \in C^\infty(\mathbb{T}^N)$ $\mathbb{P}$-a.s.;

(7) the energy inequality

$$-\int_0^T \partial_t \phi \int_{\mathbb{T}^N} \left[\frac{1}{2} \varrho |\mathbf{u}|^2 + P(\varrho) \right] \mathrm{d}x\,\mathrm{d}t + \int_0^T \phi \int_{\mathbb{T}^N} \mathbb{S}(\nabla \mathbf{u}) : \nabla \mathbf{u} \,\mathrm{d}x\,\mathrm{d}t$$

$$\leq \phi(0)\int_{\mathbb{T}^N}\left[\frac{1}{2}\varrho_0|\mathbf{u}_0|^2 + P(\varrho_0)\right]\mathrm{d}x + \frac{1}{2}\sum_{k=1}^{\infty}\int_0^T \phi\int_{\mathbb{T}^N}\varrho^{-1}|\mathbf{G}_k(\varrho,\varrho\mathbf{u})|^2\,\mathrm{d}x\,\mathrm{d}t$$
$$+\sum_{k=1}^{\infty}\int_0^T \phi\int_{\mathbb{T}^N}\mathbf{G}_k(\varrho,\varrho\mathbf{u})\cdot\mathbf{u}\,\mathrm{d}x\,\mathrm{d}W_k, \tag{3.22}$$

holds for all $\phi \in C_c^\infty([0,T))$, $\phi \geq 0$, $\mathbb{P}$-a.s.;

(8) if $b \in C^1(\mathbb{R})$ such that $b'(z) = 0$ for all $z \geq M_b$, then, for all $\phi \in C_c^\infty([0,T))$ and all $\psi \in C^\infty(\mathbb{T}^N)$, we have $\mathbb{P}$-a.s.

$$-\int_0^T \partial_t\phi\int_{\mathbb{T}^N} b(\varrho)\psi\,\mathrm{d}x\,\mathrm{d}t = \phi(0)\int_{\mathbb{T}^N} b(\varrho_0)\psi\,\mathrm{d}x + \int_0^T\phi\int_{\mathbb{T}^N} b(\varrho)\mathbf{u}\cdot\nabla\psi\,\mathrm{d}x\,\mathrm{d}t$$
$$-\int_0^T\phi\int_{\mathbb{T}^N}(b'(\varrho)\varrho - b(\varrho))\,\mathrm{div}\,\mathbf{u}\,\psi\,\mathrm{d}x\,\mathrm{d}t. \tag{3.23}$$

Equation (3.23) represents a renormalized variant of the equation of continuity. Strictly speaking, it may be omitted in the definition, however, the solution we construct will always enjoy this property.

3.4.2 Regularity properties of weak solutions

In Definition 3.4.1 we have tacitly assumed that all terms appearing in the weak formulation including the stochastic integral in the momentum equation are well-defined. Taking $\psi = 1$ in (3.20), we easily deduce the total mass conservation

$$\int_{\mathbb{T}^N}\varrho(t,x)\,\mathrm{d}x = \int_{\mathbb{T}^N}\varrho_0(x)\,\mathrm{d}x \quad \text{for any } t \geq 0.$$

Next, it follows from the energy inequality (3.22) that

$$\int_{\mathbb{T}^N}\left[\frac{1}{2}\varrho|\mathbf{u}|^2 + P(\varrho)\right](\tau)\,\mathrm{d}x + \int_0^\tau\int_{\mathbb{T}^N}\mathbb{S}(\nabla\mathbf{u}):\nabla\mathbf{u}\,\mathrm{d}x\,\mathrm{d}t$$
$$\leq \int_{\mathbb{T}^N}\left[\frac{1}{2}\varrho_0|\mathbf{u}_0|^2 + P(\varrho_0)\right]\mathrm{d}x + \frac{1}{2}\int_0^\tau\int_{\mathbb{T}^N}\sum_{k=1}^{\infty}\varrho^{-1}|\mathbf{G}_k(\varrho,\varrho\mathbf{u})|^2\,\mathrm{d}x\,\mathrm{d}t$$
$$+\sum_{k=1}^{\infty}\int_0^\tau\int_{\mathbb{T}^N}\mathbf{G}_k(\varrho,\varrho\mathbf{u})\cdot\mathbf{u}\,\mathrm{d}x\,\mathrm{d}W_k \quad \text{for a.a. } \tau \in (0,T)\ \mathbb{P}\text{-a.s.}$$

Indeed the time-dependent test functions ϕ in (3.22) may be chosen to approximate the characteristic function $1_{[0,\tau]}$; whence the above relation holds in any Lebesgue point of the function

$$\tau \mapsto \int_{\mathbb{T}^N}\left[\frac{1}{2}\varrho|\mathbf{u}|^2 + P(\varrho)\right](\tau)\,\mathrm{d}x.$$

By virtue of hypothesis (3.13), we get

$$\int_0^\tau\int_{\mathbb{T}^N}\sum_{k=1}^{\infty}\varrho^{-1}|\mathbf{G}_k(\varrho,\varrho\mathbf{u})|^2\,\mathrm{d}x\,\mathrm{d}t \lesssim \int_0^\tau\sum_{k=1}^{\infty}g_k^2\int_{\mathbb{T}^N}(\varrho + \varrho|\mathbf{u}|^2)\,\mathrm{d}x\,\mathrm{d}t$$

$$\lesssim \int_0^\tau \int_{\mathbb{T}^N} (\varrho + \varrho|\mathbf{u}|^2)\,\mathrm{d}x\,\mathrm{d}t.$$

Similarly, we apply Burkholder–Davis–Gundy's inequality to the stochastic integral, obtaining

$$\begin{aligned}
&\mathbb{E}\left[\sup_{t\in[0,\tau]}\left|\sum_{k=1}^\infty \int_0^t \int_{\mathbb{T}^N} \mathbf{G}_k(\varrho,\varrho\mathbf{u})\cdot\mathbf{u}\,\mathrm{d}x\,\mathrm{d}W_k\right|^r\right]\\
&\lesssim \mathbb{E}\left[\int_0^\tau \sum_{k=1}^\infty \left|\int_{\mathbb{T}^N} \mathbf{G}_k(\varrho,\varrho\mathbf{u})\cdot\mathbf{u}\,\mathrm{d}x\right|^2 \mathrm{d}t\right]^{r/2}\\
&\lesssim \mathbb{E}\left[\int_0^\tau \left|\int_{\mathbb{T}^N} (\varrho+\varrho|\mathbf{u}|^2)\,\mathrm{d}x\right|^2 \mathrm{d}t\right]^{r/2}.
\end{aligned}$$

Supposing $r \geq 2$ in (3.19), we use Gronwall's lemma to conclude

$$\mathbb{E}\left[\left(\sup_{t\in[0,T]}\int_{\mathbb{T}^N}\left[\frac{1}{2}\varrho|\mathbf{u}|^2 + P(\varrho)\right]\mathrm{d}x\right)^2\right] + \mathbb{E}\left[\left(\int_0^T\int_{\mathbb{T}^N}|\nabla\mathbf{u}|^2\,\mathrm{d}x\,\mathrm{d}t\right)^2\right] < \infty. \tag{3.24}$$

Going back to the equation of continuity, we deduce from (3.24)

$$\varrho \in C([0,T]; W^{-b,2}(\mathbb{T}^N)), \quad b > \frac{N}{2}, \quad \text{in particular } \varrho(0,\cdot) = \varrho_0. \tag{3.25}$$

Finally, we deduce from the momentum equation

$$\begin{aligned}
&\int_{\mathbb{T}^N} \varrho\mathbf{u}\cdot\boldsymbol{\varphi}(\tau)\,\mathrm{d}x - \int_{\mathbb{T}^N} \varrho_0\mathbf{u}_0\cdot\boldsymbol{\varphi}\,\mathrm{d}x\\
&\quad= \int_0^\tau\int_{\mathbb{T}^N} [\varrho\mathbf{u}\otimes\mathbf{u} : \nabla\boldsymbol{\varphi} + p(\varrho)\mathrm{div}\boldsymbol{\varphi}]\,\mathrm{d}x\,\mathrm{d}t - \int_0^\tau\int_{\mathbb{T}^N} \mathbb{S}(\nabla\mathbf{u}) : \nabla\boldsymbol{\varphi}\,\mathrm{d}x\,\mathrm{d}t\\
&\qquad+ \sum_{k=1}^\infty \int_0^\tau\int_{\mathbb{T}^N} \mathbf{G}_k(\varrho,\varrho\mathbf{u})\cdot\boldsymbol{\varphi}\,\mathrm{d}x\,\mathrm{d}W_k.
\end{aligned}$$

Seeing that by (3.13)

$$\mathbb{E}\left[\int_0^\tau\left(\sum_{k=1}^\infty\left(\int_{\mathbb{T}^N}\mathbf{G}_k(\varrho,\varrho\mathbf{u})\cdot\boldsymbol{\varphi}\,\mathrm{d}x\right)^2\right)^{p/2}\mathrm{d}t\right] \lesssim \|\boldsymbol{\varphi}\|^p_{L^\infty_x}\mathbb{E}\left[\int_0^\tau [\|\varrho\|_{L^1_x} + \|\varrho\mathbf{u}\|_{L^1_x}]^p\,\mathrm{d}t\right],$$

we use Lemma 2.3.9 with $p = 2r \geq 4$ to conclude

$$\varrho\mathbf{u} \in C([0,T]; W^{-b,2}(\mathbb{T}^N)), \quad b > \frac{N}{2}, \quad \text{in particular } \varrho\mathbf{u}(0,\cdot) = \varrho_0\mathbf{u}_0. \tag{3.26}$$

Remark 3.4.2. In view of the energy bounds (3.24), continuity of ϱ and $\varrho\mathbf{u}$ can be strengthened to

$$\varrho \in C_w([0,T]; L^\gamma(\mathbb{T}^N)), \quad \varrho\mathbf{u} \in C_w([0,T]; L^{\frac{2\gamma}{\gamma+1}}(\mathbb{T}^N)).$$

Note that, by Hölder's inequality,

$$\|\varrho\mathbf{u}\|_{L_x^{\frac{2\gamma}{\gamma+1}}} \le \|\varrho\|_{L_x^\gamma}^{1/2}\|\varrho|\mathbf{u}|^2\|_{L_x^1}^{1/2}.$$

As ϱ and $\mathbf{u}$ are random distributions adapted to $(\mathfrak{F}_t)_{t\ge0}$, we conclude that the time-dependent stochastic processes ϱ, $\varrho\mathbf{u}$ are $(\mathfrak{F}_t)$-progressively measurable; cf. Proposition 2.1.18.

The regularity properties of dissipative martingale solutions are summarized below.

Proposition 3.4.3. *Let $\Lambda = \Lambda(\varrho, \mathbf{q})$ be a Borel probability measure on $L^1(\mathbb{T}^N)\times L^1(\mathbb{T}^N)$ satisfying* (3.19) *for some $r \ge 2$. Suppose that $\gamma > \frac{N}{2}$ and*

$$p(\varrho) = a\varrho^\gamma, \quad P(\varrho) = \varrho\int_1^\varrho \frac{p(z)}{z^2}\,\mathrm{d}z.$$

Let $((\Omega, \mathfrak{F}, (\mathfrak{F}_t)_{t\ge0}, \mathbb{P}), \varrho, \mathbf{u}, W)$ be a dissipative martingale solution to (3.1)–(3.6) *with the initial law Λ in the sense of Definition* 3.4.1. *Then:*

(1) *the density $\varrho \in C_w([0,T]; L^\gamma(\mathbb{T}^N))$ and the momentum $\varrho\mathbf{u} \in C_w([0,T]; L^{\frac{2\gamma}{\gamma+1}}(\mathbb{T}^N))$ are $(\mathfrak{F}_t)$-progressively measurable stochastic processes, where $\mathbb{P}$-a.s.*

$$\varrho(0,\cdot) = \varrho_0, \quad \varrho\mathbf{u}(0,\cdot) = \varrho_0\mathbf{u}_0;$$

(2) *the functions ϱ, $\mathbf{u}$ enjoy the following properties:*

$$\mathbb{E}\Big[\sup_{t\in[0,T]}\|\varrho(t)\|_{L_x^\gamma}^{\gamma r}\Big] < \infty, \tag{3.27}$$

$$\mathbb{E}\bigg(\int_0^T \|\mathbf{u}\|_{W_x^{1,2}}^2\,\mathrm{d}t\bigg)^r < \infty, \tag{3.28}$$

$$\mathbb{E}\Big[\sup_{t\in[0,T]}\|\varrho\mathbf{u}(t)\|_{L_x^{\frac{2\gamma}{\gamma+1}}}^{\frac{2\gamma}{\gamma+1}r}\Big] < \infty; \tag{3.29}$$

(3) *the equation of continuity*

$$\bigg[\int_{\mathbb{T}^N}\varrho\psi\,\mathrm{d}x\bigg]_{t=0}^{t=\tau} = -\int_0^\tau\int_{\mathbb{T}^N}\varrho\mathbf{u}\cdot\nabla\psi\,\mathrm{d}x\,\mathrm{d}t$$

holds for all $\tau \in [0,T]$ and any $\psi \in C^\infty(\mathbb{T}^N)$ $\mathbb{P}$-a.s.;

(4) *the momentum equation*

$$\begin{aligned}&\bigg[\int_{\mathbb{T}^N}\varrho\mathbf{u}\cdot\boldsymbol{\varphi}\,\mathrm{d}x\bigg]_{t=0}^{t=\tau}\\ &= \int_0^\tau\int_{\mathbb{T}^N}[\varrho\mathbf{u}\otimes\mathbf{u} : \nabla\boldsymbol{\varphi} + p(\varrho)\mathrm{div}\,\boldsymbol{\varphi}]\,\mathrm{d}x\,\mathrm{d}t - \int_0^\tau\int_{\mathbb{T}^N}\mathbb{S}(\nabla\mathbf{u}) : \nabla\boldsymbol{\varphi}\,\mathrm{d}x\,\mathrm{d}t\end{aligned}$$

$$+\int_0^\tau \int_{\mathbb{T}^N} \mathbb{G}(\varrho, \varrho\mathbf{u}) \cdot \boldsymbol{\varphi} \, dx \, dW \tag{3.30}$$

holds for all $\tau \in [0, T]$ *and any* $\boldsymbol{\varphi} \in C^\infty(\mathbb{T}^N)$ $\mathbb{P}$*-a.s.*

Remark 3.4.4. Note that the pressure potential $P(\varrho) = \varrho \int_1^\varrho \frac{p(z)}{z^2} \, dz$ may be replaced by $P(\varrho) = \frac{a}{\gamma-1}\varrho^\gamma$, which differs only by a linear function of ϱ. As the total mass of the fluid is conserved, this makes no difference in the energy inequality. Of course, the conclusion of Proposition 3.4.3 remains valid for a more general pressure law

$$p(\varrho) \approx \varrho^\gamma, \quad p'(\varrho) \geq 0 \quad \text{for large } \varrho,$$

in particular monotonicity of the pressure is not needed. Under hypothesis (3.19) and due to Remark 2.3.7 and the assumptions on the coefficient $\mathbb{G}$, the stochastic integral in (3.30) as well as (3.22) is a well-defined stochastic process with values in $W^{-b,2}(\mathbb{T}^N)$ and $\mathbb{R}$, respectively. In particular, the integrands possess an $(\mathfrak{F}_t)$-progressively measurable representative. As discussed above, the conditions on ϱ and $\mathbf{u}$ already guarantee that $\mathbb{G}(\varrho, \varrho\mathbf{u})$ takes values in $L_2(\mathfrak{U}; W^{-b,2}(\mathbb{T}^N))$.

In Chapter 4, we prove existence of dissipative martingale solutions. The proof relies on a multi-layer approximation scheme whose core follows the technique developed in [FNP01] in order to deal with the deterministic counterpart. Our proof makes an essential use of the stochastic compactness method introduced in Section 2.6 and in particular of the Jakubowski–Skorokhod representation Theorem 2.7.1. However, in comparison with the existence result from [BH16], we present several modifications. Most importantly, the energy inequality (3.22) originally introduced in [BFH17] is included in the definition of a solution. This requires certain refinements of the compactness argument.

Let us stress the importance of the energy inequality (3.22) appearing in Definition 3.4.1. Indeed, (3.22) is the key ingredient in order to establish the so-called relative energy inequality presented in Chapter 6. This may be viewed as a kind of distance between a dissipative martingale solution of system (3.1)–(3.6) and a pair of arbitrary (smooth) processes. As a consequence, it is possible to compare dissipative martingale solutions with strong solutions (in the PDE sense) and to prove a weak–strong uniqueness principle in Section 8.1. In addition, the relative energy inequality is employed in Section 8.2 in order to establish an inviscid-incompressible limit of (3.1)–(3.6).

3.5 Stationary solutions

As the next step towards a better understanding of a compressible fluid flow subject to stochastic perturbation, we are concerned with the existence of stationary solutions. Generally speaking, stationary solutions of an evolutionary system provide an important piece of information concerning the behavior in the long run. For systems with a background in classical fluid mechanics, stationary solutions typically minimize the entropy production and play the role of an attractor, at least for energetically insulated fluid flows; see, e.g., [FP10].

The principal question arising in the context of randomly driven systems is the existence of a stochastic steady state solution for the system. Earlier results in this direction concern the incompressible case: Flandoli [Fla94] proved existence of an invariant measure by the "remote start" method in the two-dimensional case. This result has been extended in a few papers, for instance in Goldys–Maslowski [GM06, GM05], where existence of an invariant measure is shown by the method of embedded Markov chain theory, which also verifies the exponential speed of convergence to the invariant measure. A different approach has been adopted by Hairer–Mattingly [HM06], in which case a slightly weaker convergence result (implying, however, the uniqueness of invariant measure) has been shown under much weaker conditions on the non-degeneracy of the noise. Brzeźniak et al. [BMO15] proved the existence of an invariant measure for a two-dimensional Navier–Stokes equation on unbounded domain by a compactness method in the weak bw-topology.

In the three-dimensional case, much less is known regarding incompressible fluids. The problems here appear already on the level of the Markov property induced by the equation as uniqueness is unknown. A transition Markov semi-group has been constructed in the papers by Da Prato–Debussche [DPD03, DPD08], provided the noise term is sufficiently rough in space. A different approach was adopted by Flandoli–Romito [FR08], who used the classical Stroock–Varadhan type argument to find a suitable Markov selection and construct a semi-group. The transition semi-group is shown to be exponentially ergodic (under appropriate conditions on the noise term) by the same arguments as in Goldys–Maslowski [GM06]. However, the uniqueness of the Markov transition semi-group has not been proved so far.

In the absence of the Markov property (i.e., in the situation where the concept of invariant measure as a steady state is not well-defined), it is possible to work directly with stationary solutions, i.e., with solutions which are stationary stochastic processes. In the pioneering paper by Flandoli–Gątarek [FG95], existence of such stationary solutions has been shown in the three-dimensional incompressible case by means of finite-dimensional approximations. A generalization to less regular noise on the whole space $\mathbb{R}^N$ was given by Brzeźniak–Ferrario [BF17], where existence of an invariant measure was proved if $N = 2$ and existence of a stationary solution if $N = 3$.

Our goal is to establish the existence of global-in-time solutions to system (3.1)–(3.6) that are stationary in the stochastic sense. To this end, we use a direct method based on the multi-layer approximation scheme presented in Chapter 4. Let us recall that two notions of stationarity were introduced in Section 2.11, namely stationarity of stochastic processes and stationarity of random variables taking values in a space of space-time distributions, i.e., random distributions as introduced in Definition 2.2.1. Due to the specific structure of the Navier–Stokes system (3.1)–(3.6), the concept of stationarity must be chosen accordingly. Motivated by Definition 2.11.2, we adapt the concept introduced in the context of incompressible viscous fluids by Romito [Rom10]; cf. also the approach proposed by Itô–Nisio [IN64]. As above, $\mathcal{S}_\tau$ denotes the time shift on the space of trajectories given by $\mathcal{S}_\tau \varphi(t) = \varphi(t+\tau)$.

Definition 3.5.1. A dissipative martingale solution $[\varrho, \mathbf{u}, W]$ to (3.1)–(3.6) is called *stationary*, provided the joint law of the time shift $[\mathcal{S}_\tau \varrho, \mathcal{S}_\tau \mathbf{u}, \mathcal{S}_\tau W - W(\tau)]$ on

$$L^1_{\mathrm{loc}}(0,\infty; L^\gamma(\mathbb{T}^3)) \times L^1_{\mathrm{loc}}(0,\infty; W^{1,2}(\mathbb{T}^3)) \times C_{\mathrm{loc}}([0,\infty); \mathfrak{U}_0)$$

is independent of $\tau \geq 0$.

Remark 3.5.2. Equivalently, we may say that, for all $\ell \in \mathbb{N}$, the law of the random distribution $[\varrho, \mathbf{u}, W_1, \dots, W_\ell]$ in $\mathcal{D}'((0,\infty) \times \mathbb{T}^3; \mathbb{R}^{\ell+4})$ coincides with the law of its time shift $[\mathcal{S}_\tau \varrho, \mathcal{S}_\tau \mathbf{u}, \mathcal{S}_\tau W_1 - W_1(\tau), \dots, \mathcal{S}_\tau W_\ell - W_\ell(\tau)]$ for any $\tau \geq 0$.

Note that, trivially, $\mathcal{S}_\tau W(\cdot) - W(\tau) = W(\cdot + \tau) - W(\tau)$ is a cylindrical Wiener process. The Wiener process was included in Definition 3.5.1 to point out that the joint law of $[\varrho, \mathbf{u}, W]$ is shift invariant.

In accordance with the discussion in Section 2.11 and in particular due to Corollary 2.11.8, if $[\varrho, \mathbf{u}, W]$ is a stationary dissipative martingale solution of (3.1)–(3.6), then the stochastic process $[\varrho, \varrho\mathbf{u}]$ is stationary on $L^\gamma(\mathbb{T}^3) \times L^{\frac{2\gamma}{\gamma+1}}(\mathbb{T}^3)$ in the sense of Definition 2.11.1, whereas $\mathbf{u}$ is only weakly stationary in the sense of Definition 2.11.2.

In Chapter 7, we establish existence of a stationary dissipative martingale solution with prescribed total mass $\int_{\mathbb{T}^3} \varrho(t,x)\,\mathrm{d}x = M_0$ for all $t \in [0,\infty)$, where $M_0 > 0$ is a deterministic constant. As mentioned above, the proof is based on the approximation scheme from Chapter 4. More specifically, the stationary solutions are constructed at the very basic approximation level using the Krylov–Bogoliubov method (see Section 2.12). The final result is obtained by means of delicate global-in-time estimates and a combination of deterministic and stochastic compactness methods.

4 Global existence

This chapter is devoted to the proof of existence of global-in-time *dissipative martingale solutions* to the barotropic Navier–Stokes system

$$\mathrm{d}\varrho + \operatorname{div}(\varrho\mathbf{u})\,\mathrm{d}t = 0, \tag{4.1}$$

$$\mathrm{d}(\varrho\mathbf{u}) + \operatorname{div}(\varrho\mathbf{u}\otimes\mathbf{u})\,\mathrm{d}t + \nabla p(\varrho)\,\mathrm{d}t = \operatorname{div}\mathbb{S}(\nabla\mathbf{u})\,\mathrm{d}t + \mathbb{G}(\varrho,\varrho\mathbf{u})\,\mathrm{d}W. \tag{4.2}$$

For definiteness, we focus on the most complex and physically relevant case $N = 3$. The same proof applies to $N = 1, 2$ with obvious modifications. In addition, the critical values of certain parameters, notably the adiabatic exponent γ, can be slightly improved. The density ϱ and the velocity $\mathbf{u}$ are random variables ranging in a suitable path space of functions defined on a finite time interval $[0, T]$ and periodic with respect to the spatial variable $x \in \mathbb{T}^3$.

Equation (4.2) includes a cylindrical Wiener process $W = (W_k)_{k\in\mathbb{N}}$, defined on a complete probability space $(\Omega, \mathfrak{F}, \mathbb{P})$ with a complete right-continuous filtration $(\mathfrak{F}_t)_{t\geq 0}$. We study (4.1)–(4.2) in the context of dissipative martingale solutions introduced in Section 3.4.1. As uniqueness of solutions is an outstanding open problem even at the deterministic level, the existence of strong solutions (in the stochastic sense) defined on a given probability space is apparently out of reach for the available analytical tools. We have to content ourselves with martingale solutions, where the probability space $(\Omega, \mathfrak{F}, \mathbb{P})$ and the Wiener process W are not *a priori* given and form an integral part of the solution together with the fields ϱ, $\mathbf{u}$. Accordingly, the solutions we look for are weak in the PDE sense (satisfying the underlying equations in the sense of generalized derivatives) as well as in the probability sense. In addition, solutions will satisfy a certain form of total energy balance.

For the sake of convenience, we recall the definition of a dissipative martingale solution from Section 3.4.1.

Definition 4.0.1. Let $\Lambda = \Lambda(\varrho, \mathbf{q})$ be a Borel probability measure on $L^1(\mathbb{T}^3) \times L^1(\mathbb{T}^3)$ such that

$$\Lambda\{\varrho \geq 0\} = 1, \quad \int_{L^1_x \times L^1_x} \left| \int_{\mathbb{T}^3} \left[\frac{|\mathbf{q}|^2}{\varrho} + P(\varrho) \right] \mathrm{d}x \right|^r \mathrm{d}\Lambda(\varrho, \mathbf{q}) < \infty,$$

where the pressure potential is given by

$$P(\varrho) = \varrho \int_1^\varrho \frac{p(z)}{z^2}\,\mathrm{d}z$$

and $r \geq 1$.

The quantity $((\Omega, \mathfrak{F}, (\mathfrak{F}_t)_{t\geq 0}, \mathbb{P}), \varrho, \mathbf{u}, W)$ is called a *dissipative martingale solution* to the Navier–Stokes system (4.1)–(4.2) with the initial law Λ if:

https://doi.org/10.1515/9783110492552-004

(1) $(\Omega, \mathfrak{F}, (\mathfrak{F}_t)_{t\geq 0}, \mathbb{P})$ is a stochastic basis with a complete right-continuous filtration;
(2) W is a cylindrical $(\mathfrak{F}_t)$-Wiener process;
(3) the density ϱ and the velocity $\mathbf{u}$ are random distributions adapted to $(\mathfrak{F}_t)_{t\geq 0}$, $\varrho \geq 0$ $\mathbb{P}$-a.s.;
(4) there exists an $\mathfrak{F}_0$-measurable random variable $[\varrho_0, \mathbf{u}_0]$ such that $\Lambda = \mathcal{L}[\varrho_0, \varrho_0\mathbf{u}_0]$;
(5) the equation of continuity

$$-\int_0^T \partial_t \phi \int_{\mathbb{T}^3} \varrho\psi \,\mathrm{d}x\,\mathrm{d}t = \phi(0)\int_{\mathbb{T}^3} \varrho_0\psi\,\mathrm{d}x + \int_0^T \phi \int_{\mathbb{T}^3} \varrho\mathbf{u}\cdot\nabla\psi\,\mathrm{d}x\,\mathrm{d}x\,\mathrm{d}t$$

holds for all $\phi \in C_c^\infty([0,T))$ and all $\psi \in C^\infty(\mathbb{T}^3)$ $\mathbb{P}$-a.s.;
(6) the momentum equation

$$\begin{aligned}
&-\int_0^T \partial_t\phi \int_{\mathbb{T}^3} \varrho\mathbf{u}\cdot\boldsymbol{\varphi}\,\mathrm{d}x\,\mathrm{d}t - \phi(0)\int_{\mathbb{T}^3}\varrho_0\mathbf{u}_0\cdot\boldsymbol{\varphi}\,\mathrm{d}x \\
&= \int_0^T \phi\int_{\mathbb{T}^3} [\varrho\mathbf{u}\otimes\mathbf{u} : \nabla\boldsymbol{\varphi} + p(\varrho)\operatorname{div}\boldsymbol{\varphi}]\,\mathrm{d}x\,\mathrm{d}t - \int_0^T\phi\int_{\mathbb{T}^3}\mathbb{S}(\nabla\mathbf{u}) : \nabla\boldsymbol{\varphi}\,\mathrm{d}x\,\mathrm{d}t \\
&\quad + \sum_{k=1}^\infty \int_0^T \phi\int_{\mathbb{T}^3} \mathbf{G}_k(\varrho,\varrho\mathbf{u})\cdot\boldsymbol{\varphi}\,\mathrm{d}x\,\mathrm{d}W_k
\end{aligned} \tag{4.3}$$

holds for all $\phi \in C_c^\infty([0,T))$ and all $\boldsymbol{\varphi} \in C^\infty(\mathbb{T}^3)$ $\mathbb{P}$-a.s.;
(7) the energy inequality

$$\begin{aligned}
&-\int_0^T \partial_t\phi\int_{\mathbb{T}^3}\Big[\frac{1}{2}\varrho|\mathbf{u}|^2 + P(\varrho)\Big]\,\mathrm{d}x\,\mathrm{d}t + \int_0^T\phi\int_{\mathbb{T}^3}\mathbb{S}(\nabla\mathbf{u}) : \nabla\mathbf{u}\,\mathrm{d}x\,\mathrm{d}t \\
&\leq \phi(0)\int_{\mathbb{T}^3}\Big[\frac{1}{2}\varrho_0|\mathbf{u}_0|^2 + P(\varrho_0)\Big]\,\mathrm{d}x + \frac{1}{2}\int_0^T\phi\int_{\mathbb{T}^3}\sum_{k=1}^\infty \varrho^{-1}|\mathbf{G}_k(\varrho,\varrho\mathbf{u})|^2\,\mathrm{d}x\,\mathrm{d}t \\
&\quad + \sum_{k=1}^\infty\int_0^T\phi\int_{\mathbb{T}^3}\mathbf{G}_k(\varrho,\varrho\mathbf{u})\cdot\mathbf{u}\,\mathrm{d}x\,\mathrm{d}W_k
\end{aligned}$$

holds for all $\phi \in C_c^\infty([0,T))$, $\phi \geq 0$, $\mathbb{P}$-a.s.;
(8) if $b \in C^1(\mathbb{R})$ such that $b'(z) = 0$ for all $z \geq M_b$, then, for all $\phi \in C_c^\infty([0,T))$ and all $\psi \in C^\infty(\mathbb{T}^3)$, we have $\mathbb{P}$-a.s.

$$\begin{aligned}
-\int_0^T\partial_t\phi\int_{\mathbb{T}^3} b(\varrho)\psi\,\mathrm{d}x\,\mathrm{d}t &= \phi(0)\int_{\mathbb{T}^3} b(\varrho_0)\psi\,\mathrm{d}x + \int_0^T\phi\int_{\mathbb{T}^3} b(\varrho)\mathbf{u}\cdot\nabla\psi\,\mathrm{d}x\,\mathrm{d}t \\
&\quad - \int_0^T\phi\int_{\mathbb{T}^3}(b'(\varrho)\varrho - b(\varrho))\operatorname{div}\mathbf{u}\,\psi\,\mathrm{d}x\,\mathrm{d}t.
\end{aligned}$$

The main objective of this chapter is to show the following existence result.

Theorem 4.0.2. *Let*

$$p(\varrho) = a\varrho^\gamma, \quad a > 0,\ \gamma > \frac{3}{2}.$$

Let Λ be a Borel probability measure defined on the space $L^1(\mathbb{T}^3) \times L^1(\mathbb{T}^3)$ such that

$$\Lambda\{\varrho \geq 0\} = 1, \quad \Lambda\Big\{0 < \underline{\varrho} \leq \int_{\mathbb{T}^3} \varrho \,\mathrm{d}x \leq \overline{\varrho} < \infty\Big\} = 1,$$

for some deterministic constants $\underline{\varrho}, \overline{\varrho}$ and

$$\int_{L^1_x \times L^1_x} \left| \int_{\mathbb{T}^3} \left[\frac{1}{2}\frac{|\mathbf{q}|^2}{\varrho} + P(\varrho)\right] \mathrm{d}x \right|^r \mathrm{d}\Lambda \lesssim 1, \quad P(\varrho) = \varrho \int_1^{\varrho} \frac{p(z)}{z^2}\,\mathrm{d}z,$$

for some $r \geq 4$. Let the diffusion coefficients $\mathbb{G} = \mathbb{G}(x, \varrho, \mathbf{q})$ be continuously differentiable satisfying (3.13) *and* (3.14).

Then there is a dissipative martingale solution to (4.1)–(4.2) *in the sense of Definition* 4.0.1.

Remark 4.0.3. – The moments of the initial law are transferred in a standard way from the initial datum to the solution; see Proposition 3.4.3. The assumption $r \geq 4$ in Theorem 4.0.2 is needed in order to estimate the L^2_x-norm of $\mathbf{u}$; see estimate (4.88). Due to the periodic boundary conditions, we do not have Poincaré's inequality at hand. This drawback can be overcome when working under the no-slip boundary condition.
- As already pointed out, the specific form of the pressure is taken only for the sake of simplicity. A general pressure density state equation can be considered as long as (i) the pressure is a non-decreasing function of the density and (ii) $p(\varrho) \approx \varrho^\gamma$, $\gamma > \frac{3}{2}$ for large ϱ. As a matter of fact, monotonicity of the pressure is not really necessary; see [Fei02].
- The spatial domain is a periodic flat torus. We claim that the same result can be shown on a bounded spatial domain endowed with the no-slip boundary condition for the velocity (cf. Smith [Smi15]) and also on the whole space $\mathbb{R}^3$ (see Mensah [Men16]).
- If $N = 1, 2$, the result holds for $\gamma \geq 1$ and $\gamma > 1$, respectively.
- The driving force can be augmented by a deterministic component $\varrho(t,x)\mathbf{f}(t,x)\,\mathrm{d}t$.

The standard approach to solve non-linear PDEs starts with a finite-dimensional approximation of Galerkin-type. Unfortunately, this can only be applied to the momentum equation (4.2) since we need the density ϱ to be positive at the first level of approximation. Positivity of ϱ results from the maximum principle, where the latter is usually incompatible with a Galerkin type approximation. It seems therefore more convenient to apply the artificial viscosity method, adding diffusive terms to both (4.1) and (4.2). Thus we are led to study the following approximate system:

$$\mathrm{d}\varrho + \operatorname{div}(\varrho\mathbf{u})\,\mathrm{d}t = \varepsilon\Delta\varrho\,\mathrm{d}t, \tag{4.4}$$

$$\mathrm{d}(\varrho\mathbf{u}) + \operatorname{div}(\varrho\mathbf{u}\otimes\mathbf{u})\,\mathrm{d}t + \nabla p_\delta(\varrho)\,\mathrm{d}t = \varepsilon\Delta(\varrho\mathbf{u})\,\mathrm{d}t + \operatorname{div}\mathbb{S}(\nabla\mathbf{u})\,\mathrm{d}t + \mathbb{G}_\varepsilon(\varrho, \varrho\mathbf{u})\,\mathrm{d}W, \tag{4.5}$$

where $p_\delta(\varrho) = p(\varrho) + \delta(\varrho + \varrho^\Gamma)$ with $\Gamma \geq \max\{6, \gamma\}$. In addition, certain cut-off operators will be applied to avoid other technical difficulties in the existence proof. The term $\varepsilon\Delta(\varrho\mathbf{u})$ is added to the momentum equation to maintain the energy balance. In order to ensure that the term $\varepsilon\Delta\varrho$ in (4.4) converges to zero in the vanishing viscosity limit, the artificial pressure $\delta(\varrho + \varrho^\Gamma)$ is needed (it implies higher integrability of ϱ and also a better control near vacuum regions). It yields an estimate for $\sqrt{\varepsilon}\nabla\varrho$ which is uniformly in ε by (4.4). For technical reasons we also employ an approximation of the noise coefficient; see (4.8) below. The aim is to pass to the limit first in $\varepsilon \to 0$ and subsequently in $\delta \to 0$. However, in order to solve (4.4)–(4.5) for $\varepsilon > 0$ and $\delta > 0$ fixed, we need two additional approximation layers.

In particular, we employ a stopping time technique to establish the existence of a unique solution to a finite-dimensional approximation of (4.4)–(4.5). We gain the so-called Faedo–Galerkin approximation, on each random time interval $[0, \tau_R)$, where the stopping time τ_R cuts the norms of certain quantities if they reach the value R. It is then shown that the blow-up cannot occur in a finite time, so letting $R \to \infty$ gives a unique solution to the Faedo–Galerkin approximation on the whole time interval $[0, T]$. The solutions to the Faedo–Galerkin approximation will be constructed in the space of trigonometric polynomials of order m. To be precise, let

$$H_m = \left\{ v = \sum_{\mathbf{m}, \max_{j=1,2,3} |m_j| \leq m} [a_\mathbf{m} \cos(\pi\mathbf{m}\cdot x) + b_\mathbf{m} \sin(\pi\mathbf{m}\cdot x)] \mid a_\mathbf{m}, b_\mathbf{m} \in \mathbb{R} \right\}^3,$$

endowed with the Hilbert structure of the Lebesgue space $L^2(\mathbb{T}^3)$. Let

$$\Pi_m : L^2(\mathbb{T}^3) \to H_m$$

be the associated L^2-orthogonal projection. For $f \in L^1(\mathbb{T}^3)$, the projection $\Pi_m[f]$ represents the mth cubic partial sum of the Fourier series of f. In particular,

$$\begin{aligned} &\|\Pi_m[f]\|_{W^{k,p}_x} \leq c(k,p)\|f\|_{W^{k,p}_x}, \\ &\Pi_m[f] \to f \quad \text{in } W^{k,p}(\mathbb{T}^3) \text{ as } m \to \infty \quad k \in \mathbb{N}_0,\ 1 < p < \infty, \end{aligned} \tag{4.6}$$

whenever $f \in W^{k,p}(\mathbb{T}^3)$; see Section 1.7 and Grafakos [Gra08, Chapter 3]. As H_m is finite-dimensional, all norms are equivalent for a fixed m – a property that will be frequently used at the first level of approximation. We introduce a cut-off function,

$$\chi \in C^\infty(\mathbb{R}), \quad \chi(z) = \begin{cases} 1 & \text{for } z \leq 0, \\ \chi'(z) \leq 0 & \text{for } 0 < z < 1, \\ \chi(z) = 0 & \text{for } z \geq 1, \end{cases}$$

together with the operators

$$[\mathbf{v}]_R = \chi(\|\mathbf{v}\|_{H_m} - R)\mathbf{v}, \quad \text{defined for } \mathbf{v} \in H_m.$$

Similarly, we consider a suitable approximation of the diffusion coefficients. It is convenient to introduce $\mathbb{F} = (\mathbf{F}_k)_{k\in\mathbb{N}}$ by

$$\mathbf{F}_k(\varrho, \mathbf{v}) = \frac{\mathbf{G}_k(\varrho, \varrho\mathbf{v})}{\varrho}.$$

Note that, in accordance with hypotheses (3.13)–(3.14), the functions $\mathbf{F}_k$ satisfy

$$\mathbf{F}_k : \mathbb{T}^3 \times [0,\infty) \times \mathbb{R}^3 \to \mathbb{R}^3, \quad \mathbf{F}_k \in C^1(\mathbb{T}^3 \times (0,\infty) \times \mathbb{R}^3)$$

and there exist constants $(f_k)_{k\in\mathbb{N}} \subset [0,\infty)$ such that

$$\|\mathbf{F}_k(\cdot,\cdot,0)\|_{L^\infty_{x,\varrho}} + \|\nabla_\mathbf{v}\mathbf{F}_k\|_{L^\infty_{x,\varrho,\mathbf{v}}} \le f_k, \quad \sum_{k=1}^\infty f_k^2 < \infty. \tag{4.7}$$

Finally, we define the noise coefficient $\mathbb{G}_\varepsilon = (\mathbf{G}_{k,\varepsilon})_{k\in\mathbb{N}}$ appearing in (4.5) by

$$\mathbf{G}_{k,\varepsilon}(\varrho, \mathbf{q}) = \varrho\mathbf{F}_{k,\varepsilon}\left(\varrho, \frac{\mathbf{q}}{\varrho}\right), \tag{4.8}$$

where

$$\mathbf{F}_{k,\varepsilon}(\varrho, \mathbf{v}) = \chi\left(\frac{\varepsilon}{\varrho} - 1\right)\chi\left(|\mathbf{v}| - \frac{1}{\varepsilon}\right)\mathbf{F}_k(\varrho, \mathbf{v}). \tag{4.9}$$

Consequently, there exist constants $(f_{k,\varepsilon})_{k\in\mathbb{N}} \subset [0,\infty)$ such that

$$\|\mathbf{F}_{k,\varepsilon}\|_{L^\infty_{x,\varrho,\mathbf{v}}} + \|\nabla_{\varrho,\mathbf{v}}\mathbf{F}_{k,\varepsilon}\|_{L^\infty_{x,\varrho,\mathbf{v}}} \le f_{k,\varepsilon}, \quad \sum_{k=1}^\infty f_{k,\varepsilon}^2 < \infty, \tag{4.10}$$

with a bound depending on ε. The basic *approximate problem* reads

$$\mathrm{d}\varrho + \operatorname{div}(\varrho[\mathbf{u}]_R)\,\mathrm{d}t = \varepsilon\Delta\varrho\,\mathrm{d}t, \tag{4.11}$$

$$\begin{aligned}
&\mathrm{d}\Pi_m[\varrho\mathbf{u}] + \Pi_m[\operatorname{div}(\varrho[\mathbf{u}]_R \otimes \mathbf{u})]\,\mathrm{d}t + \Pi_m[\chi(\|\mathbf{u}\|_{H_m} - R)\nabla p_\delta(\varrho)]\,\mathrm{d}t \\
&\quad = \Pi_m[\varepsilon\Delta(\varrho\mathbf{u}) + \operatorname{div}\mathbb{S}(\nabla\mathbf{u})]\,\mathrm{d}t + \Pi_m[\varrho\Pi_m[\mathbb{F}_\varepsilon(\varrho, \mathbf{u})]]\,\mathrm{d}W,
\end{aligned} \tag{4.12}$$

where $\mathbb{F}_\varepsilon = (\mathbf{F}_{k,\varepsilon})_{k\in\mathbb{N}}$. System (4.11)–(4.12) shall be solved in the space-time cylinder $[0,T] \times \mathbb{T}^3$, with the following prescribed initial data:

$$\varrho(0) = \varrho_0 \in C^{2+\nu}(\mathbb{T}^3), \quad \varrho_0 > 0, \quad \mathbf{u}(0) = \mathbf{u}_0 \in H_m, \quad \mathbb{P}\text{-a.s.} \tag{4.13}$$

In (4.11)–(4.12) we recognize the artificial viscosity terms $\varepsilon\Delta\varrho$, $\varepsilon\Delta(\varrho\mathbf{u})$, pressure regularization $\delta(\varrho + \varrho^\Gamma)$ as well as the cut-off operators applied to various quantities. Note that equation (4.11) is deterministic, meaning it can be solved pathwise, while (4.12) involves stochastic integration.

We adopt the following strategy:

(1) The Galerkin projection applied in (4.12) reduces the problem to an ordinary stochastic differential equation. Accordingly, system (4.11)–(4.13) can be solved by means of a simple *iteration scheme*. This is the objective of Section 4.1. In addition, the approximate solutions satisfy the associated energy balance which in turn yields uniform bounds necessary to carry out the asymptotic limits $R \to \infty$, $m \to \infty$, $\varepsilon \to 0$, and $\delta \to 0$.
(2) In Section 4.2, we establish existence to the Galerkin approximation by sending $R \to \infty$. Since this parameter only governs the corresponding cut-off operators, we first solve the system on each random time interval $[0, \tau_R)$ and the passage to the limit relies on a stopping time argument showing that a blow-up cannot occur in a finite time, i.e., $\tau_R \to T$ $\mathbb{P}$-a.s.
(3) In Section 4.3, we perform the limit $m \to \infty$ while keeping the remaining parameters $\varepsilon > 0$ and $\delta > 0$ fixed. Thanks to the regularization effect of the artificial viscosity, this is a relatively routine matter based on suitable uniform bounds and application of probabilistic methods, notably the Skorokhod representation theorem.
(4) The artificial viscosity limit $\varepsilon \to 0$ is more delicate. Similarly to the deterministic case, the fundamental issue is compactness of the approximate densities for which only uniform L^p-bounds are available. The limit passage, discussed in Section 4.4, combines the deterministic method based on the analysis of the effective viscous flux and the stochastic compactness method based on the extension of Skorokhod's theorem due to Jakubowski.
(5) Finally, in Section 4.5, we eliminate the artificial pressure letting $\delta \to 0$. Similarly to the preceding steps, the method leans on a combination of the deterministic approach, here based on the concept of *oscillation defect measure*, and Jakubowski's extension of the Skorokhod representation theorem.

Remark 4.0.4. We point out that, in view of Theorem 2.7.1 and the discussion in Remark 2.3.7, the dissipative martingale solution constructed in Theorem 4.0.2 is defined on the standard probability space $([0,1], \overline{\mathfrak{B}([0,1])}, \mathfrak{L})$ with the complete right-continuous and non-anticipative filtration

$$\sigma\left(\sigma_t[\varrho] \cup \sigma_t[\mathbf{u}] \cup \bigcup_{k=1}^{\infty} \sigma_t[W]\right), \quad t \in [0,T].$$

The same applies to the martingale solutions on the approximation levels in Section 4.1, Section 4.3, and Section 4.4. To summarize, in every approximation step we may assume without loss of generality that the corresponding approximate solutions constructed on the previous level are all defined on the same probability space, there is no need to specify the filtration, and we can always consider the joint history of $[\varrho, \mathbf{u}, W]$.

In order to conclude this introductory part, let us recall the issue arising in the passage to the limit in the stochastic integral within the stochastic compactness method, namely, to show convergence of a sequence of stochastic integrals driven by a sequence of Wiener processes. One possibility is to pass to the limit directly using Lemma 2.6.6. Alternatively, one may rely on the elementary approach from Lemma 2.6.8.

4.1 Solvability of the basic approximate problem

As the first step towards the proof of our main existence result, Theorem 4.0.2, we establish existence of a unique solution to the basic approximate problem (4.11)–(4.12). The proof relies on a simple iteration scheme which approximates (4.11)–(4.12). The corresponding limit procedure then employs the stochastic compactness method (see Section 2.6). Therefore, we obtain existence of a martingale solution to (4.11)–(4.12). Furthermore, at this stage, even pathwise uniqueness holds true and as a consequence we may apply the method of Gyöngy–Krylov from Section 2.10 to deduce existence of a unique pathwise solution, that is, a solution which is strong in the probabilistic sense. However, let us point out that this argument only applies to the basic approximation level as uniqueness is lost at the subsequent limits for $\varepsilon \to 0$ and $\delta \to 0$.

To begin with, let us specify what we exactly mean by a martingale solution to (4.11)–(4.12). On this level we require smooth solutions to the continuity equation. Consequently we work with rather strong assumptions on the initial law (of ϱ). These assumptions will later be relaxed by a suitable approximation procedure.

Definition 4.1.1. Let Λ be a Borel probability measure on $C^{2+\nu}(\mathbb{T}^3) \times H_m$. Then $((\Omega, \mathfrak{F}, (\mathfrak{F}_t)_{t\geq 0}, \mathbb{P}), \varrho, \mathbf{u}, W)$ is called a *martingale solution* to (4.11)–(4.12) with the initial law Λ, provided:

(1) $(\Omega, \mathfrak{F}, (\mathfrak{F}_t)_{t\geq 0}, \mathbb{P})$ is a stochastic basis with a complete right-continuous filtration;
(2) W is a cylindrical $(\mathfrak{F}_t)$-Wiener process;
(3) the density ϱ is $(\mathfrak{F}_t)$-adapted, belongs to $C([0,T]; C^{2+\nu}(\mathbb{T}^3))$, and satisfies $\varrho > 0$ $\mathbb{P}$-a.s.;
(4) the velocity field $\mathbf{u}$ is $(\mathfrak{F}_t)$-adapted and belongs to $C([0,T]; H_m)$ $\mathbb{P}$-a.s.;
(5) there exists an $\mathfrak{F}_0$-measurable random variable $[\varrho_0, \mathbf{u}_0]$ such that $\Lambda = \mathcal{L}[\varrho_0, \mathbf{u}_0]$;
(6) the approximate equation of continuity

$$\partial_t \varrho + \operatorname{div}(\varrho[\mathbf{u}]_R) = \varepsilon \Delta \varrho \tag{4.14}$$

holds in $(0,T) \times \mathbb{T}^3$ $\mathbb{P}$-a.s. and we have $\varrho(0) = \varrho_0$ $\mathbb{P}$-a.s.;
(7) the approximate momentum equation

$$-\int_0^T \partial_t \phi \int_{\mathbb{T}^3} \varrho \mathbf{u} \cdot \boldsymbol{\varphi} \,\mathrm{d}x \,\mathrm{d}t - \phi(0) \int_{\mathbb{T}^3} \varrho_0 \mathbf{u}_0 \cdot \boldsymbol{\varphi} \,\mathrm{d}x$$

$$= \int_0^T \phi \int_{\mathbb{T}^3} [\varrho[\mathbf{u}]_R \otimes \mathbf{u} : \nabla \boldsymbol{\varphi} + \chi(\|\mathbf{u}\|_{H^m} - R) p_\delta(\varrho) \operatorname{div} \boldsymbol{\varphi}] \,\mathrm{d}x \,\mathrm{d}t$$

$$- \int_0^T \phi \int_{\mathbb{T}^3} [\mathbb{S}(\nabla \mathbf{u}) : \nabla \boldsymbol{\varphi} - \varepsilon \varrho \mathbf{u} \cdot \Delta \boldsymbol{\varphi}] \,\mathrm{d}x \,\mathrm{d}t$$

$$+ \int_0^T \phi \int_{\mathbb{T}^3} \varrho \Pi_m [\mathbb{F}_\varepsilon(\varrho, \mathbf{u})] \cdot \boldsymbol{\varphi} \,\mathrm{d}x \,\mathrm{d}W \tag{4.15}$$

holds for all $\phi \in C_c^\infty([0,T))$ and all $\boldsymbol{\varphi} \in H_m$ $\mathbb{P}$-a.s.

Remark 4.1.2. As the processes ϱ, $\mathbf{u}$ are $(\mathfrak{F}_t)$-adapted and continuous, the composition $\varrho \Pi_m [\mathbb{F}_\varepsilon(\varrho, \mathbf{u})]$ is progressively measurable as a mapping ranging in $L^2(\mathfrak{U}, W^{-b,2}(\mathbb{T}^3))$ with $b > \frac{3}{2}$ and the stochastic integral is well-defined. Indeed, by virtue of (4.10) and in accordance with (4.6),

$$\begin{aligned} \|\Pi_m[\varrho \Pi_m [\mathbf{F}_{k,\varepsilon}(\varrho, \mathbf{u})]]\|_{H_m} &\lesssim \|\varrho \Pi_m [\mathbf{F}_{k,\varepsilon}(\varrho, \mathbf{u})]\|_{W_x^{-b,2}} \\ &\lesssim \|\varrho \Pi_m [\mathbf{F}_{k,\varepsilon}(\varrho, \mathbf{u})]\|_{L_x^1} \\ &\lesssim \|\varrho\|_{L_x^\Gamma} \|\mathbf{F}_{k,\varepsilon}(\varrho, \mathbf{u})\|_{L_x^{\Gamma'}} \\ &\lesssim c(\varepsilon) \|\varrho\|_{L_x^\Gamma} f_{k,\varepsilon}, \quad \frac{1}{\Gamma} + \frac{1}{\Gamma'} = 1, \end{aligned} \tag{4.16}$$

and the proportionality constant does not depend on m. As we shall see below, the right hand side of (4.16) will be controlled by the initial data.

Our main goal in this section is to prove the following result concerning solvability of the approximate problem (4.11)–(4.12).

Theorem 4.1.3. *Let Λ be a Borel probability measure on $C^{2+\nu}(\mathbb{T}^3) \times H_m$ such that*

$$\Lambda\{0 < \underline{\varrho} \le \varrho, \ \|\varrho\|_{C_x^{2+\nu}} \le \overline{\varrho}\} = 1, \quad \int_{C_x^{2+\nu} \times H_m} \|\mathbf{v}\|_{H_m}^r \,\mathrm{d}\Lambda[\varrho, \mathbf{v}] \le \overline{u} \tag{4.17}$$

for some deterministic constants $\underline{\varrho}, \overline{\varrho}, \overline{u}$ and some $r > 2$. Then the approximate problem (4.11)–(4.12) admits a martingale solution in the sense of Definition 4.1.1. The solution satisfies in addition

$$\operatorname{esssup}_{t \in [0,T]} \left(\|\varrho(t)\|_{C_x^{2+\nu}} + \|\partial_t \varrho(t)\|_{C_x^\nu} + \|\varrho^{-1}(t)\|_{C_x^0} \right) \le c \quad \mathbb{P}\text{-a.s.}, \tag{4.18}$$

$$\mathbb{E}\Big[\sup_{\tau \in [0,T]} \|\mathbf{u}(\tau)\|_{H_m}^r\Big] \le c(1 + \mathbb{E}[\|\mathbf{u}_0\|_{H^m}^r]), \tag{4.19}$$

with a constant $c = (m, R, T, \underline{\varrho}, \overline{\varrho})$.

4.1.1 Iteration scheme

Solutions to problem (4.11)–(4.12) will be constructed by means of a modification of the Cauchy collocation method. Let $(\Omega, \mathfrak{F}, (\mathfrak{F}_t)_{t \ge 0}, \mathbb{P})$ be a stochastic basis with a complete right-continuous filtration and W be a cylindrical $(\mathfrak{F}_t)$-Wiener process. Consider an $\mathfrak{F}_0$-measurable initial datum $(\varrho_0, \mathbf{u}_0)$ with law Λ existence of which follows from Corollary 2.6.4. As a consequence of (4.17), we have

$$\begin{aligned}&\varrho_0 \geq \underline{\varrho} > 0, \quad \|\varrho_0\|_{C^{2+\nu}(\mathbb{T}^3)} \leq \overline{\varrho} \quad \mathbb{P}\text{-a.s. for some deterministic constants } \underline{\varrho},\ \overline{\varrho},\\ &\mathbb{E}\left[\|\mathbf{u}_0\|_{H_m}^r\right] \leq \overline{u} \quad \text{for some deterministic constant } \overline{u}, \text{ and some } r > 2.\end{aligned} \tag{4.20}$$

Fixing a time step $h > 0$, we set

$$\varrho(t) = \varrho_0, \quad \mathbf{u}(t) = \mathbf{u}_0 \quad \text{for } t \leq 0, \tag{4.21}$$

and define recursively

$$\begin{aligned}\partial_t \varrho + \operatorname{div}\left(\varrho(t)[\mathbf{u}(nh)]_R\right) &= \varepsilon \Delta \varrho(t), \quad t \in [nh, (n+1)h),\\ \varrho(nh) = \varrho(nh-) &:= \lim_{s \uparrow nh} \varrho(s)\end{aligned} \tag{4.22}$$

and

$$\begin{aligned}&\mathrm{d}\Pi_m(\varrho \mathbf{u}) + \Pi_m[\operatorname{div}(\varrho(t)[\mathbf{u}(nh)]_R \otimes \mathbf{u}(nh))]\, \mathrm{d}t\\ &\quad + \Pi_m[\chi(\|\mathbf{u}(nh)\|_{H_m} - R)\nabla p_\delta(\varrho(t))]\, \mathrm{d}t\\ &= \Pi_m[\varepsilon \Delta(\varrho(t)\mathbf{u}(nh)) + \operatorname{div} \mathbb{S}(\nabla \mathbf{u}(nh))]\, \mathrm{d}t\\ &\quad + \Pi_m[\varrho(t)\Pi_m[\mathbb{F}_\varepsilon(\varrho(nh), \mathbf{u}(nh))]]\, \mathrm{d}W, \quad t \in [nh, (n+1)h),\end{aligned} \tag{4.23}$$

where $\mathbf{u}(nh) = \mathbf{u}(nh-) := \lim_{s \uparrow nh} \mathbf{u}(s)$ and $n \in \{0, \dots, \lfloor h^{-1}T \rfloor\}$. Note that the velocity on the right hand side of (4.22)–(4.23) is always frozen at the time nh, whereas the density is evaluated at time t everywhere except for the stochastic integral. In order to see that (4.22)–(4.23) define a simple iteration scheme, it is convenient to rewrite (4.23) in terms of $\mathrm{d}\mathbf{u}$. To this end, we introduce (for a given function ϱ) the linear mapping $\mathcal{M}[\varrho]$,

$$\mathcal{M}[\varrho] : H_m \to H_m, \quad \mathcal{M}[\varrho](\mathbf{v}) = \Pi_m(\varrho \mathbf{v}),$$

or, equivalently,

$$\int_{\mathbb{T}^3} \mathcal{M}[\varrho]\mathbf{v} \cdot \boldsymbol{\varphi}\, \mathrm{d}x \equiv \int_{\mathbb{T}^3} \varrho \mathbf{v} \cdot \boldsymbol{\varphi}\, \mathrm{d}x \quad \text{for all } \boldsymbol{\varphi} \in H_m.$$

The operator $\mathcal{M}$ has been introduced in [FNP01, Section 2.2], where one can also find the following properties. $\mathcal{M}[\varrho]$ is invertible and we have

$$\|\mathcal{M}^{-1}[\varrho]\|_{\mathcal{L}(H_m^*, H_m)} \leq \left(\inf_{x \in \mathbb{T}^3} \varrho\right)^{-1}, \tag{4.24}$$

as long as ϱ is bounded below away from zero, and, clearly,

$$\mathcal{M}^{-1}[\varrho](\Pi_m[\varrho \mathbf{v}]) = \mathbf{v} \quad \text{for any } \mathbf{v} \in H_m.$$

Let us finally mention that

$$\|\mathcal{M}^{-1}[\varrho] - \mathcal{M}^{-1}[\varrho]\|_{\mathcal{L}(H_m^*,H_m)} \le c(m,\underline{\varrho})\|\varrho^1 - \varrho^2\|_{L_x^1}, \tag{4.25}$$

provided both ϱ^1 and ϱ^2 are bounded from below by some positive constant $\underline{\varrho}$. Accordingly, relation (4.23) can be written in the form

$$\begin{aligned}
&\mathbf{u}(t) - \mathcal{M}^{-1}[\varrho(t)]\varrho\mathbf{u}(nh)\\
&\quad + \mathcal{M}^{-1}[\varrho(t)]\int_{nh}^{t} \Pi_m[\operatorname{div}(\varrho(s)[\mathbf{u}(nh)]_R \otimes \mathbf{u}(nh))]\,\mathrm{d}s\\
&\quad + \mathcal{M}^{-1}[\varrho(t)]\int_{nh}^{t} \Pi_m[\chi(\|\mathbf{u}(nh)\|_{H_m} - R)\nabla p_\delta(\varrho(s))]\,\mathrm{d}s\\
&= \mathcal{M}^{-1}[\varrho(t)]\int_{nh}^{t} \Pi_m[\varepsilon\Delta(\varrho(s)\mathbf{u}(nh)) + \operatorname{div}\mathbb{S}(\nabla\mathbf{u}(nh))]\,\mathrm{d}s\\
&\quad + \mathcal{M}^{-1}[\varrho(t)]\int_{nh}^{t} \varrho(s)\Pi_m[\mathbb{F}_\varepsilon(\varrho(nh),\mathbf{u}(nh))]\,\mathrm{d}W, \quad t \in [nh,(n+1)h),
\end{aligned} \tag{4.26}$$

where ϱ is given by (4.22). The iteration scheme (4.22), (4.26) provides a unique solution for any initial data (4.21). Indeed, as $\mathbf{u}(nh) \in H_m$ is a smooth function, equation (4.22) admits a unique solution for the initial datum $\varrho(nh)$. Moreover, as a direct consequence of the parabolic maximum principle from Theorem A.2.5, ϱ remains positive as long as the initial datum $\varrho(nh)$ is positive. We therefore infer that (4.21)–(4.23) give rise to uniquely determined functions ϱ, $\mathbf{u}$ which are progressively $(\mathfrak{F}_t)$-measurable and satisfy

$$\varrho \in C([0,T];C^{2+\nu}(\mathbb{T}^3)), \quad \varrho > 0, \quad \mathbf{u} \in C([0,T];H_m) \quad \mathbb{P}\text{-a.s.},$$

and solve (4.21)–(4.23) for any $n \in \mathbb{N}$, $\mathbb{P}$-a.s.

4.1.2 The limit for vanishing time step

Our next goal is to let $h \to 0$ in (4.21)–(4.23) in order to obtain a solution to the approximate problem (4.11)–(4.12). This step leans essentially on suitable uniform bounds independent of h.

4.1.2.1 Regularity for the viscous approximation of the equation of continuity

For simplicity of notation, we shall write

$$[v]_h = v(nh,\cdot), \quad [v]_{h,R}(t,\cdot) = [v(nh,\cdot)]_R \quad \text{for } t \in [nh,(n+1)h),\ n \in \mathbb{N}.$$

As all norms are equivalent on the finite-dimensional space H_m, we get

$$\|[\mathbf{u}]_{h,R}\|_{W_x^{l,\infty}} \le c(l,m,R) \quad \text{uniformly for } h > 0 \text{ and } t \in [0,T].$$

Consequently, the approximate equation of continuity (4.22) admits a unique regular solution, the smoothness of which is determined by the initial data; cf. Theorem A.2.3. In particular, the solution ϱ belongs to the class

$$\varrho \in C([0,T]; C^{2+\nu}(\mathbb{T}^3)), \quad \partial_t \varrho \in L^\infty(0,T; C^\nu(\mathbb{T}^3)) \tag{4.27}$$

as soon as $\varrho_0 \in C^{2+\nu}(\mathbb{T}^3)$ for some $\nu > 0$. In addition, the standard parabolic maximum principle (Theorem A.2.5) yields

$$0 < \underline{r}(t,m,R) \min_{\mathbb{T}^3} \varrho_0 \le \varrho(t,\cdot) \le \overline{r}(t,m,R) \max_{\mathbb{T}^3} \varrho_0 \quad \text{for all } t \in [0,T], \tag{4.28}$$

and some constants $\underline{r}(t,m,R), \overline{r}(t,m,R) > 0$.

Remark 4.1.4. As the regularized velocity $[\mathbf{u}]_{h,R}$ is only piecewise continuous, the same is true for $\partial_t \varrho$. In general, we do not expect $\partial_t \varrho \in C([0,T]; C^\nu(\mathbb{T}^3))$.

Note carefully that ϱ is bounded in the aforementioned spaces only in terms of the initial datum ϱ_0, meaning that no probabilistic averaging has been applied. In particular, recalling (4.28), we infer

$$\mathrm{esssup}_{t\in[0,T]} \left(\|\varrho(t)\|_{C_x^{2+\nu}} + \|\partial_t \varrho(t)\|_{C_x^\nu} + \|\varrho^{-1}(t)\|_{C_x^0} \right) \le c, \tag{4.29}$$

with a deterministic constant $c = c(m,R,T,\underline{\varrho},\overline{\varrho})$, whenever

$$0 < \underline{\varrho} \le \varrho_0, \quad \|\varrho_0\|_{C_x^{2+\nu}} \le \overline{\varrho} \quad \mathbb{P}\text{-a.s.} \tag{4.30}$$

for some deterministic constants $\underline{\varrho}$, $\overline{\varrho}$.

4.1.2.2 Bounds on the approximate velocities

As a first step we are going to derive estimates for the velocity which are uniform in h. We will systematically use the fact that all norms are equivalent on the finite-dimensional space H_m. It follows from (4.23) and the equivalence of norms on H_m that, uniformly in h,

$$\begin{aligned} \int_{\mathbb{T}^3} \varrho \mathbf{u}(\tau) \cdot \boldsymbol{\varphi} \, \mathrm{d}x &\lesssim \|\mathbf{u}_0\|_{H_m} + \int_0^\tau \sup_{0\le s\le t} \|\mathbf{u}\|_{H_m} \, \mathrm{d}t + T \\ &\quad + \left\| \int_0^\tau \Pi_m [\varrho \Pi_m [\mathbb{F}_\varepsilon(\varrho,\mathbf{u})]_h] \, \mathrm{d}W \right\|_{H_m} \end{aligned}$$

for any $\boldsymbol{\varphi} \in H_m$, $\|\boldsymbol{\varphi}\|_{H^m} \le 1$, whenever $0 \le \tau \le T$. Here we take into account (4.29) and the definition of χ as well as the cut-off $[\cdot]_R$. Consequently, taking the supremum over $\boldsymbol{\varphi}$, we obtain

$$\begin{aligned} &\|\Pi_m[\varrho \mathbf{u}](\tau)\|_{H_m} \\ &\lesssim \|\mathbf{u}_0\|_{H_m} + \int_0^\tau \sup_{0\le s\le t} \|\mathbf{u}\|_{H_m} \, \mathrm{d}t + T + \left\| \int_0^\tau \Pi_m [\varrho \Pi_m [\mathbb{F}_\varepsilon(\varrho,\mathbf{u})]_h] \, \mathrm{d}W \right\|_{H^m} \end{aligned}$$

and

$$\|\Pi_m[\varrho\mathbf{u}](\tau)\|_{H_m}^r$$
$$\lesssim \left[\|\mathbf{u}_0\|_{H_m}^r + \int_0^\tau \sup_{0\le s\le t}\|\mathbf{u}\|_{H_m}^r \,\mathrm{d}t + 1 + \left\|\int_0^\tau \Pi_m[\varrho\Pi_m[\mathbb{F}_\varepsilon(\varrho,\mathbf{u})]_h]\,\mathrm{d}W\right\|_{H^m}^r\right]$$

uniformly in h for all $0\le\tau\le T$ and for any $r\ge 1$ with a constant $c = c(m,R,T,\underline{\varrho},\overline{\varrho},r)$. Finally, we pass to expectations and apply Burkholder–Davis–Gundy's inequality and (4.16) to control the last integral, obtaining

$$\begin{aligned}
\mathbb{E}\Big[\sup_{0\le t\le\tau}\|\Pi_m[\varrho\mathbf{u}](t)\|_{H_m}^r\Big] &\lesssim \mathbb{E}[\|\mathbf{u}_0\|_{H_m}^r] + \int_0^\tau \mathbb{E}\Big[\sup_{0\le s\le t}\|\mathbf{u}\|_{H_m}^r\Big]\,\mathrm{d}t + 1\\
&\quad + \mathbb{E}\left[\int_0^\tau \sum_{k=1}^\infty \|\Pi_m[\varrho\Pi_m[[\mathbf{F}_{k,\varepsilon}(\varrho,\mathbf{u})]_h]]\|_{H_m}^2\,\mathrm{d}t\right]^{r/2}\\
&\lesssim \mathbb{E}[\|\mathbf{u}_0\|_{H_m}^r] + \int_0^\tau \mathbb{E}\Big[\sup_{0\le s\le t}\|\mathbf{u}\|_{H_m}^r\Big]\,\mathrm{d}t + 1 + \mathbb{E}\left[\int_0^\tau \|\varrho(t)\|_{L_x^\Gamma}^2 \sum_{k=1}^\infty f_{k,\varepsilon}^2\,\mathrm{d}t\right]^{r/2}\\
&\lesssim \mathbb{E}[\|\mathbf{u}_0\|_{H_m}^r] + \int_0^\tau \mathbb{E}\Big[\sup_{0\le s\le t}\|\mathbf{u}\|_{H_m}^r\Big]\,\mathrm{d}t + 1
\end{aligned}\tag{4.31}$$

uniformly in h. Here, we have used the uniform bounds for the density obtained in (4.29) and boundedness of the approximated noise coefficients (4.10). Moreover, seeing that

$$\mathbf{u} = \mathcal{M}^{-1}[\varrho][\Pi_m[\varrho\mathbf{u}]],$$

we use again the bounds (4.29) and (4.30) to conclude

$$\|\Pi_m[\varrho\mathbf{u}]\|_{H_m} \lesssim \|\mathbf{u}\|_{H_m} \lesssim \|\Pi_m[\varrho\mathbf{u}]\|_{H_m},$$

where the constants in $\lesssim$ depend only on $\underline{\varrho}$, $\overline{\varrho}$; cf. (4.24). Consequently, a direct application of Gronwall's lemma gives rise to the following estimate:

$$\begin{aligned}
&\mathbb{E}\Big[\sup_{\tau\in[0,T]}\|\Pi_m[\varrho\mathbf{u}](\tau)\|_{H_m}^r\Big] + \mathbb{E}\Big[\sup_{\tau\in[0,T]}\|\mathbf{u}(\tau)\|_{H_m}^r\Big]\\
&\quad\lesssim c(m,R,T,\underline{\varrho},\overline{\varrho},r)(1+\mathbb{E}[\|\mathbf{u}_0\|_{H^m}^r]),\ r\ge 1,
\end{aligned}\tag{4.32}$$

uniformly in h.

4.1.2.3 Hölder continuity of approximate velocities

In addition to the uniform bound (4.32) we will need compactness of the approximate velocities in the space $C([0,T];H_m)$. Moreover, we have to control the difference $\mathbf{u}-[\mathbf{u}]_h$ uniformly in time. To this end, estimates on the modulus of continuity of $\mathbf{u}$ are needed.

Evoking (4.23), we obtain

$$\begin{aligned}
&\int_{\mathbb{T}^3} [\varrho\mathbf{u}(\tau_1,\cdot) - \varrho\mathbf{u}(\tau_2,\cdot)]\cdot\boldsymbol{\varphi}\,\mathrm{d}x \\
&= \int_{\tau_1}^{\tau_2}\int_{\mathbb{T}^3} (\varrho[\mathbf{u}]_{h,R}\otimes[\mathbf{u}]_h) : \nabla\boldsymbol{\varphi}\,\mathrm{d}x\,\mathrm{d}t \\
&+ \int_{\tau_1}^{\tau_2}\int_{\mathbb{T}^3} \chi(\|[\mathbf{u}]_h\|_{H_m} - R)p_\delta(\varrho)\mathrm{div}\,\boldsymbol{\varphi}\,\mathrm{d}x\,\mathrm{d}t \\
&+ \int_{\tau_1}^{\tau_2}\int_{\mathbb{T}^3} \varepsilon\varrho[\mathbf{u}]_h\cdot\Delta\boldsymbol{\varphi}\,\mathrm{d}x\,\mathrm{d}t - \int_{\tau_1}^{\tau_2}\int_{\mathbb{T}^3} \mathbb{S}(\nabla[\mathbf{u}]_h) : \nabla\boldsymbol{\varphi}\,\mathrm{d}x\,\mathrm{d}t \\
&+ \int_{\tau_1}^{\tau_2}\int_{\mathbb{T}^3} \varrho\Pi_m[\mathbb{F}_\varepsilon(\varrho,\mathbf{u})]_h\cdot\boldsymbol{\varphi}\,\mathrm{d}x\,\mathrm{d}W,
\end{aligned}$$

for any $\boldsymbol{\varphi}\in H_m$, $0\le\tau_1<\tau_2$. With the bound (4.32) at hand, we repeat the arguments leading to (4.31) to obtain

$$\mathbb{E}[\|\Pi_m[\varrho\mathbf{u}(\tau_1) - \varrho\mathbf{u}(\tau_2)]\|_{H_m}^r] \lesssim |\tau_1-\tau_2|^{r/2}(1+\mathbb{E}\|\mathbf{u}_0\|_{H_m}^r), \quad r\ge 1,$$

uniformly in h whenever $0\le\tau_1<\tau_2\le T$, $|\tau_1-\tau_2|\le 1$ with a constant $c=c(m,R,T,\underline{\varrho},\overline{\varrho},r)$. Thus, for $r>2$, we apply the Kolmogorov continuity criterion, Theorem 2.3.11, to conclude that (up to a modification) $\Pi_m[\varrho\mathbf{u}]$ has $\mathbb{P}$-a.s. β-Hölder continuous trajectories for all $\beta\in(0,\frac{1}{2}-\frac{1}{r})$. In addition, we have

$$\mathbb{E}[\|\Pi_m[\varrho\mathbf{u}]\|_{C_t^\beta H_m}^r] \le c(r,T)(1+\mathbb{E}\|\mathbf{u}_0\|_{H_m}^r), \quad r>2,$$

uniformly in h. Recalling the relation

$$\mathbf{u} = \mathcal{M}^{-1}[\varrho]\Pi_m[\varrho\mathbf{u}],$$

the boundedness of ϱ from (4.29) and (4.24), we infer

$$\mathbb{E}[\|\mathbf{u}\|_{C_t^\beta H_m}^r] \lesssim (1+\mathbb{E}\|\mathbf{u}_0\|_{H_m}^r) \tag{4.33}$$

uniformly in h, whenever $r>2$ and $\beta\in(0,\frac{1}{2}-\frac{1}{r})$ with a constant $c=c(m,R,T,\underline{\varrho},\overline{\varrho},r)$.

4.1.2.4 Solvability of the first level approximate problem

Let $[\varrho_h,\mathbf{u}_h,W]$ be the unique approximate solution issuing from the iteration scheme (4.21)–(4.23), with the initial data satisfying (4.20). With the estimates (4.29) and (4.33) at hand, we are ready to perform the limit $h\to 0$ in (4.21)–(4.23). This is based on the stochastic compactness method introduced in Section 2.6. The corresponding path space for the basic state variables $[\varrho_h,\mathbf{u}_h,W]$ is defined as

$$\mathcal{X} = \mathcal{X}_\varrho\times\mathcal{X}_\mathbf{u}\times\mathcal{X}_W = C^\iota([0,T];C^{2+\iota}(\mathbb{T}^3))\times C^\kappa([0,T];H_m)\times C([0,T];\mathfrak{U}_0),$$

where $\iota \in (0,\nu)$, $\kappa \in (0,\beta)$, and ν and β are the Hölder exponents in (4.29) and (4.33), respectively. Let $\mathcal{L}[\varrho_h, \mathbf{u}_h, W]$ denote the joint law of $[\varrho_h, \mathbf{u}_h, W]$ on $\mathcal{X}$, whereas $\mathcal{L}[\varrho_h]$, $\mathcal{L}[\mathbf{u}_h]$, and $\mathcal{L}[W]$ denote the corresponding marginals on $\mathcal{X}_\varrho$, $\mathcal{X}_\mathbf{u}$, and $\mathcal{X}_W$, respectively.

In view of the bounds (4.29) and (4.33), we obtain tightness of the family of joint laws associated to the triple $[\varrho_h, \mathbf{u}_h, W]$, which is the key assumption for Prokhorov's and Skorokhod's theorems needed in the sequel (cf. Theorem 2.6.1 and Theorem 2.6.2).

Proposition 4.1.5. *The set $\{\mathcal{L}[\varrho_h, \mathbf{u}_h, W];\, h \in (0,1)\}$ is tight on $\mathcal{X}$.*

Proof. First of all, we observe that it follows directly from (4.29) that the set $\{\mathcal{L}[\varrho_h];\, h \in (0,1)\}$ is tight on $\mathcal{X}_\varrho$. Indeed, for any $L > 0$, the set

$$B_L = \{\varrho \in C^\nu([0,T]; C^{2+\nu}(\mathbb{T}^3)) \cap W^{1,\infty}([0,T]; C^\nu(\mathbb{T}^3));\, \|\varrho\|_{C_t C_x^{2+\nu}} + \|\varrho\|_{W_t^{1,\infty} C_x^\nu} \le L\}$$

is, due to the compact embedding,

$$C^\nu([0,T]; C^{2+\nu}(\mathbb{T}^3)) \cap W^{1,\infty}([0,T]; C^\nu(\mathbb{T}^3)) \overset{c}{\hookrightarrow} C^\iota([0,T]; C^{2+\iota}(\mathbb{T}^3))$$

relatively compact in $\mathcal{X}_\varrho$. In view of (4.29) we obtain

$$\mathcal{L}[\varrho_h](B_L^c) = \mathbb{P}(\|\varrho_h\|_{C_t C_x^{2+\nu}} + \|\varrho_h\|_{W_t^{1,\infty} C_x^\nu} > L) = 0,$$

provided L is sufficiently large, i.e., bigger than the constant $c(m, T, R, \underline{\varrho}, \overline{\varrho})$ on the right hand side of (4.29).

Similarly, we obtain tightness of the set $\{\mathcal{L}[\mathbf{u}_h];\, h \in (0,1)\}$ on $\mathcal{X}_\mathbf{u}$. More precisely, for $\kappa \in (0,\beta)$ we consider the set

$$B_L = \{\mathbf{u} \in C^\beta([0,T]; H_m);\, \|\mathbf{u}\|_{C_t^\beta H_m} \le L\},$$

which is relatively compact in $\mathcal{X}_\mathbf{u}$ due to Arzelà–Ascoli's theorem. Due to Chebyshev's inequality and (4.33), we have

$$\mathcal{L}[\mathbf{u}_h](B_R^c) = \mathbb{P}(\|\mathbf{u}_h\|_{C_t^\beta H_m} > L) \le \frac{1}{L^r} \mathbb{E}\|\mathbf{u}_h\|_{C_t^\beta H_m}^r \le \frac{C}{L^r}.$$

Choosing L sufficiently large yields the claim.

Finally, $\mathcal{L}[W]$ is tight on $\mathcal{X}_W$ since it is a Radon measure on a Polish space.

To conclude, let $\theta \in (0,1)$ be given. Then there exist compact sets $K_\varrho \subset \mathcal{X}_\varrho$, $K_\mathbf{u} \subset \mathcal{X}_\mathbf{u}$, $K_W \subset \mathcal{X}_W$ such that

$$\mathcal{L}[\varrho_h](K_\varrho) \ge 1 - \frac{\theta}{3}, \quad \mathcal{L}[\mathbf{u}_h](K_\mathbf{u}) \ge 1 - \frac{\theta}{3}, \quad \mathcal{L}[W](K_W) \ge 1 - \frac{\theta}{3}.$$

Because of Tychonoff's Theorem $K_\varrho \times K_{\mathbf{u}} \times K_W$ is compact in $\mathcal{X}$ and using de Morgan's law $A \cap B = (A^c \cup B^c)^c$, we get

$$\begin{aligned}\mathcal{L}[\varrho_h, \mathbf{u}_h, W](K_\varrho \times K_{\mathbf{u}} \times K_W) &= \mathbb{P}(\varrho_h \in K_\varrho, \mathbf{u}_h \in K_{\mathbf{u}}, W \in K_W) \\ &= 1 - \mathbb{P}([\varrho_h \notin K_\varrho] \cup [\mathbf{u}_h \notin K_{\mathbf{u}}] \cup [W \notin K_W]) \\ &\geq 1 - \mathbb{P}(\varrho_h \notin K_\varrho) - \mathbb{P}(\mathbf{u}_h \notin K_{\mathbf{u}}) - \mathbb{P}(W \notin K_W) \geq 1 - \theta,\end{aligned}$$

which completes the proof. □

Accordingly, we may apply Prokhorov's and Skorokhod's theorems (Theorem 2.6.1 and Theorem 2.6.2). To be more precise, since $\mathcal{X}$ is a Polish space, due to Theorem 2.6.1, there exists a subsequence, still denoted by $\mathcal{L}[\varrho_h, \mathbf{u}_h, W]$, which converges weakly, in the sense of probability measures on $\mathcal{X}$, to a probability measure $\mathcal{L}$. Using Theorem 2.6.2, we infer the following result.

Proposition 4.1.6. *There exists a complete probability space $(\tilde\Omega, \tilde{\mathfrak{F}}, \tilde{\mathbb{P}})$ with $\mathcal{X}$-valued Borel measurable random variables $(\tilde\varrho_h, \tilde{\mathbf{u}}_h, \tilde W_h)$, $h \in (0,1)$, and $(\tilde\varrho, \tilde{\mathbf{u}}, \tilde W)$ such that (up to a subsequence):*

(1) *the law of $(\tilde\varrho_h, \tilde{\mathbf{u}}_h, \tilde W_h)$ on $\mathcal{X}$ is given by $\mathcal{L}[\varrho_h, \mathbf{u}_h, W]$, $h \in (0,1)$;*
(2) *the law of $(\tilde\varrho, \tilde{\mathbf{u}}, \tilde W)$ on $\mathcal{X}$ is a Radon measure;*
(3) *$(\tilde\varrho_h, \tilde{\mathbf{u}}_h, \tilde W_h)$ converges $\tilde{\mathbb{P}}$-almost surely to $(\tilde\varrho, \tilde{\mathbf{u}}, \tilde W)$ in the topology of $\mathcal{X}$, i.e.,*

$$\begin{aligned}\tilde\varrho_h &\to \tilde\varrho \quad \text{in } C^\iota([0,T]; C^{2+\iota}(\mathbb{T}^3))\ \tilde{\mathbb{P}}\text{-a.s.}, \\ \tilde{\mathbf{u}}_h &\to \tilde{\mathbf{u}} \quad \text{in } C^\kappa([0,T]; H_m)\ \tilde{\mathbb{P}}\text{-a.s.}, \\ \tilde W_h &\to \tilde W \quad \text{in } C([0,T]; \mathfrak{U}_0)\ \tilde{\mathbb{P}}\text{-a.s.}\end{aligned} \tag{4.34}$$

Note that at this stage of our construction, $\tilde\varrho$ and $\tilde{\mathbf{u}}$ are stochastic processes in the classical sense; cf. Definition 2.1.11. Since their trajectories are $\tilde{\mathbb{P}}$-a.s. continuous, progressive measurability with respect to their respective canonical filtrations follows from Proposition 2.1.18. Consequently, they are progressively measurable with respect to the canonical filtration generated by $[\tilde\varrho, \tilde{\mathbf{u}}, \tilde W]$, namely,

$$\tilde{\mathfrak{F}}_t := \sigma\left(\sigma_t[\tilde\varrho] \cup \sigma_t[\tilde{\mathbf{u}}] \cup \bigcup_{k=1}^\infty \sigma_t[\tilde W_k]\right), \quad t \in [0,T].$$

In view of Lemma 2.1.35, the process $\tilde W$ is a cylindrical Wiener processes with respect to its canonical filtration. In order to show that $\tilde W$ is a cylindrical Wiener process with respect to $(\tilde{\mathfrak{F}}_t)_{t\geq 0}$, we intend to apply Corollary 2.1.36. Hence we need to show that the filtration is non-anticipative with respect to $\tilde W$. To this end, we first recall Theorem 2.9.1 and deduce that, for every $h \in (0,1)$, $\tilde W_h = \sum_{k=1}^\infty e_k \tilde W_{h,k}$ is a cylindrical Wiener process with respect to

$$\sigma\left(\sigma_t[\tilde\varrho_h] \cup \sigma_t[\tilde{\mathbf{u}}_h] \cup \bigcup_{k=1}^\infty \sigma_t[\tilde W_{h,k}]\right), \quad t \in [0,T].$$

In other words, this filtration is non-anticipative with respect to $\tilde{W}_h$. Lemma 2.9.3, together with Proposition 4.1.6, then yields the claim.

First, we show that $[\tilde{\varrho},\tilde{\mathbf{u}}]$ solves the approximate continuity equation.

Lemma 4.1.7. *The process* $[\tilde{\varrho},\tilde{\mathbf{u}}]$ *satisfies* (4.11) *a.e. in* $(0,T)\times\mathbb{T}^3$, $\tilde{\mathbb{P}}$*-a.s.*

Proof. As a consequence of the equality of laws from Proposition 4.1.6 and Theorem 2.9.1, we see that the approximate continuity equation (4.22) is satisfied on the new probability space by $[\tilde{\varrho}_h,\tilde{\mathbf{u}}_h]$. Moreover, the uniform bounds (4.29) and (4.33) hold true also for $[\tilde{\varrho}_h,\tilde{\mathbf{u}}_h]$. Hence by Proposition 4.1.6 and Vitali's convergence theorem we pass to the limit in (4.22) and deduce that $[\tilde{\varrho},\tilde{\mathbf{u}}]$ is a weak solution to the approximate continuity equation (4.11). Furthermore, the bounds (4.29) and (4.33) are also valid for the limit process $[\tilde{\varrho},\tilde{\mathbf{u}}]$. Consequently, (4.11) is satisfied a.e. in $(0,T)\times\mathbb{T}^3$, $\tilde{\mathbb{P}}$-a.s. □

As the next step, we are now going to show that the triple $[\tilde{\varrho},\tilde{\mathbf{u}},\tilde{W}]$ solves the approximate momentum equation (4.12).

Proposition 4.1.8. *The process* $[\tilde{\varrho},\tilde{\mathbf{u}},\tilde{W}]$ *satisfies* (4.15) *for all* $\phi\in C^\infty([0,T]))$, $\boldsymbol{\varphi}\in H_m$ *and* $t\in[0,T]$, $\tilde{\mathbb{P}}$*-a.s.*

Proof. Slightly modifying the proof, the result of Theorem 2.9.1 remains valid in the current setting. Hence as a consequence of the equality of laws from Proposition 4.1.6, the approximate momentum equation (4.23) is satisfied on the new probability space by $[\tilde{\varrho}_h,\tilde{\mathbf{u}}_h,\tilde{W}_h]$. It suffices to pass to the limit with respect to h.

We observe

$$\|[\tilde{\mathbf{u}}_h]_h(t)-\tilde{\mathbf{u}}_h(t)\|_{H_m}\lesssim h^\beta\|\tilde{\mathbf{u}}_h\|_{C_t^\beta H_m}$$

and similarly

$$\|[\tilde{\varrho}_h]_h(t)-\tilde{\varrho}_h(t)\|_{C_x^{2+\nu}}\lesssim h^\nu\|\tilde{\varrho}_h\|_{C_t^\nu C_x^{2+\nu}}.$$

Now, with the convergences (4.34), the bounds (4.29), (4.33), and the assumption (4.10) at hand, we may pass to the limit in the approximate momentum equation (4.23). The only term that needs explanation is the stochastic integral. By the uniform convergence of $\tilde{\varrho}_h$ and $\tilde{\mathbf{u}}_h$ (recall Proposition 4.1.6), the continuity of the coefficients $\mathbf{F}_{k,\varepsilon}$, and the continuity of Π_m on $L^q(\mathbb{T}^3)$ for any $1<q<\infty$ from (4.6), it is easy to see that $\tilde{\mathbb{P}}$-a.s.

$$\Pi_m[\tilde{\varrho}_h\Pi_m[\mathbf{F}_{k,\varepsilon}([\tilde{\varrho}_h]_h,[\tilde{\mathbf{u}}_h]_h)]]\to\Pi_m[\tilde{\varrho}\Pi_m[\mathbf{F}_{k,\varepsilon}(\tilde{\varrho},\tilde{\mathbf{u}})]]\quad\text{in }L^q((0,T)\times\mathbb{T}^3),\tag{4.35}$$

for all $k\in\mathbb{N}$ and all $q<\infty$. On the other hand, we have

$$\tilde{\mathbb{E}}\int_0^T\|\Pi_m[\tilde{\varrho}_h\Pi_m[\mathbb{F}_\varepsilon([\tilde{\varrho}_h]_h,[\tilde{\mathbf{u}}_h]_h)]]\|^2_{L_2(\mathfrak{U};L^2_x)}\,\mathrm{d}t$$

$$
\begin{aligned}
&= \sum_{k=1}^{\infty} \tilde{\mathbb{E}} \int_0^T \|\Pi_m[\tilde{\varrho}_h \Pi_m[\mathbf{F}_{k,\varepsilon}([\tilde{\varrho}_h]_h, [\tilde{\mathbf{u}}_h]_h)]]\|_{L^2_x}^2 \,\mathrm{d}t \\
&\leq \sum_{k=1}^{\infty} \tilde{\mathbb{E}} \int_0^T \|[\tilde{\varrho}_h \Pi_m[\mathbf{F}_{k,\varepsilon}([\tilde{\varrho}_h]_h, [\tilde{\mathbf{u}}_h]_h)]]\|_{L^2_x}^2 \,\mathrm{d}t \\
&\leq \|\tilde{\varrho}_h\|_{L^\infty_{\omega,t,x}}^2 \sum_{k=1}^{\infty} \tilde{\mathbb{E}} \int_0^T \|\Pi_m[\mathbf{F}_{k,\varepsilon}([\tilde{\varrho}_h]_h, [\tilde{\mathbf{u}}_h]_h)]\|_{L^2_x}^2 \,\mathrm{d}t \\
&\leq \|\tilde{\varrho}_h\|_{L^\infty_{\omega,t,x}}^2 \sum_{k=1}^{\infty} \tilde{\mathbb{E}} \int_0^T \|\mathbf{F}_{k,\varepsilon}([\tilde{\varrho}_h]_h, [\tilde{\mathbf{u}}_h]_h)\|_{L^2_x}^2 \,\mathrm{d}t \\
&\leq \|\tilde{\varrho}_h\|_{L^\infty_{\omega,t,x}}^2 \sum_{k=1}^{\infty} f_{k,\varepsilon}^2 \lesssim c,
\end{aligned}
$$

using (4.6), (4.10), and (4.29). Hence, for any $\kappa > 0$,

$$
\tilde{\mathbb{E}} \sum_{k=N+1}^{\infty} \int_0^t \left(\int_{\mathbb{T}^3} \tilde{\varrho}_h \Pi_m[\mathbf{F}_{k,\varepsilon}([\tilde{\varrho}_h]_h, [\tilde{\mathbf{u}}_h]_h)] \cdot \boldsymbol{\varphi} \,\mathrm{d}x \right)^2 \mathrm{d}s \leq \kappa
$$

uniformly in h, provided $N \geq N_0(\kappa)$. Consequently, we strengthen (4.35) to

$$
\Pi_m[\tilde{\varrho}_h \Pi_m[\mathbb{F}_\varepsilon([\tilde{\varrho}_h]_h, [\tilde{\mathbf{u}}_h]_h)]] \to \Pi_m[\tilde{\varrho} \Pi_m[\mathbb{F}_\varepsilon(\tilde{\varrho}, \tilde{\mathbf{u}})]] \quad \text{in } L^2(0,T; L_2(\mathfrak{U}; L^2(\mathbb{T}^3))), \tag{4.36}
$$

$\tilde{\mathbb{P}}$-a.s. Combining this with the convergence of $\tilde{W}_h$ from Proposition 4.1.6 we may apply Lemma 2.6.6 to pass to the limit in the stochastic integral and hence complete the proof. □

The proof of Theorem 4.1.3 is hereby complete.

4.1.3 Pathwise uniqueness

In this section, we show that solutions of the approximate problem (4.11)–(4.12), defined on the same probability space with the same Wiener process W, are uniquely determined by the initial data. Such a result is called *pathwise uniqueness.*

Proposition 4.1.9. *Let* $[\varrho^1, \mathbf{u}^1, W]$, $[\varrho^2, \mathbf{u}^2, W]$ *be two martingale solutions of problem* (4.11)–(4.12) *in the sense of Definition* 4.1.1 *satisfying* (4.18) *and* (4.19). *Suppose they are defined on the same probability space* $(\Omega, \mathfrak{F}, \mathbb{P})$ *such that*

$$
[\varrho^1, \mathbf{u}^1](0) = [\varrho^2, \mathbf{u}^2](0) \quad \text{in } C^{2+\nu}(\mathbb{T}^3) \times H_m \ \mathbb{P}\text{-a.s.}
$$

Then

$$
[\varrho^1, \mathbf{u}^1] = [\varrho^2, \mathbf{u}^2] \quad \text{in } C([0,T]; C^{2+\nu}(\mathbb{T}^3) \times H_m) \ \mathbb{P}\text{-a.s.}
$$

Proof. As all non-linearities appearing in the system (4.11)–(4.12) are locally Lipschitz continuous in the phase space $C^{2+\nu}(\mathbb{T}^3) \times H_m$, it is convenient to introduce the stopping times

$$\tau_M^i = \inf\{t \in [0,T] \mid \|\varrho^i(t)\|_{C_x^{2+\nu}} + \|(\varrho^i)^{-1}(t)\|_{C_x^0} + \|\mathbf{u}^i(t)\|_{H^m} > M\}, \quad i = 1,2,$$

with the convention $\inf \emptyset = T$, as well as $\tau_M = \tau_M^1 \wedge \tau_M^2$. Note that the stopping times are well-defined due to the continuity (in time) of the involved quantities and the bounds in (4.18) and (4.19). Moreover, we have $\tau_M \le \tau_L$ whenever $M \le L$ and as a consequence of the *a priori* estimates (4.18) and (4.19), $\tau_M \to T$ holds a.s. Indeed,

$$\begin{aligned}
\mathbb{P}\Big(\sup_{M\in\mathbb{N}} \tau_M = T\Big) &= 1 - \mathbb{P}\Big(\sup_{M\in\mathbb{N}} \tau_M < T\Big) \\
&\ge 1 - \sum_{i=1,2} \mathbb{P}\Big(\sup_{t\in[0,T]} [\|\varrho^i(t)\|_{C_x^{2+\nu}} + \|(\varrho^i)^{-1}(t)\|_{C_x^0} + \|\mathbf{u}^i(t)\|_{H^m}] > M,\Big) \\
&\ge 1 - \frac{C}{M} \to 1 \quad \text{as } M \to \infty.
\end{aligned}$$

As the next step, we recall

$$\mathrm{d}\Pi_m(\varrho\mathbf{u}) = \Pi_m(\mathrm{d}\varrho\mathbf{u}) + \Pi_m(\varrho\,\mathrm{d}\mathbf{u}) = \Pi_m(\partial_t\varrho\mathbf{u})\,\mathrm{d}t + \Pi_m(\varrho\,\mathrm{d}\mathbf{u});$$

whence (4.12) can be rewritten in the form

$$\begin{aligned}
&\mathrm{d}\mathbf{u} + \mathcal{M}^{-1}[\varrho]\Pi_m[\mathrm{d}\varrho\mathbf{u}] \\
&\quad + \mathcal{M}^{-1}[\varrho]\Pi_m[\operatorname{div}(\varrho[\mathbf{u}]_R \otimes \mathbf{u}) + \chi(\|\mathbf{u}\|_{H_m} - R)\nabla p_\delta(\varrho)]\,\mathrm{d}t \\
&= \mathcal{M}^{-1}[\varrho]\Pi_m[\varepsilon\Delta(\varrho\mathbf{u}) + \operatorname{div}\mathbb{S}(\nabla\mathbf{u})]\,\mathrm{d}t + \Pi_m[\mathbb{F}_\varepsilon(\varrho,\mathbf{u})]\,\mathrm{d}W.
\end{aligned} \tag{4.37}$$

Taking the difference of equations (4.11) and (4.37), we obtain

$$\partial_t(\varrho^1 - \varrho^2) - \varepsilon\Delta(\varrho^1 - \varrho^2) = -\operatorname{div}(\varrho^1[\mathbf{u}^1]_R - \varrho^2[\mathbf{u}^2]_R) \tag{4.38}$$

and

$$\begin{aligned}
\mathrm{d}(\mathbf{u}^1 - \mathbf{u}^2) &= (\mathcal{M}^{-1}[\varrho^2] - \mathcal{M}^{-1}[\varrho^1])\Pi_m[\partial_t\varrho^1\mathbf{u}^1]\,\mathrm{d}t + \mathcal{M}^{-1}[\varrho^2]\Pi_m[\partial_t\varrho^2\mathbf{u}^2 - \partial_t\varrho^1\mathbf{u}^1]\,\mathrm{d}t \\
&\quad - \mathcal{M}^{-1}[\varrho^1]\Pi_m[\operatorname{div}(\varrho^1[\mathbf{u}^1]_R \otimes \mathbf{u}^1) + \chi(\|\mathbf{u}^1\|_{H_m} - R)\nabla p_\delta(\varrho^1)]\,\mathrm{d}t \\
&\quad + \mathcal{M}^{-1}[\varrho^2]\Pi_m[\operatorname{div}(\varrho^2[\mathbf{u}^2]_R \otimes \mathbf{u}^2) + \chi(\|\mathbf{u}^2\|_{H_m} - R)\nabla p_\delta(\varrho^2)]\,\mathrm{d}t \\
&\quad + (\mathcal{M}^{-1}[\varrho^1] - \mathcal{M}^{-1}[\varrho^2])\Pi_m[\varepsilon\Delta(\varrho^1\mathbf{u}^1) + \operatorname{div}\mathbb{S}(\nabla\mathbf{u}^1)]\,\mathrm{d}t \\
&\quad + \mathcal{M}^{-1}[\varrho^2]\Pi_m[\varepsilon\Delta(\varrho^1\mathbf{u}^1 - \varrho^2\mathbf{u}^2) + \operatorname{div}\mathbb{S}(\nabla\mathbf{u}^1 - \nabla\mathbf{u}^2)]\,\mathrm{d}t \\
&\quad + \Pi_m[\mathbb{F}_\varepsilon(\varrho^1,\mathbf{u}^1) - \mathbb{F}_\varepsilon(\varrho^2,\mathbf{u}^2)]\,\mathrm{d}W.
\end{aligned} \tag{4.39}$$

Furthermore, Itô's product rule applied to (4.39) yields

$$\begin{aligned}
\frac{1}{2}\mathrm{d}|\mathbf{u}^1 - \mathbf{u}^2|^2 &= (\mathcal{M}^{-1}[\varrho^2] - \mathcal{M}^{-1}[\varrho^1])(\mathbf{u}^1 - \mathbf{u}^2) \cdot \Pi_m[\partial_t\varrho^1\mathbf{u}^1]\,\mathrm{d}t \\
&\quad + \mathcal{M}^{-1}[\varrho^2]\Pi_m[\partial_t\varrho^2\mathbf{u}^2 - \partial_t\varrho^1\mathbf{u}^1] \cdot (\mathbf{u}^1 - \mathbf{u}^2)\,\mathrm{d}t
\end{aligned}$$

$$
\begin{aligned}
&- \mathcal{M}^{-1}[\varrho^1]\Pi_m[\operatorname{div}(\varrho^1[\mathbf{u}^1]_R \otimes \mathbf{u}^1) + \chi(\|\mathbf{u}^1\|_{H_m} - R)\nabla p_\delta(\varrho^1)] \cdot (\mathbf{u}^1 - \mathbf{u}^2)\,\mathrm{d}t \\
&+ \mathcal{M}^{-1}[\varrho^2]\Pi_m[\operatorname{div}(\varrho^2[\mathbf{u}^2]_R \otimes \mathbf{u}^2) + \chi(\|\mathbf{u}^2\|_{H_m} - R)\nabla p_\delta(\varrho^2)] \cdot (\mathbf{u}^1 - \mathbf{u}^2)\,\mathrm{d}t \\
&+ (\mathcal{M}^{-1}[\varrho^1] - \mathcal{M}^{-1}[\varrho^2])\Pi_m[\varepsilon\Delta(\varrho^1\mathbf{u}^1) + \operatorname{div}\mathbb{S}(\nabla\mathbf{u}^1)] \cdot (\mathbf{u}^1 - \mathbf{u}^2)\,\mathrm{d}t \\
&+ \mathcal{M}^{-1}[\varrho^2]\Pi_m[\varepsilon\Delta(\varrho^1\mathbf{u}^1 - \varrho^2\mathbf{u}^2) + \operatorname{div}\mathbb{S}(\nabla\mathbf{u}^1 - \mathbf{u}^2)] \cdot (\mathbf{u}^1 - \mathbf{u}^2)\,\mathrm{d}t \\
&+ \frac{1}{2}\sum_{k=1}^{\infty} |\Pi_m[\mathbf{F}_{k,\varepsilon}(\varrho^1,\mathbf{u}^1) - \mathbf{F}_{k,\varepsilon}(\varrho^2,\mathbf{u}^2)]|^2\,\mathrm{d}t \\
&+ \Pi_m[\mathbb{F}_\varepsilon(\varrho^1,\mathbf{u}^1) - \mathbb{F}_\varepsilon(\varrho^2,\mathbf{u}^2)] \cdot (\mathbf{u}^1 - \mathbf{u}^2)\,\mathrm{d}W.
\end{aligned} \tag{4.40}
$$

Now, we integrate over $\mathbb{T}^3$, take the supremum in time, and apply expectations. Making use of the definition of τ_M, (4.25), and the Lipschitz continuity of the noise coefficients (recall (4.10)), we infer, for any $\kappa > 0$,

$$
\begin{aligned}
&\mathbb{E}\Big[\sup_{s\in[0,T]} \|(\mathbf{u}^1 - \mathbf{u}^2)(s\wedge\tau_M)\|_{H^m}^2\Big] \\
&\quad\le \mathbb{E}[\|(\mathbf{u}^1 - \mathbf{u}^2)(0)\|_{H^m}^2] + c(M)\mathbb{E}\Big[\int_0^{T\wedge\tau_M} \|\varrho^1 - \varrho^2\|_{C_x^{2+\nu}} \|\mathbf{u}^1 - \mathbf{u}^2\|_{H^m}\,\mathrm{d}s\Big] \\
&\qquad + c\mathbb{E}\Big[\int_0^{T\wedge\tau_M} \big(\|\mathbf{u}^1 - \mathbf{u}^2\|_{H^m}^2 + \|\varrho^1 - \varrho^2\|_{L_x^2}^2\big)\,\mathrm{d}s\Big] + \mathbb{E}\Big[\sup_{s\in[0,T]} |\mathfrak{M}_{s\wedge\tau_M}|\Big] \\
&\quad\le \kappa\mathbb{E}\Big[\sup_{s\in[0,T]} \|(\varrho^1 - \varrho^2)(s\wedge\tau_M)\|_{C_x^{2+\nu}}^2\Big] + \mathbb{E}[\|(\mathbf{u}^1 - \mathbf{u}^2)(0)\|_{H^m}^2] \\
&\qquad + c(M,\kappa)\mathbb{E}\Big[\int_0^{T\wedge\tau_M} \big(\|\mathbf{u}^1 - \mathbf{u}^2\|_{H^m}^2 + \|\varrho^1 - \varrho^2\|_{L_x^2}^2\big)\,\mathrm{d}s\Big] + \mathbb{E}\Big[\sup_{s\in[0,T]} |\mathfrak{M}_{s\wedge\tau_M}|\Big],
\end{aligned} \tag{4.41}
$$

where $\mathfrak{M}$ is the stochastic integral

$$
\mathfrak{M} = \int_0^{\cdot} \int_{\mathbb{T}^3} \Pi_m[\mathbb{F}_\varepsilon(\varrho^1,\mathbf{u}^1) - \mathbb{F}_\varepsilon(\varrho^2,\mathbf{u}^2)] \cdot (\mathbf{u}^1 - \mathbf{u}^2)\,\mathrm{d}x\,\mathrm{d}W.
$$

By Burkholder–Davis–Gundy's inequality and (4.10), we estimate $\mathfrak{M}$ similarly by

$$
\begin{aligned}
\mathbb{E}\Big[\sup_{s\in[0,T]} |\mathfrak{M}_{s\wedge\tau_M}|\Big] &\le c\,\mathbb{E}\Big[\int_0^{T\wedge\tau_M} \sum_{k=1}^{\infty}\Big(\int_{\mathbb{T}^3} [\mathbf{F}_{k,\varepsilon}(\varrho^1,\mathbf{u}^1) - \mathbf{F}_{k,\varepsilon}(\varrho^2,\mathbf{u}^2)] \cdot (\mathbf{u}^1 - \mathbf{u}^2)\,\mathrm{d}x\Big)^2\,\mathrm{d}t\Big]^{1/2} \\
&\le c\,\mathbb{E}\Big[\int_0^{T\wedge\tau_M} \Big(\int_{\mathbb{T}^3} (|\varrho^1 - \varrho^2| + |\mathbf{u}^1 - \mathbf{u}^2|)|\mathbf{u}^1 - \mathbf{u}^2|\,\mathrm{d}x\Big)^2\,\mathrm{d}t\Big]^{1/2} \\
&\le \kappa\,\mathbb{E}\Big[\sup_{s\in[0,T]} \int_{\mathbb{T}^3} (|(\varrho^1 - \varrho^2)(s\wedge\tau_M)|^2 + |(\mathbf{u}^1 - \mathbf{u}^2)(s\wedge\tau_M)|^2)\,\mathrm{d}x\Big] \\
&\quad + c(\kappa)\,\mathbb{E}\Big[\int_0^{T\wedge\tau_M} \int_{\mathbb{T}^3} |\mathbf{u}^1 - \mathbf{u}^2|^2\,\mathrm{d}x\,\mathrm{d}t\Big],
\end{aligned}
$$

where $\kappa > 0$ is arbitrary. Using this in (4.41) implies

$$
\begin{aligned}
&\mathbb{E}\Big[\sup_{s\in[0,T]} \|(\mathbf{u}^1 - \mathbf{u}^2)(s\wedge\tau_M)\|_{H^m}^2\Big] \\
&\quad\le \kappa\mathbb{E}\Big[\sup_{s\in[0,T]} \|(\varrho^1 - \varrho^2)(s\wedge\tau_M)\|_{C_x^{2+\nu}}^2\Big] + \mathbb{E}[\|(\mathbf{u}^1 - \mathbf{u}^2)(0)\|_{H^m}^2]
\end{aligned}
$$

$$+ c(M,\kappa)\mathbb{E}\bigg[\int_0^{T\wedge\tau_M} (\|\mathbf{u}^1 - \mathbf{u}^2\|_{H_m}^2 + \|\varrho^1 + \varrho^2\|_{L_x^2}^2)\,\mathrm{d}s\bigg], \tag{4.42}$$

for any $\kappa > 0$. On the other hand, the standard parabolic regularity theory from Theorem A.2.3 applied to (4.38) provides the estimate

$$\begin{aligned}&\sup_{0\leq s\leq \tau} (\|(\varrho^1 - \varrho^2)(s)\|_{C_x^{2+\nu}} + \|\partial_t(\varrho^1 - \varrho^2)(s)\|_{C_x^\nu})\\ &\quad\lesssim \sup_{0\leq s\leq \tau} \|\operatorname{div}(\varrho^1[\mathbf{u}^1]_R - \varrho^2[\mathbf{u}^2]_R)(s)\|_{C_x^\nu} + \|(\varrho^1 - \varrho^2)(0)\|_{C_x^{2+\nu}},\end{aligned}$$

such that, using the definition of $[\mathbf{u}]_R$,

$$\begin{aligned}&\sup_{s\in[0,\tau]} (\|(\varrho^1 - \varrho^2)(s)\|_{C_x^{2+\nu}} + \|\partial_t(\varrho^1 - \varrho^2)(s)\|_{C_x^\nu})\\ &\quad\lesssim \sup_{s\in[0,\tau]} \|(\varrho^1 - \varrho^2)(s)\|_{C_x^{1+\nu}} + \sup_{s\in[0,\tau]} \|(\mathbf{u}^1 - \mathbf{u}^2)(s)\|_{H_m}\\ &\quad\quad + \|(\varrho^1 - \varrho^2)(0)\|_{C_x^{2+\nu}}.\end{aligned} \tag{4.43}$$

It is easy to show (for instance by contradiction) that, for every $\kappa > 0$, there is some $c(\kappa)$ such that

$$\sup_{s\in[0,\tau]} \|v\|_{C^{1+\nu}} \leq \kappa\Big(\sup_{s\in[0,\tau]} \|v\|_{C_x^{2+\nu}} + \sup_{s\in[0,\tau]} \|\partial_t v\|_{C_x^\nu}\Big) + c(\kappa)\bigg(\int_0^\tau \|v\|_{C_x^\nu}^2\,dt\bigg)^{\frac{1}{2}},$$

for all $v \in W^{1,\infty}(0,\tau; C^\nu(\mathbb{T}^3)) \cap L^\infty(0,\tau; C^{2+\nu}(\mathbb{T}^3))$. Using this in (4.43) and choosing κ samll enough yields

$$\begin{aligned}&\sup_{s\in[0,\tau]} (\|(\varrho^1 - \varrho^2)(s)\|_{C_x^{2+\nu}} + \|\partial_t(\varrho^1 - \varrho^2)(s)\|_{C_x^\nu})\\ &\quad\lesssim \bigg(\int_0^\tau \|\varrho^1 - \varrho^2\|_{C_x^{2+\nu}}^2\,\mathrm{d}t\bigg)^{\frac{1}{2}} + \sup_{s\in[0,\tau]} \|(\mathbf{u}^1 - \mathbf{u}^2)(s)\|_{H_m}\\ &\quad\quad + \|(\varrho^1 - \varrho^2)(0)\|_{C_x^{2+\nu}}.\end{aligned} \tag{4.44}$$

Next, we combine (4.42) and (4.44) with $\tau = T \wedge \tau_M$ and choose κ small enough to get

$$\begin{aligned}&\mathbb{E}\Big[\sup_{s\in[0,T]} (\|(\mathbf{u}^1 - \mathbf{u}^2)(s\wedge\tau_M)\|_{H^m}^2 + \|(\varrho^1 - \varrho^2)(s\wedge\tau_M)\|_{C_x^{2+\nu}}^2)\Big]\\ &\quad\lesssim \mathbb{E}[\|(\mathbf{u}^1 - \mathbf{u}^2)(0)\|_{H^m}^2 + \|(\varrho^1 - \varrho^2)(0)\|_{C_x^{2+\nu}}^2]\\ &\quad\quad + \mathbb{E}\bigg[\int_0^{T\wedge\tau_M} (\|\mathbf{u}^1 - \mathbf{u}^2\|_{H_m}^2 + \|\varrho^1 - \varrho^2\|_{C_x^{2+\nu}}^2)\,\mathrm{d}s\bigg],\end{aligned}$$

with a constant depending on M. Finally, a direct application of Gronwall's lemma yields

$$\mathbb{E}\Big[\sup_{s\in[0,T]} (\|(\mathbf{u}^1 - \mathbf{u}^2)(s\wedge\tau_M)\|_{H^m}^2 + \|(\varrho^1 - \varrho^2)(s\wedge\tau_M)\|_{C_x^{2+\nu}}^2)\Big]$$

$$\lesssim \mathbb{E}\big[\|(\mathbf{u}^1 - \mathbf{u}^2)(0)\|_{H^m}^2 + \|(\varrho^1 - \varrho^2)(0)\|_{C_x^{2+\nu}}^2\big],$$

with a constant depending on M. Now, as the initial data coincide, the desired conclusion follows by sending $M \to \infty$. □

It can be seen from the proof of Proposition 4.1.9 that we have the following Corollary.

Corollary 4.1.10. *Under the assumption of Proposition* 4.1.9, *we have the estimate*

$$\mathbb{E}\Big[\sup_{s\in[0,T]} \big(\|(\mathbf{u}^1 - \mathbf{u}^2)(s \wedge \tau_M)\|_{H^m}^2 + \|(\varrho^1 - \varrho^2)(s \wedge \tau_M)\|_{C_x^{2+\nu}}^2\big)\Big]$$
$$\lesssim \mathbb{E}\big[\|(\mathbf{u}^1 - \mathbf{u}^2)(0)\|_{H^m}^2 + \|(\varrho^1 - \varrho^2)(0)\|_{C_x^{2+\nu}}^2\big]$$

with a constant depending of M, *where* $\tau_M = \tau_M^1 \wedge \tau_M^2$ *with*

$$\tau_M^i = \inf\{t \in [0,T] \mid \|\varrho^i(t)\|_{C_x^{2+\nu}} + \|(\varrho^i)^{-1}(t)\|_{C_x^0} + \|\mathbf{u}^i(t)\|_{H^m} > M\}, \quad i = 1, 2.$$

4.1.4 Strong solutions

Thanks to the pathwise uniqueness established in Proposition 4.1.9, we are able to show existence of a unique pathwise solution to (4.11)–(4.12). To be more precise, we solve (4.11)–(4.12) in the context of the following definition.

Definition 4.1.11. Let $(\Omega, \mathfrak{F}, (\mathfrak{F}_t)_{t\geq 0}, \mathbb{P})$ be a given stochastic basis with a complete right-continuous filtration, let W be a cylindrical $(\mathfrak{F}_t)$-Wiener process, and let $(\varrho_0, \mathbf{u}_0)$ be an $\mathfrak{F}_0$-measurable random variable taking values in $C^{2+\nu}(\mathbb{T}^3) \times H_m$. Then $(\varrho, \mathbf{u})$ is called a *pathwise solution* to (4.11)–(4.12) with the initial condition $(\varrho_0, \mathbf{u}_0)$, provided:
(1) the density ϱ is $(\mathfrak{F}_t)$-adapted, belongs to $C([0,T]; C^{2+\nu}(\mathbb{T}^3))$, and satisfies $\varrho > 0$ $\mathbb{P}$-a.s.;
(2) the velocity field $\mathbf{u}$ is $(\mathfrak{F}_t)$-adapted and belongs to $C([0,T]; H_m)$ $\mathbb{P}$-a.s.;
(3) $(\varrho(0), \mathbf{u}(0)) = (\varrho_0, \mathbf{u}_0)$ $\mathbb{P}$-a.s.;
(4) the approximate equation of continuity

$$\partial_t \varrho + \operatorname{div}(\varrho[\mathbf{u}]_R) = \varepsilon \Delta \varrho \tag{4.45}$$

holds in $(0,T) \times \mathbb{T}^3$ $\mathbb{P}$-a.s.;
(5) the approximate momentum equation

$$-\int_0^T \partial_t \phi \int_{\mathbb{T}^3} \varrho \mathbf{u} \cdot \boldsymbol{\varphi} \, dx \, dt - \phi(0) \int_{\mathbb{T}^3} \varrho_0 \mathbf{u}_0 \cdot \boldsymbol{\varphi} \, dx$$
$$= \int_0^T \phi \int_{\mathbb{T}^3} [\varrho[\mathbf{u}]_R \otimes \mathbf{u} : \nabla \boldsymbol{\varphi} + \chi(\|\mathbf{u}\|_{H^m} - R) p_\delta(\varrho) \operatorname{div} \boldsymbol{\varphi}] \, dx \, dt$$

$$- \int_0^T \phi \int_{\mathbb{T}^3} [\mathbb{S}(\nabla \mathbf{u}) : \nabla \boldsymbol{\varphi} - \varepsilon \varrho \mathbf{u} \cdot \Delta \boldsymbol{\varphi}] \,\mathrm{d}x \,\mathrm{d}t$$

$$+ \int_0^T \phi \int_{\mathbb{T}^3} \varrho \Pi_m [\mathbb{F}_\varepsilon(\varrho, \mathbf{u})] \cdot \boldsymbol{\varphi} \,\mathrm{d}x \,\mathrm{d}W \tag{4.46}$$

holds for all $\phi \in C_c^\infty([0,T))$ and all $\boldsymbol{\varphi} \in H_m$ $\mathbb{P}$-a.s.

The main result of this section reads as follows.

Theorem 4.1.12. *Let $\varrho_0 \in C^{2+\nu}(\mathbb{T}^3)$, $\mathbf{u}_0 \in H_m$ be an $\mathfrak{F}_0$-measurable random variables satisfying*

$$\mathbb{P}\{0 < \underline{\varrho} \le \varrho_0,\ \|\varrho_0\|_{C_x^{2+\nu}} \le \overline{\varrho}\} = 1, \quad \mathbb{E}[\|\mathbf{u}_0\|_{H_m}^r] \le \overline{u}, \tag{4.47}$$

for some deterministic constants $\underline{\varrho}, \overline{\varrho}, \overline{u}, r > 2$. Then the approximate problem (4.11)–(4.12) *admits a unique pathwise solution $[\varrho, \mathbf{u}]$ in the sense of Definition* 4.1.11. *The solution satisfies in addition*

$$\operatorname{esssup}_{t\in[0,T]} \big(\|\varrho(t)\|_{C_x^{2+\nu}} + \|\partial_t \varrho(t)\|_{C_x^\nu} + \|\varrho^{-1}(t)\|_{C_x^0}\big) \le c \quad \mathbb{P}\text{-a.s.}, \tag{4.48}$$

$$\mathbb{E}\Big[\sup_{\tau\in[0,T]} \|\mathbf{u}(\tau)\|_{H_m}^r\Big] \le c(1 + \mathbb{E}[\|\mathbf{u}_0\|_{H^m}^r]), \tag{4.49}$$

with a constant $c = (m, R, T, \underline{\varrho}, \overline{\varrho})$.

Proof. Consider a family of approximate solutions

$$\{[\varrho_\ell, \mathbf{u}_\ell];\ \ell \in \mathbb{N}\} := \{[\varrho_{h_\ell}, \mathbf{u}_{h_\ell}];\ \ell \in \mathbb{N}\},$$

constructed on the original probability space $(\Omega, \mathfrak{F}, (\mathfrak{F}_t)_{t\ge0}, \mathbb{P})$ by means of the iteration procedure (4.21)–(4.23). Similarly to Section 4.1.2, we apply the Skorokhod representation theorem to the joint law generated by the random variables

$$\{[\varrho_{n_k}, \mathbf{u}_{n_k}, \varrho_{m_k}, \mathbf{u}_{m_k}, W];\ k \in \mathbb{N}\}$$

in the space

$$[C^\iota([0,T]; C^{2+\iota}(\mathbb{T}^3)) \times C^\kappa([0,T]; H_m)]^2 \times C([0,T]; \mathfrak{U}_0),$$

for suitable $\iota, \kappa \in (0,1)$. Consequently, we obtain a subsequence

$$\{[\tilde\varrho_{n_l}, \tilde{\mathbf{u}}_{n_l}, \tilde\varrho_{m_l}, \tilde{\mathbf{u}}_{m_l}, \tilde W_l];\ l \in \mathbb{N}\}$$

defined on a new probability space $(\tilde\Omega, \tilde{\mathfrak{F}}, \tilde{\mathbb{P}})$ converging $\tilde{\mathbb{P}}$-a.s. to some random variable $[\tilde\varrho^1, \tilde{\mathbf{u}}^1, \tilde\varrho^2, \tilde{\mathbf{u}}^2, \tilde W]$. Applying the arguments used in the proof of Proposition 4.1.3, we conclude

$$[\tilde\varrho_{n_l}, \tilde{\mathbf{u}}_{n_l}, \tilde W_l] \to [\tilde\varrho^1, \tilde{\mathbf{u}}^1, \tilde W] \quad \tilde{\mathbb{P}}\text{-a.s.}, \quad [\tilde\varrho_{m_l}, \tilde{\mathbf{u}}_{m_l}, \tilde W_l] \to [\tilde\varrho^2, \tilde{\mathbf{u}}^2, \tilde W] \quad \tilde{\mathbb{P}}\text{-a.s.},$$

where $[\tilde{\varrho}^i, \tilde{\mathbf{u}}^i, \tilde{W}]$, $i = 1, 2$, are solutions of problem (4.11)–(4.12) on the same probability space and with the initial data

$$[\tilde{\varrho}^1(0), \tilde{\mathbf{u}}^1(0)] = [\tilde{\varrho}^2(0), \tilde{\mathbf{u}}^2(0)] \quad \tilde{\mathbb{P}}\text{-a.s.}$$

As a consequence of Proposition 4.1.9, the two solutions $[\tilde{\varrho}^i, \tilde{\mathbf{u}}^i]$ coincide $\tilde{\mathbb{P}}$-a.s. Therefore, the family $\{[\varrho_\ell, \mathbf{u}_\ell]; \ell \in \mathbb{N}\}$ satisfies the hypothesis of Gyöngy–Krylov's result, Lemma 2.10.1. Passing to a subsequence if necessary, we conclude that $[\varrho_\ell, \mathbf{u}_\ell]$ converges $\mathbb{P}$-a.s. and therefore gives rise to a solution $[\varrho, \mathbf{u}]$ defined on the original probability space $(\Omega, \mathfrak{F}, \mathbb{P})$. Note that the corresponding passage to the limit is significantly easier than the one performed in Proposition 4.1.8; the approximate equations are satisfied for the same Wiener process. □

4.1.5 General initial data

Our ultimate goal in this section is to relax the restriction (4.47). This step leans essentially on the uniqueness property established in Proposition 4.1.9. We will show the following.

Corollary 4.1.13. *Let $(\Omega, \mathfrak{F}, (\mathfrak{F}_t)_{t\geq 0}, \mathbb{P})$ be a stochastic basis endowed with a complete right-continuous filtration $(\mathfrak{F}_t)_{t\geq 0}$ and let W be a cylindrical $(\mathfrak{F}_t)$-Wiener process. Let $[\varrho_0, \mathbf{u}_0]$ be a given $\mathfrak{F}_0$-measurable initial datum satisfying*

$$\varrho_0 \in C^{2+\nu}(\mathbb{T}^3), \quad 0 < \underline{\varrho} \leq \varrho_0 \leq \overline{\varrho}, \quad \mathbf{u}_0 \in H_m, \quad \mathbb{P}\text{-a.s.}, \tag{4.50}$$

for some deterministic constants $\underline{\varrho}$, $\overline{\varrho}$. Then the approximate problem (4.11)–(4.12) *admits a unique pathwise solution $[\varrho, \mathbf{u}]$ in the sense of Definition* 4.1.11.

Proof. We consider the general initial data $[\varrho_0, \mathbf{u}_0]$ satisfying (4.50). For a given $M > 0$, we consider the set

$$I_M = \{(\varrho, \mathbf{v}) \mid \|\varrho\|_{C_x^{2+\nu}} \leq M, \ \|\mathbf{v}\|_{H_m} \leq M\} \subset C^{2+\nu}(\mathbb{T}^3) \times H_m.$$

Next, we modify the initial data introducing

$$[\varrho_{0,M}, \mathbf{u}_{0,M}] = \begin{cases} [\varrho_0, \mathbf{u}_0] & \text{if } [\varrho_0, \mathbf{u}_0] \in I_M, \\ \varrho_0 = M + 1, \ \mathbf{u}_0 = 0 & \text{otherwise.} \end{cases}$$

In accordance with Theorem 4.1.12, problem (4.11)–(4.12) admits a unique pathwise solution $[\varrho_M, \mathbf{u}_M]$ starting from the initial condition $[\varrho_{0,M}, \mathbf{u}_{0,M}]$, for any $M > 0$. Moreover, in accordance with the uniqueness result stated in Proposition 4.1.9,

$$1_{[\varrho_0,\mathbf{u}_0]^{-1}[I_N]}[\varrho_N, \mathbf{u}_N] = 1_{[\varrho_0,\mathbf{u}_0]^{-1}[I_N]}[\varrho_M, \mathbf{u}_M] \quad \text{whenever } M \geq N.$$

Seeing that

$$\mathbb{P}\left([\varrho_0, \mathbf{u}_0] \in \bigcup_{N=1}^{\infty} \bigcap_{M \geq N} I_M \right) = 1,$$

we define $[\varrho, \mathbf{u}]$ – the unique solution of the approximate problem (4.11)–(4.12) – as

$$[\varrho, \mathbf{u}] = [\varrho_M, \mathbf{u}_M] \quad \text{whenever } [\varrho_0, \mathbf{u}_0] \in I_M.$$ □

4.1.6 Energy balance

As the next step, we show that any solution of the approximate problem (4.11)–(4.12) in the sense of Definition 4.1.11 satisfies a variant of the energy balance.

Proposition 4.1.14. *Let* $[\varrho, \mathbf{u}]$ *be a pathwise solution to* (4.11)–(4.12) *in the sense of Definition* 4.1.11. *Then the energy balance*

$$\begin{aligned} &-\int_0^T \partial_t \phi \int_{\mathbb{T}^3} \left[\frac{1}{2} \varrho |\mathbf{u}|^2 + P_\delta(\varrho) \right] \mathrm{d}x \, \mathrm{d}t \\ &\quad + \int_0^T \phi \int_{\mathbb{T}^3} [\mathbb{S}(\nabla \mathbf{u}) : \nabla \mathbf{u} + \varepsilon \varrho |\nabla \mathbf{u}|^2 + \varepsilon P_\delta''(\varrho) |\nabla \varrho|^2] \, \mathrm{d}x \, \mathrm{d}t \\ &= \frac{1}{2} \sum_{k=1}^{\infty} \int_0^T \phi \int_{\mathbb{T}^3} \varrho |\Pi_m[\mathbf{F}_{k,\varepsilon}(\varrho, \mathbf{u})]|^2 \, \mathrm{d}x \, \mathrm{d}t + \int_0^T \phi \int_{\mathbb{T}^3} \varrho \Pi_m[\mathbb{F}_\varepsilon(\varrho, \mathbf{u})] \cdot \mathbf{u} \, \mathrm{d}x \, \mathrm{d}W, \\ &\quad + \phi(0) \int_{\mathbb{T}^3} \left[\frac{1}{2} \varrho_0 |\mathbf{u}_0|^2 + P_\delta(\varrho_0) \right] \mathrm{d}x \end{aligned} \tag{4.51}$$

holds for all $\phi \in C_c^\infty([0, T))$ $\mathbb{P}$*-a.s. with the approximate pressure potential*

$$P_\delta(\varrho) = \varrho \int_1^\varrho \frac{p_\delta(z)}{z^2} \, \mathrm{d}z.$$

Proof. To this end, we formally test (4.12) by $\mathbf{u}$ and integrate the resulting expression by parts. More precisely, we apply Itô's formula to system (4.11)–(4.12) and the functional

$$f : L^2(\mathbb{T}^3) \times H_m^* \to \mathbb{R}, \quad (\varrho, \mathbf{q}) \mapsto \frac{1}{2} \int_{\mathbb{T}^3} \mathbf{q} \cdot \mathcal{M}^{-1}[\varrho] \mathbf{q} \, \mathrm{d}x.$$

We observe that

$$\partial_{\mathbf{q}} f(\varrho, \mathbf{q}) = \mathcal{M}^{-1}[\varrho] \mathbf{q} \in H_m, \quad \partial_{\mathbf{q}}^2 f(\varrho, \mathbf{q}) = \mathcal{M}^{-1}[\varrho] \in \mathcal{L}(H_m, H_m^*)$$

and

$$\partial_\varrho f(\varrho, \mathbf{q}) = -\frac{1}{2} \langle \mathbf{q}, \mathcal{M}^{-1}[\varrho] \mathcal{M}[\cdot] \mathcal{M}^{-1}[\varrho] \mathbf{q} \rangle \in \mathcal{L}(L^2(\mathbb{T}^3), \mathbb{R}).$$

Therefore

$$
\begin{aligned}
&f(\varrho,\varrho\mathbf{u}) = \frac{1}{2}\int_{\mathbb{T}^3} \varrho|\mathbf{u}|^2\,\mathrm{d}x,\\
&\partial_{\mathbf{q}} f(\varrho,\varrho\mathbf{u}) = \mathbf{u}, \quad \partial^2_{\mathbf{q}} f(\varrho,\varrho\mathbf{u}) = \mathcal{M}^{-1}[\varrho],\\
&\partial_\varrho f(\varrho,\varrho\mathbf{u}) = -\frac{1}{2}|\mathbf{u}|^2.
\end{aligned}
$$

Hence we deduce

$$
\begin{aligned}
\mathrm{d}\int_{\mathbb{T}^3} \frac{1}{2}\varrho|\mathbf{u}|^2\,\mathrm{d}x = &-\int_{\mathbb{T}^3} [\operatorname{div}(\varrho[\mathbf{u}]_R\otimes\mathbf{u}) + \chi(\|\mathbf{u}\|_{H_m} - R)\nabla p_\delta(\varrho)]\cdot\mathbf{u}\,\mathrm{d}x\,\mathrm{d}t\\
&+\int_{\mathbb{T}^3} [\varepsilon\Delta(\varrho\mathbf{u}) + \operatorname{div}\mathbb{S}(\nabla\mathbf{u})]\cdot\mathbf{u}\,\mathrm{d}x\,\mathrm{d}t - \frac{1}{2}\int_{\mathbb{T}^3} |\mathbf{u}|^2\,\mathrm{d}\varrho\,\mathrm{d}x\\
&+\frac{1}{2}\sum_{k=1}^{\infty}\int_{\mathbb{T}^3} \varrho|\Pi_m[\mathbf{F}_{k,\varepsilon}(\varrho,\mathbf{u})]|^2\,\mathrm{d}x\,\mathrm{d}t + \int_{\mathbb{T}^3} \varrho\Pi_m[\mathbb{F}_\varepsilon(\varrho,\mathbf{u})]\cdot\mathbf{u}\,\mathrm{d}x\,\mathrm{d}W.
\end{aligned}
\tag{4.52}
$$

Furthermore, equation (4.11) tells us that

$$
\frac{1}{2}\int_{\mathbb{T}^3} |\mathbf{u}|^2\,\mathrm{d}\varrho\,\mathrm{d}x = \frac{1}{2}\int_{\mathbb{T}^3} \varepsilon|\mathbf{u}|^2\Delta\varrho\,\mathrm{d}x\,\mathrm{d}t - \frac{1}{2}\int_{\mathbb{T}^3} \operatorname{div}(\varrho[\mathbf{u}]_R)|\mathbf{u}|^2\,\mathrm{d}x\,\mathrm{d}t,
$$

while

$$
\int_{\mathbb{T}^3} \operatorname{div}(\varrho[\mathbf{u}]_R\otimes\mathbf{u})\cdot\mathbf{u}\,\mathrm{d}x = -\frac{1}{2}\int_{\mathbb{T}^3} \varrho[\mathbf{u}]_R\cdot\nabla|\mathbf{u}|^2\,\mathrm{d}x = \frac{1}{2}\int_{\mathbb{T}^3} \operatorname{div}(\varrho[\mathbf{u}]_R)|\mathbf{u}|^2\,\mathrm{d}x
$$

and

$$
\varepsilon\int_{\mathbb{T}^3} \Delta(\varrho\mathbf{u})\cdot\mathbf{u}\,\mathrm{d}x = -\varepsilon\int_{\mathbb{T}^3} \varrho|\nabla\mathbf{u}|^2\,\mathrm{d}x + \frac{1}{2}\int_{\mathbb{T}^3} \varepsilon|\mathbf{u}|^2\Delta\varrho\,\mathrm{d}x.
$$

Consequently, relation (4.52) reduces to

$$
\begin{aligned}
&\mathrm{d}\int_{\mathbb{T}^3} \frac{1}{2}\varrho|\mathbf{u}|^2\,\mathrm{d}x + \int_{\mathbb{T}^3} \mathbb{S}(\nabla\mathbf{u}) : \nabla\mathbf{u}\,\mathrm{d}x\,\mathrm{d}t + \varepsilon\int_{\mathbb{T}^3} \varrho|\nabla\mathbf{u}|^2\,\mathrm{d}x\,\mathrm{d}t\\
&\quad = \int_{\mathbb{T}^3} \chi(\|\mathbf{u}\|_{H_m} - R)p_\delta(\varrho)\operatorname{div}\mathbf{u}\,\mathrm{d}x\,\mathrm{d}t\\
&\qquad + \frac{1}{2}\sum_{k=1}^{\infty}\int_{\mathbb{T}^3} \varrho|\Pi_m[\mathbf{F}_{k,\varepsilon}(\varrho,\mathbf{u})]|^2\,\mathrm{d}x\,\mathrm{d}t + \int_{\mathbb{T}^3} \varrho\Pi_m[\mathbb{F}_\varepsilon(\varrho,\mathbf{u})]\cdot\mathbf{u}\,\mathrm{d}x\,\mathrm{d}W.
\end{aligned}
\tag{4.53}
$$

Finally, we multiply equation (4.11) by $b'(\varrho)$ to obtain its *renormalized* formulation as follows:

$$
\mathrm{d}b(\varrho) + \operatorname{div}(b(\varrho)[\mathbf{u}]_R)\,\mathrm{d}t + (b'(\varrho)\varrho - b(\varrho))\operatorname{div}[\mathbf{u}]_R\,\mathrm{d}t = \varepsilon b'(\varrho)\Delta\varrho\,\mathrm{d}t.
\tag{4.54}
$$

Seeing that

$$
\chi(\|\mathbf{u}\|_{H_m} - R)p_\delta(\varrho)\operatorname{div}\mathbf{u} = p_\delta(\varrho)\operatorname{div}[\mathbf{u}]_R,
$$

we rewrite the energy balance (4.53) in its final form

$$\begin{aligned}
&\mathrm{d}\int_{\mathbb{T}^3}\left[\frac{1}{2}\varrho|\mathbf{u}|^2+P_\delta(\varrho)\right]\mathrm{d}x\\
&\quad+\int_{\mathbb{T}^3}\mathbb{S}(\nabla\mathbf{u}):\nabla\mathbf{u}\,\mathrm{d}x\,\mathrm{d}t+\varepsilon\int_{\mathbb{T}^3}\varrho|\nabla\mathbf{u}|^2\,\mathrm{d}x\,\mathrm{d}t+\varepsilon\int_{\mathbb{T}^3}P_\delta''(\varrho)|\nabla\varrho|^2\,\mathrm{d}x\,\mathrm{d}t\\
&=\frac{1}{2}\sum_{k=1}^{\infty}\int_{\mathbb{T}^3}\varrho|\Pi_m[\mathbf{F}_{k,\varepsilon}(\varrho,\mathbf{u})]|^2\,\mathrm{d}x\,\mathrm{d}t+\int_{\mathbb{T}^3}\varrho\Pi_m[\mathbb{F}_\varepsilon(\varrho,\mathbf{u})]\cdot\mathbf{u}\,\mathrm{d}x\,\mathrm{d}W. \qquad (4.55)
\end{aligned}$$

Consistently with the weak formulation of the field equations in Definition 4.1.1, we finally rewrite (4.55) in the form of a variational equality (4.51) with a deterministic test function $\phi\in C_c^\infty([0,T))$. □

Remark 4.1.15. In (4.55), the stochastic integral ranges in $\mathbb{R}$. More precisely, the expression

$$\int_{\mathbb{T}^3}\varrho\Pi_m[\mathbb{F}_\varepsilon(\varrho,\mathbf{u})]\cdot\mathbf{u}\,\mathrm{d}x=\left(\int_{\mathbb{T}^3}\varrho\Pi_m[\mathbf{F}_{k,\varepsilon}(\varrho,\mathbf{u})]\cdot\mathbf{u}\,\mathrm{d}x\right)_{k\in\mathbb{N}}$$

must be interpreted as an object in $L_2(\mathfrak{U};\mathbb{R})$.

Remark 4.1.16. Solutions satisfying, in addition, some form of the global energy balance will be called *dissipative solutions*.

4.2 Solvability of the Galerkin approximation

Our next goal is to let $R\to\infty$ in the approximate system (4.11)–(4.12). Accordingly, our target problem reads

$$\mathrm{d}\varrho+\operatorname{div}(\varrho\mathbf{u})\,\mathrm{d}t=\varepsilon\Delta\varrho\,\mathrm{d}t, \qquad (4.56)$$

$$\begin{aligned}
&\mathrm{d}\Pi_m[\varrho\mathbf{u}]+\Pi_m[\operatorname{div}(\varrho\mathbf{u}\otimes\mathbf{u})]\,\mathrm{d}t+\Pi_m[\nabla p_\delta(\varrho)]\,\mathrm{d}t\\
&\quad=\Pi_m[\varepsilon\Delta(\varrho\mathbf{u})+\operatorname{div}\mathbb{S}(\nabla\mathbf{u})]\,\mathrm{d}t+\Pi_m[\varrho\Pi_m[\mathbb{F}_\varepsilon(\varrho,\mathbf{u})]]\,\mathrm{d}W. \qquad (4.57)
\end{aligned}$$

It will be solved in the context of strong pathwise solutions given by the following Definition.

Definition 4.2.1. Let $(\Omega,\mathfrak{F},(\mathfrak{F}_t)_{t\geq0},\mathbb{P})$ be a given stochastic basis with a complete right-continuous filtration and let W be a cylindrical $(\mathfrak{F}_t)$-Wiener process. Let ϱ_0 and $\mathbf{u}_0$ be $\mathfrak{F}_0$-measurable random variables with values in $C^{2+\nu}(\mathbb{T}^3)$ and H_m, respectively. Then $(\varrho,\mathbf{u})$ is called a *strong pathwise solution* to (4.56)–(4.57) with initial datum $(\varrho_0,\mathbf{u}_0)$, provided the following hold:

(1) the density ϱ is $(\mathfrak{F}_t)$-progressively measurable and satisfies

$$\varrho\in C([0,T];C^{2+\nu}(\mathbb{T}^3)),\quad \varrho>0,\ \partial_t\varrho\in C([0,T];C^\nu(\mathbb{T}^3)),\quad \mathbb{P}\text{-a.s.};$$

(2) the velocity $\mathbf{u}$ is $(\mathfrak{F}_t)$-progressively measurable and satisfies

$$\mathbf{u} \in C([0,T]; H_m) \quad \mathbb{P}\text{-a.s.};$$

(3) we have $\mathbb{P}$-a.s.

$$(\varrho(0), \mathbf{u}(0)) = (\varrho_0, \mathbf{u}_0);$$

(4) the approximate continuity equation

$$\partial_t \varrho + \operatorname{div}(\varrho \mathbf{u}) = \varepsilon \Delta \varrho \tag{4.58}$$

holds in $(0,T) \times \mathbb{T}^3$ $\mathbb{P}$-a.s.;

(5) the approximate momentum equation

$$\begin{aligned}
&- \int_0^T \partial_t \phi \int_{\mathbb{T}^3} \varrho \mathbf{u} \cdot \boldsymbol{\varphi} \, dx \, dt - \phi(0) \int_{\mathbb{T}^3} \varrho_0 \mathbf{u}_0 \cdot \boldsymbol{\varphi} \, dx \\
&= \int_0^T \phi \int_{\mathbb{T}^3} [\varrho \mathbf{u} \otimes \mathbf{u} : \nabla \boldsymbol{\varphi} + p_\delta(\varrho) \operatorname{div} \boldsymbol{\varphi}] \, dx \, dt \\
&- \int_0^T \phi \int_{\mathbb{T}^3} [\mathbb{S}(\nabla \mathbf{u}) : \nabla \boldsymbol{\varphi} - \varepsilon \varrho \mathbf{u} \cdot \Delta \boldsymbol{\varphi}] \, dx \, dt \\
&+ \int_0^T \phi \int_{\mathbb{T}^3} \varrho \Pi_m [\mathbb{F}_\varepsilon(\varrho, \mathbf{u})] \cdot \boldsymbol{\varphi} \, dx \, dW
\end{aligned} \tag{4.59}$$

holds for all $\phi \in C_c^\infty([0,T))$ and all $\boldsymbol{\varphi} \in H_m$ $\mathbb{P}$-a.s.;

(6) the energy equality

$$\begin{aligned}
&- \int_0^T \partial_t \phi \int_{\mathbb{T}^3} \left[\frac{1}{2} \varrho |\mathbf{u}|^2 + P_\delta(\varrho) \right] dx \, dt - \phi(0) \int_{\mathbb{T}^3} \left[\frac{1}{2} \varrho_0 |\mathbf{u}_0|^2 + P_\delta(\varrho_0) \right] dx \\
&+ \int_0^T \phi \int_{\mathbb{T}^3} [\mathbb{S}(\nabla \mathbf{u}) : \nabla \mathbf{u} + \varepsilon \varrho |\nabla \mathbf{u}|^2 + \varepsilon P_\delta''(\varrho) |\nabla \varrho|^2] \, dx \, dt \\
&= \frac{1}{2} \sum_{k=1}^\infty \int_0^T \phi \int_{\mathbb{T}^3} \varrho |\Pi_m [\mathbf{F}_{k,\varepsilon}(\varrho, \mathbf{u})]|^2 \, dx \, dt + \int_0^T \phi \int_{\mathbb{T}^3} \varrho \Pi_m [\mathbb{F}_\varepsilon(\varrho, \mathbf{u})] \cdot \mathbf{u} \, dx \, dW
\end{aligned} \tag{4.60}$$

holds for all $\phi \in C_c^\infty([0,T))$ $\mathbb{P}$-a.s.

Theorem 4.2.2. *Let $(\Omega, \mathfrak{F}, \mathbb{P})$ be a complete probability space, endowed with a cylindrical Wiener process W relative to a complete right-continuous filtration $(\mathfrak{F}_t)_{t \geq 0}$. Let $[\varrho_0, \mathbf{u}_0]$ be a given initial datum which is $\mathfrak{F}_0$-measurable such that*

$$\varrho_0 \in C^{2+\nu}(\mathbb{T}^3), \quad 0 < \underline{\varrho} \leq \varrho_0 \leq \overline{\varrho}, \quad \mathbf{u}_0 \in H_m, \quad \mathbb{P}\text{-a.s.}, \tag{4.61}$$

for some deterministic constants $\underline{\varrho}$, $\overline{\varrho}$, and

$$\mathbb{E}\left[\int_{\mathbb{T}^3} \left[\frac{1}{2} \varrho_0 |\mathbf{u}_0|^2 + P_\delta(\varrho_0) \right] dx \right]^r < \infty, \tag{4.62}$$

for some $r > 2$. Then there exists a unique strong pathwise solution $[\varrho, \mathbf{u}]$ to (4.56)–(4.57) *in the sense of Definition* 4.2.1.

In order to prove Theorem 4.2.2 we adopt the following strategy:
(1) Using the energy balance from Proposition 4.1.14 we derive uniform bounds independent of the parameter R.
(2) We perform the limit $R \to \infty$. This is an easy task as the velocity field remains bounded in H_m; whence in any desired topology. The proof is based on a suitable stopping time argument.

4.2.1 Uniform energy bounds

Let us begin with the uniform energy bound. In fact, this estimate holds true uniformly in both parameters R, m.

Proposition 4.2.3. *Let* $[\varrho, \mathbf{u}]$ *be a solution to* (4.11)–(4.12) *in the sense of Definition* 4.1.11. *Then we have, uniformly in* R *and* m,

$$\begin{aligned}
&\mathbb{E}\left[\left|\sup_{\tau\in[0,T]}\int_{\mathbb{T}^3}\left[\frac{1}{2}\varrho|\mathbf{u}|^2+P_\delta(\varrho)\right]\mathrm{d}x\right|^r\right] \\
&\quad+\mathbb{E}\left[\left|\int_0^T\int_{\mathbb{T}^3}[\mathbb{S}(\nabla\mathbf{u}):\nabla\mathbf{u}+\varepsilon\varrho|\nabla\mathbf{u}|^2+\varepsilon P_\delta''(\varrho)|\nabla\varrho|^2]\,\mathrm{d}x\,\mathrm{d}t\right|^r\right] \\
&\quad\le c\,\mathbb{E}\left[\left|\int_{\mathbb{T}^3}\left[\frac{1}{2}\varrho_0|\mathbf{u}_0|^2+P_\delta(\varrho_0)\right]\mathrm{d}x\right|^r+1\right] \quad \textit{whenever } r\ge 2,
\end{aligned} \tag{4.63}$$

with a constant $c = c(r, T, \varepsilon)$.

Proof. The energy balance from Proposition 4.1.14, written in its integral form, reads

$$\begin{aligned}
&\int_{\mathbb{T}^3}\left[\frac{1}{2}\varrho|\mathbf{u}|^2+P_\delta(\varrho)\right](\tau)\,\mathrm{d}x-\int_{\mathbb{T}^3}\left[\frac{1}{2}\varrho_0|\mathbf{u}_0|^2+P_\delta(\varrho_0)\right]\mathrm{d}x \\
&\quad+\int_0^\tau\int_{\mathbb{T}^3}[\mathbb{S}(\nabla\mathbf{u}):\nabla\mathbf{u}+\varepsilon\varrho|\nabla\mathbf{u}|^2+\varepsilon P_\delta''(\varrho)|\nabla\varrho|^2]\,\mathrm{d}x\,\mathrm{d}t \\
&=\frac{1}{2}\sum_{k=1}^\infty\int_0^\tau\int_{\mathbb{T}^3}\varrho|\Pi_m[\mathbf{F}_{k,\varepsilon}(\varrho,\mathbf{u})]|^2\,\mathrm{d}x\,\mathrm{d}t+\int_0^\tau\int_{\mathbb{T}^3}\varrho\Pi_m[\mathbb{F}_\varepsilon(\varrho,\mathbf{u})]\cdot\mathbf{u}\,\mathrm{d}x\,\mathrm{d}W,
\end{aligned} \tag{4.64}$$

for any $0 \le \tau \le T$ $\mathbb{P}$-a.s. Keeping ε, $\delta \in (0,1)$ fixed we are able to derive bounds independent of the parameters R, $m \in \mathbb{N}$. As the projections Π_m are bounded in L^p uniformly in m (cf. (4.6)), (4.10) yields

$$\begin{aligned}
\sum_{k=1}^\infty\int_{\mathbb{T}^3}\varrho|\Pi_m[\mathbf{F}_{k,\varepsilon}(\varrho,\mathbf{u})]|^2\,\mathrm{d}x &\lesssim \sum_{k=1}^\infty\|\varrho\|_{L_x^\Gamma}\|\mathbf{F}_{k,\varepsilon}(\varrho,\mathbf{u})\|_{L_x^q}^2 \\
&\lesssim \sum_{k=1}^\infty\|\varrho\|_{L_x^\Gamma}\|\mathbf{F}_{k,\varepsilon}(\varrho,\mathbf{u})\|_{L_x^\infty}^2 \lesssim \|\varrho\|_{L_x^\Gamma}
\end{aligned} \tag{4.65}$$

uniformly in R and m, where q is chosen such that $\frac{1}{\Gamma}+\frac{2}{q}=1$. Note that the above bound depends on ε as a consequence of (4.10). Next, by means of the Burkholder–Davis–Gundy inequality,

$$\mathbb{E}\left[\sup_{0\le t\le \tau}\left|\int_0^t\int_{\mathbb{T}^3}\varrho\Pi_m[\mathbb{F}_\varepsilon(\varrho,\mathbf{u})]\cdot\mathbf{u}\,\mathrm{d}x\,\mathrm{d}W\right|^r\right]$$
$$\lesssim \mathbb{E}\left[\int_0^\tau\sum_{k=1}^\infty\left|\int_{\mathbb{T}^3}\varrho\Pi_m[\mathbf{F}_{k,\varepsilon}(\varrho,\mathbf{u})]\cdot\mathbf{u}\,\mathrm{d}x\right|^2\mathrm{d}t\right]^{r/2},\quad r\ge 1, \tag{4.66}$$

uniformly in R and m. Furthermore, using once more (4.6) and (4.10), we deduce

$$\begin{aligned}\left|\int_{\mathbb{T}^3}\varrho\Pi_m[\mathbf{F}_{k,\varepsilon}(\varrho,\mathbf{u})]\cdot\mathbf{u}\,\mathrm{d}x\right|^2 &\lesssim \big|\|\sqrt{\varrho}\|_{L_x^{2\Gamma}}\|\sqrt{\varrho}\mathbf{u}\|_{L_x^2}\|\Pi_m[\mathbf{F}_{k,\varepsilon}(\varrho,\mathbf{u})]\|_{L_x^q}\big|^2\\ &\lesssim \big|\|\sqrt{\varrho}\|_{L_x^{2\Gamma}}\|\sqrt{\varrho}\mathbf{u}\|_{L_x^2}\|\mathbf{F}_{k,\varepsilon}(\varrho,\mathbf{u})\|_{L_x^\infty}\big|^2\\ &\lesssim f_{k,\varepsilon}^2\big|\|\sqrt{\varrho}\|_{L_x^{2\Gamma}}\|\sqrt{\varrho}\mathbf{u}\|_{L_x^2}\big|^2,\end{aligned} \tag{4.67}$$

uniformly in R and m, where $\frac{1}{2\Gamma}+\frac{1}{2}+\frac{1}{q}=1$. Finally, we estimate

$$\big|\|\sqrt{\varrho}\|_{L_x^{2\Gamma}}\|\sqrt{\varrho}\mathbf{u}\|_{L_x^2}\big|^2\lesssim\left(\int_{\mathbb{T}^3}\varrho|\mathbf{u}|^2\,\mathrm{d}x\right)^2+\left(\int_{\mathbb{T}^3}\varrho^\Gamma\,\mathrm{d}x\right)^{\frac{2}{\Gamma}}$$

uniformly in R and m. Consequently, passing to expectations in (4.64), we apply Gronwall's inequality to deduce (4.63). □

Remark 4.2.4. A refined analysis would give (4.63) also for $1\le r<2$. However, we content ourselves with $r\ge 2$ requiring more integrability of the initial data.

Finally, we apply Korn's inequality on $\mathbb{T}^3$ (see Theorem A.1.8) to deduce, from (4.63),

$$\mathbb{E}\left[\left|\int_0^T\int_{\mathbb{T}^3}|\nabla\mathbf{u}|^2\,\mathrm{d}x\,\mathrm{d}t\right|^r\right]\lesssim\mathbb{E}\left[\left|\int_{\mathbb{T}^3}\left[\frac{1}{2}\varrho_0|\mathbf{u}_0|^2+P_\delta(\varrho_0)\right]\mathrm{d}x\right|^r+1\right]$$

uniformly in R and m. From this we obtain, using the equivalence of all norms on H_m,

$$\mathbb{E}\left[\left|\int_0^T\|\nabla\mathbf{u}\|_{W_x^{k,\infty}}^2\,\mathrm{d}t\right|^r\right]\le c(m,k,r,T)\mathbb{E}\left[\left|\int_{\mathbb{T}^3}\left[\frac{1}{2}\varrho_0|\mathbf{u}_0|^2+P_\delta(\varrho_0)\right]\mathrm{d}x\right|^r+1\right], \tag{4.68}$$

for any $k\ge 1$, $r\ge 2$ uniformly in R.

4.2.2 Passage to the limit

Finally, everything is in hand to pass to the limit as $R\to\infty$. This is based on a simple application of a stopping time argument.

Proof of Theorem 4.2.2. Let $[\varrho_0, \mathbf{u}_0]$ be the initial condition from the statement of the theorem, that is, satisfying (4.61) and (4.62). Let us denote by $(\varrho_R, \mathbf{u}_R)$ the unique pathwise solution to (4.11)–(4.12) starting from $[\varrho_0, \mathbf{u}_0]$, which was constructed in Corollary 4.1.13. Let us consider

$$\tau_R = \inf\{t \in [0,T] \mid \|\mathbf{u}_R(t)\|_{H_m} > R\},$$

with the convention $\inf \emptyset = T$. Note that, since $\mathbf{u}_R$ has continuous trajectories in H_m, τ_R defines an $(\mathfrak{F}_t)$-stopping time. Besides, due to uniqueness, if $R' > R$, then $\tau_{R'} \geq \tau_R$ and $(\varrho_{R'}, \mathbf{u}_{R'}) = (\varrho_R, \mathbf{u}_R)$ on $[0, \tau_R)$. In addition, $(\varrho_R, \mathbf{u}_R)$ is a solution to (4.56)–(4.57) on $[0, \tau_R)$. Therefore, we define $(\varrho, \mathbf{u})$ by $(\varrho, \mathbf{u}) := (\varrho_R, \mathbf{u}_R)$ on $[0, \tau_R)$. In order to make sure that $(\varrho, \mathbf{u})$ solves (4.56)–(4.57) on the whole time interval $[0, T]$, i.e., the blow-up cannot occur in a finite time, we are going to show that

$$\mathbb{P}\Big(\sup_{R\in\mathbb{N}} \tau_R = T\Big) = 1. \tag{4.69}$$

This relies on the uniform bounds from the last section; cf. Proposition 4.2.3. As $(\varrho_R, \mathbf{u}_R)$ is a smooth solution of the deterministic equation (4.11), the standard maximum principle from Theorem A.2.5, together with the assumption on the initial datum (4.61), yields

$$\underline{\varrho} \exp\left(- \int_0^\tau \|\mathrm{div}\,[\mathbf{u}_R]_R\|_{L^\infty_x} \,\mathrm{d}t \right) \leq \varrho_R(\tau, x) \leq \overline{\varrho} \exp\left(\int_0^\tau \|\mathrm{div}\,[\mathbf{u}_R]_R\|_{L^\infty_x} \,\mathrm{d}t \right);$$

whence, due to (4.68) and (4.62),

$$\underline{\varrho} \exp\left(- \int_0^\tau \|\nabla \mathbf{u}_R\|_{L^\infty_x} \,\mathrm{d}t \right) \leq \varrho_R(\tau, x) \leq \overline{\varrho} \exp\left(\int_0^\tau \|\nabla \mathbf{u}_R\|_{L^\infty_x} \,\mathrm{d}t \right), \tag{4.70}$$

for all $\tau \in [0, T]$, $x \in \mathbb{T}^3$. Note carefully that the proportional constants in both (4.68) and (4.70) are independent of R. Since all norms on H_m are equivalent, the above left hand side can be further estimated from below by

$$\underline{\varrho} \exp\left(-T - c \int_0^\tau \|\nabla \mathbf{u}_R\|^2_{L^2_x} \,\mathrm{d}s \right) \leq \varrho_R(\tau, x).$$

Plugging this into (4.63), we infer

$$\mathbb{E}\left[\exp\left(-c \int_0^\tau \|\nabla \mathbf{u}_R\|^2_{L^2_x} \,\mathrm{d}s \right) \sup_{0\leq t\leq \tau} \|\mathbf{u}_R\|^2_{L^2_x} \right] \leq \tilde{c}. \tag{4.71}$$

Next, let us fix two increasing sequences (a_R) and (b_R) such that $a_R, b_R \to \infty$ and $a_R\, \mathrm{e}^{b_R} = R$ for each $R \in \mathbb{N}$. As in [FM12], we introduce the following events:

$$A = \left[\exp\left(-c \int_0^\tau \|\nabla \mathbf{u}_R\|^2_{L^2_x} \,\mathrm{d}t \right) \sup_{0\leq t\leq T} \|\mathbf{u}_R\|^2_{L^2_x} \leq a_R \right],$$

$$B = \left[c \int_0^T \|\nabla \mathbf{u}_R\|_{L_x^2}^2 \, \mathrm{d}t \le b_R \right],$$

$$C = \left[\sup_{0 \le t \le T} \|\mathbf{u}_R\|_{L_x^2}^2 \le a_R \, \mathrm{e}^{b_R} \right].$$

Then $A \cap B \subset C$ because about $A \cap B$ we know

$$\begin{aligned} \sup_{0 \le t \le T} \|\mathbf{u}_R\|_{L_x^2}^2 &= \mathrm{e}^{b_R} \mathrm{e}^{-b_R} \sup_{0 \le t \le T} \|\mathbf{u}_R\|_{L_x^2}^2 \\ &\le \mathrm{e}^{b_R} \exp\left(-c \int_0^T \|\nabla \mathbf{u}_R\|_{L_x^2}^2 \, \mathrm{d}t \right) \sup_{0 \le t \le T} \|\mathbf{u}_R\|_{L_x^2}^2 \le \mathrm{e}^{b_R} a_R. \end{aligned}$$

Furthermore, according to (4.68), (4.71), and Chebyshev's inequality,

$$\mathbb{P}(A) \ge 1 - \frac{c}{a_R}, \quad \mathbb{P}(B) \ge 1 - \frac{c}{b_R}.$$

Due to the general inequality for probabilities $\mathbb{P}(C) \ge \mathbb{P}(A) + \mathbb{P}(B) - 1$ we deduce

$$\mathbb{P}(C) \ge 1 - \frac{c}{a_R} - \frac{c}{b_R} \longrightarrow 1, \quad R \to \infty.$$

This in turn implies (4.69).

Summarizing the previous discussion, $(\varrho, \mathbf{u})$ satisfies (4.58) and (4.59) together with the corresponding energy balance (4.60). Finally, uniqueness follows from the uniqueness of (4.11)–(4.12) (cf. Theorem 4.1.12) and (4.69). Thus we have proved Theorem 4.2.2. □

4.3 The limit in the Galerkin approximation scheme

Our next goal is to let $m \to \infty$ in the approximate system (4.56)–(4.57). Accordingly, our target problem is given, formally, by

$$\mathrm{d}\varrho + \operatorname{div}(\varrho \mathbf{u}) \, \mathrm{d}t = \varepsilon \Delta \varrho \, \mathrm{d}t, \tag{4.72}$$

$$\begin{aligned} \mathrm{d}(\varrho \mathbf{u}) + \operatorname{div}(\varrho \mathbf{u} \otimes \mathbf{u}) \, \mathrm{d}t + \nabla p_\delta(\varrho) \, \mathrm{d}t &= \varepsilon \Delta(\varrho \mathbf{u}) \, \mathrm{d}t + \operatorname{div} \mathbb{S}(\nabla \mathbf{u}) \, \mathrm{d}t \\ &\quad + \varrho \mathbb{F}_\varepsilon(\varrho, \mathbf{u}) \, \mathrm{d}W. \end{aligned} \tag{4.73}$$

A rigorous formulation reads as follows.

Definition 4.3.1. Let Λ be a Borel probability measure on $C^{2+\nu}(\mathbb{T}^3) \times L^1(\mathbb{T}^3)$. Then $((\Omega, \mathfrak{F}, (\mathfrak{F}_t)_{t \ge 0}, \mathbb{P}), \varrho, \mathbf{u}, W)$ is called a *dissipative martingale solution* to (4.72)–(4.73) with the initial law Λ, provided:

(1) $(\Omega, \mathfrak{F}, (\mathfrak{F}_t)_{t \ge 0}, \mathbb{P})$ is a stochastic basis with a complete right-continuous filtration;
(2) W is a cylindrical $(\mathfrak{F}_t)$-Wiener process;

(3) the density ϱ is $(\mathfrak{F}_t)$-adapted and satisfies $\varrho \geq 0$ and $t \mapsto \langle \varrho(t), \psi \rangle \in C([0,T])$ for any $\psi \in C^\infty(\mathbb{T}^3)$ $\mathbb{P}$-a.s.;
(4) the velocity field $\mathbf{u}$ is a random distribution adapted to $(\mathfrak{F}_t)_{t\geq 0}$ and $\mathbf{u} \in L^2(0,T, W^{1,2}(\mathbb{T}^3))$ $\mathbb{P}$-a.s.;
(5) there exists an $\mathfrak{F}_0$-measurable random variable $[\varrho_0, \mathbf{u}_0]$ such that $\Lambda = \mathcal{L}[\varrho_0, \varrho_0\mathbf{u}_0]$;
(6) the approximate equation of continuity

$$\partial_t \varrho + \operatorname{div}(\varrho \mathbf{u}) = \varepsilon \Delta \varrho \tag{4.74}$$

holds in $(0,T) \times \mathbb{T}^3$ $\mathbb{P}$-a.s. and $\varrho(0) = \varrho_0$ $\mathbb{P}$-a.s.;
(7) the approximate momentum equation

$$\begin{aligned}
&-\int_0^T \partial_t \phi \int_{\mathbb{T}^3} \varrho \mathbf{u} \cdot \boldsymbol{\varphi} \, \mathrm{d}x \, \mathrm{d}t - \phi(0) \int_{\mathbb{T}^3} \varrho_0 \mathbf{u}_0 \cdot \boldsymbol{\varphi} \, \mathrm{d}x \\
&= \int_0^T \phi \int_{\mathbb{T}^3} [\varrho \mathbf{u} \otimes \mathbf{u} : \nabla \boldsymbol{\varphi} + p_\delta(\varrho) \operatorname{div} \boldsymbol{\varphi}] \, \mathrm{d}x \, \mathrm{d}t \\
&\quad - \int_0^T \phi \int_{\mathbb{T}^3} [\mathbb{S}(\nabla \mathbf{u}) : \nabla \boldsymbol{\varphi} - \varepsilon \varrho \mathbf{u} \cdot \Delta \boldsymbol{\varphi}] \, \mathrm{d}x \, \mathrm{d}t \\
&\quad + \int_0^T \phi \int_{\mathbb{T}^3} \varrho \mathbb{F}_\varepsilon(\varrho, \mathbf{u}) \cdot \boldsymbol{\varphi} \, \mathrm{d}x \, \mathrm{d}W
\end{aligned} \tag{4.75}$$

holds for all $\phi \in C_c^\infty([0,T))$ and all $\boldsymbol{\varphi} \in C^\infty(\mathbb{T}^3)$ $\mathbb{P}$-a.s.;
(8) the energy inequality

$$\begin{aligned}
&-\int_0^T \partial_t \phi \int_{\mathbb{T}^3} \left[\frac{1}{2} \varrho |\mathbf{u}|^2 + P_\delta(\varrho)\right] \mathrm{d}x \, \mathrm{d}t - \phi(0) \int_{\mathbb{T}^3} \left[\frac{1}{2} \varrho_0 |\mathbf{u}_0|^2 + P_\delta(\varrho_0)\right] \mathrm{d}x \\
&+ \int_0^T \phi \int_{\mathbb{T}^3} [\mathbb{S}(\nabla \mathbf{u}) : \nabla \mathbf{u} + \varepsilon \varrho |\nabla \mathbf{u}|^2 + \varepsilon P_\delta''(\varrho) |\nabla \varrho|^2] \, \mathrm{d}x \, \mathrm{d}t \\
&\leq \frac{1}{2} \sum_{k=1}^\infty \int_0^T \phi \int_{\mathbb{T}^3} \varrho |\mathbf{F}_{k,\varepsilon}(\varrho, \mathbf{u})|^2 \, \mathrm{d}x \, \mathrm{d}t + \int_0^T \phi \int_{\mathbb{T}^3} \varrho \mathbb{F}_\varepsilon(\varrho, \mathbf{u}) \cdot \mathbf{u} \, \mathrm{d}x \, \mathrm{d}W
\end{aligned} \tag{4.76}$$

holds for all $\phi \in C_c^\infty([0,T))$, $\phi \geq 0$, $\mathbb{P}$-a.s.

Note that at this point the velocity field $\mathbf{u}$ may not have continuous trajectories. Consequently, it is not a stochastic process in the classical sense and the role of a given initial state is transferred from $[\varrho(0), \mathbf{u}(0)]$ to $[\varrho(0), \varrho\mathbf{u}(0)]$.

Theorem 4.3.2. *Let Λ be a Borel probability measure on $C^{2+\nu}(\mathbb{T}^3) \times L^1(\mathbb{T}^3)$, satisfying*

$$\Lambda\{0 < \underline{\varrho} \leq \varrho \leq \overline{\varrho}\} = 1, \tag{4.77}$$

for some deterministic constants $\underline{\varrho}$ and $\overline{\varrho}$. Also, assume that

$$\int_{C_x^{2+\nu} \times L_x^1} \left(\left| \int_{\mathbb{T}^3} \left[\frac{1}{2} \frac{|\mathbf{q}|^2}{\varrho} + P_\delta(\varrho) \right] \mathrm{d}x \right|^{2r} + \|\varrho\|_{C_x^{2+\nu}}^r \right) \mathrm{d}\Lambda(\varrho, \mathbf{q}) \lesssim 1, \tag{4.78}$$

for some $r \geq 2$. *Then problem* (4.72)–(4.73) *admits a dissipative martingale solution in the sense of Definition* 4.3.1.

Remark 4.3.3. As a matter of fact, the solution constructed in this section belongs to the following class:

$$\begin{aligned} &\mathbb{E}\Big[\sup_{t\in[0,T]}\Big|\int_{\mathbb{T}^3}\Big[\frac{1}{2}\varrho|\mathbf{u}|^2+P_\delta(\varrho)\Big]\,\mathrm{d}x\Big|^r\Big]\lesssim 1,\\ &\mathbb{E}\Big[\Big|\int_0^T\int_{\mathbb{T}^3}[\mathbb{S}(\nabla\mathbf{u}):\nabla\mathbf{u}+\varepsilon\varrho|\nabla\mathbf{u}|^2+\varepsilon P_\delta''(\varrho)|\nabla\varrho|^2]\,\mathrm{d}x\,\mathrm{d}t\Big|^r\Big]\lesssim 1, \end{aligned} \tag{4.79}$$

with the same $r \geq 2$ as in (4.78). In particular, we have

$$\mathbb{E}\Big[\sup_{t\in[0,T]}\|\varrho(t)\|_{L_x^\Gamma}^{\Gamma p}\Big]<\infty \quad \text{for some } p\in(1,\infty), \tag{4.80}$$

$$\mathbb{E}\Big(\int_0^T\|\mathbf{u}\|_{W_x^{1,2}}^2\,\mathrm{d}t\Big)^p<\infty \quad \text{for some } p\in(1,\infty), \tag{4.81}$$

$$\mathbb{E}\Big[\sup_{t\in[0,T]}\|\varrho\mathbf{u}(t)\|_{L_x^{\frac{2\Gamma}{\Gamma+1}}}^{\frac{2\Gamma}{\Gamma+1}p}\Big]<\infty \quad \text{for some } p\in(1,\infty). \tag{4.82}$$

In order to prove Theorem 4.3.2 we adopt a strategy similar to the preceding section:

(1) Using the energy balance we derive uniform bounds independent of the parameter m.

(2) We let $m \to \infty$ with the help of the stochastic compactness method based on the Jakubowski–Skorokhod representation theorem, Theorem 2.7.1. In contrast with Section 4.1, however, we get only *martingale solutions* as uniqueness is lost already at this stage of the approximation procedure.

Compared to the procedure performed in Section 4.1.2, the limit $m \to \infty$ is more delicate as (i) the system (4.57) of stochastic ODEs becomes a system of stochastic PDEs (4.73) and (ii) the lower bound on the density will be lost in the asymptotic limit. The latter phenomenon is obviously related to the (hypothetical) presence of *vacuum zones* pertaining to compressible fluid models.

4.3.1 Uniform bounds

In this section, the approximation parameters $\varepsilon, \delta \in (0,1)$ are kept fixed. Accordingly, we call *uniform* the estimates that are independent of m but may depend on ε, δ as well as $T > 0$. Denote by $[\varrho, \mathbf{u}]$ the solution of problem (4.56)–(4.57) starting from initial condition $[\varrho_0, \mathbf{u}_0]$, the existence of which is guaranteed by Theorem 4.2.2. Recall that the energy bound (4.63) holds true uniformly in m. It gives rise to the following estimates which are independent of m:

$$\mathbb{E}\Big[\Big|\sup_{t\in[0,T]}\|\varrho\|_{L_x^\Gamma}^\Gamma\Big|^r\Big]+\mathbb{E}[\|\varrho^{-\frac{1}{2}}\nabla\varrho\|_{L_t^2L_x^2}^{2r}]+\mathbb{E}[\|\varrho^{\frac{\Gamma}{2}-1}\nabla\varrho\|_{L_t^2L_x^2}^{2r}]\lesssim c(r), \tag{4.83}$$

$$\mathbb{E}\Big[\Big|\sup_{t\in[0,T]}\|\varrho|\mathbf{u}|^2\|_{L_x^1}\Big|^r+\Big|\sup_{t\in[0,T]}\|\varrho\mathbf{u}\|_{L_x^{\frac{2\Gamma}{\Gamma+1}}}^{\frac{2\Gamma}{\Gamma+1}}\Big|^r\Big]\lesssim c(r), \tag{4.84}$$

$$\mathbb{E}[\|\nabla\mathbf{u}\|_{L_t^2L_x^2}^{2r}]\lesssim c(r), \tag{4.85}$$

where

$$c(r)\approx 1+\mathbb{E}\Big[\Big|\int_{\mathbb{T}^3}\Big[\frac{1}{2}\varrho_0|\mathbf{u}_0|^2+P_\delta(\varrho_0)\Big]\,\mathrm{d}x\Big|^r\Big],\quad r\geq 2. \tag{4.86}$$

Moreover, as a direct consequence of (4.58) and the non-negativity of ϱ, we have

$$\|\varrho(\tau)\|_{L_x^1}=\|\varrho_0\|_{L_x^1}\leq c(\overline{\varrho}),\quad \tau\in[0,T]. \tag{4.87}$$

Apparently, none of these bounds directly controls the amplitude of $\mathbf{u}_m$. The following holds by Sobolev's embedding and Poincaré's inequality, because $\gamma>\frac{3}{2}$:

$$\begin{aligned}\|\varrho_0\|_{L_x^1}|(\mathbf{u})_{\mathbb{T}^3}|=\Big|\int_{\mathbb{T}^3}\varrho(\mathbf{u})_{\mathbb{T}^3}\,\mathrm{d}x\Big|&\leq\int_{\mathbb{T}^3}\varrho|(\mathbf{u})_{\mathbb{T}^3}-\mathbf{u}|\,\mathrm{d}x+\int_{\mathbb{T}^3}\varrho|\mathbf{u}|\,\mathrm{d}x\\&\lesssim\|\varrho\|_{L_x^\gamma}\|(\mathbf{u})_{\mathbb{T}^3}-\mathbf{u}\|_{L_x^{\gamma'}}+\|\sqrt{\varrho}\|_{L_x^2}\|\sqrt{\varrho}\mathbf{u}\|_{L_x^2}\\&\lesssim\|\varrho\|_{L_x^\gamma}\|\nabla\mathbf{u}\|_{L_x^2}+\|\varrho\|_{L_x^1}+\|\varrho|\mathbf{u}|^2\|_{L_x^1}.\end{aligned}$$

Consequently,

$$\begin{aligned}\|\varrho_0\|_{L_x^1}^2\int_0^\tau|(\mathbf{u})_{\mathbb{T}^3}|^2\,\mathrm{d}t\lesssim&\sup_{t\in[0,\tau]}\|\varrho\|_{L_x^\gamma}^2\int_0^\tau\|\nabla\mathbf{u}\|_{L_x^2}^2\,\mathrm{d}t\\&+\tau\sup_{t\in[0,\tau]}(\|\varrho\|_{L_x^1}^2+\|\varrho|\mathbf{u}|^2\|_{L_x^1}^2).\end{aligned} \tag{4.88}$$

In view of the bounds established in (4.83)–(4.86) and the assumptions on the initial law in (4.77), we obtain the desired estimate

$$\mathbb{E}[\|\mathbf{u}\|_{L_t^2W_x^{1,2}}^{2r}]\lesssim c_2(r), \tag{4.89}$$

with

$$c_2(r)\approx 1+\mathbb{E}\Big[\Big|\int_{\mathbb{T}^3}\Big[\frac{1}{2}\varrho_0|\mathbf{u}_0|^2+P_\delta(\varrho_0)\Big]\,\mathrm{d}x\Big|^{2r}\Big],\quad r\geq 2.$$

Next, we recall the standard parabolic maximal regularity estimates (cf. Theorem A.2.1 and Theorem A.2.2), applied to (4.58):

$$\begin{aligned}\|\partial_t\varrho\|_{L_t^pL_x^q}+\|\varrho\|_{L_t^pW_x^{2,q}}&\lesssim\|\operatorname{div}(\varrho\mathbf{u})\|_{L_t^pL_x^q}+\|\varrho_0\|_{C_x^{2+\nu}},\\\|\varrho\|_{L_t^pW_x^{1,q}}&\lesssim\|\varrho\mathbf{u}\|_{L_t^pL_x^q}+\|\varrho_0\|_{C_x^{2+\nu}},\end{aligned} \tag{4.90}$$

for $1<p,q<\infty$.

Remark 4.3.4. In (4.90), the regularity of the initial data can be considerably weakened. However, such generality is not needed here.

Now observe that (4.83), together with $\Gamma \geq 6$, (4.85), and the standard Sobolev embedding $W^{1,2}(\mathbb{T}^3) \hookrightarrow L^6(\mathbb{T}^3)$, gives rise to

$$\begin{aligned}
\mathbb{E}\big[\big|\|\varrho\mathbf{u}\|_{L^1_t L^3_x}\big|^r\big] &\leq \mathbb{E}\bigg[\bigg|\int_0^T \|\varrho\|_{L^6_x}\|\mathbf{u}\|_{L^6_x}\,\mathrm{d}t\bigg|^r\bigg] \\
&\lesssim \mathbb{E}\bigg[\bigg|\sup_{t\in[0,T]}\|\varrho\|_{L^6_x}\int_0^T \|\mathbf{u}\|_{L^6_x}\,\mathrm{d}t\bigg|^r\bigg] \\
&\lesssim \mathbb{E}\Big[\Big|\sup_{t\in[0,T]}\|\varrho\|^2_{L^\Gamma_x}\Big|^r\Big] + \mathbb{E}\bigg[\bigg|\int_0^T \|\mathbf{u}\|^2_{W^{1,2}_x}\,\mathrm{d}t\bigg|^r\bigg] \\
&\lesssim c_2(r).
\end{aligned}$$

Interpolating this with (4.84) yields

$$\mathbb{E}\big[\big|\|\varrho\mathbf{u}\|_{L^p_t L^p_x}\big|^r\big] \lesssim c_2(r) \quad \text{for a certain } p > 2, \tag{4.91}$$

which, plugged in the right hand side of (4.90), implies

$$\mathbb{E}\big[\big|\|\varrho\|_{L^p_t W^{1,p}_x}\big|^r\big] \lesssim \tilde{c}_2(r) \quad \text{for a certain } p > 2. \tag{4.92}$$

Finally, the estimates (4.89) and (4.92) can be used again in (4.90) to conclude

$$\mathbb{E}\big[\big|\|\partial_t\varrho\|_{L^p_t L^p_x} + \|\varrho\|_{L^p_t W^{2,p}_x}\big|^r\big] \lesssim \tilde{c}_2(r), \tag{4.93}$$

for some $p > 1$, where

$$\tilde{c}_2(r) \approx 1 + \mathbb{E}\bigg[\bigg|\int_{\mathbb{T}^3}\Big[\frac{1}{2}\varrho_0|\mathbf{u}_0|^2 + P_\delta(\varrho_0)\Big]\,\mathrm{d}x\bigg|^{2r} + \big|\|\varrho_0\|_{C^{2+\nu}_x}\big|^r\bigg], \quad r \geq 2. \tag{4.94}$$

4.3.2 Asymptotic limit

With the uniform bounds established in the preceding part at hand, we are ready to perform the limit $m \to \infty$ in a similar way as in Section 4.1.2. Note that the available estimates are considerably weaker than those obtained in Section 4.1.2, making the choice of the appropriate *path space* more delicate. In particular, the role of “leading” phase variables is here and hereafter transferred from $[\varrho, \mathbf{u}]$ to $[\varrho, \varrho\mathbf{u}]$ as it is the latter that enjoys a certain path continuity in time. Another new aspect is the necessity to work with *weak topologies* that are in general not Polish. Accordingly, we use the Jakubowski modification of the classical Skorokhod representation theorem (Theorem 2.7.1).

Let Λ be a probability measure on $C^{2+\nu}(\mathbb{T}^3) \times L^1(\mathbb{T}^3)$ satisfying (4.77) and (4.78). Let $(\Omega, \mathfrak{F}, (\mathfrak{F}_t)_{t\geq 0}, \mathbb{P})$ be a stochastic basis with a complete right-continuous filtration and let W be a cylindrical Wiener process relative to $(\mathfrak{F}_t)_{t\geq 0}$. Finally, let $(\varrho_0, \mathbf{q}_0)$ be $\mathfrak{F}_0$-measurable random variables with values in $C^{2+\nu}(\mathbb{T}^3) \times L^1(\mathbb{T}^3)$ and law Λ. The existence of $(\varrho_0, \mathbf{q}_0)$ is guaranteed by Corollary 2.6.4. Since $\varrho_0 \geq \underline{\varrho} > 0$, we can set $\mathbf{u}_0 = \frac{\mathbf{q}_0}{\varrho_0}$. Note that, due to assumption (4.78), we have $\mathbf{u}_0 \in L^2(\mathbb{T}^3)$ $\mathbb{P}$-a.s. We define the initial velocities $\mathbf{u}_{0,m} = \Pi_m \mathbf{u}_0$ and observe that the assumptions on the initial condition in Theorem 4.2.2 are satisfied by $(\varrho_0, \mathbf{u}_{0,m})$. In addition, setting $\Lambda_m = \mathbb{P} \circ (\varrho_0, \mathbf{u}_{0,m})^{-1}$, the corresponding version of (4.78) holds true for Λ_m and uniformly in m. To be more precise, we have

$$\int_{C_x^{2+\nu} \times L_x^2} \left(\left| \int_{\mathbb{T}^3} \left[\frac{1}{2} \varrho |\mathbf{v}|^2 + P_\delta(\varrho) \right] \mathrm{d}x \right|^{2r} + \|\varrho\|_{C_x^{2+\nu}}^r \right) \mathrm{d}\Lambda_m(\varrho, \mathbf{v}) \leq c, \tag{4.95}$$

for some $r > 2$ with a constant independent of m. Finally, note that we obtain

$$\begin{aligned} &\int_{C_x^{2+\nu} \times L_x^2} \left(\left| \int_{\mathbb{T}^3} \left[\frac{1}{2} \varrho |\mathbf{v}|^2 + P_\delta(\varrho) \right] \mathrm{d}x \right|^{2r} + \|\varrho\|_{C_x^{2+\nu}}^r \right) \mathrm{d}\Lambda_m(\varrho, \mathbf{v}) \\ &\quad \longrightarrow \int_{C_x^{2+\nu} \times L_x^1} \left(\left| \int_{\mathbb{T}^3} \left[\frac{1}{2} \frac{|\mathbf{q}|^2}{\varrho} + P_\delta(\varrho) \right] \mathrm{d}x \right|^{2r} + \|\varrho\|_{C_x^{2+\nu}}^r \right) \mathrm{d}\Lambda(\varrho, \mathbf{q}), \end{aligned} \tag{4.96}$$

as $m \to \infty$. Relation (4.96) follows immediately, if we rewrite the quantities as expectations and use the definition of $\mathbf{u}_0$ and $\mathbf{u}_{0,m}$. To be precise, we have

$$\begin{aligned} &\mathbb{E}\left[\left| \int_{\mathbb{T}^3} \left[\frac{1}{2} \varrho_0 |\mathbf{u}_{0,m}|^2 + P_\delta(\varrho_0) \right] \mathrm{d}x \right|^{2r} + \|\varrho_0\|_{C_x^{2+\nu}}^r \right] \\ &\quad \longrightarrow \mathbb{E}\left[\left| \int_{\mathbb{T}^3} \left[\frac{1}{2} \varrho_0 |\mathbf{u}_0|^2 + P_\delta(\varrho_0) \right] \mathrm{d}x \right|^{2r} + \|\varrho_0\|_{C_x^{2+\nu}}^r \right], \end{aligned}$$

using majorized convergences (recall that $\varrho_0 \leq \overline{\varrho}$ and $\|\Pi_m \mathbf{u}_0\|_{L_x^2} \leq \|\mathbf{u}_0\|_{L_x^2}$).

Finally, let $(\varrho_m, \mathbf{u}_m)$ be the solution of problem (4.56)–(4.57) obtained in Theorem 4.2.2 and starting from $(\varrho_0, \mathbf{u}_{0,m})$. As a consequence of (4.95), all the estimates in Section 4.3.1 hold true for $(\varrho_m, \mathbf{u}_m)$ uniformly in m.

In accordance with the uniform bounds derived in the preceding part of this section, we choose the path space

$$\mathcal{X} = \mathcal{X}_{\varrho_0} \times \mathcal{X}_{\mathbf{u}_0} \times \mathcal{X}_{\varrho} \times \mathcal{X}_{\varrho\mathbf{u}} \times \mathcal{X}_{\mathbf{u}} \times \mathcal{X}_W,$$

where

$$\begin{aligned} \mathcal{X}_{\varrho_0} &= C(\mathbb{T}^3), \\ \mathcal{X}_{\mathbf{u}_0} &= L^2(\mathbb{T}^3), \\ \mathcal{X}_{\varrho} &= L^p(0,T; W^{1,p}(\mathbb{T}^3)) \cap (W^{1,q}(0,T; L^q(\mathbb{T}^3)) \cap L^q(0,T; W^{2,q}(\mathbb{T}^3)), w) \\ &\quad \cap C_w([0,T]; L^\Gamma(\mathbb{T}^3)), \end{aligned}$$

$$
\begin{aligned}
\mathcal{X}_{\varrho\mathbf{u}} &= C([0,T]; W^{-k,2}(\mathbb{T}^3)) \cap C_w([0,T]; L^{\frac{2\Gamma}{\Gamma+1}}(\mathbb{T}^3)),\\
\mathcal{X}_{\mathbf{u}} &= (L^2(0,T; W^{1,2}(\mathbb{T}^3)), w),\\
\mathcal{X}_W &= C([0,T]; \mathfrak{U}_0),
\end{aligned}
$$

for certain $p > 2$, $q > 1$ and $k \in \mathbb{N}$. Note that the initial condition was included in the path space in order to be able to pass to the limit in the energy inequality.

Now we claim that the parameters p, q, k can be adjusted in such a way that the family of joint laws

$$
\{\mathcal{L}[\varrho_0, \mathbf{u}_{0,m}, \varrho_m, \Pi_m(\varrho_m \mathbf{u}_m), \mathbf{u}_m, W];\ m \in \mathbb{N}\} \quad \text{is tight on } \mathcal{X}.
$$

To see this, we proceed in several steps. First of all, we observe that the law $\mathcal{L}[\varrho_0]$ is tight, being a Radon measure on the Polish space $C(\mathbb{T}^3)$. Next, we show tightness of the initial conditions $\mathbf{u}_{0,m}$.

Proposition 4.3.5. *The set $\{\mathcal{L}[\mathbf{u}_{0,m}];\ m \in \mathbb{N}\}$ is tight on $\mathcal{X}_{\mathbf{u}_0}$.*

Proof. By definition, $\mathbf{u}_{0,m} = \Pi_m \mathbf{u}_0$. Hence, due to continuity of the projections Π_m, we obtain $\mathbf{u}_{0,m} \to \mathbf{u}_0$ a.s. Consequently, the convergence in law follows and Prokhorov's theorem, Theorem 2.6.1, implies tightness of the corresponding laws. □

As the next step, we prove tightness of the marginals corresponding to $\mathbf{u}_m$.

Proposition 4.3.6. *The set $\{\mathcal{L}[\mathbf{u}_m];\ m \in \mathbb{N}\}$ is tight on $\mathcal{X}_{\mathbf{u}}$.*

Proof. The proof follows directly from (4.89). Indeed, for any $L > 0$, the set

$$
B_L = \{\mathbf{u} \in L^?(0,T; W^{1,2}(\mathbb{T}^3));\ \|\mathbf{u}\|_{L_t^2 W_x^{1,2}} \le L\}
$$

is relatively compact in $\mathcal{X}_{\mathbf{u}}$ and

$$
\mathcal{L}[\mathbf{u}_m](B_R^C) = \mathbb{P}(\|\mathbf{u}_m\|_{L_t^2 W_x^{1,2}} \ge L) \le \frac{1}{L} \mathbb{E}\|\mathbf{u}_m\|_{L_t^2 W_x^{1,2}} \le \frac{C}{L},
$$

which yields the claim by choosing L sufficiently large. The same argument was already employed in Proposition 4.1.5. □

As the next step, we establish tightness of the marginals corresponding to ϱ_m.

Proposition 4.3.7. *The set $\{\mathcal{L}[\varrho_m];\ m \in \mathbb{N}\}$ is tight on $\mathcal{X}_\varrho$.*

Proof. Using (4.84) we obtain

$$
\mathbb{E}\Big[\Big|\sup_{t\in[0,T]} \|\operatorname{div}(\varrho_m \mathbf{u}_m)\|_{W_x^{-1,\frac{2\Gamma}{\Gamma+1}}}^{\frac{2\Gamma}{\Gamma+1}}\Big|^r\Big] \lesssim c(r). \tag{4.97}
$$

Similarly, due to (4.83), we have

$$\mathbb{E}\Big[\Big|\sup_{t\in[0,T]}\|\varepsilon\Delta\varrho_m\|^{\Gamma}_{W_x^{-2,\Gamma}}\Big|^r\Big]\lesssim c(r).$$

As a consequence of the continuity equation (4.56),

$$\mathbb{E}\|\varrho_m\|_{C_t^{0,1}W_x^{-2,\frac{2\Gamma}{\Gamma+1}}}\leq C.$$

Now, the required tightness in $C_w([0,T];L^{\Gamma}(\mathbb{T}^3))$ follows by a similar reasoning as in Proposition 4.3.6 due to the following compact embedding (see Theorem 1.8.5):

$$L^{\infty}(0,T;L^{\Gamma}(\mathbb{T}^3))\cap C^{0,1}([0,T];W^{-2,\frac{2\Gamma}{\Gamma+1}}(\mathbb{T}^3))\overset{c}{\hookrightarrow} C_w([0,T];L^{\Gamma}(\mathbb{T}^3)).$$

Tightness in $(W^{1,q}(0,T;L^q(\mathbb{T}^3))\cap L^q(0,T;W^{2,q}(\mathbb{T}^3)),w)$ is a direct consequence of (4.93) due to the fact that balls are relatively compact with respect to the weak topology.

In order to show tightness in $L^p(0,T;W^{1,p}(\mathbb{T}^3))$ for some $p>2$, we observe that, in view of compactness of the embedding

$$W^{1,q}(0,T;L^q(\mathbb{T}^3))\cap L^q(0,T;W^{2,q}(\mathbb{T}^3))\overset{c}{\hookrightarrow} L^r(0,T;W^{1,r}(\mathbb{T}^3)),$$

which holds true for a certain $r(q)>1$ by Aubin–Lions' theorem, we obtain tightness in $L^r(0,T;W^{1,r}(\mathbb{T}^3))$ for $r>1$. In order to increase integrability, we use an interpolation argument. Namely, we observe that, if $q>1$ and $r'>2$, the set

$$\begin{aligned}B_L=\{\varrho\in W^{1,q}(0,T;L^q(\mathbb{T}^3))\cap L^q(0,T;W^{2,q}(\mathbb{T}^3))\cap L^{r'}(0,T;W^{1,r'}(\mathbb{T}^3));\\ \|\varrho\|_{W_t^{1,q}L_x^q}+\|\varrho\|_{L_t^qW_x^{2,q}}+\|\varrho\|_{L_t^{r'}W_x^{1,r'}}\leq L\}\end{aligned}\tag{4.98}$$

is relatively compact in $L^p(0,T;W^{1,p}(\mathbb{T}^3))$ for some $p>2$. Hence the desired tightness can be deduced from (4.92) and (4.93). □

Proposition 4.3.8. *The set* $\{\mathcal{L}[\Pi_m(\varrho_m\mathbf{u}_m)];\ m\in\mathbb{N}\}$ *is tight on* $\mathcal{X}_{\varrho\mathbf{u}}$.

Proof. First of all, we prove time regularity of $\Pi_m(\varrho_m\mathbf{u}_m)$, which holds true uniformly in m. Let us start with the deterministic part of (4.57), namely,

$$\begin{aligned}Y_m(t)=\Pi_m(\varrho_m\mathbf{u}_m)(0)-\int_0^t\Pi_m[\operatorname{div}(\varrho_m\mathbf{u}_m\otimes\mathbf{u}_m)]\,\mathrm{d}s+\int_0^t\Pi_m[\operatorname{div}\mathbb{S}(\nabla\mathbf{u}_m)]\,\mathrm{d}s\\ -\int_0^t\Pi_m[\nabla p_\delta(\varrho_m)]\,\mathrm{d}s+\varepsilon\int_0^t\Pi_m\Delta(\varrho_m\mathbf{u}_m)\,\mathrm{d}s.\end{aligned}$$

We will show that there exist $\kappa\in(0,1)$ and $l\in\mathbb{N}$ such that

$$\mathbb{E}\|Y_m\|_{C^{\kappa}([0,T];W_x^{-l,2})}\leq C.\tag{4.99}$$

To this end, we observe that, according to (4.84) and (4.89), we have

$$\varrho_m \mathbf{u}_m \otimes \mathbf{u}_m \in L^r(\Omega; L^2(0,T; L^{\frac{6\Gamma}{4\Gamma+3}}(\mathbb{T}^3))) \tag{4.100}$$

for a suitable $r > 1$ uniformly in m. Thus

$$\{\Pi_m \operatorname{div}(\varrho_m \mathbf{u}_m \otimes \mathbf{u}_m)\} \quad \text{is bounded in } L^r(\Omega; L^2(0,T; W^{-1,\frac{6\Gamma}{4\Gamma+3}}(\mathbb{T}^3))), \tag{4.101}$$

by continuity of Π_m; cf. (4.6). Similarly,

$$\{\Pi_m \operatorname{div} \mathbb{S}(\nabla \mathbf{u}_m)\} \quad \text{is bounded in } L^r(\Omega; L^2(0,T; W^{-1,2}(\mathbb{T}^3))). \tag{4.102}$$

The correct space for $p_\delta(\varrho_m)$ is $L^\infty(0,T; L^1(\mathbb{T}^3))$, so choosing $l > \frac{5}{2}$ Sobolev's embedding yields

$$\{\nabla p_\delta(\varrho_m)\} \quad \text{is bounded in } L^r(\Omega; L^p(0,T; W^{-l,2}(\mathbb{T}^3))), \tag{4.103}$$

for all $p \in (1,\infty)$. Finally, as a consequence of (4.91), we obtain

$$\{\Pi_m \Delta(\varrho_m \mathbf{u}_m)\} \quad \text{is bounded in } L^r(\Omega; L^p(0,T; W^{-2,p}(\mathbb{T}^3))), \tag{4.104}$$

for some $p > 2$. Plugging all together yields boundedness of Y_m in

$$L^r(\Omega; W^{1,2}(0,T; W^{-l,2}(\mathbb{T}^3))),$$

provided l is large enough. Now, (4.99) follows if $\kappa < \frac{1}{2}$.

Similarly to Section 4.1.2 and in view of the bound established in (4.99), it suffices to check the time regularity of the stochastic integral. Applying the Burkholder–Davis–Gundy inequality, we obtain

$$\begin{aligned}
&\mathbb{E}\left[\left\| \int_{\tau_1}^{\tau_2} \varrho_m \Pi_m[\mathbb{F}_\varepsilon(\varrho_m, \mathbf{u}_m)] \, \mathrm{d}W \right\|_{L^2_x}^r\right] \\
&\quad \lesssim |\tau_1 - \tau_2|^{r/2} \mathbb{E}\left[\sup_{0 \le t \le T} \sum_{k=1}^{\infty} \|\varrho_m \Pi_m[\mathbf{F}_{k,\varepsilon}(\varrho_m, \mathbf{u}_m)]\|_{L^2_x}^2\right]^{r/2} \quad \text{for any } r \ge 1.
\end{aligned}$$

Next, by Hölder's inequality, (4.6), (4.10), and $\Gamma \ge 4$,

$$\begin{aligned}
\|\varrho_m \Pi_m[\mathbf{F}_{k,\varepsilon}(\varrho_m, \mathbf{u}_m)]\|_{L^2_x} &\le \|\varrho_m\|_{L^4_x} \|\Pi_m[\mathbf{F}_{k,\varepsilon}(\varrho_m, \mathbf{u}_m)]\|_{L^4_x} \\
&\lesssim \|\varrho_m\|_{L^\Gamma_x} \|\mathbf{F}_{k,\varepsilon}(\varrho_m, \mathbf{u}_m)\|_{L^\infty_x} \lesssim f_{k,\varepsilon} \|\varrho_m\|_{L^\Gamma_x}.
\end{aligned} \tag{4.105}$$

Next, we use the bounds (4.83) and apply Theorem 2.3.11 to obtain (lowering κ if necessary)

$$\mathbb{E}\left[\left\| \int_0^{\cdot} \varrho_m \Pi_m[\mathbb{F}_\varepsilon(\varrho_m, \mathbf{u}_m)] \, \mathrm{d}W \right\|_{C^\kappa_t L^2_x}^r\right] \lesssim c(r) \quad \text{for any } r > 2. \tag{4.106}$$

In combination with (4.99), this implies

$$\mathbb{E}\|\Pi_m(\varrho_m\mathbf{u}_m)\|_{C_t^\kappa W_x^{-l,2}} \le \tilde{c}_2(r) \tag{4.107}$$

uniformly for $m \to \infty$, as soon as the initial data satisfy (4.94) for some $r > 2$.

Accordingly, due to (4.107), tightness in $C([0,T]; W^{-k,2}(\mathbb{T}^3))$ follows from the compact embedding

$$C^\kappa([0,T]; W^{-l,2}(\mathbb{T}^3)) \overset{c}{\hookrightarrow} C([0,T]; W^{-k,2}(\mathbb{T}^3)),$$

which is valid, provided $k > l$. In addition, tightness in $C_w([0,T]; L^{\frac{2\Gamma}{\Gamma+1}}(\mathbb{T}^3))$ is obtained from (4.84) and (4.107) with the help of the following compact embedding (see Theorem 1.8.5):

$$L^\infty(0,T; L^{\frac{2\Gamma}{\Gamma+1}}(\mathbb{T}^3)) \cap C^\kappa([0,T]; W^{-l,2}(\mathbb{T}^3)) \overset{c}{\hookrightarrow} C_w([0,T]; L^{\frac{2\Gamma}{\Gamma+1}}(\mathbb{T}^3)).$$ □

The singleton $\mathcal{L}[W]$ on $C([0,T]; \mathfrak{U}_0)$, being a Radon measure on a Polish space, is tight. Consequently, as in Proposition 4.1.5, we obtain the following.

Corollary 4.3.9. *The set* $\{\mathcal{L}[\varrho_0, \mathbf{u}_{0,m}, \varrho_m, \Pi_m(\varrho_m\mathbf{u}_m), \mathbf{u}_m, W]; m \in \mathbb{N}\}$ *is tight on* $\mathcal{X}$.

Since weak topologies are generally not metrizable, the path space $\mathcal{X}$ is not a Polish space. Nevertheless, it can be seen that $\mathcal{X}$ belongs to the class of sub-Polish spaces, introduced in Definition 2.1.3. Accordingly, our compactness argument is based on the Jakubowski–Skorokhod representation theorem instead of the classical Skorokhod representation theorem; see Theorem 2.7.1. To be more precise, we infer the following result.

Proposition 4.3.10. *There exists a complete probability space* $(\tilde{\Omega}, \tilde{\mathfrak{F}}, \tilde{\mathbb{P}})$ *with* $\mathcal{X}$*-valued Borel measurable random variables* $[\tilde{\varrho}_{0,m}, \tilde{\mathbf{u}}_{0,m}, \tilde{\varrho}_m, \tilde{\mathbf{q}}_m, \tilde{\mathbf{u}}_m, \tilde{W}_m]$, $m \in \mathbb{N}$, *and* $[\tilde{\varrho}_0, \tilde{\mathbf{u}}_0, \tilde{\varrho}, \tilde{\mathbf{q}}, \tilde{\mathbf{u}}, \tilde{W}]$ *such that (up to a subsequence):*

(1) *for all* $m \in \mathbb{N}$, *the laws of* $[\tilde{\varrho}_{0,m}, \tilde{\mathbf{u}}_{0,m}, \tilde{\varrho}_m, \tilde{\mathbf{q}}_m, \tilde{\mathbf{u}}_m, \tilde{W}_m]$ *and* $[\varrho_0, \mathbf{u}_{0,m}, \varrho_m, \Pi_m(\varrho_m\mathbf{u}_m), \mathbf{u}_m, W]$ *coincide on* $\mathcal{X}$;

(2) *the law of* $[\tilde{\varrho}_0, \tilde{\mathbf{u}}_0, \tilde{\varrho}, \tilde{\mathbf{q}}, \tilde{\mathbf{u}}, \tilde{W}]$ *on* $\mathcal{X}$ *is a Radon measure;*

(3) $[\tilde{\varrho}_{0,m}, \tilde{\mathbf{u}}_{0,m}, \tilde{\varrho}_m, \tilde{\mathbf{q}}_m, \tilde{\mathbf{u}}_m, \tilde{W}_m]$ *converges in the topology of* $\mathcal{X}$ $\tilde{\mathbb{P}}$*-a.s. to* $[\tilde{\varrho}_0, \tilde{\mathbf{u}}_0, \tilde{\varrho}, \tilde{\mathbf{q}}, \tilde{\mathbf{u}}, \tilde{W}]$, *i.e.,*

$$\begin{aligned}
\tilde{\varrho}_{0,m} &\to \tilde{\varrho}_0 \quad \text{in } C(\mathbb{T}^3),\\
\tilde{\mathbf{u}}_{0,m} &\to \tilde{\mathbf{u}}_0 \quad \text{in } L^2(\mathbb{T}^3),\\
\tilde{\varrho}_m &\to \tilde{\varrho} \quad \text{in } L^p(0,T; W^{1,p}(\mathbb{T}^3)),\\
\tilde{\varrho}_m &\to \tilde{\varrho} \quad \text{in } W^{1,q}(0,T; L^q(\mathbb{T}^3)) \cap L^q(0,T; W^{2,q}(\mathbb{T}^3)),\\
\tilde{\varrho}_m &\to \tilde{\varrho} \quad \text{in } C_w([0,T]; L^\Gamma(\mathbb{T}^3)),
\end{aligned} \tag{4.108}$$

$$\begin{aligned}
\tilde{\mathbf{q}}_m &\to \tilde{\mathbf{q}} \quad \textit{in } C([0,T]; W^{-k,2}(\mathbb{T}^3)),\\
\tilde{\mathbf{q}}_m &\to \tilde{\mathbf{q}} \quad \textit{in } C_w([0,T]; L^{\frac{2\Gamma}{\Gamma+1}}(\mathbb{T}^3)),\\
\tilde{\mathbf{u}}_m &\rightharpoonup \tilde{\mathbf{u}} \quad \textit{in } L^2(0,T; W^{1,2}(\mathbb{T}^3)),\\
\tilde{W}_m &\to \tilde{W} \quad \textit{in } C([0,T]; \mathfrak{U}_0),
\end{aligned}$$

$\tilde{\mathbb{P}}$-a.s. for certain $p > 2$, $q > 1$, and $k \in \mathbb{N}$.

We observe that at this stage of the proof it becomes convenient to work with random distributions as introduced in Section 2.2. Indeed, it is seen from the above compactness result that the limit velocity $\tilde{\mathbf{u}}$ is not a stochastic process in the classical sense; cf. Definition 2.1.11. As discussed in Section 2.3, the stochastic integration theory relies on the associated "arrow of time", which is represented by the progressive measurability of the corresponding integrands. Recall that, for random distributions that are adapted in the sense of Definition 2.2.13 and satisfy a suitable integrability assumption, Lemma 2.2.18 guarantees the existence of a progressively measurable stochastic process belonging to the same class of equivalence.

As a consequence, it was discussed in Remark 2.3.7 that the minimal assumption on integrands, under which the stochastic integral is well-defined, is the non-anticipativity of the corresponding joint canonical filtration with respect to the driving Wiener process. In particular, we define

$$\tilde{\mathfrak{F}}_t := \sigma\left(\sigma_t[\tilde{\varrho}] \cup \sigma_t[\tilde{\mathbf{u}}] \cup \bigcup_{k=1}^{\infty} \sigma_t[\tilde{W}_k] \right), \quad t \in [0,T],$$

and need to check that $\tilde{\mathfrak{F}}_t$ is independent of $\sigma(\tilde{W}(s) - \tilde{W}(t))$ for all $s > t$. This can be done by following the same arguments as in the discussion after Proposition 4.1.6.

More precisely, we first recall Theorem 2.9.1 and deduce that, for every $m \in \mathbb{N}$, $\tilde{W}_m = \sum_{k=1}^{\infty} e_k \tilde{W}_{m,k}$ is a cylindrical Wiener process with respect to

$$\sigma\left(\sigma_t[\tilde{\varrho}_m] \cup \sigma_t[\tilde{\mathbf{u}}_m] \cup \bigcup_{k=1}^{\infty} \sigma_t[\tilde{W}_{m,k}] \right), \quad t \in [0,T].$$

In other words, this filtration is non-anticipative with respect to $\tilde{W}_m$. Lemma 2.9.3, together with Proposition 4.3.10, then allows one to pass to the limit as $m \to \infty$ and the non-anticipativity of $(\tilde{\mathfrak{F}}_t)_{t\geq 0}$ with respect to $\tilde{W}$ follows. Finally, due to Lemma 2.1.35 and Corollary 2.1.36, the process $\tilde{W}$ is a cylindrical Wiener processes with respect to $(\tilde{\mathfrak{F}}_t)_{t\geq 0}$.

We are immediately able to identify $\tilde{\mathbf{q}}_m$, $m \in \mathbb{N}$, and $\tilde{\mathbf{q}}$.

Lemma 4.3.11. *We have*

$$\tilde{\mathbf{q}}_m = \Pi_m(\tilde{\varrho}_m \tilde{\mathbf{u}}_m), \quad \tilde{\mathbf{q}} = \tilde{\varrho}\tilde{\mathbf{u}}, \quad \tilde{\mathbb{P}}\text{-a.s.}$$

Proof. The first statement follows from the equality of joint laws of

$$(\varrho_m, \mathbf{u}_m, \Pi_m(\varrho_m \mathbf{u}_m)) \quad \text{and} \quad (\tilde{\varrho}_m, \tilde{\mathbf{u}}_m, \tilde{\mathbf{q}}_m).$$

In order to identify the limit $\tilde{\mathbf{q}}$, note that

$$\tilde{\varrho}_m \tilde{\mathbf{u}}_m \rightharpoonup \tilde{\varrho}\tilde{\mathbf{u}} \quad \text{in } L^1(0,T;L^1(\mathbb{T}^3)) \ \tilde{\mathbb{P}}\text{-a.s.}$$

as a consequence of the convergence of $\tilde{\varrho}_m$ and $\tilde{\mathbf{u}}_m$ in $\mathcal{X}_\varrho$ and $\mathcal{X}_\mathbf{u}$, respectively. Clearly, this also identifies the limit of $\Pi_m(\tilde{\varrho}_m \tilde{\mathbf{u}}_m)$ with $\tilde{\varrho}\tilde{\mathbf{u}}$. □

Next, we observe that due to equality of laws and weak lower semi-continuity of the involved norms, the uniform bounds from Section 4.3.1 hold true also for $[\tilde{\varrho}_m, \tilde{\mathbf{u}}_m]$. Consequently, based on Proposition 4.3.10, we obtain the following.

Corollary 4.3.12. *The following convergence holds true $\tilde{\mathbb{P}}$-a.s.:*

$$\tilde{\varrho}_m \tilde{\mathbf{u}}_m \otimes \tilde{\mathbf{u}}_m \rightharpoonup \tilde{\varrho}\tilde{\mathbf{u}} \otimes \tilde{\mathbf{u}} \quad \text{in } L^1(0,T;L^1(\mathbb{T}^3)). \tag{4.109}$$

Proof. From Proposition 4.3.10 and Lemma 4.3.11, we gain

$$\begin{aligned} \|\sqrt{\tilde{\varrho}_m}\tilde{\mathbf{u}}_m\|^2_{L^2_t L^2_x} &= \int_0^T \int_{\mathbb{T}^3} \Pi_m(\tilde{\varrho}_m \tilde{\mathbf{u}}_m) \cdot \tilde{\mathbf{u}}_m \, \mathrm{d}x \, \mathrm{d}t \\ &\to \int_0^T \int_{\mathbb{T}^3} \tilde{\varrho}\tilde{\mathbf{u}} \cdot \tilde{\mathbf{u}} \, \mathrm{d}x \, \mathrm{d}t = \|\sqrt{\tilde{\varrho}}\tilde{\mathbf{u}}\|^2_{L^2_t L^2_x} \quad \tilde{\mathbb{P}}\text{-a.s.} \end{aligned} \tag{4.110}$$

Here we used the compact embedding

$$L^{\frac{2\Gamma}{\Gamma+1}}(\mathbb{T}^3) \overset{C}{\hookrightarrow} W^{-1,2}(\mathbb{T}^3),$$

which implies, together with Proposition 4.3.10 and Lemma 4.3.11,

$$\Pi_m(\tilde{\varrho}_m \tilde{\mathbf{u}}_m) \to \tilde{\varrho}\tilde{\mathbf{u}} \quad \text{in } L^2(0,T;W^{-1,2}(\mathbb{T}^3)) \ \tilde{\mathbb{P}}\text{-a.s.}$$

In accordance with (4.110), we infer that, for almost every ω, the sequence $(\sqrt{\tilde{\varrho}_m}\tilde{\mathbf{u}}_m(\omega))$ is bounded in $L^2(0,T;L^2(\mathbb{T}^3))$. Hence the combination of weak and strong convergence from Proposition 4.3.10 implies

$$\sqrt{\tilde{\varrho}_m}\tilde{\mathbf{u}}_m \rightharpoonup \sqrt{\tilde{\varrho}}\tilde{\mathbf{u}} \quad \text{in } L^2(0,T;L^2(\mathbb{T}^3)) \ \tilde{\mathbb{P}}\text{-a.s.} \tag{4.111}$$

So (4.109) follows by combining (4.110) and (4.111). □

Similarly to Lemma 4.1.7, the continuity equation (4.72) is satisfied by $[\tilde{\varrho}, \tilde{\mathbf{u}}]$ on the new probability space. Note that, by virtue of (4.108), the limit $\tilde{\varrho}$, $\tilde{\mathbf{u}}$ is regular enough for the equation to be satisfied a.e. in $(0,T) \times \mathbb{T}^3$.

Lemma 4.3.13. *The random distribution $[\tilde{\varrho}, \tilde{\mathbf{u}}]$ satisfies (4.74) a.e. in $(0,T) \times \mathbb{T}^3$, $\tilde{\mathbb{P}}$-a.s.*

Next, we perform the limit $m \to \infty$ in the momentum equation (4.57).

Proposition 4.3.14. *The random distribution* $[\tilde{\varrho}, \tilde{\mathbf{u}}, \tilde{W}]$ *satisfies* (4.75) *for all* $\phi \in C_c^\infty([0,T))$ *and* $\boldsymbol{\varphi} \in C^\infty(\mathbb{T}^3)$ $\tilde{\mathbb{P}}$*-a.s.*

Proof. Similarly to Proposition 4.1.8, we apply Theorem 2.9.1 to show that the approximate momentum equation (4.57) is satisfied on the new probability space by $[\tilde{\varrho}_m, \tilde{\mathbf{u}}_m, \tilde{W}_m]$. In order to let $m \to \infty$ in (4.57), we use Proposition 4.3.10. The limit passage in the deterministic terms follows immediately. Let us now discuss in detail the convergence of the terms coming from the stochastic integral. Here, the bulk of the proof is to show

$$\mathbf{F}_{k,\varepsilon}(\tilde{\varrho}_m, \tilde{\mathbf{u}}_m) \to \mathbf{F}_{k,\varepsilon}(\tilde{\varrho}, \tilde{\mathbf{u}}) \quad \text{in } L^2(\tilde{\Omega} \times (0,T) \times \mathbb{T}^3) \tag{4.112}$$

as $m \to \infty$ for any $k \in \mathbb{N}$. This implies

$$\Pi_m[\mathbf{F}_{k,\varepsilon}(\tilde{\varrho}_m, \tilde{\mathbf{u}}_m)] \to \mathbf{F}_{k,\varepsilon}(\tilde{\varrho}, \tilde{\mathbf{u}}) \quad \text{in } L^2(\tilde{\Omega} \times (0,T) \times \mathbb{T}^3) \tag{4.113}$$

by properties of the projection Π_m; cf. (4.6). Combining (4.113) with (4.108), we finally gain

$$\tilde{\varrho}_m \Pi_m[\mathbf{F}_{k,\varepsilon}(\tilde{\varrho}_m, \tilde{\mathbf{u}}_m)] \to \tilde{\varrho}\mathbf{F}_{k,\varepsilon}(\tilde{\varrho}, \tilde{\mathbf{u}}) \quad \text{in } L^1(\tilde{\Omega} \times (0,T) \times \mathbb{T}^3). \tag{4.114}$$

The proof of (4.112) requires strong convergence of $\tilde{\varrho}_m$ and $\tilde{\mathbf{u}}_m$. The strong convergence of $\tilde{\varrho}_m$ follows directly from (4.108). However, as we will see below, strong convergence of $\tilde{\mathbf{u}}_m$ can only be shown outside the vacuum set $\{\tilde{\varrho} = 0\}$. Fortunately, by the definition of $\mathbb{F}_\varepsilon$ in (4.10), the vacuum set is not seen for fixed ε. Note that, as a consequence of (4.110) and (4.111), we have (up to a subsequence)

$$\sqrt{\tilde{\varrho}_m}\tilde{\mathbf{u}}_m \to \sqrt{\tilde{\varrho}}\tilde{\mathbf{u}} \quad \text{a.e. in } (0,T) \times \mathbb{T}^3 \ \tilde{\mathbb{P}}\text{-a.s.}$$

Let us fix some arbitrary $0 < \kappa < \frac{\varepsilon}{2}$. By Egorov's theorem and (4.108) there exists a measurable set $\mathcal{O}_\kappa \subset \tilde{\Omega} \times (0,T) \times \mathbb{T}^3$ such that $\tilde{\mathbb{P}} \otimes \mathcal{L}^4([\tilde{\Omega} \times (0,T) \times \mathbb{T}^3] \setminus \mathcal{O}_\kappa) < \kappa$ and

$$\sqrt{\tilde{\varrho}_m}\tilde{\mathbf{u}}_m \to \sqrt{\tilde{\varrho}}\tilde{\mathbf{u}}, \quad \tilde{\varrho}_m \to \tilde{\varrho} \text{ uniformly in } \mathcal{O}_\kappa. \tag{4.115}$$

Finally, we consider the sets

$$\mathcal{O}_\kappa^1 = \{(\omega, t, x) \in \mathcal{O}_\kappa : \tilde{\varrho} < \kappa\},$$
$$\mathcal{O}_\kappa^2 = \{(\omega, t, x) \in \mathcal{O}_\kappa : \tilde{\varrho} \geq \kappa\}.$$

Using (4.115), we choose m large enough such that

$$\tilde{\varrho}_m \leq 2\kappa \quad \text{in } \mathcal{O}_\kappa^1, \quad \tilde{\varrho}_m \geq \frac{\kappa}{2} \quad \text{in } \mathcal{O}_\kappa^2.$$

We remark that $\mathbf{F}_{k,\varepsilon}(\tilde{\varrho}_m, \tilde{\mathbf{u}}_m) = \mathbf{F}_{k,\varepsilon}(\tilde{\varrho}, \tilde{\mathbf{u}}) = 0$ in $\mathcal{O}^1_\kappa$. With these preparations at hand, we gain

$$\begin{aligned}
&\tilde{\mathbb{E}} \int_0^T \int_{\mathbb{T}^3} |\mathbf{F}_{k,\varepsilon}(\tilde{\varrho}_m, \tilde{\mathbf{u}}_m) - \mathbf{F}_{k,\varepsilon}(\tilde{\varrho}, \tilde{\mathbf{u}})|^2 \,\mathrm{d}x \,\mathrm{d}t \\
&= \int_{\mathcal{O}^c_\kappa} |\mathbf{F}_{k,\varepsilon}(\tilde{\varrho}_m, \tilde{\mathbf{u}}_m) - \mathbf{F}_{k,\varepsilon}(\tilde{\varrho}, \tilde{\mathbf{u}})|^2 \,\mathrm{d}x \,\mathrm{d}t \,\mathrm{d}\tilde{\mathbb{P}} \\
&\quad + \int_{\mathcal{O}^2_\kappa} \left| \mathbf{F}_{k,\varepsilon}\left(\tilde{\varrho}_m, \frac{\sqrt{\tilde{\varrho}_m}\tilde{\mathbf{u}}_m}{\sqrt{\tilde{\varrho}_m}}\right) - \mathbf{F}_{k,\varepsilon}\left(\tilde{\varrho}, \frac{\sqrt{\tilde{\varrho}}\tilde{\mathbf{u}}}{\sqrt{\tilde{\varrho}}}\right)\right|^2 \,\mathrm{d}x \,\mathrm{d}t \,\mathrm{d}\tilde{\mathbb{P}} \\
&= (I)^1_\kappa + (I)^2_\kappa .
\end{aligned}$$

Due to the boundedness of $\mathbf{F}_{k,\varepsilon}$ from (4.10), the first integral is bounded by κ. On the other hand, by (4.115) and continuity of $\mathbf{F}_{k,\varepsilon}$ (and the lower bounds for $\tilde{\varrho}$ and $\tilde{\varrho}_m$ in $\mathcal{O}^2_\kappa$) the last integral vanishes as $m \to \infty$. Since κ was arbitrary, we obtain (4.112) and (4.114) follows.

In order to apply Lemma 2.6.6 we have to deal with the infinite sum as well as the additional square. Due to (4.6), (4.10), and (4.83), we obtain, for all $q \in (1, \Gamma]$,

$$\tilde{\mathbb{E}} \int_0^t \left(\int_{\mathbb{T}^3} \tilde{\varrho}_m \Pi_m[\mathbf{F}_{k,\varepsilon}(\tilde{\varrho}_m, \tilde{\mathbf{u}}_m)] \cdot \boldsymbol{\varphi} \,\mathrm{d}x \right)^2 \mathrm{d}s \lesssim f^2_{k,\varepsilon} \tilde{\mathbb{E}} \int_0^T \left(\int_{\mathbb{T}^3} \tilde{\varrho}^q_m \,\mathrm{d}x \right)^2 \mathrm{d}s \lesssim f^2_{k,\varepsilon}.$$

Hence, for any $\kappa > 0$,

$$\tilde{\mathbb{E}} \sum_{k=N+1}^{\infty} \int_0^t \left(\int_{\mathbb{T}^3} \varrho \Pi_m[\mathbf{F}_{k,\varepsilon}(\tilde{\varrho}_m, \tilde{\mathbf{u}}_m)] \cdot \boldsymbol{\varphi} \,\mathrm{d}x \right)^2 \mathrm{d}s \leq \kappa,$$

provided $N \geq N_0(\kappa)$. In addition, using (4.6), (4.10), and (4.83) again, we strengthen (4.114) to

$$\tilde{\varrho}_m \Pi_m[\mathbf{F}_{k,\varepsilon}(\tilde{\varrho}_m, \tilde{\mathbf{u}}_m)] \to \tilde{\varrho} \mathbf{F}_{k,\varepsilon}(\tilde{\varrho}, \tilde{\mathbf{u}}) \quad \text{in } L^r(\tilde{\Omega}; L^p(0, T; L^q(\mathbb{T}^3))), \tag{4.116}$$

for some $r > 2$, all $p \in [1, \infty)$, and all $q \in [1, \Gamma)$. Consequently, we have

$$\tilde{\varrho}_m \Pi_m[\mathbb{F}_\varepsilon(\tilde{\varrho}_m, \tilde{\mathbf{u}}_m)] \to \tilde{\varrho} \mathbb{F}_\varepsilon(\tilde{\varrho}, \tilde{\mathbf{u}}) \quad \text{in } L^2(0, T; L_2(\mathfrak{U}; L^2(\mathbb{T}^3))) \tag{4.117}$$

$\tilde{\mathbb{P}}$-a.s. Combining this with the convergence of $\tilde{W}_m$ from Proposition 4.3.10, we may apply Lemma 2.6.6 to pass to the limit in the stochastic integral and hence complete the proof. □

As the next step, we pass to the limit in the stochastic integral appearing in the energy inequality.

Proposition 4.3.15. *We have*

$$\int_0^T \int_{\mathbb{T}^3} \tilde{\varrho}_m \Pi_m[\mathbb{F}_\varepsilon(\tilde{\varrho}_m, \tilde{\mathbf{u}}_m)] \cdot \tilde{\mathbf{u}}_m \,\mathrm{d}x \,\mathrm{d}\tilde{W}_m \to \int_0^T \int_{\mathbb{T}^3} \tilde{\varrho} \mathbb{F}_\varepsilon(\tilde{\varrho}, \tilde{\mathbf{u}}) \cdot \tilde{\mathbf{u}} \,\mathrm{d}x \,\mathrm{d}\tilde{W} \quad \textit{in } L^2(0, T)$$

in probability.

Proof. First, we observe that (4.113) can be strengthened to

$$\Pi_m[\mathbf{F}_{k,\varepsilon}(\tilde{\varrho}_m,\tilde{\mathbf{u}}_m)] \to \mathbf{F}_{k,\varepsilon}(\tilde{\varrho},\tilde{\mathbf{u}}) \quad \text{in } L^q((0,T)\times\mathbb{T}^3) \tag{4.118}$$

$\tilde{\mathbb{P}}$-a.s., for all $q \in [1,\infty)$ and all $k \in \mathbb{N}$. This can be combined with the convergence of $\Pi_m(\tilde{\varrho}_m\tilde{\mathbf{u}}_m)$ from Proposition 4.3.10 such that

$$\int_{\mathbb{T}^3} \tilde{\varrho}_m \Pi_m[\mathbf{F}_{k,\varepsilon}(\tilde{\varrho}_m,\tilde{\mathbf{u}}_m)]\cdot\tilde{\mathbf{u}}_m \,\mathrm{d}x \to \int_{\mathbb{T}^3} \tilde{\varrho}\mathbf{F}_{k,\varepsilon}(\tilde{\varrho},\tilde{\mathbf{u}})\cdot\tilde{\mathbf{u}}\,\mathrm{d}x \quad \text{in } L^2(0,T) \tag{4.119}$$

$\tilde{\mathbb{P}}$-a.s., for all $k \in \mathbb{N}$. On the other hand we obtain, from (4.6), (4.10), and (4.84),

$$\begin{aligned}
&\tilde{\mathbb{E}}\int_0^T \left\| \int_{\mathbb{T}^3} \tilde{\varrho}_m\Pi_m[\mathbb{F}_{k,\varepsilon}(\tilde{\varrho}_m,\tilde{\mathbf{u}}_m)]\cdot\tilde{\mathbf{u}}_m\,\mathrm{d}x\right\|^2_{L_2(\mathfrak{U},\mathbb{R})} \mathrm{d}t \\
&= \tilde{\mathbb{E}}\int_0^T \sum_{k=1}^\infty \left(\int_{\mathbb{T}^3} \tilde{\varrho}_m\Pi_m[\mathbf{F}_{k,\varepsilon}(\tilde{\varrho}_m,\tilde{\mathbf{u}}_m)]\cdot\tilde{\mathbf{u}}_m\,\mathrm{d}x\right)^2 \mathrm{d}s \\
&\lesssim \sum_{k=1}^\infty f_{k,\varepsilon}^2 \tilde{\mathbb{E}}\int_0^T \left(\int_{\mathbb{T}^3} |\tilde{\varrho}_m\tilde{\mathbf{u}}_m|^q\,\mathrm{d}x\right)^2 \lesssim \sum_{k=1}^\infty f_{k,\varepsilon}^2,
\end{aligned}$$

for all $q \in (1, \frac{2\Gamma}{\Gamma+1}]$. Thus, (4.119) implies

$$\int_{\mathbb{T}^3} \tilde{\varrho}_m \Pi_m[\mathbb{F}_{\varepsilon}(\tilde{\varrho}_m,\tilde{\mathbf{u}}_m)]\cdot\tilde{\mathbf{u}}_m\,\mathrm{d}x \to \int_{\mathbb{T}^3} \tilde{\varrho}\mathbb{F}_{\varepsilon}(\tilde{\varrho},\tilde{\mathbf{u}})\cdot\tilde{\mathbf{u}}\,\mathrm{d}x \quad \text{in } L^2(0,T;L_2(\mathfrak{U},\mathbb{R})) \tag{4.120}$$

$\tilde{\mathbb{P}}$-a.s. Combining this with the convergence of $\tilde{W}_m$ from Proposition 4.3.10, we apply Lemma 2.6.6 and the claim follows. □

Lemma 4.3.16. *The energy inequality* (4.76) *is satisfied by* $[\tilde{\varrho},\tilde{\mathbf{u}},\tilde{W}]$ *for all test functions* $\phi \in C_c^\infty([0,T))$, $\phi \geq 0$, $\tilde{\mathbb{P}}$*-a.s.*

Proof. Our goal is to perform the asymptotic limit in the total energy balance (4.60). Note that (4.60) is also satisfied by $[\tilde{\varrho}_m,\tilde{\mathbf{u}}_m,\tilde{W}_m]$ on the new probability space with the initial condition

$$\int_{\mathbb{T}^3}\left[\frac{1}{2}\tilde{\varrho}_{0,m}|\tilde{\mathbf{u}}_{0,m}|^2 + P_\delta(\tilde{\varrho}_{0,m})\right]\mathrm{d}x.$$

This follows from the equality of laws from Proposition 4.3.10 and Theorem 2.9.1. Using Proposition 4.3.10, we pass to the limit in the initial condition to obtain

$$\int_{\mathbb{T}^3}\left[\frac{1}{2}\tilde{\varrho}_0|\tilde{\mathbf{u}}_0|^2 + P_\delta(\tilde{\varrho}_0)\right]\mathrm{d}x.$$

Using Proposition 4.3.15, we pass to the limit in the stochastic integral. Using (4.6), (4.10), (4.83), and (4.118), as well as the convergence of $\tilde{\varrho}_m$ from (4.108), we may pass to the limit in the Itô correction term. Note that the integral

$$\int_0^T\int_{\mathbb{T}^3}[\mathbb{S}(\nabla\tilde{\mathbf{u}}):\nabla\tilde{\mathbf{u}} + \varepsilon\tilde{\varrho}|\nabla\tilde{\mathbf{u}}|^2 + \varepsilon P_\delta''(\tilde{\varrho})|\nabla\tilde{\varrho}|^2]\,\mathrm{d}x\,\mathrm{d}t$$

is only lower semi-continuous with respect to the topologies in (4.108). As a result, only energy inequality (4.76) is recovered in the limit $m \to \infty$. □

The proof of Theorem 4.3.2 is hereby complete.

4.4 Vanishing viscosity limit

Our next goal is to let $\varepsilon \to 0$ in the approximate system (4.72)–(4.73). Accordingly, our target problem is given formally by

$$\mathrm{d}\varrho + \operatorname{div}(\varrho \mathbf{u})\,\mathrm{d}t = 0, \tag{4.121}$$

$$\mathrm{d}(\varrho \mathbf{u}) + \operatorname{div}(\varrho \mathbf{u} \otimes \mathbf{u})\,\mathrm{d}t + \nabla p_\delta(\varrho)\,\mathrm{d}t = \operatorname{div}\mathbb{S}(\nabla \mathbf{u})\,\mathrm{d}t + \mathbb{G}(\varrho, \varrho \mathbf{u})\,\mathrm{d}W. \tag{4.122}$$

A rigorous formulation reads as follows.

Definition 4.4.1. Let Λ be a Borel probability measure on $L^1(\mathbb{T}^3) \times L^1(\mathbb{T}^3)$. Then $((\Omega, \mathfrak{F}, (\mathfrak{F}_t)_{t\geq 0}, \mathbb{P}), \varrho, \mathbf{u}, W)$ is called a *dissipative martingale solution* to (4.121)–(4.122) with the initial law Λ, provided:

(1) $(\Omega, \mathfrak{F}, (\mathfrak{F}_t)_{t\geq 0}, \mathbb{P})$ is a stochastic basis with a complete right-continuous filtration;
(2) W is a cylindrical $(\mathfrak{F}_t)$-Wiener process;
(3) the density ϱ and the velocity $\mathbf{u}$ are random distributions adapted to $(\mathfrak{F}_t)_{t\geq 0}$, $\varrho \geq 0$ $\mathbb{P}$-a.s.;
(4) there exists an $\mathfrak{F}_0$-measurable random variable $[\varrho_0, \mathbf{u}_0]$ such that $\Lambda = \mathcal{L}[\varrho_0, \varrho_0 \mathbf{u}_0]$;
(5) the equation of continuity

$$-\int_0^T \partial_t \phi \int_{\mathbb{T}^3} \varrho \psi \,\mathrm{d}x\,\mathrm{d}t = \phi(0) \int_{\mathbb{T}^3} \varrho_0 \psi \,\mathrm{d}x + \int_0^T \phi \int_{\mathbb{T}^3} \varrho \mathbf{u} \cdot \nabla \psi \,\mathrm{d}x\,\mathrm{d}t \tag{4.123}$$

holds for all $\phi \in C_c^\infty([0,T))$ and all $\psi \in C^\infty(\mathbb{T}^3)$ $\mathbb{P}$-a.s.;
(6) the approximate momentum equation

$$\begin{aligned} -\int_0^T \partial_t \phi \int_{\mathbb{T}^3} \varrho \mathbf{u} \cdot \boldsymbol{\varphi} \,\mathrm{d}x\,\mathrm{d}t - \phi(0) \int_{\mathbb{T}^3} \varrho_0 \mathbf{u}_0 \cdot \boldsymbol{\varphi} \,\mathrm{d}x \\ = \int_0^T \phi \int_{\mathbb{T}^3} [\varrho \mathbf{u} \otimes \mathbf{u} : \nabla \boldsymbol{\varphi} + p_\delta(\varrho) \operatorname{div} \boldsymbol{\varphi}] \,\mathrm{d}x\,\mathrm{d}t \\ - \int_0^T \phi \int_{\mathbb{T}^3} \mathbb{S}(\nabla \mathbf{u}) : \nabla \boldsymbol{\varphi} \,\mathrm{d}x\,\mathrm{d}t + \int_0^T \phi \int_{\mathbb{T}^3} \mathbb{G}(\varrho, \varrho \mathbf{u}) \cdot \boldsymbol{\varphi} \,\mathrm{d}x\,\mathrm{d}W \end{aligned} \tag{4.124}$$

holds for all $\boldsymbol{\varphi} \in C^\infty(\mathbb{T}^3)$ and all $\phi \in C_c^\infty([0,T))$ $\mathbb{P}$-a.s.;
(7) the energy inequality

$$-\int_0^T \partial_t \phi \int_{\mathbb{T}^3} \left[\frac{1}{2} \varrho |\mathbf{u}|^2 + P_\delta(\varrho)\right] \mathrm{d}x\,\mathrm{d}t - \phi(0) \int_{\mathbb{T}^3} \left[\frac{1}{2} \varrho_0 |\mathbf{u}_0|^2 + P_\delta(\varrho_0)\right] \mathrm{d}x$$

$$+\int_0^T \phi \int_{\mathbb{T}^3} \mathbb{S}(\nabla \mathbf{u}) : \nabla \mathbf{u} \, dx \, dt$$
$$\leq \frac{1}{2} \sum_{k=1}^{\infty} \int_0^T \phi \int_{\mathbb{T}^3} \varrho^{-1} |\mathbf{G}_k(\varrho, \varrho \mathbf{u})|^2 \, dx \, dt + \int_0^T \phi \int_{\mathbb{T}^3} \mathbb{G}(\varrho, \varrho \mathbf{u}) \cdot \mathbf{u} \, dx \, dW \tag{4.125}$$

holds for all $\phi \in C_c^\infty([0,T))$, $\phi \geq 0$, $\mathbb{P}$-a.s.

The main result of this section is the following.

Theorem 4.4.2. *Let $\Gamma \geq 6$. Let Λ be a Borel probability measure defined on the space $L^1(\mathbb{T}^3) \times L^1(\mathbb{T}^3)$ such that*

$$\Lambda\{\varrho > 0\} = 1, \quad \Lambda\left\{0 < \underline{\varrho} \leq \int_{\mathbb{T}^3} \varrho \, dx \leq \overline{\varrho} < \infty\right\} = 1, \tag{4.126}$$

where $\underline{\varrho}$, $\overline{\varrho}$ are two deterministic constants. Assume that

$$\int_{L_x^1 \times L_x^1} \left| \int_{\mathbb{T}^3} \left[\frac{1}{2} \frac{|\mathbf{q}|^2}{\varrho} + P_\delta(\varrho) \right] dx \right|^r d\Lambda \lesssim 1, \tag{4.127}$$

for some $r \geq 4$. Then problem (4.121)–(4.122) *admits a dissipative martingale solution in the sense of Definition* 4.4.1.

Remark 4.4.3. As in Remark 4.3.3, we have

$$\begin{aligned} &\mathbb{E}\left[\sup_{t \in [0,T]} \left| \int_{\mathbb{T}^3} \left[\frac{1}{2} \varrho |\mathbf{u}|^2 + P_\delta(\varrho) \right] dx \right|^r \right] \lesssim 1, \\ &\mathbb{E}\left[\left| \int_0^T \int_{\mathbb{T}^3} [\mathbb{S}(\nabla \mathbf{u}) : \nabla \mathbf{u}] \, dx \, dt \right|^r \right] \lesssim 1, \end{aligned} \tag{4.128}$$

such that again

$$\mathbb{E}\left[\sup_{t \in [0,T]} \|\varrho(t)\|_{L_x^\Gamma}^{\Gamma p} \right] < \infty \quad \text{for some } p \in (1,\infty), \tag{4.129}$$

$$\mathbb{E}\left(\int_0^T \|\mathbf{u}\|_{W_x^{1,2}}^2 \, dt \right)^p < \infty \quad \text{for some } p \in (1,\infty), \tag{4.130}$$

$$\mathbb{E}\left[\sup_{t \in [0,T]} \|\varrho \mathbf{u}(t)\|_{L_x^{\frac{2\Gamma}{\Gamma+1}}}^{\frac{2\Gamma}{\Gamma+1} p} \right] < \infty \quad \text{for some } p \in (1,\infty). \tag{4.131}$$

The proof of Theorem 4.4.2 requires the full strength of the method developed in the context of the deterministic Navier–Stokes system. Possible oscillations of the density are ruled out thanks to the weak compactness of a quantity called *effective viscous flux*,

$$(\eta + \mu) \operatorname{div} \mathbf{u} - p(\varrho),$$

where $\eta = \frac{\mu}{3} + \lambda$ and $\mu > 0$ and $\lambda \geq 0$ are the viscosity coefficients, namely, we recall that

$$\mathbb{S}(\nabla \mathbf{u}) = \mu\left(\nabla \mathbf{u} + \nabla^t \mathbf{u} - \frac{2}{3}\operatorname{div}\mathbf{u}\mathbb{I}\right) + \lambda \operatorname{div}\mathbf{u}\mathbb{I}.$$

Remark 4.4.4. It follows from the DiPerna–Lions theory (cf. Theorem A.3.1) that the martingale solution constructed in Theorem 4.4.2 satisfies $\mathbb{P}$-a.s. the renormalized equation of continuity

$$\begin{aligned} -\int_0^T \partial_t \phi \int_{\mathbb{T}^3} b(\varrho)\psi \,\mathrm{d}x\,\mathrm{d}t = \phi(0) \int_{\mathbb{T}^3} b(\varrho_0)\psi \,\mathrm{d}x + \int_0^T \phi \int_{\mathbb{T}^3} b(\varrho)\mathbf{u}\cdot\nabla\psi \,\mathrm{d}x\,\mathrm{d}t \\ - \int_0^T \phi \int_{\mathbb{T}^3} (b'(\varrho)\varrho - b(\varrho))\operatorname{div}\mathbf{u}\,\psi \,\mathrm{d}x\,\mathrm{d}t, \end{aligned} \tag{4.132}$$

for all $\phi \in C_c^\infty([0,T))$, all $\psi \in C^\infty(\mathbb{T}^3)$, and any $b \in C^1([0,\infty))$ with $b'(\varrho) = 0$ for $\varrho \gg 1$. Moreover, it can be shown that

$$\varrho \in C([0,T]; L^1(\mathbb{T}^3)) \quad \mathbb{P}\text{-a.s.}; \tag{4.133}$$

see Lemma A.3.3.

We start with uniform bounds, independent of the parameter ε, that can be derived directly from the energy inequality (4.74); see Section 4.4.1. In addition, the absence of the regularization effect of the diffusion term in the asymptotic limit of the continuity equation must be compensated by a new type of estimates of the pressure term, obtained in Section 4.4.2. Finally, the asymptotic limit for $\varepsilon \to 0$ is performed in Section 4.4.3 by means of an adaptation of deterministic techniques to the stochastic setting.

4.4.1 Uniform energy bounds

In this section, the parameter $\delta \in (0,1)$ is kept fixed and we derive bounds that are uniform in ε and may depend on δ as well as $T > 0$. Denote by $[\varrho, \mathbf{u}]$ the solution of problem (4.72)–(4.73) with the initial law Λ, the existence of which is guaranteed by Theorem 4.3.2. The integral form of the energy inequality (4.76) reads

$$\begin{aligned} &\int_{\mathbb{T}^3} \left[\frac{1}{2}\varrho|\mathbf{u}|^2 + P_\delta(\varrho)\right](\tau)\,\mathrm{d}x \\ &\quad + \int_0^\tau \int_{\mathbb{T}^3} [\mathbb{S}(\nabla\mathbf{u}) : \nabla\mathbf{u} + \varepsilon\varrho|\nabla\mathbf{u}|^2 + \varepsilon P_\delta''(\varrho)|\nabla\varrho|^2]\,\mathrm{d}x\,\mathrm{d}t \\ &\leq \int_{\mathbb{T}^3} \left[\frac{1}{2}\varrho_0|\mathbf{u}_0|^2 + P_\delta(\varrho_0)\right]\mathrm{d}x + \frac{1}{2}\sum_{k=1}^\infty \int_0^\tau \int_{\mathbb{T}^3} \varrho|\mathbf{F}_{k,\varepsilon}(\varrho,\mathbf{u})|^2\,\mathrm{d}x\,\mathrm{d}t \\ &\quad + \int_0^\tau \int_{\mathbb{T}^3} \varrho\mathbb{F}_\varepsilon(\varrho,\mathbf{u})\cdot\mathbf{u}\,\mathrm{d}x\,\mathrm{d}W \quad \mathbb{P}\text{-a.s.} \end{aligned} \tag{4.134}$$

This can easily be seen by approximating the characteristic function of the time interval $[0,\tau]$ by a family of smooth test functions $\phi \in C^\infty([0,T])$. Such a procedure yields (4.134) for a.e. τ. However, the weak lower semi-continuity of the energy yields (4.134) for any $\tau \in [0,T]$ $\mathbb{P}$-a.s. Note that the instantaneous values $\varrho(\tau)$, $\varrho\mathbf{u}(\tau)$ are well-defined as these processes are at least weakly continuous in time.

Now the energy estimates can be obtained in the same way as in Section 4.2.1. Using (4.7), the integrals on the right hand side of (4.134) can be controlled by the energy as

$$\int_{\mathbb{T}^3} \varrho |\mathbf{F}_{k,\varepsilon}(\varrho,\mathbf{u})|^2 \,\mathrm{d}x \lesssim f_k^2 \int_{\mathbb{T}^3} (\varrho + \varrho|\mathbf{u}|^2)\,\mathrm{d}x$$

and, by the Burkholder–Davis–Gundy inequality,

$$\begin{aligned}
&\mathbb{E}\left[\sup_{0\le t\le \tau}\left|\int_0^t\int_{\mathbb{T}^3} \varrho \mathbb{F}_\varepsilon(\varrho,\mathbf{u})\cdot\mathbf{u}\,\mathrm{d}x\,\mathrm{d}W\right|^r\right] \\
&\quad\lesssim \mathbb{E}\left[\int_0^\tau \sum_{k=1}^\infty \left|\int_{\mathbb{T}^3} \varrho \mathbf{F}_{k,\varepsilon}(\varrho,\mathbf{u})\cdot\mathbf{u}\,\mathrm{d}x\right|^2\right]^{r/2} \lesssim \mathbb{E}\left[\int_0^\tau \left|\int_{\mathbb{T}^3} (\varrho+\varrho|\mathbf{u}|^2)\,\mathrm{d}x\right|^2\right]^{r/2}.
\end{aligned}$$

Note that, unlike their counterparts in Section 4.2.1, the present estimates are *independent* of ε.

Exactly as in Section 4.2.2, we obtain the uniform bounds:

$$\|\varrho(\tau)\|_{L^1_x} = \|\varrho_0\|_{L^1_x} \le \overline{\varrho}, \quad \tau \in [0,T],$$

$$\mathbb{E}\left[\left|\sup_{t\in[0,T]} \|\varrho\|_{L^\Gamma_x}^\Gamma\right|^r\right] + \mathbb{E}\left[\|\sqrt{\varepsilon}\nabla\varrho\|_{L^2_tL^2_x}^{2r}\right] \lesssim c_1(r), \tag{4.135}$$

$$\mathbb{E}\left[\left|\sup_{t\in[0,T]} \|\varrho|\mathbf{u}|^2\|_{L^1(\mathbb{T}^3)}\right|^r + \left|\sup_{t\in[0,T]} \|\varrho\mathbf{u}\|_{L_x^{\frac{2\Gamma}{\Gamma+1}}}^{\frac{2\Gamma}{\Gamma+1}}\right|^r\right] \lesssim c_1(r), \tag{4.136}$$

$$\mathbb{E}\left[\|\nabla\mathbf{u}\|_{L^2_tL^2_x}^{2r}\right] \lesssim c_1(r), \tag{4.137}$$

where

$$c_1(r) \approx 1 + \mathbb{E}\left[\left|\int_{\mathbb{T}^3}\left[\frac{1}{2}\frac{|\varrho\mathbf{u}(0)|^2}{\varrho(0)} + P_\delta(\varrho(0))\right]\mathrm{d}x\right|^r\right], \quad r\ge 2. \tag{4.138}$$

It remains to deduce a bound for $\|\mathbf{u}\|_{L^2}$. We argue similarly to (4.89). By the standard Poincaré inequality,

$$\|\mathbf{u} - (\mathbf{u})_{\mathbb{T}^3}\|_{L^2_x} \lesssim \|\nabla\mathbf{u}\|_{L^2_x}.$$

On the other hand, by virtue of (4.135), we have, similarly to (4.88),

$$\begin{aligned}
\|\varrho(0)\|_{L^1_x}^2 \int_0^\tau |(\mathbf{u})_{\mathbb{T}^3}|^2\,\mathrm{d}t \lesssim{}& \sup_{t\in[0,\tau]} \|\varrho\|_{L^\gamma_x}^2 \int_0^\tau \|\nabla\mathbf{u}\|_{L^2_x}^2\,\mathrm{d}t \\
&+ \tau \sup_{t\in[0,\tau]} \left(\|\varrho\|_{L^1_x}^2 + \|\varrho|\mathbf{u}|^2\|_{L^1_x}^2\right).
\end{aligned}$$

In view of the bounds established in (4.135)–(4.138), we obtain the desired estimate

$$\mathbb{E}\big[\|\mathbf{u}\|^{2r}_{L^2_t W^{1,2}_x}\big] \lesssim c_2(r), \tag{4.139}$$

with

$$c_2(r) \approx 1 + \mathbb{E}\bigg[\bigg|\int_{\mathbb{T}^3}\Big[\frac{1}{2}\frac{|\varrho\mathbf{u}(0)|^2}{\varrho(0)} + P_\delta(\varrho(0))\Big]\,\mathrm{d}x\bigg|^{2r}\bigg], \quad r \geq 2, \tag{4.140}$$

taking into account (4.126). The above estimates are not strong enough to control the pressure term proportional to ϱ^Γ that is now bounded only in the non-reflexive space L^1. The adequate bounds will be derived in the next section.

4.4.2 Pressure estimates

Throughout this section, we again denote by $[\varrho, \mathbf{u}]$ the solution of problem (4.72)–(4.73) with the initial law Λ, the existence of which is guaranteed by Theorem 4.3.2. In view of Section 4.4.1, we derive refined estimates for the pressure. We stress that all the estimates in this section hold true uniformly in ε.

The idea is to use the quantity

$$\nabla\Delta^{-1}[\varrho - (\varrho)_{\mathbb{T}^3}] = \Delta^{-1}\nabla\varrho,$$

with Δ^{-1} defined on $\mathbb{T}^3$ for functions of zero mean, as a test function in the momentum balance (4.75). Note that this is not straightforward as the legal test functions allowed in (4.75) have the form $\phi(t)\boldsymbol{\varphi}(x)$, where both ϕ and $\boldsymbol{\varphi}$ are smooth and deterministic. Nevertheless, such a procedure can be rigorously justified by the application of a suitable version of the generalized Itô formula, Theorem A.4.1, to the functional

$$(\varrho, \mathbf{q}) \mapsto \int_{\mathbb{T}^3} \mathbf{q}\cdot\Delta^{-1}\nabla\varrho\,\mathrm{d}x.$$

We rewrite (4.75) in the following differential form:

$$\begin{aligned}
&\mathrm{d}\int_{\mathbb{T}^3}\varrho\mathbf{u}\cdot\boldsymbol{\varphi}\,\mathrm{d}x - \int_{\mathbb{T}^3}[\varrho\mathbf{u}\otimes\mathbf{u} : \nabla\boldsymbol{\varphi} + p_\delta(\varrho)\mathrm{div}\,\boldsymbol{\varphi}]\,\mathrm{d}x\,\mathrm{d}t \\
&= -\int_{\mathbb{T}^3}[\mathbb{S}(\nabla\mathbf{u}) : \nabla\boldsymbol{\varphi} - \varepsilon\varrho\mathbf{u}\cdot\Delta\boldsymbol{\varphi}]\,\mathrm{d}x\,\mathrm{d}t + \int_{\mathbb{T}^3}\varrho\mathbb{F}_\varepsilon(\varrho,\mathbf{u})\cdot\boldsymbol{\varphi}\,\mathrm{d}x\,\mathrm{d}W.
\end{aligned}$$

Seeing that, in accordance with (4.74),

$$\mathrm{d}(\nabla\Delta^{-1}[\varrho - (\varrho)_{\mathbb{T}^3}]) = -\nabla\Delta^{-1}\mathrm{div}\,(\varrho\mathbf{u})\,\mathrm{d}t + \varepsilon\nabla\varrho\,\mathrm{d}t,$$

Theorem A.4.1 implies

$$\begin{aligned}
&\int_{\mathbb{T}^3} p_\delta(\varrho)[\varrho - (\varrho)_{\mathbb{T}^3}]\,\mathrm{d}x\,\mathrm{d}t \\
&= \mathrm{d}\int_{\mathbb{T}^3}\varrho\mathbf{u}\cdot\Delta^{-1}[\nabla\varrho]\,\mathrm{d}x - \int_{\mathbb{T}^3}\varrho\mathbf{u}\otimes\mathbf{u} : \nabla\Delta^{-1}\nabla\varrho\,\mathrm{d}x\,\mathrm{d}t
\end{aligned}$$

$$+\int_{\mathbb{T}^3} \mathbb{S}(\nabla\mathbf{u}) : \nabla\Delta^{-1}\nabla\varrho \,dx\,dt - \int_{\mathbb{T}^3} \varrho\mathbb{F}_\varepsilon(\varrho,\mathbf{u})\cdot\Delta^{-1}\nabla\varrho\,dx\,dW$$
$$+\varepsilon\int_{\mathbb{T}^3} \varrho^2 \operatorname{div}\mathbf{u}\,dx\,dt + \int_{\mathbb{T}^3} \varrho\mathbf{u}\cdot\nabla\Delta^{-1}\operatorname{div}(\varrho\mathbf{u})\,dx\,dt. \tag{4.141}$$

Relation (4.141) integrated in time gives rise to the desired pressure bounds. Indeed, we obtain

$$\int_0^\tau\int_{\mathbb{T}^3} p_\delta(\varrho)[\varrho - (\varrho)_{\mathbb{T}^3}]\,dx\,dt$$
$$= \left[\int_{\mathbb{T}^3} \varrho\mathbf{u}\cdot\Delta^{-1}[\nabla\varrho]\,dx\right]_{t=0}^{t=\tau} - \int_0^\tau\int_{\mathbb{T}^3} \varrho\mathbf{u}\otimes\mathbf{u} : \nabla\Delta^{-1}\nabla\varrho\,dx\,dt$$
$$+\int_0^\tau\int_{\mathbb{T}^3} \mathbb{S}(\nabla\mathbf{u}) : \nabla\Delta^{-1}\nabla\varrho\,dx\,dt + \int_0^\tau\int_{\mathbb{T}^3} \varrho\mathbf{u}\cdot\nabla\Delta^{-1}\operatorname{div}(\varrho\mathbf{u})\,dx\,dt$$
$$+\varepsilon\int_0^\tau\int_{\mathbb{T}^3} \varrho^2\operatorname{div}\mathbf{u}\,dx\,dt - \int_0^\tau\int_{\mathbb{T}^3} \varrho\mathbb{F}_\varepsilon(\varrho,\mathbf{u})\cdot\Delta^{-1}\nabla\varrho\,dx\,dW \equiv \sum_{i=1}^6 I_i. \tag{4.142}$$

We will show in the following that all integrals on the right hand side are controlled by the energy bounds derived in the preceding section. Moreover, as $\Gamma > 3$, we may use the standard elliptic estimates to deduce

$$\|\Delta^{-1}\nabla\varrho\|_{L^\infty_x} \lesssim \|\varrho\|_{L^\Gamma_x}.$$

Performing a bit tedious but straightforward manipulation, we obtain the following estimates:

(1) By Hölder's inequality we have

$$|I_1| \le \sup_{t\in[0,\tau]}\left|\int_{\mathbb{T}^3} \varrho\mathbf{u}\cdot\Delta^{-1}[\nabla\varrho]\,dx\right|$$
$$\lesssim \sup_{t\in[0,\tau]}\|\sqrt{\varrho}\|_{L^2_x}\sup_{t\in[0,\tau]}\|\sqrt{\varrho}\mathbf{u}\|_{L^2_x}\sup_{t\in[0,\tau]}\|\Delta^{-1}[\nabla\varrho]\|_{L^\infty_x}$$
$$\lesssim \Big(\sup_{t\in[0,\tau]}\|\varrho\|_{L^1_x}\Big)^{1/2}\sup_{t\in[0,\tau]}\|\varrho|\mathbf{u}|^2\|_{L^1_x}^{1/2}\sup_{t\in[0,\tau]}\|\varrho\|_{L^\Gamma_x}.$$

Consequently, in view of the energy bounds (4.135) and (4.136) as well as (4.126),

$$\mathbb{E}[|I_1|^r] \lesssim c_1(r), \tag{4.143}$$

with $c_1(r)$ given by (4.138).

(2) By virtue of the standard properties of Δ^{-1},

$$|I_2| \le \sup_{t\in[0,\tau]}\|\nabla\Delta^{-1}[\nabla\varrho]\|_{L^\Gamma_x}\int_0^\tau\|\varrho|\mathbf{u}|^2\|_{L^{\Gamma'}_x}\,dt$$
$$\lesssim \sup_{t\in[0,\tau]}\|\nabla\Delta^{-1}[\nabla\varrho]\|_{L^\Gamma_x}\sup_{t\in[0,\tau]}\|\varrho\|_{L^\Gamma}\int_0^\tau\|\mathbf{u}\|^2_{L^q_x}\,dt$$
$$\lesssim \sup_{t\in[0,\tau]}\|\varrho\|^2_{L^\Gamma}\int_0^\tau\|\mathbf{u}\|^2_{L^q_x}\,dt,$$

where

$$\frac{2}{\Gamma} + \frac{2}{q} = 1.$$

Thus if $\Gamma \geq 4$, we take $q \leq 4$, and thanks to (4.135) and (4.139) combined with the embedding $W^{1,2} \hookrightarrow L^4$ we conclude

$$\mathbb{E}\big[|I_2|^r\big] \lesssim c_2(r), \tag{4.144}$$

where $c_2(r)$ is as in (4.140).

(3) As

$$\|\nabla\Delta^{-1}\nabla\varrho\|_{L^2_x} \lesssim \|\varrho\|_{L^2_x},$$

we get

$$|I_3| \lesssim \int_0^\tau \int_{\mathbb{T}^3} |\nabla \mathbf{u}|^2 \,\mathrm{d}x\,\mathrm{d}t + \tau \sup_{t\in[0,\tau]} \|\varrho\|^2_{L^\Gamma_x}.$$

In view of (4.135) and (4.137), we find

$$\mathbb{E}\big[|I_3|^r\big] \lesssim c_1(r), \tag{4.145}$$

where $c_1(r)$ is given by (4.142).

(4) Similarly to the previous step,

$$\begin{aligned} |I_4| &\lesssim \int_0^\tau \|\varrho\mathbf{u}\|^2_{L^2_x}\,\mathrm{d}t \lesssim \sup_{t\in[0,\tau]} \|\varrho\|^2_{L^4_x} \int_0^\tau \|\mathbf{u}\|^2_{L^4_x}\,\mathrm{d}t \\ &\lesssim \sup_{t\in[0,\tau]} \|\varrho\|^4_{L^4_x} + \left(\int_0^\tau \|\mathbf{u}\|^2_{L^4_x}\,\mathrm{d}t\right)^2. \end{aligned}$$

If $\Gamma \geq 4$, the first term is controlled by the energy (cf. (4.135)) while the second one can be estimated by virtue of (4.139) together with the standard embedding $W^{1,2} \hookrightarrow L^4$. Thus we obtain

$$\mathbb{E}\big[|I_4|^r\big] \lesssim c_2(r). \tag{4.146}$$

(5) The term I_5 can easily be bounded by

$$|I_5| \lesssim \int_0^\tau \int_{\mathbb{T}^3} \varrho^4 \,\mathrm{d}x\,\mathrm{d}t + \int_0^\tau \int_{\mathbb{T}^3} |\nabla\mathbf{u}|^2 \,\mathrm{d}x\,\mathrm{d}t.$$

Consequently, we have

$$\mathbb{E}\big[|I_5|^r\big] \lesssim c_1(r)$$

using (4.135) and (4.137) together with $\Gamma \geq 4$.

(6) Finally, the stochastic integral can be handled as follows using the Burkholder–Davis–Gundy inequality:

$$\begin{aligned}\mathbb{E}[|I_6|^r] &= \mathbb{E}\bigg[\bigg|\int_0^\tau \int_{\mathbb{T}^3} \varrho \mathbb{F}_\varepsilon(\varrho,\mathbf{u}) \cdot \Delta^{-1}\nabla\varrho \,\mathrm{d}x\,\mathrm{d}W\bigg|^r\bigg] \\ &\leq \mathbb{E}\bigg[\sup_{t\in[0,\tau]}\bigg|\int_0^t \int_{\mathbb{T}^3} \varrho \mathbb{F}_\varepsilon(\varrho,\mathbf{u}) \cdot \Delta^{-1}\nabla\varrho \,\mathrm{d}x\,\mathrm{d}W\bigg|^r\bigg] \\ &\lesssim \mathbb{E}\bigg[\int_0^\tau \sum_{k=1}^\infty \bigg|\int_{\mathbb{T}^3} \varrho \mathbf{F}_{k,\varepsilon}(\varrho,\mathbf{u}) \cdot \Delta^{-1}\nabla\varrho \,\mathrm{d}x\bigg|^2 \mathrm{d}t\bigg]^{r/2},\end{aligned}$$

where, due to (4.7) and the properties of Δ^{-1},

$$\begin{aligned}\bigg|\int_{\mathbb{T}^3} \varrho \mathbf{F}_{k,\varepsilon}(\varrho,\mathbf{u}) \cdot \Delta^{-1}\nabla\varrho \,\mathrm{d}x\bigg| &\lesssim f_k \|\Delta^{-1}\nabla\varrho\|_{L^\infty_x} \int_{\mathbb{T}^3} (\varrho + \varrho|\mathbf{u}|)\,\mathrm{d}x \\ &\lesssim f_k \|\varrho\|_{L^\Gamma_x} \int_{\mathbb{T}^3} (\varrho + \varrho|\mathbf{u}|)\,\mathrm{d}x \\ &\lesssim f_k \|\varrho\|^2_{L^\Gamma_x} + f_k\bigg(\int_{\mathbb{T}^3} (\varrho + \varrho|\mathbf{u}|^2)\,\mathrm{d}x\bigg)^2.\end{aligned}$$

Hence we conclude

$$\mathbb{E}[|I_6|^r] \lesssim c_2(r), \tag{4.147}$$

with $c_2(r)$ given by (4.140).

Summing up the estimates (4.142)–(4.147), we obtain the desired bound for the pressure. We have

$$\mathbb{E}\bigg[\bigg|\int_0^T \int_{\mathbb{T}^3} (p(\varrho) + \delta\varrho^\Gamma)\varrho \,\mathrm{d}x\,\mathrm{d}t\bigg|^r\bigg] \lesssim c_2(r). \tag{4.148}$$

Note that the term $\int_0^T \int_{\mathbb{T}^3} p_\delta(\varrho)(\varrho)_{\mathbb{T}^3}\,\mathrm{d}x\,\mathrm{d}t$ can be handled using (4.135) and (4.126).

4.4.3 Limit $\varepsilon \to 0$

The uniform bounds derived in the previous section are optimal in view of the energy method. We are ready to perform the limit $\varepsilon \to 0$. We proceed in two steps. First, we use Jakubowski's extension of the Skorokhod representation theorem and change the probability space to obtain compactness in probability. Then we adapt the method known for the deterministic case to show compactness of the densities, which is the main issue here.

Assume that Λ is the initial law given by Theorem 4.4.2, that is, (4.126) and (4.127) are satisfied. We need to approximate Λ by Borel probability measures Λ_ε which fulfill the assumptions of Theorem 4.3.2, namely, (4.77) and (4.78), such that (4.126) and

(4.127) hold true uniformly in ε. To this end, consider a random variable $[\varrho_0, \mathbf{q}_0]$ with law Λ which exists due to Corollary 2.6.4. Then one can find random variables $\varrho_{0,\varepsilon}$ with values in $C^{2+\kappa}(\mathbb{T}^3)$, for some $\kappa > 0$, such that $\mathbb{P}$-a.s.

$$0 < \varepsilon \leq \varrho_{0,\varepsilon} \leq \frac{1}{\varepsilon}, \quad \frac{\underline{\varrho}}{2} \leq \int_{\mathbb{T}^3} \varrho_{0,\varepsilon}\,\mathrm{d}x \leq 2\overline{\varrho},$$

as well as

$$\varrho_{0,\varepsilon} \to \varrho_0 \quad \text{in } L^p(\Omega; L^\Gamma(\mathbb{T}^3)) \quad \forall p \in [1, r\Gamma]. \tag{4.149}$$

Next, setting

$$\hat{\mathbf{q}}_{0,\varepsilon} = \begin{cases} \mathbf{q}_0 \sqrt{\frac{\varrho_{0,\varepsilon}}{\varrho_0}}, & \text{if } \varrho_0 > 0, \\ 0, & \text{if } \varrho_0 = 0, \end{cases}$$

it follows from the assumptions on Λ that

$$\frac{|\hat{\mathbf{q}}_{0,\varepsilon}|^2}{\varrho_{0,\varepsilon}} \in L^p(\Omega; L^1(\mathbb{T}^3)) \quad \forall p \in [1, r]$$

uniformly in ε. Moreover, by mollification we can find random variables h_ε with values in $C^2(\mathbb{T}^3)$ such that

$$\frac{\hat{\mathbf{q}}_{0,\varepsilon}}{\sqrt{\varrho_{0,\varepsilon}}} - h_\varepsilon \to 0 \quad \text{in } L^p(\Omega; L^2(\mathbb{T}^3)) \quad \forall p \in [1, 2r].$$

Let $\mathbf{q}_{0,\varepsilon} = h_\varepsilon \sqrt{\varrho_{0,\varepsilon}}$. Then

$$\frac{|\mathbf{q}_{0,\varepsilon}|^2}{\varrho_{0,\varepsilon}} \in L^p(\Omega; L^1(\mathbb{T}^3)) \quad \forall p \in [1, r]$$

uniformly in ε and

$$\mathbf{q}_{0,\varepsilon} \to \mathbf{q}_0 \quad \text{in } L^p(\Omega; L^1(\mathbb{T}^3)) \quad \forall p \in [1, r] \tag{4.150}$$

$$\frac{\mathbf{q}_{0,\varepsilon}}{\sqrt{\varrho_{0,\varepsilon}}} \to \frac{\mathbf{q}_0}{\sqrt{\varrho_0}} \quad \text{in } L^p(\Omega; L^2(\mathbb{T}^3)) \quad \forall p \in [1, 2r]. \tag{4.151}$$

Note that (4.150) also uses (4.149). Finally, we define $\Lambda_\varepsilon = \mathbb{P} \circ (\varrho_{0,\varepsilon}, \mathbf{q}_{0,\varepsilon})^{-1}$. Note that (4.149) and (4.150) imply in particular $\Lambda_\varepsilon \overset{*}{\rightharpoonup} \Lambda$ in the sense of measures on $L^1(\mathbb{T}^3) \times L^1(\mathbb{T}^3)$.

As a consequence of the above, Theorem 4.3.2 yields existence of $[\varrho_\varepsilon, \mathbf{u}_\varepsilon]$, which is a dissipative martingale solution to (4.72)–(4.73) with the initial law Λ_ε. To be precise, Theorem 4.3.2 yields for every $\varepsilon \in (0, 1)$ a multiplet

$$((\Omega^\varepsilon, \mathfrak{F}^\varepsilon, (\mathfrak{F}^\varepsilon_t), \mathbb{P}^\varepsilon), \varrho_\varepsilon, \mathbf{u}_\varepsilon, W_\varepsilon),$$

which is a weak martingale solution to (4.72)–(4.73). In view of Remark 4.0.4 we may assume without loss of generality that

$$(\Omega^\varepsilon, \mathfrak{F}^\varepsilon, \mathbb{P}^\varepsilon) = ([0,1], \overline{\mathfrak{B}([0,1])}, \mathfrak{L}) \quad \varepsilon \in (0,1)$$

and that

$$\mathfrak{F}_t^\varepsilon = \sigma\left(\sigma_t[\varrho_\varepsilon] \cup \sigma_t[\mathbf{u}_\varepsilon] \cup \bigcup_{k=1}^{\infty} \sigma_t[W_\varepsilon] \right), \quad t \in [0,T].$$

Moreover, we assume without loss of generality that there exists one common Wiener process W for all ε. Indeed, this can be achieved by performing the compactness argument in Section 4.3 for the parameters from any chosen subsequence $(\varepsilon_n)_{n\in\mathbb{N}}$ at once.

Since (4.126) and (4.127) are satisfied by Λ_ε uniformly in ε, we obtain the uniform bounds for the quantities $c_1(r)$ and $c_2(r)$ appearing in Section 4.4.1 and Section 4.4.2. This in turn implies the corresponding uniform bounds for $[\varrho_\varepsilon, \mathbf{u}_\varepsilon]$.

4.4.3.1 Stochastic compactness method

Due to the lack of smoothness, the principal fields ϱ_ε, $\mathbf{u}_\varepsilon$ will be treated in the framework of weak topologies. It is convenient to use the extension of Jakubowski's theorem stated in Theorem 2.7.1, where the family $[\varrho_\varepsilon, \mathbf{u}_\varepsilon, \nabla\mathbf{u}_\varepsilon]$ is considered along with the associated Young measure, which allows one to pass to the limit in compositions with non-linear functions; see Section 2.8. In our context, it will be used to pass to the limit in the pressure as well as in the stochastic integral. Similarly to Section 4.3.2, we first fix the path space the solutions live in. As the present uniform bounds are considerably weaker than those in Section 4.3.1, everything must be done more carefully. It is convenient to include the energy

$$E_\varepsilon := E(\varrho_\varepsilon, \mathbf{u}_\varepsilon) = \frac{1}{2}\varrho_\varepsilon|\mathbf{u}_\varepsilon|^2 + P_\delta(\varrho_\varepsilon).$$

As we have shown in Section 4.4.1, the energy E_ε is bounded in expectation in the space $L^\infty(0,T;L^1(\mathbb{T}^3))$. This embeds into $L^\infty(0,T;\mathcal{M}_b(\mathbb{T}^3))$, considered as the dual to the separable Banach space $L^1(0,T;C(\mathbb{T}^3))$. Here $\mathcal{M}_b(\mathbb{T}^3)$ denotes the space of bounded Borel measures on $\mathbb{T}^3$. Consequently, $L^\infty(0,T;\mathcal{M}_b(\mathbb{T}^3))$ is a sub-Polish space and the Jakubowski–Skorokhod theorem, Theorem 2.7.1, is applicable.

Finally, we also include the Young measure corresponding to $[\varrho_\varepsilon, \mathbf{u}_\varepsilon, \nabla\mathbf{u}_\varepsilon]$. The reader is referred to Section 2.8 for an introduction. Let ν_ε be the canonical Young measure associated to $[\varrho_\varepsilon, \mathbf{u}_\varepsilon, \nabla\mathbf{u}_\varepsilon]$. To be more precise, ν_ε is the weakly-$*$ measurable mapping

$$\nu_\varepsilon : [0,T] \times \mathbb{T}^3 \to \mathcal{P}(\mathbb{R} \times \mathbb{R}^3 \times \mathbb{R}^{3\times3}) \simeq \mathcal{P}(\mathbb{R}^{13}),$$

given by

$$\nu_{\varepsilon,t,x}(\cdot) = \delta_{[\varrho_\varepsilon, \mathbf{u}_\varepsilon, \nabla\mathbf{u}_\varepsilon](t,x)}(\cdot).$$

Note that there is an additional dependence on the randomness variable ω. In view of the discussion in Section 2.8, ν_ε can be regarded as a random variable taking values in the space of Young measures which we denoted by

$$(L^\infty_{w^*}((0,T)\times\mathbb{T}^3;\mathcal{P}(\mathbb{R}^{13})),w^*).$$

We recall that the weak-∗ topology on this space is determined by functionals

$$L^\infty_{w^*}((0,T)\times\mathbb{T}^3;\mathcal{P}(\mathbb{R}^{13}))\to\mathbb{R},\quad \nu\mapsto\int_0^T\int_{\mathbb{T}^3}\psi(t,x)\int_{\mathbb{R}^{13}}\phi(\xi)\,\mathrm{d}\nu_{t,x}(\xi)\,\mathrm{d}x\,\mathrm{d}t,$$

where $\psi\in L^1((0,T)\times\mathbb{T}^3)$ and $\phi\in C_b(\mathbb{R}^{13})$. It was discussed in Section 2.8 that this topology is finer than the weak-∗ topology on $L^\infty_{w^*}((0,T)\times\mathbb{T}^3;\mathcal{M}_b(\mathbb{R}^{13}))$, which is the topological dual of $L^1((0,T)\times\mathbb{T}^3;C_0(\mathbb{R}^{13}))$. Consequently, the space of Young measures $(L^\infty_{w^*}((0,T)\times\mathbb{T}^3;\mathcal{P}(\mathbb{R}^{13})),w^*)$ belongs to the class of sub-Polish spaces.

As the next step, we define the corresponding path space. We have

$$\mathcal{X}=\mathcal{X}_{\varrho_0}\times\mathcal{X}_{\mathbf{q}_0}\times\mathcal{X}_{\frac{\mathbf{q}_0}{\sqrt{\varrho_0}}}\times\mathcal{X}_\varrho\times\mathcal{X}_{\varrho\mathbf{u}}\times\mathcal{X}_{\mathbf{u}}\times\mathcal{X}_W\times\mathcal{X}_E\times\mathcal{X}_\nu,$$

where

$$\begin{aligned}
\mathcal{X}_{\varrho_0}&=L^\Gamma(\mathbb{T}^3),\quad \mathcal{X}_{\mathbf{q}_0}=L^1(\mathbb{T}^3),\quad \mathcal{X}_{\frac{\mathbf{q}_0}{\sqrt{\varrho_0}}}=L^2(\mathbb{T}^3),\\
\mathcal{X}_\varrho&=(L^{\Gamma+1}((0,T)\times\mathbb{T}^3),w)\cap C_w([0,T];L^\Gamma(\mathbb{T}^3)),\\
\mathcal{X}_{\varrho\mathbf{u}}&=C_w([0,T];L^{\frac{2\Gamma}{\Gamma+1}}(\mathbb{T}^3))\cap C([0,T];W^{-k,2}(\mathbb{T}^3)),\quad k>\frac{5}{2},\\
\mathcal{X}_{\mathbf{u}}&=(L^2(0,T;W^{1,2}(\mathbb{T}^3)),w),\\
\mathcal{X}_W&=C([0,T];\mathfrak{U}_0),\\
\mathcal{X}_E&=(L^\infty(0,T;\mathcal{M}_b(\mathbb{T}^3)),w^*),\\
\mathcal{X}_\nu&=(L^\infty_{w^*}((0,T)\times\mathbb{T}^3;\mathcal{P}(\mathbb{R}^{13})),w^*).
\end{aligned}$$

To proceed, it is necessary to establish tightness of the set

$$\left\{\mathcal{L}\left[\varrho_{0,\varepsilon},\mathbf{q}_{0,\varepsilon},\frac{\mathbf{q}_{0,\varepsilon}}{\sqrt{\varrho_{0,\varepsilon}}},\varrho_\varepsilon,\varrho_\varepsilon\mathbf{u}_\varepsilon,\mathbf{u}_\varepsilon,W,E_\delta(\varrho_\varepsilon,\mathbf{u}_\varepsilon),\nu_\varepsilon\right];\varepsilon\in(0,1)\right\}$$

on $\mathcal{X}$. To this end, we observe that tightness of the initial laws

$$\left\{\mathcal{L}\left[\varrho_{0,\varepsilon},\mathbf{q}_{0,\varepsilon},\frac{\mathbf{q}_{0,\varepsilon}}{\sqrt{\varrho_{0,\varepsilon}}}\right];\varepsilon\in(0,1)\right\}$$

follows from (4.149)–(4.151) and Prokhorov's theorem (Theorem 2.6.1). Tightness of $\{\mathcal{L}[\mathbf{u}_\varepsilon];\varepsilon\in(0,1)\}$ can be shown as in Proposition 4.3.6 using (4.139), while tightness of $\{\mathcal{L}[\varrho_\varepsilon];\varepsilon\in(0,1)\}$ is as in Proposition 4.3.7 using (4.135), (4.136), and (4.148). Tightness of μ_W is immediate and was discussed just before Corollary 4.3.9. To show tightness for $\{\mathcal{L}[\varrho_\varepsilon\mathbf{u}_\varepsilon];\varepsilon\in(0,1)\}$, we observe that the proof of Proposition 4.3.8 requires some modifications.

Proposition 4.4.5. *The set $\{\mathcal{L}[\varrho_\varepsilon \mathbf{u}_\varepsilon];\, \varepsilon \in (0,1)\}$ is tight on $\mathcal{X}_{\varrho\mathbf{u}}$.*

Proof. We proceed similarly to Proposition 4.3.8 and decompose $\varrho_\varepsilon \mathbf{u}_\varepsilon$ into two parts, namely, $\varrho_\varepsilon \mathbf{u}_\varepsilon(t) = Y^\varepsilon(t) + Z^\varepsilon(t)$, where

$$\begin{aligned} Y^\varepsilon(t) &= (\varrho_\varepsilon \mathbf{u}_\varepsilon)(0) - \int_0^t \operatorname{div}(\varrho_\varepsilon \mathbf{u}_\varepsilon \otimes \mathbf{u}_\varepsilon)\,\mathrm{d}s - \int_0^t \operatorname{div} \mathbb{S}(\nabla \mathbf{u}_\varepsilon)\,\mathrm{d}s \\ &\quad + \int_0^t \nabla p_\delta(\varrho_\varepsilon)\,\mathrm{d}s - \varepsilon \int_0^t \Delta(\varrho_\varepsilon \mathbf{u}_\varepsilon)\,\mathrm{d}s, \\ Z^\varepsilon(t) &= \int_0^t \mathbb{G}_\varepsilon(\varrho_\varepsilon, \varrho_\varepsilon \mathbf{u}_\varepsilon)\,\mathrm{d}W. \end{aligned}$$

By a similar approach as in Proposition 4.3.8, we obtain Hölder continuity of Y^ε, namely, there exist $\kappa > 0$ and $l > 5/2$ such that

$$\mathbb{E}\|Y^\varepsilon\|_{C_t^\kappa W_x^{-l,2}} \le C.$$

This is a consequence of the *a priori* estimates (4.135)–(4.139). Concerning the stochastic integral Z^ε, we obtain due to (3.13) as well as (4.135) and (4.136) (similarly to (3.15))

$$\begin{aligned} \mathbb{E}\left\| \int_s^t \mathbb{G}_\varepsilon(\varrho_\varepsilon, \varrho_\varepsilon \mathbf{u}_\varepsilon)\,\mathrm{d}W \right\|_{W_x^{-l,2}}^r &\lesssim \mathbb{E}\left(\int_s^t \sum_{k=1}^\infty \|\mathbf{G}_{k,\varepsilon}(\varrho_\varepsilon, \varrho_\varepsilon \mathbf{u}_\varepsilon)\|_{W_x^{-l,2}}^2\,\mathrm{d}\tau \right)^{r/2} \\ \lesssim \mathbb{E}\left(\int_s^t \sum_{k=1}^\infty \|\mathbf{G}_{k,\varepsilon}(\varrho_\varepsilon, \varrho_\varepsilon \mathbf{u}_\varepsilon)\|_{L_x^1}^2\,\mathrm{d}\tau \right)^{r/2} &\lesssim \mathbb{E}\left(\int_s^t \int_{\mathbb{T}^3} (\varrho_\varepsilon + \varrho_\varepsilon |\mathbf{u}_\varepsilon|^2)\,\mathrm{d}x\,\mathrm{d}\tau \right)^{r/2} \\ \lesssim |t-s|^{r/2} \Big(1 + \mathbb{E} \sup_{0 \le t \le T} \|\sqrt{\varrho_\varepsilon} \mathbf{u}_\varepsilon\|_{L^2}^r \Big) &\le C|t-s|^{r/2} \end{aligned}$$

and the Kolmogorov continuity criterion applies. We can now complete the proof as in Proposition 4.3.8. □

Proposition 4.4.6. *The set $\{\mathcal{L}[E_\varepsilon];\, \varepsilon \in (0,1)\}$ is tight on $\mathcal{X}_E$.*

Proof. The claim follows immediately from the bounds in Section 4.4.1, namely, (4.134)–(4.136), and the fact that bounded sets in $L^\infty(0,T;\mathcal{M}_b(\mathbb{T}^3))$ are relatively compact with respect to the weak-$*$ topology. □

Proposition 4.4.7. *The set $\{\mathcal{L}[\nu_\varepsilon];\, \varepsilon \in (0,1)\}$ is tight on $\mathcal{X}_\nu$.*

Proof. The aim is to apply the compactness criterion in $(L^\infty_{w^*}((0,T) \times \mathbb{T}^3; \mathcal{P}(\mathbb{R}^{13})), w^*)$ established in Corollary 2.8.6. To this end, define the set

$$B_R := \Big\{ \nu \in L^\infty_{w^*}((0,T) \times \mathbb{T}^3; \mathcal{P}(\mathbb{R}^{13}));$$

$$\int_0^T \int_{\mathbb{T}^3} \int_{\mathbb{R}^{13}} \left(|\xi_1|^{\Gamma+1} + \sum_{i=2}^{13} |\xi_i|^2 \right) \mathrm{d}\nu_{t,x}(\xi)\, \mathrm{d}x\, \mathrm{d}t \le R \Big\},$$

which is relatively compact in $(L^\infty_{w^*}((0,T)\times\mathbb{T}^3;\mathcal{P}(\mathbb{R}^{13})), w^*)$. From (4.137), (4.139), and (4.148) we deduce

$$\begin{aligned}
\mathcal{L}[\nu_\varepsilon](B_R^c) &= \mathbb{P}\left(\int_0^T \int_{\mathbb{T}^3} \int_{\mathbb{R}^{13}} \left(|\xi_1|^{\Gamma+1} + \sum_{i=2}^{13} |\xi_i|^2 \right) \mathrm{d}\nu_{\varepsilon,t,x}(\xi)\, \mathrm{d}x\, \mathrm{d}t > R \right) \\
&= \mathbb{P}\left(\int_0^T \int_{\mathbb{T}^3} |\varrho_\varepsilon|^{\Gamma+1} + |\mathbf{u}_\varepsilon|^2 + |\nabla \mathbf{u}_\varepsilon|^2\, \mathrm{d}x\, \mathrm{d}t > R \right) \\
&\le \frac{1}{R}\mathbb{E}[\|\varrho_\varepsilon\|^{\Gamma+1}_{L^{\Gamma+1}_{t,x}} + \|\mathbf{u}_\varepsilon\|^2_{L^2_{t,x}} + \|\nabla \mathbf{u}_\varepsilon\|^2_{L^2_{t,x}}] \le \frac{C}{R}.
\end{aligned}$$

The proof is complete. □

The desired conclusion follows.

Corollary 4.4.8. *The set*

$$\left\{ \mathcal{L}\left[\varrho_{0,\varepsilon}, \mathbf{q}_{0,\varepsilon}, \frac{\mathbf{q}_{0,\varepsilon}}{\sqrt{\varrho_{0,\varepsilon}}}, \varrho_\varepsilon, \varrho_\varepsilon \mathbf{u}_\varepsilon, \mathbf{u}_\varepsilon, W, E(\varrho_\varepsilon, \mathbf{u}_\varepsilon), \nu_\varepsilon \right]; \varepsilon \in (0,1) \right\}$$

is tight on $\mathcal{X}$.

Consequently, we are allowed to apply the Jakubowski–Skorokhod representation theorem (Theorem 2.7.1).

Proposition 4.4.9. *There exists a complete probability space $(\tilde{\Omega}, \tilde{\mathfrak{F}}, \tilde{\mathbb{P}})$ with $\mathcal{X}$-valued Borel measurable random variables $[\tilde{\varrho}_{0,\varepsilon}, \tilde{\mathbf{q}}_{0,\varepsilon}, \tilde{\mathbf{k}}_{0,\varepsilon}, \tilde{\varrho}_\varepsilon, \tilde{\mathbf{q}}_\varepsilon, \tilde{\mathbf{u}}_\varepsilon, \tilde{W}_\varepsilon, \tilde{E}_\varepsilon, \tilde{\nu}_\varepsilon]$, $\varepsilon \in (0,1)$, as well as $[\tilde{\varrho}_0, \tilde{\mathbf{q}}_0, \tilde{\mathbf{k}}_0, \tilde{\varrho}, \tilde{\mathbf{q}}, \tilde{\mathbf{u}}, \tilde{W}, \tilde{E}, \tilde{\nu}]$ such that (up to a subsequence):*

(1) *for all $\varepsilon \in (0,1)$, we know that $\mathcal{L}[\tilde{\varrho}_{0,\varepsilon}, \tilde{\mathbf{q}}_{0,\varepsilon}, \tilde{\mathbf{k}}_{0,\varepsilon}, \tilde{\varrho}_\varepsilon, \tilde{\mathbf{q}}_\varepsilon, \tilde{\mathbf{u}}_\varepsilon, \tilde{W}_\varepsilon, \tilde{E}_\varepsilon, \tilde{\nu}_\varepsilon]$ and $\mathcal{L}[\varrho_{0,\varepsilon}, \mathbf{q}_{0,\varepsilon}, \frac{\mathbf{q}_{0,\varepsilon}}{\sqrt{\varrho_{0,\varepsilon}}}, \varrho_\varepsilon, \varrho_\varepsilon \mathbf{u}_\varepsilon, \mathbf{u}_\varepsilon, W, E_\varepsilon, \nu_\varepsilon]$ coincide on $\mathcal{X}$. In particular,*

$$\begin{aligned}
&\tilde{\varrho}_{0,\varepsilon} = \tilde{\varrho}_\varepsilon(0), \quad \tilde{\mathbf{q}}_{0,\varepsilon} = \tilde{\varrho}_\varepsilon \tilde{\mathbf{u}}_\varepsilon(0), \quad \tilde{\mathbf{k}}_{0,\varepsilon} = \frac{\tilde{\mathbf{q}}_{0,\varepsilon}}{\sqrt{\tilde{\varrho}_{0,\varepsilon}}} = \frac{\tilde{\varrho}_\varepsilon \tilde{\mathbf{u}}_\varepsilon(0)}{\sqrt{\tilde{\varrho}_\varepsilon(0)}}, \\
&\tilde{\mathbf{q}}_\varepsilon = \tilde{\varrho}_\varepsilon \tilde{\mathbf{u}}_\varepsilon, \quad \tilde{E}_\varepsilon = E(\tilde{\varrho}_\varepsilon, \tilde{\mathbf{u}}_\varepsilon), \quad \tilde{\nu}_\varepsilon = \delta_{[\tilde{\varrho}_\varepsilon, \tilde{\mathbf{u}}_\varepsilon, \nabla \tilde{\mathbf{u}}_\varepsilon]},
\end{aligned}$$

$\tilde{\mathbb{P}}$-a.s., as well as

$$\mathbb{E}\left[\left| \sup_{t\in[0,T]} \int_{\mathbb{T}^3} \left[\frac{1}{2} \tilde{\varrho}_\varepsilon |\tilde{\mathbf{u}}_\varepsilon|^2 + P_\delta(\tilde{\varrho}_\varepsilon) \right] \mathrm{d}x \right|^r \right] \le c_2(r), \tag{4.152}$$

with $c_2(r)$ given by (4.140);

(2) *the law of $[\tilde{\varrho}_0, \tilde{\mathbf{q}}_0, \tilde{\mathbf{k}}_0, \tilde{\varrho}, \tilde{\mathbf{q}}, \tilde{\mathbf{u}}, \tilde{W}, \tilde{E}, \tilde{\nu}]$ on $\mathcal{X}$ is a Radon measure;*

(3) $[\tilde{\varrho}_{0,\varepsilon}, \tilde{\mathbf{q}}_{0,\varepsilon}, \tilde{\mathbf{k}}_{0,\varepsilon}, \tilde{\varrho}_\varepsilon, \tilde{\varrho}_\varepsilon \tilde{\mathbf{u}}_\varepsilon, \tilde{\mathbf{u}}_\varepsilon, \tilde{W}_\varepsilon, \tilde{E}_\varepsilon, \tilde{\nu}_\varepsilon]$ *converges in the topology of* $\mathcal{X}$ $\tilde{\mathbb{P}}$*-a.s. to* $[\tilde{\varrho}_0, \tilde{\mathbf{q}}_0, \tilde{\mathbf{k}}_0, \tilde{\varrho}, \tilde{\mathbf{q}}, \tilde{\mathbf{u}}, \tilde{W}, \tilde{E}, \tilde{\nu}]$, *i.e.,*

$$\begin{aligned}
\tilde{\varrho}_{0,\varepsilon} &\to \tilde{\varrho}_0 \quad \text{in } L^\Gamma(\mathbb{T}^3),\\
\tilde{\mathbf{q}}_{0,\varepsilon} &\to \tilde{\mathbf{q}}_0 \quad \text{in } L^1(\mathbb{T}^3),\\
\tilde{\mathbf{k}}_{0,\varepsilon} &\to \tilde{\mathbf{k}}_0 \quad \text{in } L^2(\mathbb{T}^3),\\
\tilde{\varrho}_\varepsilon &\to \tilde{\varrho} \quad \text{in } C_w([0,T]; L^\Gamma(\mathbb{T}^3)),\\
\tilde{\varrho}_\varepsilon &\rightharpoonup \tilde{\varrho} \quad \text{in } L^{\Gamma+1}((0,T)\times\mathbb{T}^3),\\
\tilde{\varrho}_\varepsilon \tilde{\mathbf{u}}_\varepsilon &\to \tilde{\varrho}\tilde{\mathbf{u}} \quad \text{in } C_w([0,T]; L^{\frac{2\Gamma}{\Gamma+1}}(\mathbb{T}^3)),\\
\tilde{\mathbf{u}}_\varepsilon &\rightharpoonup \tilde{\mathbf{u}} \quad \text{in } L^2(0,T; W^{1,2}(\mathbb{T}^3)),\\
\tilde{W}_\varepsilon &\to \tilde{W} \quad \text{in } C([0,T]; \mathfrak{U}_0),\\
E(\tilde{\varrho}_\varepsilon, \tilde{\mathbf{u}}_\varepsilon) &\overset{*}{\rightharpoonup} \tilde{E} \quad \text{in } L^\infty(0,T; \mathcal{M}_b(\mathbb{T}^3)),\\
\delta_{[\tilde{\varrho}_\varepsilon, \tilde{\mathbf{u}}_\varepsilon, \nabla\tilde{\mathbf{u}}_\varepsilon]} &\overset{*}{\rightharpoonup} \tilde{\nu} \quad \text{in } L^\infty((0,T)\times\mathbb{T}^3; \mathcal{P}(\mathbb{R}^{13})),
\end{aligned} \tag{4.153}$$

as $\varepsilon \to 0$ $\tilde{\mathbb{P}}$*-a.s.;*

(4) *for any Carathéodory function* $H = H(t,x,\varrho,\mathbf{v},\mathbf{V})$ *where* $(t,x) \in (0,T)\times\mathbb{T}^3$, $(\varrho,\mathbf{v},\mathbf{V}) \in \mathbb{R}^{13}$, *satisfying for some* $q_1, q_2, q_3 > 0$ *the growth condition*

$$|H(t,x,\varrho,\mathbf{v},\mathbf{V})| \lesssim 1 + |\varrho|^{q_1} + |\mathbf{v}|^{q_2} + |\mathbf{V}|^{q_2},$$

uniformly in (t,x), *denote* $\overline{H(\tilde{\varrho},\tilde{\mathbf{u}},\nabla\tilde{\mathbf{u}})}(t,x) = \langle \tilde{\nu}_{t,x}, H\rangle$. *Then we have*

$$\begin{aligned}
H(\tilde{\varrho}_\varepsilon, \tilde{\mathbf{u}}_\varepsilon, \nabla\tilde{\mathbf{u}}_\varepsilon) &\rightharpoonup \overline{H(\tilde{\varrho},\tilde{\mathbf{u}},\nabla\tilde{\mathbf{u}})} \quad \text{in } L^r((0,T)\times\mathbb{T}^3)\\
&\text{for all } 1 < r \le \frac{\Gamma+1}{q_1} \wedge \frac{2}{q_2},
\end{aligned} \tag{4.154}$$

as $\varepsilon \to 0$ $\tilde{\mathbb{P}}$*-a.s.*

Proof. The parts (1)–(3) follow from Theorem 2.7.1. The remaining point was obtained a consequence of Corollary 2.8.3. □

For the same reasons as in Section 4.3.2, we work in the context of random distributions introduced in Section 2.2 and follow the same arguments as in the discussion after Proposition 4.3.10. To be more precise, in view of Remark 2.3.7 and Corollary 2.1.36, it is necessary to show that joint canonical filtration given by

$$\tilde{\mathfrak{F}}_t := \sigma\left(\sigma_t[\tilde{\varrho}] \cup \sigma_t[\tilde{\mathbf{u}}] \cup \bigcup_{k=1}^\infty \sigma_t[\tilde{W}_k]\right), \quad t \in [0,T]$$

is non-anticipative with respect to the limit Wiener process $\tilde{W}$.

This is obtained as follows. According to Theorem 2.9.1, for every $\varepsilon \in (0,1)$, $\tilde{W}_\varepsilon = \sum_{k=1}^\infty e_k \tilde{W}_{\varepsilon,k}$ is a cylindrical Wiener process with respect to

$$\sigma\Big(\sigma_t[\tilde{\varrho}_\varepsilon] \cup \sigma_t[\tilde{\mathbf{u}}_\varepsilon] \cup \bigcup_{k=1}^\infty \sigma_t[\tilde{W}_{\varepsilon,k}]\Big), \quad t \in [0,T].$$

In other words, this filtration is non-anticipative with respect to $\tilde{W}_\varepsilon$. Lemma 2.9.3 together with Proposition 4.4.9 then allows one to pass to the limit as $\varepsilon \to 0$ and the non-anticipativity of $(\tilde{\mathfrak{F}}_t)_{t\geq 0}$ with respect to $\tilde{W}$ follows. Finally, due to Lemma 2.1.35 and Corollary 2.1.36, the process $\tilde{W}$ is a cylindrical Wiener processes with respect to $(\tilde{\mathfrak{F}}_t)_{t\geq 0}$ and is given by the formal expansion $\tilde{W} = \sum_{k=1}^\infty e_k \tilde{W}_k$ for some sequence of mutually independent real-valued $(\tilde{\mathfrak{F}}_t)$-Wiener processes $\tilde{W}_k$, $k \in \mathbb{N}$.

As in Corollary 4.3.12 we obtain the following as a direct consequence of Proposition 4.4.9.

Corollary 4.4.10. *The following convergence holds true $\tilde{\mathbb{P}}$-a.s.:*

$$\tilde{\varrho}_\varepsilon \tilde{\mathbf{u}}_\varepsilon \otimes \tilde{\mathbf{u}}_\varepsilon \rightharpoonup \tilde{\varrho}\tilde{\mathbf{u}} \otimes \tilde{\mathbf{u}} \quad \text{in } L^1(0,T;L^1(\mathbb{T}^3)). \tag{4.155}$$

Similar to Lemma 4.1.7 the continuity equation (4.123) is satisfied by $[\tilde{\varrho}, \tilde{\mathbf{u}}]$ on the new probability space. Note, however, that we only have a weak solution.

Lemma 4.4.11. *The random distribution $[\tilde{\varrho}, \tilde{\mathbf{u}}]$ satisfies (4.123) for all $\phi \in C_c^\infty([0,T))$ and $\psi \in C^\infty(\mathbb{T}^3)$ $\tilde{\mathbb{P}}$-a.s.*

As a consequence of Theorem 2.9.1 we see that $[\tilde{\varrho}_\varepsilon, \tilde{\mathbf{u}}_\varepsilon]$ satisfies the continuity equation

$$\partial_t \tilde{\varrho}_\varepsilon + \operatorname{div}(\tilde{\varrho}_\varepsilon \tilde{\mathbf{u}}_\varepsilon) = \varepsilon \Delta \tilde{\varrho}_\varepsilon \tag{4.156}$$

a.e. in $(0,T) \times \mathbb{T}^3$ $\tilde{\mathbb{P}}$-a.s. We may therefore multiply (4.156) by $\tilde{\varrho}_\varepsilon$ and integrate by parts, obtaining

$$\frac{\mathrm{d}}{\mathrm{d}t} \int_{\mathbb{T}^3} \frac{1}{2} |\tilde{\varrho}_\varepsilon|^2 \, \mathrm{d}x + \varepsilon \int_{\mathbb{T}^3} |\nabla \tilde{\varrho}_\varepsilon|^2 \, \mathrm{d}x = -\frac{1}{2} \int_{\mathbb{T}^3} |\tilde{\varrho}_\varepsilon|^2 \operatorname{div} \tilde{\mathbf{u}}_\varepsilon \, \mathrm{d}x.$$

In accordance with (4.135) and (4.137) (which continue to hold on the new probability space due to Proposition 4.4.9), the integral on the right hand side is bounded in expectation and we obtain

$$\varepsilon \mathbb{E} \int_0^T \|\nabla \tilde{\varrho}_\varepsilon\|_{L^2_x}^2 \, \mathrm{d}t \lesssim c_1(r), \tag{4.157}$$

where $c_1(r)$ is given in (4.138). In particular, for a suitable subsequence, $\tilde{\mathbb{P}}$-a.s.,

$$\varepsilon \Delta \tilde{\varrho}_\varepsilon \to 0 \quad \text{in } L^2(0,T; W^{-1,2}(\mathbb{T}^3)).$$

Next, we perform the limit $\varepsilon \to 0$ in the momentum equation (4.73). To this end, we apply Proposition 4.4.9, part (4), to compositions $p_\delta(\tilde{\varrho}_\varepsilon)$ and $\tilde{\varrho}_\varepsilon \mathbf{F}_k(\tilde{\varrho}_\varepsilon, \tilde{\mathbf{u}}_\varepsilon)$, $k \in \mathbb{N}$. Specifically, there exist $\overline{p_\delta(\tilde{\varrho})}$ and $\overline{\tilde{\varrho}\mathbf{F}_k(\tilde{\varrho}, \tilde{\mathbf{u}})}$, $k \in \mathbb{N}$, such that

$$p_\delta(\tilde{\varrho}_\varepsilon) \rightharpoonup \overline{p_\delta(\tilde{\varrho})} \quad \text{in } L^q((0,T) \times \mathbb{T}^3), \tag{4.158}$$

$$\tilde{\varrho}_\varepsilon \mathbf{F}_k(\tilde{\varrho}_\varepsilon, \tilde{\mathbf{u}}_\varepsilon) \rightharpoonup \overline{\tilde{\varrho}\mathbf{F}_k(\tilde{\varrho}, \tilde{\mathbf{u}})} \quad \text{in } L^q((0,T) \times \mathbb{T}^3), \tag{4.159}$$

for some $q > 1$ $\tilde{\mathbb{P}}$-a.s. Let us now define the Hilbert–Schmidt operator $\overline{\tilde{\varrho}\mathbb{F}(\tilde{\varrho}, \tilde{\mathbf{u}})}$ by

$$\overline{\tilde{\varrho}\mathbb{F}(\tilde{\varrho}, \tilde{\mathbf{u}})} e_k := \overline{\tilde{\varrho}\mathbf{F}_k(\tilde{\varrho}, \tilde{\mathbf{u}})} \quad k \in \mathbb{N}.$$

We obtain the following result.

Proposition 4.4.12. *The random distribution $[\tilde{\varrho}, \tilde{\mathbf{u}}, \tilde{W}]$ satisfies*

$$\begin{aligned} &-\int_0^T \partial_t \phi \int_{\mathbb{T}^3} \tilde{\varrho}\tilde{\mathbf{u}} \cdot \boldsymbol{\varphi} \,\mathrm{d}x\,\mathrm{d}t - \phi(0) \int_{\mathbb{T}^3} \tilde{\varrho}\tilde{\mathbf{u}}_0 \cdot \boldsymbol{\varphi}\,\mathrm{d}x \\ &= \int_0^T \phi \int_{\mathbb{T}^3} [\tilde{\varrho}\tilde{\mathbf{u}} \otimes \tilde{\mathbf{u}} : \nabla \boldsymbol{\varphi} + \overline{p_\delta(\tilde{\varrho})} \operatorname{div} \boldsymbol{\varphi}]\,\mathrm{d}x\,\mathrm{d}t \\ &\quad - \int_0^T \phi \int_{\mathbb{T}^3} \mathbb{S}(\nabla \tilde{\mathbf{u}}) : \nabla \boldsymbol{\varphi}\,\mathrm{d}x\,\mathrm{d}t + \int_0^T \phi \int_{\mathbb{T}^3} \overline{\tilde{\varrho}\mathbb{F}(\tilde{\varrho}, \tilde{\mathbf{u}})} \cdot \boldsymbol{\varphi}\,\mathrm{d}x\,\mathrm{d}\tilde{W}, \end{aligned} \tag{4.160}$$

for all $\phi \in C_c^\infty([0,T))$ and all $\boldsymbol{\varphi} \in C^\infty(\mathbb{T}^3)$ $\mathbb{P}$-a.s.

Proof. Similarly to Proposition 4.3.14 we apply Theorem 2.9.1 to show that the random distributions $[\tilde{\varrho}_\varepsilon, \tilde{\mathbf{u}}_\varepsilon, \tilde{W}_\varepsilon]$ solve the momentum equation (4.75). Before we proceed we have to explain how to pass to the limit in the pressure and the stochastic integral. Note that at the current stage we are only able to show that a certain limit exists but we are unable to identify it. For the identification, strong convergence of the approximate densities $\tilde{\varrho}_\varepsilon$ is necessary, which is the main goal of Section 4.4.3.2. The identification of the limit will then be completed at the very end of this chapter.

In accordance with (4.158), we pass to the limit in the pressure term. The remaining part of the proof is devoted to the passage to the limit in the stochastic integral. In particular, we are going to show that, for $l > \frac{3}{2}$,

$$\tilde{\varrho}_\varepsilon \mathbf{F}_{k,\varepsilon}(\tilde{\varrho}_\varepsilon, \tilde{\mathbf{u}}_\varepsilon) \to \overline{\tilde{\varrho}\mathbf{F}_k(\tilde{\varrho}, \tilde{\mathbf{u}})} \quad \text{in } L^2(0,T; W^{-l,2}(\mathbb{T}^3)) \tag{4.161}$$

$\tilde{\mathbb{P}}$-a.s., for any $\boldsymbol{\varphi} \in C^\infty(\mathbb{T}^3)$ and any $k \in \mathbb{N}$, where $\overline{\tilde{\varrho}\mathbf{F}_k(\tilde{\varrho}, \tilde{\mathbf{u}})}$ was constructed in (4.159) as the weak limit of $\tilde{\varrho}_\varepsilon \mathbf{F}_k(\tilde{\varrho}_\varepsilon, \tilde{\mathbf{u}}_\varepsilon)$. Hence (4.161) means that $\tilde{\varrho}_\varepsilon \mathbf{F}_{k,\varepsilon}(\tilde{\varrho}_\varepsilon, \tilde{\mathbf{u}}_\varepsilon)$ converges to the same limit and the convergence is even strong in $L^2(0,T; W^{-l,2}(\mathbb{T}^3))$, which is needed in order to pass to the limit in the stochastic integral; cf. Lemma 2.6.6. First observe that, by (4.9) and (4.7),

$$\begin{aligned} &\int_{\mathbb{T}^3} \tilde{\varrho}_\varepsilon |\mathbf{F}_{k,\varepsilon}(\tilde{\varrho}_\varepsilon, \tilde{\mathbf{u}}_\varepsilon) - \mathbf{F}_k(\tilde{\varrho}_\varepsilon, \tilde{\mathbf{u}}_\varepsilon)|\,\mathrm{d}x \\ &\quad \lesssim \int_{\tilde{\varrho}_\varepsilon < \varepsilon} \tilde{\varrho}_\varepsilon |\mathbf{F}_k(\tilde{\varrho}_\varepsilon, \tilde{\mathbf{u}}_\varepsilon)|\,\mathrm{d}x + \int_{|\tilde{\mathbf{u}}_\varepsilon| > \frac{1}{\varepsilon}} \tilde{\varrho}_\varepsilon |\mathbf{F}_k(\tilde{\varrho}_\varepsilon, \tilde{\mathbf{u}}_\varepsilon)|\,\mathrm{d}x \end{aligned}$$

$$\lesssim f_k\Big[\int_{\tilde{\varrho}_\varepsilon<\varepsilon}(\tilde{\varrho}_\varepsilon+\tilde{\varrho}_\varepsilon|\tilde{\mathbf{u}}_\varepsilon|)\,\mathrm{d}x+\int_{|\tilde{\mathbf{u}}_\varepsilon|>\frac{1}{\varepsilon}}(\tilde{\varrho}_\varepsilon+\tilde{\varrho}_\varepsilon|\tilde{\mathbf{u}}_\varepsilon|)\,\mathrm{d}x\Big]$$
$$\lesssim f_k\Big[\varepsilon\int_{\mathbb{T}^3}1+|\tilde{\mathbf{u}}_\varepsilon|\,\mathrm{d}x+\int_{|\tilde{\mathbf{u}}_\varepsilon|>\frac{1}{\varepsilon}}(\tilde{\varrho}_\varepsilon|\tilde{\mathbf{u}}_\varepsilon|)\,\mathrm{d}x\Big]$$
$$\lesssim \varepsilon f_k\int_{\mathbb{T}^3}(1+|\tilde{\mathbf{u}}_\varepsilon|+\tilde{\varrho}_\varepsilon|\tilde{\mathbf{u}}_\varepsilon|^2)\,\mathrm{d}x.$$

Hence, due to (4.7) and the uniform *a priori* estimates (4.136) and (4.139) (which continue to hold on the new probability space by Proposition 4.4.9), we obtain

$$\tilde{\varrho}_\varepsilon\mathbf{F}_{k,\varepsilon}(\tilde{\varrho}_\varepsilon,\tilde{\mathbf{u}}_\varepsilon)-\tilde{\varrho}_\varepsilon\mathbf{F}_k(\tilde{\varrho}_\varepsilon,\tilde{\mathbf{u}}_\varepsilon)\to 0\quad\text{in } L^2(0,T;L^1(\mathbb{T}^3))$$

$\tilde{\mathbb{P}}$-a.s., as well as

$$\tilde{\varrho}_\varepsilon\mathbf{F}_{k,\varepsilon}(\tilde{\varrho}_\varepsilon,\tilde{\mathbf{u}}_\varepsilon)-\tilde{\varrho}_\varepsilon\mathbf{F}_k(\tilde{\varrho}_\varepsilon,\tilde{\mathbf{u}}_\varepsilon)\to 0\quad\text{in } L^2(0,T;W^{-l,2}(\mathbb{T}^3))\tag{4.162}$$

$\tilde{\mathbb{P}}$-a.s. for any $k\in\mathbb{N}$. To see (4.161), we first write

$$\int_{\mathbb{T}^3}\tilde{\varrho}_\varepsilon\mathbf{F}_k(\tilde{\varrho}_\varepsilon,\tilde{\mathbf{u}}_\varepsilon)\cdot\boldsymbol{\varphi}\,\mathrm{d}x=\int_{\mathbb{T}^3}\tilde{\varrho}_\varepsilon(\mathbf{F}_k(\tilde{\varrho}_\varepsilon,\tilde{\mathbf{u}}_\varepsilon)-\mathbf{F}_k(\tilde{\varrho}_\varepsilon,\tilde{\mathbf{u}}))\cdot\boldsymbol{\varphi}\,\mathrm{d}x+\int_{\mathbb{T}^3}\tilde{\varrho}_\varepsilon\mathbf{F}_k(\tilde{\varrho}_\varepsilon,\tilde{\mathbf{u}})\cdot\boldsymbol{\varphi}\,\mathrm{d}x.$$

By Hölder's inequality, we have

$$\int_{\mathbb{T}^3}\tilde{\varrho}_\varepsilon(\mathbf{F}_k(\tilde{\varrho}_\varepsilon,\tilde{\mathbf{u}}_\varepsilon)-\mathbf{F}_k(\tilde{\varrho}_\varepsilon,\tilde{\mathbf{u}}))\cdot\boldsymbol{\varphi}\,\mathrm{d}x\lesssim f_k\|\boldsymbol{\varphi}\|_{L^\infty_x}\int_{\mathbb{T}^3}\tilde{\varrho}_\varepsilon|\tilde{\mathbf{u}}_\varepsilon-\tilde{\mathbf{u}}|\,\mathrm{d}x$$
$$\lesssim f_k\|\boldsymbol{\varphi}\|_{L^\infty_x}\|\sqrt{\tilde{\varrho}_\varepsilon}\|_{L^2_x}\|\tilde{\varrho}_\varepsilon|\tilde{\mathbf{u}}_\varepsilon-\tilde{\mathbf{u}}|^2\|_{L^1_x}^{1/2}.$$

It follows from Proposition 4.4.9 and Corollary 4.4.10 that $\tilde{\mathbb{P}}$-a.s.

$$\|\tilde{\varrho}_\varepsilon|\tilde{\mathbf{u}}_\varepsilon-\tilde{\mathbf{u}}|^2\|_{L^1_x}=\int_{\mathbb{T}^3}\tilde{\varrho}_\varepsilon|\tilde{\mathbf{u}}_\varepsilon|^2\,\mathrm{d}x-2\int_{\mathbb{T}^3}\tilde{\varrho}_\varepsilon\tilde{\mathbf{u}}_\varepsilon\cdot\tilde{\mathbf{u}}\,\mathrm{d}x+\int_{\mathbb{T}^3}\tilde{\varrho}_\varepsilon|\tilde{\mathbf{u}}|^2\,\mathrm{d}x$$
$$\to 0\quad\text{in } L^2(0,T).\tag{4.163}$$

This implies

$$\tilde{\varrho}_\varepsilon(\mathbf{F}_k(\tilde{\varrho}_\varepsilon,\tilde{\mathbf{u}}_\varepsilon)-\mathbf{F}_k(\tilde{\varrho}_\varepsilon,\tilde{\mathbf{u}}))\to 0\quad\text{in } L^2(0,T;W^{-l,2}(\mathbb{T}^3)),$$

for any $\boldsymbol{\varphi}\in C^\infty(\mathbb{T}^3)$ $\tilde{\mathbb{P}}$-a.s. Thus, (4.161) follows if we show that

$$\tilde{\varrho}_\varepsilon\mathbf{F}_k(\tilde{\varrho}_\varepsilon,\tilde{\mathbf{u}})\to\overline{\tilde{\varrho}\mathbf{F}_k(\tilde{\varrho},\tilde{\mathbf{u}})}\quad\text{in } L^2(0,T;W^{-l,2}(\mathbb{T}^3))\tag{4.164}$$

$\tilde{\mathbb{P}}$-a.s. Finally, to see (4.164), we need the following renormalized form of the equation of continuity:

$$\partial_t b(\tilde{\varrho}_\varepsilon)+\operatorname{div}(b(\tilde{\varrho}_\varepsilon)\tilde{\mathbf{u}}_\varepsilon)+(b'(\tilde{\varrho}_\varepsilon)\tilde{\varrho}_\varepsilon-b(\tilde{\varrho}_\varepsilon))\operatorname{div}\tilde{\mathbf{u}}_\varepsilon$$
$$=\varepsilon\operatorname{div}(b'(\tilde{\varrho}_\varepsilon)\nabla\tilde{\varrho}_\varepsilon)-\varepsilon b''(\tilde{\varrho}_\varepsilon)|\nabla\tilde{\varrho}_\varepsilon|^2,\tag{4.165}$$

for any b having at most quadratic growth,

$$|b''(\varrho)| \lesssim 1 \quad \text{for all } \varrho \geq 0.$$

Equation (4.165) follows easily by multiplying (4.156) by $b'(\tilde{\varrho}_\varepsilon)$. Now, with the help of (4.154) and (4.157), we deduce from (4.165)

$$b(\tilde{\varrho}_\varepsilon) \to \overline{b(\tilde{\varrho})} \quad \text{in } C_w([0,T];L^2(\mathbb{T}^3)) \text{ as } \varepsilon \to 0. \tag{4.166}$$

Note carefully that we do not have to pass to a subsequence as the limit is uniquely determined by (4.154).

The function b in (4.166) depends solely on ϱ. To obtain a similar statement for the Carathéodory composition,

$$\tilde{\varrho}_\varepsilon \mathbf{F}_k(\tilde{\varrho}_\varepsilon, t, x) = \tilde{\varrho}_\varepsilon \mathbf{F}_k(\tilde{\varrho}_\varepsilon, \tilde{\mathbf{u}}(t,x)),$$

we use an approximation argument. First, we recall the compact embedding

$$C_w([0,T];L^2(\mathbb{T}^3)) \overset{c}{\hookrightarrow} L^2(0,T;W^{-1,2}).$$

Consequently, we deduce from (4.166) and the fact that $\tilde{\mathbf{u}} \in L^2(0,T;W^{1,2}(\mathbb{T}^3))$ $\tilde{\mathbb{P}}$-a.s. that

$$b(\tilde{\varrho}_\varepsilon)B(\tilde{\mathbf{u}}) \to \overline{b(\tilde{\varrho})B(\tilde{\mathbf{u}})} \quad \text{in } L^2(0,T;W^{-l,2}(\mathbb{T}^3)),$$

for any globally Lipschitz b and B. Thus (4.164) follows by a density argument via approximation of $\varrho\mathbf{F}_k(\varrho,\mathbf{u})$ by finite sums $\sum_i b_i(\varrho)B_i(\mathbf{u})$. Note again that the result holds unconditionally for the original sequence, meaning we do not need to take a subsequence in ω. We have shown (4.161).

For $l > \frac{3}{2}$, we have, by Sobolev's embedding (cf. (3.15)),

$$\begin{aligned}
&\tilde{\mathbb{E}}\int_0^T \|\tilde{\varrho}_\varepsilon \mathbb{F}_\varepsilon(\tilde{\varrho}_\varepsilon, \tilde{\mathbf{u}}_\varepsilon)\|^2_{L_2(\mathfrak{U};W_x^{-l,2})}\,\mathrm{d}t \\
&\lesssim \tilde{\mathbb{E}}\int_0^T (\tilde{\varrho}_\varepsilon)_{\mathbb{T}^3}\int_{\mathbb{T}^3}\left(\sum_{k=1}^\infty \tilde{\varrho}_\varepsilon^{-1}|\mathbf{G}_{k,\varepsilon}(\tilde{\varrho}_\varepsilon, \tilde{\varrho}_\varepsilon\tilde{\mathbf{u}}_\varepsilon)|^2\right)\mathrm{d}x\,\mathrm{d}t \\
&\lesssim \tilde{\mathbb{E}}\int_0^T (\tilde{\varrho}_\varepsilon)_{\mathbb{T}^3}\int_{\mathbb{T}^3}(\tilde{\varrho}_\varepsilon + \tilde{\varrho}_\varepsilon|\tilde{\mathbf{u}}_\varepsilon|^2)\,\mathrm{d}x\,\mathrm{d}t \lesssim c_1(r),
\end{aligned} \tag{4.167}$$

using also (4.9), (4.7), (4.135), and (4.136). Consequently, (4.161) implies

$$\tilde{\varrho}_\varepsilon \mathbb{F}_\varepsilon(\tilde{\varrho}_\varepsilon, \tilde{\mathbf{u}}_\varepsilon) \to \overline{\tilde{\varrho}\mathbb{F}(\tilde{\varrho}, \tilde{\mathbf{u}})} \quad \text{in } L^2(0,T;L_2(\mathfrak{U};W^{-l,2}(\mathbb{T}^3)))$$

$\tilde{\mathbb{P}}$-a.s. Combining this with the convergence of $\tilde{W}_\varepsilon$ from Proposition 4.4.9, we may apply Lemma 2.6.6 to pass to the limit in the stochastic integral and hence complete the proof. □

4.4.3.2 Deterministic compactness method

Our ultimate goal is to show strong (pointwise a.e.) convergence of the approximate densities. We adapt the method used in the deterministic case based on weak continuity of the *effective viscous flux*. Evoking the procedure yielding (4.142), we use the quantity

$$\nabla\Delta^{-1}[\tilde{\varrho} - (\tilde{\varrho})_{\mathbb{T}^3}] = \Delta^{-1}\nabla\tilde{\varrho}$$

as a test function in (4.160). Repeating step by step the arguments used in Section 4.4.2, we obtain

$$\begin{aligned}
&\int_0^\tau \int_{\mathbb{T}^3} \overline{p_\delta(\tilde{\varrho})}[\tilde{\varrho} - (\tilde{\varrho})_{\mathbb{T}^3}]\,\mathrm{d}x\,\mathrm{d}t \\
&\quad = \left[\int_{\mathbb{T}^3} \tilde{\varrho}\tilde{\mathbf{u}} \cdot \Delta^{-1}[\nabla\tilde{\varrho}]\,\mathrm{d}x\right]_{t=0}^{t=\tau} - \int_0^\tau \int_{\mathbb{T}^3} \tilde{\varrho}\tilde{\mathbf{u}} \otimes \tilde{\mathbf{u}} : \nabla\Delta^{-1}\nabla\tilde{\varrho}\,\mathrm{d}x\,\mathrm{d}t \\
&\qquad + \int_0^\tau \int_{\mathbb{T}^3} \mathbb{S}(\nabla\tilde{\mathbf{u}}) : \nabla\Delta^{-1}\nabla\tilde{\varrho}\,\mathrm{d}x\,\mathrm{d}t + \int_0^\tau \int_{\mathbb{T}^3} \tilde{\varrho}\tilde{\mathbf{u}} \cdot \nabla\Delta^{-1}\mathrm{div}\,(\tilde{\varrho}\tilde{\mathbf{u}})\,\mathrm{d}x\,\mathrm{d}t \\
&\qquad - \int_0^\tau \int_{\mathbb{T}^3} \overline{\tilde{\varrho}\mathbb{F}(\tilde{\varrho},\tilde{\mathbf{u}})} \cdot \Delta^{-1}\nabla\tilde{\varrho}\,\mathrm{d}x\,\mathrm{d}\tilde{W}.
\end{aligned} \tag{4.168}$$

As $\Gamma \geq 4$, we deduce from Proposition 4.4.9, the standard properties of Δ^{-1}, and the embedding $W^{1,\Gamma}(\mathbb{T}^3) \overset{C}{\hookrightarrow} C(\mathbb{T}^3)$

$$\nabla\Delta^{-1}[\tilde{\varrho}_\varepsilon - (\tilde{\varrho}_\varepsilon)_{\mathbb{T}^3}] \to \nabla\Delta^{-1}[\tilde{\varrho} - (\tilde{\varrho})_{\mathbb{T}^3}], \tag{4.169}$$

in $C([0,T] \times \mathbb{T}^3)$ $\tilde{\mathbb{P}}$-a.s. Next, it is also convenient to rewrite

$$\int_0^\tau \int_{\mathbb{T}^3} \mathbb{S}(\nabla\tilde{\mathbf{u}}_\varepsilon) : \nabla\Delta^{-1}\nabla\tilde{\varrho}_\varepsilon\,\mathrm{d}x\,\mathrm{d}t = \int_0^\tau \int_{\mathbb{T}^3} (\eta + \mu)\mathrm{div}\,\tilde{\mathbf{u}}_\varepsilon\,\tilde{\varrho}_\varepsilon\,\mathrm{d}x\,\mathrm{d}t$$

and, similarly,

$$\int_0^\tau \int_{\mathbb{T}^3} \mathbb{S}(\nabla\tilde{\mathbf{u}}) : \nabla\Delta^{-1}\nabla\tilde{\varrho}\,\mathrm{d}x\,\mathrm{d}t = \int_0^\tau \int_{\mathbb{T}^3} (\eta + \mu)\mathrm{div}\,\tilde{\mathbf{u}}\,\tilde{\varrho}\,\mathrm{d}x\,\mathrm{d}t.$$

Using the compactness properties of the family $[\tilde{\varrho}_\varepsilon, \tilde{\mathbf{u}}_\varepsilon, \tilde{W}_\varepsilon]$, established in this section, it is not difficult to let $\varepsilon \to 0$ in the corresponding version of (4.142) and to compare the limit with (4.168). For the convergence of the stochastic integrals, we note that, after the change of probability space, the approximate stochastic integrals in the corresponding version of (4.142) are driven by $\tilde{W}_\varepsilon$, not by a single process. Consequently, in order to pass to the limit, we apply Lemma 2.6.6 together with (4.161) and (4.169). We obtain

$$\lim_{\varepsilon\to 0}\left[\int_0^\tau \int_{\mathbb{T}^3} p_\delta(\tilde{\varrho}_\varepsilon)\tilde{\varrho}_\varepsilon\,\mathrm{d}x\,\mathrm{d}t - \int_0^\tau \int_{\mathbb{T}^3} (\eta + \mu)\mathrm{div}\,\tilde{\mathbf{u}}_\varepsilon\,\tilde{\varrho}_\varepsilon\,\mathrm{d}x\,\mathrm{d}t\right]$$

$$
\begin{aligned}
&-\int_0^\tau\int_{\mathbb{T}^3}\overline{p_\delta(\tilde\varrho)\tilde\varrho}\,\mathrm{d}x\,\mathrm{d}t+\int_0^\tau\int_{\mathbb{T}^3}(\eta+\mu)\operatorname{div}\tilde{\mathbf{u}}\,\tilde\varrho\,\mathrm{d}x\,\mathrm{d}t\\
&=\lim_{\varepsilon\to0}\int_0^\tau\int_{\mathbb{T}^3}\big[\tilde\varrho_\varepsilon\tilde{\mathbf{u}}_\varepsilon\cdot\nabla\Delta^{-1}\operatorname{div}(\tilde\varrho_\varepsilon\tilde{\mathbf{u}}_\varepsilon)-\tilde\varrho_\varepsilon\tilde{\mathbf{u}}_\varepsilon\otimes\tilde{\mathbf{u}}_\varepsilon:\nabla\Delta^{-1}\nabla\tilde\varrho_\varepsilon\big]\,\mathrm{d}x\,\mathrm{d}t\\
&\quad-\int_0^\tau\int_{\mathbb{T}^3}\big[\tilde\varrho\tilde{\mathbf{u}}\cdot\nabla\Delta^{-1}\operatorname{div}(\tilde\varrho\tilde{\mathbf{u}})-\tilde\varrho\tilde{\mathbf{u}}\otimes\tilde{\mathbf{u}}:\nabla\Delta^{-1}\nabla\tilde\varrho\big]\,\mathrm{d}x\,\mathrm{d}t\quad\tilde{\mathbb{P}}\text{-a.s.}
\end{aligned}
\tag{4.170}
$$

Next, we show that the right hand side of (4.170) actually vanishes. We first rewrite

$$
\begin{aligned}
&\int_0^\tau\int_{\mathbb{T}^3}\big[\tilde\varrho_\varepsilon\tilde{\mathbf{u}}_\varepsilon\cdot\nabla\Delta^{-1}\operatorname{div}(\tilde\varrho_\varepsilon\tilde{\mathbf{u}}_\varepsilon)-\tilde\varrho_\varepsilon\tilde{\mathbf{u}}_\varepsilon\otimes\tilde{\mathbf{u}}_\varepsilon:\nabla\Delta^{-1}\nabla\tilde\varrho_\varepsilon\big]\,\mathrm{d}x\,\mathrm{d}t\\
&\quad=\int_0^\tau\int_{\mathbb{T}^3}\tilde{\mathbf{u}}_\varepsilon\cdot\big[\tilde\varrho_\varepsilon\nabla\Delta^{-1}\operatorname{div}(\tilde\varrho_\varepsilon\tilde{\mathbf{u}}_\varepsilon)-\tilde\varrho_\varepsilon\tilde{\mathbf{u}}_\varepsilon\cdot\nabla\Delta^{-1}\nabla\tilde\varrho_\varepsilon\big]\,\mathrm{d}x\,\mathrm{d}t.
\end{aligned}
$$

Using the convergence result from Proposition 4.4.9 and Lemma A.1.11, we obtain $\tilde{\mathbb{P}}$-a.s.

$$
\begin{aligned}
&\tilde\varrho_\varepsilon\nabla\Delta^{-1}\operatorname{div}(\tilde\varrho_\varepsilon\tilde{\mathbf{u}}_\varepsilon)-\tilde\varrho_\varepsilon\tilde{\mathbf{u}}_\varepsilon\cdot\nabla\Delta^{-1}\nabla\tilde\varrho_\varepsilon\\
&\qquad\to\tilde\varrho\nabla\Delta^{-1}\operatorname{div}(\tilde\varrho\tilde{\mathbf{u}})-\tilde\varrho\tilde{\mathbf{u}}\cdot\nabla\Delta^{-1}\nabla\tilde\varrho
\end{aligned}
$$

weakly in $L^r(\mathbb{T}^3)$ for any $t\in[0,T]$, where

$$
\frac{1}{r}=\frac{1}{\Gamma}+\frac{\Gamma+1}{2\Gamma}.
$$

Choosing

$$
\Gamma\geq5\quad\text{so that } r=\frac{2\Gamma}{\Gamma+3}>\frac{6}{5},
$$

we get $(L^\Gamma(\mathbb{T}^3),w)\overset{c}{\hookrightarrow}W^{-1,2}(\mathbb{T}^3)$ and, consequently,

$$
\begin{aligned}
&\tilde\varrho_\varepsilon\nabla\Delta^{-1}\operatorname{div}(\tilde\varrho_\varepsilon\tilde{\mathbf{u}}_\varepsilon)-\tilde\varrho_\varepsilon\tilde{\mathbf{u}}_\varepsilon\cdot\nabla\Delta^{-1}\nabla\tilde\varrho_\varepsilon\\
&\qquad\to\tilde\varrho\nabla\Delta^{-1}\operatorname{div}(\tilde\varrho\tilde{\mathbf{u}})-\tilde\varrho\tilde{\mathbf{u}}\cdot\nabla\Delta^{-1}\nabla\tilde\varrho
\end{aligned}
$$

strongly in $L^2(0,T;W^{-1,2}(\mathbb{T}^3))$ $\tilde{\mathbb{P}}$-a.s. This, combined with (4.153), yields the desired conclusion

$$
\begin{aligned}
&\lim_{\varepsilon\to0}\int_0^\tau\int_{\mathbb{T}^3}\big[\tilde\varrho_\varepsilon\tilde{\mathbf{u}}_\varepsilon\cdot\nabla\Delta^{-1}\operatorname{div}(\tilde\varrho_\varepsilon\tilde{\mathbf{u}}_\varepsilon)-\tilde\varrho_\varepsilon\tilde{\mathbf{u}}_\varepsilon\otimes\tilde{\mathbf{u}}_\varepsilon:\nabla\Delta^{-1}\nabla\tilde\varrho_\varepsilon\big]\,\mathrm{d}x\,\mathrm{d}t\\
&\quad-\int_0^\tau\int_{\mathbb{T}^3}\big[\tilde\varrho\tilde{\mathbf{u}}\cdot\nabla\Delta^{-1}\operatorname{div}(\tilde\varrho\tilde{\mathbf{u}})-\tilde\varrho\tilde{\mathbf{u}}\otimes\tilde{\mathbf{u}}:\nabla\Delta^{-1}\nabla\tilde\varrho\big]\,\mathrm{d}x\,\mathrm{d}t=0.
\end{aligned}
$$

Consequently, relation (4.170) gives rise to

$$
\begin{aligned}
&\lim_{\varepsilon\to0}\int_0^\tau\int_{\mathbb{T}^3}\big[p_\delta(\tilde\varrho_\varepsilon)\tilde\varrho_\varepsilon-(\eta+\mu)\operatorname{div}\tilde{\mathbf{u}}_\varepsilon\,\tilde\varrho_\varepsilon\big]\,\mathrm{d}x\,\mathrm{d}t\\
&\quad=\int_0^\tau\int_{\mathbb{T}^3}\big[\overline{p_\delta(\tilde\varrho)\tilde\varrho}-(\eta+\mu)\operatorname{div}\tilde{\mathbf{u}}\,\tilde\varrho\big]\,\mathrm{d}x\,\mathrm{d}t\quad\tilde{\mathbb{P}}\text{-a.s.}
\end{aligned}
\tag{4.171}
$$

To exploit the piece of information hidden in (4.171), we let $\varepsilon \to 0$ in the renormalized equation (4.165) with $b(\tilde{\varrho}_\varepsilon) = \tilde{\varrho}_\varepsilon \log(\tilde{\varrho}_\varepsilon)$. Using convexity of b together with Proposition 4.4.9, part (4), we get

$$\partial_t \overline{\tilde{\varrho} \log(\tilde{\varrho})} + \operatorname{div}\left(\overline{\tilde{\varrho}\log(\tilde{\varrho})}\tilde{\mathbf{u}}\right) + \overline{\tilde{\varrho}\operatorname{div}\tilde{\mathbf{u}}} \le 0,$$

or, more precisely,

$$\begin{aligned}\int_{\mathbb{T}^3} \overline{\tilde{\varrho}\log(\tilde{\varrho})}\psi(\tau)\,\mathrm{d}x + \int_0^\tau \int_{\mathbb{T}^3} \overline{\tilde{\varrho}\operatorname{div}\tilde{\mathbf{u}}}\psi\,\mathrm{d}x\,\mathrm{d}t &\le \int_{\mathbb{T}^3} \tilde{\varrho}_0 \log(\tilde{\varrho}_0)\psi(0)\,\mathrm{d}x \\ &+ \int_0^\tau \int_{\mathbb{T}^3} [\overline{\tilde{\varrho}\log(\tilde{\varrho})}\partial_t\psi + \overline{\tilde{\varrho}\log(\tilde{\varrho})}\tilde{\mathbf{u}}\cdot\nabla\psi]\,\mathrm{d}x\,\mathrm{d}t, \end{aligned} \tag{4.172}$$

for any $\psi \in C_c^\infty([0,\tau)\times\mathbb{T}^3)$, $\psi \ge 0$.

Finally, we apply the DiPerna–Lions theory of renormalized solutions (cf. Theorem A.3.1) to the limit equation from Lemma 4.4.11. Since $\tilde{\mathbf{u}} \in L^2(0,T;W^{1,2}(\mathbb{T}^3))$ $\tilde{\mathbb{P}}$-a.s. and $\tilde{\varrho} \in L^\infty(0,T;L^\Gamma(\mathbb{T}^3))$ $\tilde{\mathbb{P}}$-a.s. where $\Gamma \ge 2$, equation (4.123) is satisfied also in the renormalized sense. We have

$$\begin{aligned}\int_{\mathbb{T}^3} b(\tilde{\varrho}(\tau))\psi(\tau)\,\mathrm{d}x + \int_0^\tau \int_{\mathbb{T}^3} (b'(\tilde{\varrho})\tilde{\varrho} - b(\tilde{\varrho}))\operatorname{div}\mathbf{u}\psi\,\mathrm{d}x\,\mathrm{d}t &= \int_{\mathbb{T}^3} b(\tilde{\varrho}_0)\psi(0)\,\mathrm{d}x \\ &+ \int_0^\tau \int_{\mathbb{T}^3} [b(\tilde{\varrho})\partial_t\psi + b(\tilde{\varrho})\tilde{\mathbf{u}}\cdot\nabla\psi]\,\mathrm{d}x\,\mathrm{d}t, \end{aligned} \tag{4.173}$$

for any $\psi \in C_c^\infty([0,T]\times\mathbb{T}^3)$. In particular,

$$\begin{aligned}\int_{\mathbb{T}^3} \tilde{\varrho}\log(\tilde{\varrho})\psi(\tau)\,\mathrm{d}x + \int_0^\tau \int_{\mathbb{T}^3} \tilde{\varrho}\operatorname{div}\tilde{\mathbf{u}}\psi\,\mathrm{d}x\,\mathrm{d}t &= \int_{\mathbb{T}^3} \tilde{\varrho}_0 \log(\tilde{\varrho}_0)\psi(0)\,\mathrm{d}x \\ &+ \int_0^\tau \int_{\mathbb{T}^3} [\tilde{\varrho}\log(\tilde{\varrho})\partial_t\psi + \tilde{\varrho}\log(\tilde{\varrho})\tilde{\mathbf{u}}\cdot\nabla\psi]\,\mathrm{d}x\,\mathrm{d}t. \end{aligned} \tag{4.174}$$

Subtracting (4.174) from (4.172) and taking ψ as a smooth approximation of the indicator function of $[0,\tau]$, we obtain

$$\int_{\mathbb{T}^3} [\overline{\tilde{\varrho}\log(\tilde{\varrho})} - \tilde{\varrho}\log(\tilde{\varrho})](\tau)\,\mathrm{d}x + \int_0^\tau \int_{\mathbb{T}^3} [\overline{\tilde{\varrho}\operatorname{div}\tilde{\mathbf{u}}} - \tilde{\varrho}\operatorname{div}\tilde{\mathbf{u}}]\,\mathrm{d}x\,\mathrm{d}t \le 0, \tag{4.175}$$

for any $\tau \in [0,T]$. As the pressure is a non-decreasing function of the density, we get

$$\lim_{\varepsilon\to 0} \int_0^\tau \int_{\mathbb{T}^3} p_\delta(\tilde{\varrho}_\varepsilon)\tilde{\varrho}_\varepsilon\,\mathrm{d}x\,\mathrm{d}t \ge \int_0^\tau \int_{\mathbb{T}^3} \overline{p_\delta(\tilde{\varrho})}\tilde{\varrho}\,\mathrm{d}x\,\mathrm{d}t.$$

Thus relation (4.171) yields

$$\int_0^\tau \int_{\mathbb{T}^3} [\overline{\tilde{\varrho}\operatorname{div}\tilde{\mathbf{u}}} - \tilde{\varrho}\operatorname{div}\tilde{\mathbf{u}}]\,\mathrm{d}x\,\mathrm{d}t \ge 0$$

and (4.175) reduces to

$$\int_{\mathbb{T}^3} [\overline{\tilde{\varrho}\log(\tilde{\varrho})} - \tilde{\varrho}\log(\tilde{\varrho})](\tau)\,\mathrm{d}x \le 0 \quad \text{for any } \tau \in [0,T]. \tag{4.176}$$

As the function $\tilde{\varrho} \mapsto \tilde{\varrho}\log(\tilde{\varrho})$ is strictly convex, relation (4.176) implies the strong convergence of $\tilde{\varrho}_\varepsilon$. Specifically,

$$\tilde{\varrho}_\varepsilon \to \tilde{\varrho} \quad \text{in } L^q(0,T;L^1(\mathbb{T}^3)) \quad \text{for any } 1 \le q < \infty \ \tilde{\mathbb{P}}\text{-a.s.} \tag{4.177}$$

With the strong convergence (4.177) at hand we are now able to identify the limit in the stochastic integral. In view of (4.162) and (4.163) we only need the $\tilde{\mathbb{P}}$-a.s. convergence

$$\int_{\mathbb{T}^3} \tilde{\varrho}_\varepsilon \mathbf{F}_k(\tilde{\varrho}_\varepsilon, \tilde{\mathbf{u}}) \cdot \boldsymbol{\varphi} \,\mathrm{d}x \to \int_{\mathbb{T}^3} \tilde{\varrho}\mathbf{F}_k(\tilde{\varrho}, \tilde{\mathbf{u}}) \cdot \boldsymbol{\varphi} \,\mathrm{d}x \quad \text{a.e. in } (0,T) \tag{4.178}$$

for any $\boldsymbol{\varphi} \in C^\infty(\mathbb{T}^3)$. Obviously (4.178) follows immediately from (4.177) and the Lipschitz continuity of $\mathbf{F}_k$. Hence we infer

$$\overline{\tilde{\varrho}\mathbf{F}_k(\tilde{\varrho}, \tilde{\mathbf{u}})} = \tilde{\varrho}\mathbf{F}_k(\tilde{\varrho}, \tilde{\mathbf{u}}) \quad \text{a.e. in } \tilde{\Omega} \times (0,T) \times \mathbb{T}^3.$$

We are now going to pass to the limit in the stochastic integral appearing in the energy inequality.

Proposition 4.4.13. *We have*

$$\int_0^T \int_{\mathbb{T}^3} \tilde{\varrho}_\varepsilon \, \mathbb{F}_\varepsilon(\tilde{\varrho}_\varepsilon, \tilde{\mathbf{u}}_\varepsilon) \cdot \tilde{\mathbf{u}}_\varepsilon \,\mathrm{d}x \,\mathrm{d}\tilde{W}_\varepsilon \to \int_0^T \int_{\mathbb{T}^3} \tilde{\varrho}\, \mathbb{F}(\tilde{\varrho}, \tilde{\mathbf{u}}) \cdot \tilde{\mathbf{u}} \,\mathrm{d}x \,\mathrm{d}\tilde{W} \quad \textit{in } L^2(0,T)$$

in probability.

Proof. We proceed similarly as in Proposition 4.3.15 and employ Lemma 2.6.6. It yields the claimed convergence, provided we can show

$$\int_{\mathbb{T}^3} \tilde{\varrho}_\varepsilon \, \mathbb{F}_\varepsilon(\tilde{\varrho}_\varepsilon, \tilde{\mathbf{u}}_\varepsilon) \cdot \tilde{\mathbf{u}}_\varepsilon \,\mathrm{d}x \to \int_{\mathbb{T}^3} \tilde{\varrho}\, \mathbb{F}(\tilde{\varrho}, \tilde{\mathbf{u}}) \cdot \tilde{\mathbf{u}} \,\mathrm{d}x \quad \text{in } L^2(0,T;L_2(\mathfrak{U};\mathbb{R})) \tag{4.179}$$

$\tilde{\mathbb{P}}$-a.s. First, let us denote the approximate stochastic integral by $\tilde{\mathscr{M}}_\varepsilon$. We observe, by (4.7) and (4.9),

$$\begin{aligned}\mathbb{E}\|\tilde{\mathscr{M}}_\varepsilon\|^2_{L^2(0,T;L_2(\mathfrak{U};\mathbb{R}))} &= \mathbb{E}\left[\sum_{k=1}^\infty \int_0^T \left(\int_{\mathbb{T}^3} \tilde{\varrho}_\varepsilon \, \mathbf{F}_{k,\varepsilon}(\tilde{\varrho}_\varepsilon, \tilde{\mathbf{u}}_\varepsilon) \cdot \tilde{\mathbf{u}}_\varepsilon \,\mathrm{d}x\right)^2 \mathrm{d}t\right] \\ &\lesssim f_k^2\, \mathbb{E}\left[\int_0^T \left(\int_{\mathbb{T}^3} (\tilde{\varrho}_\varepsilon + \tilde{\varrho}_\varepsilon |\tilde{\mathbf{u}}_\varepsilon|^2) \,\mathrm{d}x\right)^2 \mathrm{d}t\right] \lesssim f_k^2,\end{aligned}$$

using (4.135) and (4.136). Due to the summability of f_k, the convergence (4.179) follows, provided we can show

$$\int_{\mathbb{T}^3} \tilde{\varrho}_\varepsilon \, \mathbf{F}_{k,\varepsilon}(\tilde{\varrho}_\varepsilon, \tilde{\mathbf{u}}_\varepsilon) \cdot \tilde{\mathbf{u}}_\varepsilon \,\mathrm{d}x \to \int_{\mathbb{T}^3} \tilde{\varrho}\, \mathbf{F}_k(\tilde{\varrho}, \tilde{\mathbf{u}}) \cdot \tilde{\mathbf{u}} \,\mathrm{d}x \quad \text{in } L^2(0,T) \tag{4.180}$$

$\tilde{\mathbb{P}}$-a.s. for all $k \in \mathbb{N}$. Because of (4.135) and (4.136), this can be relaxed to

$$\int_{\mathbb{T}^3} \tilde{\varrho}_\varepsilon \mathbf{F}_{k,\varepsilon}(\tilde{\varrho}_\varepsilon, \tilde{\mathbf{u}}_\varepsilon) \cdot \tilde{\mathbf{u}}_\varepsilon \, dx \to \int_{\mathbb{T}^3} \tilde{\varrho} \mathbf{F}_k(\tilde{\varrho}, \tilde{\mathbf{u}}) \cdot \tilde{\mathbf{u}} \, dx \quad \text{for a.e. } t \in (0,T) \tag{4.181}$$

$\tilde{\mathbb{P}}$-a.s. for all $k \in \mathbb{N}$. In order to show (4.181) we observe

$$\begin{aligned}
&\int_{\mathbb{T}^3} \tilde{\varrho}_\varepsilon |\tilde{\mathbf{u}}_\varepsilon| |\mathbf{F}_{k,\varepsilon}(\tilde{\varrho}_\varepsilon, \tilde{\mathbf{u}}_\varepsilon) - \mathbf{F}_k(\tilde{\varrho}_\varepsilon, \tilde{\mathbf{u}}_\varepsilon)| \, dx \\
&\quad \lesssim \int_{\tilde{\varrho}_\varepsilon < \varepsilon} \tilde{\varrho}_\varepsilon |\tilde{\mathbf{u}}_\varepsilon| |\mathbf{F}_k(\tilde{\varrho}_\varepsilon, \tilde{\mathbf{u}}_\varepsilon)| \, dx + \int_{|\tilde{\mathbf{u}}_\varepsilon| > \frac{1}{\varepsilon}} \tilde{\varrho}_\varepsilon |\tilde{\mathbf{u}}_\varepsilon| |\mathbf{F}_k(\tilde{\varrho}_\varepsilon, \tilde{\mathbf{u}}_\varepsilon)| \, dx \\
&\quad \lesssim f_k \Big[\int_{\tilde{\varrho}_\varepsilon < \varepsilon} (\tilde{\varrho}_\varepsilon |\tilde{\mathbf{u}}_\varepsilon| + \tilde{\varrho}_\varepsilon |\tilde{\mathbf{u}}_\varepsilon|^2) \, dx + \int_{|\tilde{\mathbf{u}}_\varepsilon| > \frac{1}{\varepsilon}} (\tilde{\varrho}_\varepsilon |\tilde{\mathbf{u}}_\varepsilon| + \tilde{\varrho}_\varepsilon |\tilde{\mathbf{u}}_\varepsilon|^2) \, dx \Big] \\
&\quad \lesssim f_k \varepsilon \int_{\mathbb{T}^3} (1 + |\tilde{\mathbf{u}}_\varepsilon|^2) \, dx + f_k \Big(\mathfrak{L}^3 \Big(|\tilde{\mathbf{u}}_\varepsilon| > \frac{1}{\varepsilon} \Big) \Big)^{\frac{\gamma-1}{2\gamma}} \Big(\int_{\mathbb{T}^3} |\tilde{\varrho}_\varepsilon \tilde{\mathbf{u}}_\varepsilon|^{\frac{2\gamma}{\gamma+1}} \, dx \Big)^{\frac{\gamma+1}{2\gamma}} \\
&\qquad + f_k \Big(\mathfrak{L}^3 \Big(|\tilde{\mathbf{u}}_\varepsilon| > \frac{1}{\varepsilon} \Big) \Big)^{\frac{2\gamma-3}{6\gamma}} \Big(\int_{\mathbb{T}^3} (\tilde{\varrho}_\varepsilon |\tilde{\mathbf{u}}_\varepsilon|^2)^{\frac{6\gamma}{4\gamma+3}} \, dx \Big)^{\frac{4\gamma+3}{6\gamma}},
\end{aligned}$$

similarly to (4.161). Due to (4.135), (4.136), and Chebyshev's inequality, the right hand side vanishes as $\varepsilon \to 0$ $\tilde{\mathbb{P}}$-a.s. for a.e. $t \in (0,T)$. Consequently, it suffices to prove

$$\int_{\mathbb{T}^3} \tilde{\varrho}_\varepsilon \mathbf{F}_k(\tilde{\varrho}_\varepsilon, \tilde{\mathbf{u}}_\varepsilon) \cdot \tilde{\mathbf{u}}_\varepsilon \, dx \to \int_{\mathbb{T}^3} \tilde{\varrho} \mathbf{F}_k(\tilde{\varrho}, \tilde{\mathbf{u}}) \cdot \tilde{\mathbf{u}} \, dx \quad \text{for a.e. } t \in (0,T) \tag{4.182}$$

$\tilde{\mathbb{P}}$-a.s. for all $k \in \mathbb{N}$. We proceed similarly to the proof of (4.112). Note that, as a consequence of Corollary 4.4.10, we have (up to a subsequence)

$$\sqrt{\tilde{\varrho}_\varepsilon} \tilde{\mathbf{u}}_\varepsilon \to \sqrt{\tilde{\varrho}} \tilde{\mathbf{u}} \quad \text{a.e. in } (0,T) \times \mathbb{T}^3 \ \tilde{\mathbb{P}}\text{-a.s.}$$

Let us fix some arbitrary $\kappa > 0$. By Egorov's theorem and (4.177) there exists a measurable set $\mathcal{O}_\kappa \subset \tilde{\Omega} \times (0,T) \times \mathbb{T}^3$ such that $\tilde{\mathbb{P}} \otimes \mathfrak{L}^4([\tilde{\Omega} \times (0,T) \times \mathbb{T}^3] \setminus \mathcal{O}_\kappa) < \kappa$ and

$$\sqrt{\tilde{\varrho}_\varepsilon} \tilde{\mathbf{u}}_\varepsilon \to \sqrt{\tilde{\varrho}} \tilde{\mathbf{u}}, \quad \tilde{\varrho}_\varepsilon \to \tilde{\varrho} \text{ uniformly in } \mathcal{O}_\kappa. \tag{4.183}$$

Finally, we consider the sets

$$\begin{aligned}
\mathcal{O}_\kappa^1 &= \{(\omega, t, x) \in \mathcal{O}_\kappa : \tilde{\varrho} < \kappa\}, \\
\mathcal{O}_\kappa^2 &= \{(\omega, t, x) \in \mathcal{O}_\kappa : \tilde{\varrho} \geq \kappa\}.
\end{aligned}$$

As a consequence of (4.183), we can choose m large enough such that

$$\tilde{\varrho}_\varepsilon \leq 2\kappa \quad \text{in } \mathcal{O}_\kappa^1, \quad \tilde{\varrho}_\varepsilon \geq \frac{\kappa}{2} \quad \text{in } \mathcal{O}_\kappa^2.$$

With these preparations at hand we gain

$$\tilde{\mathbb{E}} \int_0^T \int_{\mathbb{T}^3} |\tilde{\varrho}_\varepsilon \mathbf{F}_k(\tilde{\varrho}_\varepsilon, \tilde{\mathbf{u}}_\varepsilon) \cdot \tilde{\mathbf{u}}_\varepsilon - \tilde{\varrho} \mathbf{F}_k(\tilde{\varrho}, \tilde{\mathbf{u}}) \cdot \tilde{\mathbf{u}}| \, dx \, dt$$

$$
\begin{aligned}
&= \int_{\mathcal{O}_\kappa^c} |\tilde{\varrho}_\varepsilon \mathbf{F}_k(\tilde{\varrho}_\varepsilon, \tilde{\mathbf{u}}_\varepsilon) \cdot \tilde{\mathbf{u}}_\varepsilon - \tilde{\varrho}\mathbf{F}_k(\tilde{\varrho}, \tilde{\mathbf{u}}) \cdot \tilde{\mathbf{u}}| \,\mathrm{d}x\,\mathrm{d}t\,\mathrm{d}\tilde{\mathbb{P}} \\
&\quad + \int_{\mathcal{O}_\kappa^1} |\tilde{\varrho}_\varepsilon \mathbf{F}_k(\tilde{\varrho}_\varepsilon, \tilde{\mathbf{u}}_\varepsilon) \cdot \tilde{\mathbf{u}}_\varepsilon - \tilde{\varrho}\mathbf{F}_k(\tilde{\varrho}, \tilde{\mathbf{u}}) \cdot \tilde{\mathbf{u}}| \,\mathrm{d}x\,\mathrm{d}t\,\mathrm{d}\tilde{\mathbb{P}} \\
&\quad + \int_{\mathcal{O}_\kappa^2} \left| \tilde{\varrho}_\varepsilon \mathbf{F}_k\Big(\tilde{\varrho}_\varepsilon, \frac{\sqrt{\tilde{\varrho}_\varepsilon}\tilde{\mathbf{u}}_\varepsilon}{\sqrt{\tilde{\varrho}_\varepsilon}}\Big) \cdot \frac{\sqrt{\tilde{\varrho}_\varepsilon}\tilde{\mathbf{u}}_\varepsilon}{\sqrt{\tilde{\varrho}_\varepsilon}} - \tilde{\varrho}\mathbf{F}_k\Big(\tilde{\varrho}, \frac{\sqrt{\tilde{\varrho}}\tilde{\mathbf{u}}}{\sqrt{\tilde{\varrho}}}\Big) \cdot \frac{\sqrt{\tilde{\varrho}}\tilde{\mathbf{u}}}{\sqrt{\tilde{\varrho}}} \right| \mathrm{d}x\,\mathrm{d}t\,\mathrm{d}\tilde{\mathbb{P}} \\
&= (I)_\kappa^1 + (I)_\kappa^2 + (I)_\kappa^3.
\end{aligned}
$$

By Hölder's inequality we have

$$
\begin{aligned}
(I)_\kappa^1 &\lesssim \int_{\mathcal{O}_\kappa^c} (\tilde{\varrho}_\varepsilon|\tilde{\mathbf{u}}_\varepsilon| + \tilde{\varrho}_\varepsilon|\tilde{\mathbf{u}}_\varepsilon|^2 + \tilde{\varrho}|\tilde{\mathbf{u}}| + \tilde{\varrho}|\tilde{\mathbf{u}}|^2) \,\mathrm{d}x\,\mathrm{d}t\,\mathrm{d}\tilde{\mathbb{P}} \lesssim \int_{\mathcal{O}_\kappa^c} (1 + \tilde{\varrho}_\varepsilon|\tilde{\mathbf{u}}_\varepsilon|^2 + \tilde{\varrho}|\tilde{\mathbf{u}}|^2) \,\mathrm{d}x\,\mathrm{d}t\,\mathrm{d}\tilde{\mathbb{P}} \\
&\lesssim (\tilde{\mathbb{P}} \otimes \mathfrak{L}^4(\mathcal{O}_\kappa^c))^{\frac{2\gamma-3}{6\gamma}} \left(\mathbb{E} \int_0^T \int_{\mathbb{T}^3} (1 + (\tilde{\varrho}_\varepsilon|\tilde{\mathbf{u}}_\varepsilon|^2)^{\frac{6\gamma}{4\gamma+3}} + (\tilde{\varrho}|\tilde{\mathbf{u}}|^2)^{\frac{6\gamma}{4\gamma+3}}) \,\mathrm{d}x\,\mathrm{d}t \right)^{\frac{4\gamma+3}{6\gamma}} \lesssim \kappa^{\frac{2\gamma-3}{6\gamma}},
\end{aligned}
$$

due to the assumption on the $\mathbf{F}_k$ from (4.7), (4.135), and (4.136). The second integral can be bounded by

$$
(I)_\kappa^2 \lesssim \kappa\,\mathbb{E} \int_0^T \int_{\mathbb{T}^3} (1 + |\tilde{\mathbf{u}}_\varepsilon|^2) \,\mathrm{d}x\,\mathrm{d}t + \kappa\,\mathbb{E} \int_0^T \int_{\mathbb{T}^3} (1 + |\tilde{\mathbf{u}}|^2) \,\mathrm{d}x\,\mathrm{d}t \lesssim \kappa,
$$

using (4.135) and (4.136). On the other hand, by (4.177), (4.115), and the continuity of $\mathbf{F}_k$ away from the vacuum (and the lower bounds for $\tilde{\varrho}$ and $\tilde{\varrho}_\varepsilon$ in $\mathcal{O}_\kappa^2$), the last integral vanishes as $\varepsilon \to 0$. Since κ was arbitrary we obtain (4.181) and, consequently, also (4.179). Finally, we obtain the claim by Lemma 2.6.6. □

Due to the equality of laws Proposition 4.4.9 and Theorem 2.9.1, the energy inequality (4.76) continues to hold on the new probability space. By Proposition 4.4.9, (4.177), and Proposition 4.4.13 it is a routine matter to perform the limit. Note in particular the convergence of the initial data (recall (4.153) and (4.149)–(4.151)).

The proof of Theorem 4.4.2 is hereby complete.

4.5 Vanishing artificial pressure limit

Our ultimate goal is to let $\delta \to 0$ in (4.121)–(4.122) and to gain

$$
\mathrm{d}\varrho + \operatorname{div}(\varrho\mathbf{u})\,\mathrm{d}t = 0, \tag{4.184}
$$

$$
\mathrm{d}(\varrho\mathbf{u}) + \operatorname{div}(\varrho\mathbf{u} \otimes \mathbf{u})\,\mathrm{d}t + \nabla p(\varrho)\,\mathrm{d}t = \operatorname{div}\mathbb{S}(\nabla\mathbf{u})\,\mathrm{d}t + \mathbb{G}(\varrho, \varrho\mathbf{u})\,\mathrm{d}W. \tag{4.185}
$$

As in the previous two sections, it is the momentum $\varrho\mathbf{u}$ rather than the velocity $\mathbf{u}$ that is continuous in time. Therefore, it is more natural to specify the initial values in terms of $[\varrho(0), \varrho\mathbf{u}(0)]$. We consider the space $L^1(\mathbb{T}^3) \times L^1(\mathbb{T}^3)$ along with a Borel probability measure Λ such that

$$\Lambda\{\varrho \geq 0\} = 1, \quad \Lambda\left\{0 < \underline{\varrho} \leq \int_{\mathbb{T}^3} \varrho \,\mathrm{d}x \leq \overline{\varrho} < \infty\right\} = 1, \tag{4.186}$$

for some deterministic constants $\underline{\varrho}$, $\overline{\varrho}$, and

$$\int_{L^1_x \times L^1_x} \left| \int_{\mathbb{T}^3} \left[\frac{1}{2} \frac{|\mathbf{q}|^2}{\varrho} + P(\varrho)\right] \mathrm{d}x \right|^r \mathrm{d}\Lambda \lesssim 1, \tag{4.187}$$

for some $r \geq 4$, where

$$P(\varrho) = \varrho \int_1^{\varrho} \frac{p(z)}{z^2} \,\mathrm{d}z$$

is the pressure potential.

Remark 4.5.1. Note that condition (4.187) requires

$$\mathbf{q} = 0 \quad \text{on the vacuum set } \{\varrho = 0\} \ \Lambda\text{-a.s.}$$

We recall the definition of a dissipative martingale solution given in Definition 3.4.1.

Definition 4.5.2. Let $\Lambda = \Lambda(\varrho, \mathbf{q})$ be a Borel probability measure on $L^1(\mathbb{T}^N) \times L^1(\mathbb{T}^N)$ such that (4.187) holds. The quantity $((\Omega, \mathfrak{F}, (\mathfrak{F}_t)_{t\geq 0}, \mathbb{P}), \varrho, \mathbf{u}, W)$ is called a *dissipative martingale solution* to (4.184)–(4.185) with the initial law Λ if:

(1) $(\Omega, \mathfrak{F}, (\mathfrak{F}_t)_{t\geq 0}, \mathbb{P})$ is a stochastic basis with a complete right-continuous filtration;
(2) W is a cylindrical $(\mathfrak{F}_t)$-Wiener process;
(3) the density ϱ and the velocity $\mathbf{u}$ are random distributions adapted to $(\mathfrak{F}_t)_{t\geq 0}$, $\varrho \geq 0$ $\mathbb{P}$-a.s.;
(4) there exists an $\mathfrak{F}_0$-measurable random variable $[\varrho_0, \mathbf{u}_0]$ such that $\Lambda = \mathcal{L}[\varrho_0, \varrho_0\mathbf{u}_0]$;
(5) the equation of continuity

$$-\int_0^T \partial_t \phi \int_{\mathbb{T}^3} \varrho \psi \,\mathrm{d}x \,\mathrm{d}t = \phi(0) \int_{\mathbb{T}^3} \varrho_0 \psi \,\mathrm{d}x + \int_0^T \phi \int_{\mathbb{T}^3} \varrho \mathbf{u} \cdot \nabla \psi \,\mathrm{d}x \,\mathrm{d}x \,\mathrm{d}t \tag{4.188}$$

holds for all $\phi \in C_c^\infty([0,T))$ and all $\psi \in C^\infty(\mathbb{T}^3)$ $\mathbb{P}$-a.s.;
(6) the momentum equation

$$\begin{aligned}
&-\int_0^T \partial_t \phi \int_{\mathbb{T}^3} \varrho \mathbf{u} \cdot \boldsymbol{\varphi} \,\mathrm{d}x \,\mathrm{d}t - \phi(0) \int_{\mathbb{T}^3} \varrho_0 \mathbf{u}_0 \cdot \boldsymbol{\varphi} \,\mathrm{d}x \\
&= \int_0^T \phi \int_{\mathbb{T}^3} [\varrho \mathbf{u} \otimes \mathbf{u} : \nabla \boldsymbol{\varphi} + p(\varrho) \operatorname{div} \boldsymbol{\varphi}] \,\mathrm{d}x \,\mathrm{d}t - \int_0^T \phi \int_{\mathbb{T}^3} \mathbb{S}(\nabla \mathbf{u}) : \nabla \boldsymbol{\varphi} \,\mathrm{d}x \,\mathrm{d}t \\
&\quad + \int_0^T \phi \int_{\mathbb{T}^3} \mathbb{G}(\varrho, \varrho \mathbf{u}) \cdot \boldsymbol{\varphi} \,\mathrm{d}x \,\mathrm{d}W
\end{aligned} \tag{4.189}$$

holds for all $\phi \in C_c^\infty([0,T))$ and all $\boldsymbol{\varphi} \in C^\infty(\mathbb{T}^3)$ $\mathbb{P}$-a.s.;

(7) the energy inequality

$$-\int_0^T \partial_t \phi \int_{\mathbb{T}^3} \left[\frac{1}{2}\varrho|\mathbf{u}|^2 + P(\varrho)\right] \mathrm{d}x\,\mathrm{d}t + \int_0^T \phi \int_{\mathbb{T}^3} \mathbb{S}(\nabla\mathbf{u}) : \nabla\mathbf{u}\,\mathrm{d}x\,\mathrm{d}t$$
$$\leq \phi(0) \int_{\mathbb{T}^3} \left[\frac{1}{2}\varrho_0|\mathbf{u}_0|^2 + P(\varrho_0)\right] \mathrm{d}x + \frac{1}{2}\int_0^T \phi \int_{\mathbb{T}^3} \sum_{k=1}^{\infty} \varrho^{-1}|\mathbf{G}_k(\varrho,\varrho\mathbf{u})|^2\,\mathrm{d}x\,\mathrm{d}t$$
$$+ \sum_{k=1}^{\infty} \int_0^T \phi \int_{\mathbb{T}^3} \mathbf{G}_k(\varrho,\varrho\mathbf{u}) \cdot \mathbf{u}\,\mathrm{d}x\,\mathrm{d}W_k, \tag{4.190}$$

holds for all $\phi \in C_c^\infty([0,T))$, $\phi \geq 0$, $\mathbb{P}$-a.s.;

(8) if $b \in C^1(\mathbb{R})$ such that $b'(z) = 0$ for all $z \geq M_b$, then, for all $\phi \in C_c^\infty([0,T))$ and all $\psi \in C^\infty(\mathbb{T}^3)$, we have $\mathbb{P}$-a.s.

$$-\int_0^T \partial_t \phi \int_{\mathbb{T}^3} b(\varrho)\psi\,\mathrm{d}x\,\mathrm{d}t = \phi(0)\int_{\mathbb{T}^3} b(\varrho_0)\psi\,\mathrm{d}x + \int_0^T \phi \int_{\mathbb{T}^3} b(\varrho)\mathbf{u}\cdot\nabla\psi\,\mathrm{d}x\,\mathrm{d}t$$
$$- \int_0^T \phi \int_{\mathbb{T}^3} (b'(\varrho)\varrho - b(\varrho))\operatorname{div}\mathbf{u}\,\psi\,\mathrm{d}x\,\mathrm{d}t.$$

And we obtain the following result.

Theorem 4.5.3. *Let $\gamma > \frac{3}{2}$. Let Λ be a Borel probability measure defined on the space $L^1(\mathbb{T}^3) \times L^1(\mathbb{T}^3)$ such that* (4.186) *and* (4.187) *hold true for some $\overline{\varrho} \geq \underline{\varrho} > 0$ and $r \geq 4$. Then there exists a dissipative martingale solution to* (4.184)–(4.185) *in the sense of Definition* 4.5.2.

As in the preceding sections, the proof consists of (i) showing uniform bounds independent of δ, (ii) applying the stochastic compactness method based on Jakubowski's theorem, and (iii) showing compactness of the density by means of deterministic arguments.

4.5.1 Uniform energy bounds

We start with the energy estimates that basically mimic those obtained in Section 4.4.1. Let $[\varrho, \mathbf{u}]$ be a dissipative martingale solution to (4.121)–(4.122) constructed by means of Theorem 4.4.2. We intend to derive estimates which hold true uniformly in δ.

The energy inequality (4.125) in its integral form reads

$$\int_{\mathbb{T}^3} \left[\frac{1}{2}\varrho|\mathbf{u}|^2 + P_\delta(\varrho)\right](\tau)\,\mathrm{d}x + \int_0^\tau \int_{\mathbb{T}^3} \mathbb{S}(\nabla\mathbf{u}) : \nabla\mathbf{u}\,\mathrm{d}x\,\mathrm{d}t$$
$$\leq \int_{\mathbb{T}^3} \left[\frac{1}{2}\varrho_0|\mathbf{u}_0|^2 + P_\delta(\varrho_0)\right] \mathrm{d}x$$
$$+ \frac{1}{2}\sum_{k=1}^{\infty} \int_0^\tau \int_{\mathbb{T}^3} \varrho|\mathbf{F}_k(\varrho,\mathbf{u})|^2\,\mathrm{d}x\,\mathrm{d}t + \int_0^\tau \int_{\mathbb{T}^3} \varrho\mathbb{F}(\varrho,\mathbf{u})\cdot\mathbf{u}\,\mathrm{d}x\,\mathrm{d}W. \tag{4.191}$$

In view of hypotheses (4.208)–(4.210), the inequality (4.191) provides uniform bounds that can be obtained exactly as in Section 4.4.1:

$$\mathbb{E}\Big[\Big|\sup_{t\in[0,T]}\|\varrho\|_{L_x^\gamma}^\gamma\Big|^r\Big]+\mathbb{E}\Big[\Big|\sup_{t\in[0,T]}\delta\|\varrho\|_{L_x^\Gamma}^\Gamma\Big|^r\Big]\lesssim c_1(r), \tag{4.192}$$

$$\mathbb{E}\Big[\Big|\sup_{t\in[0,T]}\|\varrho|\mathbf{u}|^2\|_{L_x^1}\Big|^r+\Big|\sup_{t\in[0,T]}\|\varrho\mathbf{u}\|_{L_x^{\frac{2\gamma}{\gamma+1}}}^{\frac{2\gamma}{\gamma+1}}\Big|^r\Big]\lesssim c_1(r), \tag{4.193}$$

$$\mathbb{E}\big[\|\mathbf{u}\|_{L_t^2W_x^{1,2}}^{2r}\big]\lesssim c_2(r), \tag{4.194}$$

with

$$\begin{aligned} c_1(r)&\approx 1+\mathbb{E}\Big[\Big|\int_{\mathbb{T}^3}\Big[\frac{1}{2}\varrho_0|\mathbf{u}_0|^2+P_\delta(\varrho_0)\Big]\,\mathrm{d}x\Big|^r\Big],\\ c_2(r)&\approx 1+\mathbb{E}\Big[\Big|\int_{\mathbb{T}^3}\Big[\frac{1}{2}\varrho_0|\mathbf{u}_0|^2+P_\delta(\varrho_0)\Big]\,\mathrm{d}x\Big|^{2r}\Big]. \end{aligned} \tag{4.195}$$

Finally, it follows from (4.123) that

$$\|\varrho(\tau)\|_{L_x^1}=\|\varrho_0\|_{L_x^1}\leq\overline{\varrho},\quad \tau\in[0,T]. \tag{4.196}$$

4.5.2 Pressure estimates

In order to derive refined estimates of the pressure, we apply a method similar to Section 4.4.2. We consider

$$\nabla\Delta^{-1}\big[b(\varrho)-(b(\varrho))_{\mathbb{T}^3}\big]=\Delta^{-1}\big[\nabla b(\varrho)\big]$$

as test function in the momentum equation (4.124). Here b is a smooth function with moderate growth specified below.

As ϱ_δ satisfies the renormalized continuity equation (see Theorem A.3.1 and Lemma A.3.2), we get

$$\begin{aligned} \mathrm{d}\Delta^{-1}\big[\nabla b(\varrho)\big]&=-\nabla\Delta^{-1}\mathrm{div}\,(b(\varrho)\mathbf{u})\,\mathrm{d}t\\ &\quad+\Delta^{-1}\big[\nabla((b(\varrho)-b'(\varrho)\varrho)\mathrm{div}\,\mathbf{u})\big]\,\mathrm{d}t. \end{aligned} \tag{4.197}$$

Exactly as in Section 4.4.2, we deduce

$$\begin{aligned} &\int_0^\tau\int_{\mathbb{T}^3}p_\delta(\varrho)\big[b(\varrho)-(b(\varrho))_{\mathbb{T}^3}\big]\,\mathrm{d}x\,\mathrm{d}t\\ &=\Big[\int_{\mathbb{T}^3}\varrho\mathbf{u}\cdot\Delta^{-1}\big[\nabla b(\varrho)\big]\,\mathrm{d}x\Big]_{t=0}^{t=\tau}-\int_0^\tau\int_{\mathbb{T}^3}\varrho\mathbf{u}\otimes\mathbf{u}:\nabla\Delta^{-1}\nabla b(\varrho)\,\mathrm{d}x\,\mathrm{d}t\\ &\quad+\int_0^\tau\int_{\mathbb{T}^3}\mathbb{S}(\nabla\mathbf{u}):\nabla\Delta^{-1}\nabla b(\varrho)\,\mathrm{d}x\,\mathrm{d}t+\int_0^\tau\int_{\mathbb{T}^3}\varrho\mathbf{u}\cdot\nabla\Delta^{-1}\mathrm{div}\,\big[b(\varrho)\mathbf{u}\big]\,\mathrm{d}x\,\mathrm{d}t\\ &\quad+\int_0^\tau\int_{\mathbb{T}^3}\varrho\mathbf{u}\cdot\Delta^{-1}\nabla\big[(b'(\varrho)\varrho-b(\varrho))\mathrm{div}\,\mathbf{u}\big]\,\mathrm{d}x\,\mathrm{d}t\\ &\quad-\int_0^\tau\int_{\mathbb{T}^3}\varrho\mathbb{F}(\varrho,\mathbf{u})\cdot\Delta^{-1}\nabla b(\varrho)\,\mathrm{d}x\,\mathrm{d}W\equiv\sum_{i=1}^6 I_i. \end{aligned} \tag{4.198}$$

Now, consider $b(\varrho) = \varrho^\beta$, where $0 < \beta < \frac{1}{3}$ will be chosen below. Although b is not be differentiable at $\varrho = 0$, relation (4.198) can easily be justified by a density argument. As a consequence of (4.196), standard L^q-estimates for the inverse Laplacian, and the embedding relation $W^{1,q}(\mathbb{T}^3) \hookrightarrow L^\infty(\mathbb{T}^3)$ for $q > 3$,

$$\sup_{t\in[0,T]} \|\Delta^{-1}\nabla b(\varrho)(t)\|_{L^\infty_x} \lesssim 1 \quad \mathbb{P}\text{-a.s.,} \tag{4.199}$$

where the norm is controlled by a deterministic constant proportional to $\overline{\varrho}$.

The integrals on the right hand side of (4.198) are estimated in a similar way as in Section 4.4.2:

(1) In accordance with (4.199), we have

$$\begin{aligned} |I_1| &\le \sup_{t\in[0,\tau]} \left| \int_{\mathbb{T}^3} \varrho\mathbf{u} \cdot \Delta^{-1}[\nabla\varrho^\beta]\,\mathrm{d}x \right| \\ &\lesssim \sup_{t\in[0,\tau]} \|\sqrt{\varrho}\|_{L^2_x} \sup_{t\in[0,\tau]} \|\sqrt{\varrho}\mathbf{u}\|_{L^2_x} \sup_{t\in[0,\tau]} \|\Delta^{-1}[\nabla\varrho^\beta]\|_{L^\infty_x} \\ &\lesssim c(\overline{\varrho}) \sup_{t\in[0,\tau]} \|\varrho|\mathbf{u}|^2\|^{1/2}_{L^1_x}. \end{aligned}$$

Consequently, by virtue of (4.193),

$$\mathbb{E}\big[|I_1|^r\big] \lesssim c_1(r)c(\overline{\varrho}). \tag{4.200}$$

(2) We use the elliptic estimates for Δ^{-1} and Hölder's inequality to obtain

$$\begin{aligned} |I_2| &\le \sup_{t\in[0,\tau]} \|\nabla\Delta^{-1}[\nabla\varrho^\beta]\|_{L^q_x} \sup_{t\in[0,\tau]} \|\varrho\|_{L^\gamma_x} \int_0^\tau \|\mathbf{u}\|^2_{L^p_x}\,\mathrm{d}t \\ &\lesssim \sup_{t\in[0,\tau]} \|\varrho^\beta\|_{L^q_x} \sup_{t\in[0,\tau]} \|\varrho\|_{L^\gamma_x} \int_0^\tau \|\mathbf{u}\|^2_{L^p_x}\,\mathrm{d}t, \end{aligned}$$

where

$$\frac{1}{q} + \frac{1}{\gamma} + \frac{2}{p} = 1.$$

Because of $\gamma > \frac{3}{2}$, $W^{1,2}(\mathbb{T}^3) \hookrightarrow L^6(\mathbb{T}^3)$, and $\beta \le \frac{2}{3}\gamma - 1$, we gain

$$\mathbb{E}\big[|I_2|^r\big] \lesssim c(\overline{\varrho})c_2(r), \tag{4.201}$$

using (4.192), (4.194), and (4.196).

(3) As

$$\|\nabla\Delta^{-1}\nabla\varrho^\beta\|_{L^2_x} \lesssim \|\varrho^\beta\|_{L^2_x},$$

we get

$$|I_3| \lesssim \int_0^\tau \int_{\mathbb{T}^3} |\nabla\mathbf{u}|^2\,\mathrm{d}x\,\mathrm{d}t + \tau \sup_{t\in[0,\tau]} \|\varrho^\beta\|^2_{L^2_x}.$$

In view of (4.194) and (4.196) together with $\beta \le 1$, we obtain

$$\mathbb{E}[|I_3|^r] \lesssim c_1(r). \tag{4.202}$$

(4) As in the previous steps,

$$\begin{aligned}|I_4| &\lesssim \sup_{t\in[0,\tau]} \|\varrho\|_{L_x^\gamma} \int_0^\tau \|\mathbf{u}\|_{L_x^q} \|\nabla\Delta^{-1}\operatorname{div}(\varrho^\beta \mathbf{u})\|_{L_x^q}\,\mathrm{d}t \\ &\lesssim \sup_{t\in[0,\tau]} \|\varrho\|_{L_x^\gamma} \int_0^\tau \|\mathbf{u}\|_{L_x^q} \|\varrho^\beta \mathbf{u}\|_{L_x^q}\,\mathrm{d}t \\ &\lesssim \sup_{t\in[0,\tau]} \|\varrho\|_{L_x^\gamma} \sup_{t\in[0,\tau]} \|\varrho^\beta\|_{L_x^p} \int_0^\tau \|\mathbf{u}\|_{L_x^q} \|\mathbf{u}\|_{L_x^r}\,\mathrm{d}t,\end{aligned}$$

where

$$\frac{1}{\gamma} + \frac{2}{q} = 1, \qquad \frac{1}{q} = \frac{1}{p} + \frac{1}{r}.$$

We observe that, as $\gamma > \frac{3}{2}$, we can choose $q < 6$ and $r \le 6$ such that

$$|I_4| \lesssim \sup_{t\in[0,\tau]} \|\varrho\|_{L_x^\gamma} \sup_{t\in[0,\tau]} \|\varrho^\beta\|_{L_x^p} \int_0^\tau \|\mathbf{u}\|^2_{W_x^{1,2}}\,\mathrm{d}t.$$

Using (4.192), (4.194), and (4.196), we find $\beta > 0$ small enough so that

$$\mathbb{E}[|I_4|^r] \lesssim c(\overline{\varrho}) c_2(r). \tag{4.203}$$

(5) By Hölder's inequality,

$$|I_5| \lesssim \sup_{t\in[0,T]} \|\varrho\|_{L_x^\gamma} \int_0^\tau \|\mathbf{u}\|_{L_x^p} \|\Delta^{-1}\nabla(\varrho^\beta \operatorname{div}\mathbf{u})\|_{L_x^q}\,\mathrm{d}t, \qquad \frac{1}{\gamma} + \frac{1}{p} + \frac{1}{q} = 1.$$

Moreover, the elliptic regularity estimates $\Delta^{-1}\nabla : L^r(\mathbb{T}^3) \to W^{1,r}(\mathbb{T}^3)$, for any $1 < r < \infty$, and the embedding

$$W^{1,r}(\mathbb{T}^3) \hookrightarrow L^q(\mathbb{T}^3), \quad q \le \frac{3r}{3-r} \text{ if } r < 3 \tag{4.204}$$

give rise to

$$\|\Delta^{-1}\nabla(\varrho^\beta \operatorname{div}\mathbf{u})\|_{L_x^q} \lesssim \|\varrho^\beta \operatorname{div}\mathbf{u}\|_{L_x^r} \lesssim \|\varrho^\beta\|_{L_x^s} \|\nabla\mathbf{u}\|_{L_x^2},$$

where r is as in (4.204) and $s = \frac{2r}{2-r}$. Note that we can choose $q < 6$, $p \le 6$ and $r < 2$. This leaves a gap for $\beta > 0$ to be chosen small enough, so that

$$\sup_{t\in[0,T]} \|\varrho^\beta\|_{L_x^s} \le c(\overline{\varrho});$$

cf. (4.196). We obtain

$$
\begin{aligned}
|I_5| &\lesssim \sup_{t\in[0,T]} \|\varrho\|_{L_x^\gamma} \int_0^T \|\mathbf{u}\|_{L_x^p} \|\nabla \mathbf{u}\|_{L_x^2}\, dt \\
&\lesssim \sup_{t\in[0,T]} \|\varrho\|_{L_x^\gamma} \int_0^T \|\mathbf{u}\|_{W_x^{1,2}}^2\, dt
\end{aligned}
$$

and, finally,

$$
\mathbb{E}\big[|I_5|^r\big] \lesssim c(\overline{\varrho}) c_2(r) \tag{4.205}
$$

on account of (4.192) and (4.194).

(6) As for the stochastic integral, we have, by the Burkholder–Davis–Gundy inequality,

$$
\begin{aligned}
\mathbb{E}\big[|I_6|^r\big] &= \mathbb{E}\left[\left|\int_0^T \int_{\mathbb{T}^3} \varrho \mathbb{F}(\varrho, \mathbf{u}) \cdot \Delta^{-1} \nabla \varrho^\beta \,\mathrm{d}x \,\mathrm{d}W\right|^r\right] \\
&\leq \mathbb{E}\left[\sup_{t\in[0,T]} \left|\int_0^t \int_{\mathbb{T}^3} \varrho \mathbb{F}(\varrho, \mathbf{u}) \cdot \Delta^{-1} \nabla \varrho^\beta \,\mathrm{d}x \,\mathrm{d}W\right|^r\right] \\
&\lesssim \mathbb{E}\left[\int_0^T \sum_{k=1}^\infty \left|\int_{\mathbb{T}^3} \varrho \mathbf{F}_k(\varrho, \mathbf{u}) \cdot \Delta^{-1} \nabla \varrho^\beta \,\mathrm{d}x\right|^2 \mathrm{d}t\right]^{r/2},
\end{aligned}
$$

where, due to (4.7), (4.199), and (4.196),

$$
\begin{aligned}
\left|\int_{\mathbb{T}^3} \varrho \mathbf{F}_k(\varrho, \mathbf{u}) \cdot \Delta^{-1} \nabla \varrho^\beta \,\mathrm{d}x\right| &\leq f_k \|\Delta^{-1} \nabla \varrho^\beta\|_{L_x^\infty} \int_{\mathbb{T}^3} (\varrho + \varrho|\mathbf{u}|)\,\mathrm{d}x \\
&\lesssim c(\overline{\varrho}) f_k \int_{\mathbb{T}^3} (\varrho + \varrho|\mathbf{u}|)\,\mathrm{d}x.
\end{aligned}
$$

We conclude

$$
\mathbb{E}\big[|I_6|^r\big] \lesssim c(\overline{\varrho}) c_2(r), \tag{4.206}
$$

as a consequence of (4.192) and (4.193).

Summarizing the previous estimates and going back to (4.198), we infer

$$
\mathbb{E}\left[\left|\int_0^T \int_{\mathbb{T}^3} (p(\varrho) + \delta(\varrho + \varrho^\Gamma)) \varrho^\beta \,\mathrm{d}x \,\mathrm{d}t\right|^r\right] \lesssim c(\overline{\varrho})(c_1(r) + c_2(r)), \tag{4.207}
$$

for a certain $\beta > 0$, with $c_1(r)$ and $c_2(r)$ given by (4.195).

4.5.3 Limit $\delta \to 0$ – stochastic compactness method

In order to proceed with passage to the limit, it is necessary to construct approximate initial laws Λ_δ such that $\Lambda_\delta \overset{*}{\rightharpoonup} \Lambda$ in the sense of measures on $L^1(\mathbb{T}^3) \times L^1(\mathbb{T}^3)$, in order for the assumptions of Theorem 4.5.3 to hold true uniformly in δ and, in addition, the assumptions of Theorem 4.4.2 to be satisfied, namely, (4.126) and (4.127). This can be done similarly to the construction of the approximate initial laws at the beginning of Section 4.4.3. More precisely, let $[\varrho_0, \mathbf{q}_0]$ be a random variable having the law Λ which exists due to Corollary 2.6.4. It is a routine matter to find a sequence $([\varrho_{0,\delta}, \mathbf{q}_{0,\delta}])_{\delta\in(0,1)}$ such that

$$\varrho_{0,\delta} \in L^\Gamma(\mathbb{T}^3), \quad \varrho_{0,\delta} > 0, \quad 0 < \frac{\underline{\varrho}}{2} \leq \int_{\mathbb{T}^3} \varrho_{0,\delta}\,\mathrm{d}x \leq 2\overline{\varrho}, \ \mathbb{P}\text{-a.s.},$$

$$\varrho_{0,\delta} \to \varrho_0 \quad \text{in } L^\gamma(\mathbb{T}^3), \quad \mathbf{q}_{0,\delta} \to \mathbf{q} \quad \text{in } L^1(\mathbb{T}^3) \ \mathbb{P}\text{-a.s.}, \tag{4.208}$$

$$\mathbb{E}\Big[\Big|\int_{\mathbb{T}^3}\Big[\frac{1}{2}\frac{|\mathbf{q}_{0,\delta}|^2}{\varrho_{0,\delta}} + P_\delta(\varrho_{0,\delta})\Big]\,\mathrm{d}x\Big|^r\Big] \lesssim 1 \tag{4.209}$$

for some $r \geq 4$ uniformly for $\delta \to 0$. Also,

$$\int_{\mathbb{T}^3}\Big[\frac{1}{2}\frac{|\mathbf{q}_{0,\delta}|^2}{\varrho_{0,\delta}} + P_\delta(\varrho_{0,\delta})\Big]\,\mathrm{d}x \to \int_{\mathbb{T}^3}\Big[\frac{1}{2}\frac{|\mathbf{q}_0|^2}{\varrho_0} + P(\varrho_0)\Big]\,\mathrm{d}x, \tag{4.210}$$

as $\delta \to 0$ $\mathbb{P}$-a.s. Finally, note that (4.210) implies

$$\frac{\mathbf{q}_{0,\delta}}{\sqrt{\varrho_{0,\delta}}} \to \frac{\mathbf{q}_0}{\sqrt{\varrho_0}} \quad \text{in } L^2(\mathbb{T}^3) \ \mathbb{P}\text{-a.s.} \tag{4.211}$$

Under these circumstances, Theorem 4.4.2 provides a solution $[\varrho_\delta, \mathbf{u}_\delta, W_\delta]$ of problem (4.121)–(4.122), with the initial law $\Lambda_\delta = \mathcal{L}[\varrho_{0,\delta}, \mathbf{q}_{0,\delta}]$. As discussed in Section 4.4.3, without any loss of generality we suppose that, for every $\delta \in (0,1)$, there exists

$$((\Omega, \mathfrak{F}, (\mathfrak{F}_t^\delta)_{t\geq 0}, \mathbb{P}), \varrho_\delta, \mathbf{u}_\delta, W),$$

which is a solution to (4.121)–(4.122). Finally, since (4.186) and (4.187) are satisfied by Λ_δ uniformly in δ, the bounds established in Section 4.5.1 and Section 4.5.2 give rise to the necessary uniform estimates on $[\varrho_\delta, \mathbf{u}_\delta]$. In the remaining part of this section we show that these approximate solutions generate a dissipative martingale solution of our target problem (4.184)–(4.185).

Finally, we are ready to apply the stochastic compactness method, replacing the original random variables by their Jakubowski–Skorokhod representation. Similarly to Section 4.4.3, we consider the path space

$$\mathcal{X} = \mathcal{X}_{\varrho_0} \times \mathcal{X}_{\mathbf{q}_0} \times \mathcal{X}_{\frac{\mathbf{q}_0}{\sqrt{\varrho_0}}} \times \mathcal{X}_\varrho \times \mathcal{X}_{\varrho\mathbf{u}} \times \mathcal{X}_\mathbf{u} \times \mathcal{X}_W \times \mathcal{X}_E \times \mathcal{X}_\nu,$$

where

$$\mathcal{X}_{\varrho_0} = L^\gamma(\mathbb{T}^3), \quad \mathcal{X}_{\mathbf{q}_0} = L^1(\mathbb{T}^3), \quad \mathcal{X}_{\frac{\mathbf{q}_0}{\sqrt{\varrho_0}}} = L^2(\mathbb{T}^3),$$

$$\mathcal{X}_\varrho = (L^{\gamma+\beta}((0,T)\times\mathbb{T}^3), w) \cap C_w([0,T]; L^\gamma(\mathbb{T}^3)),$$

$$
\begin{aligned}
\mathcal{X}_{\varrho\mathbf{u}} &= C_w([0,T]; L^{\frac{2\gamma}{\gamma+1}}(\mathbb{T}^3)) \cap C([0,T]; W^{-k,2}(\mathbb{T}^3)), \quad k > \frac{3}{2},\\
\mathcal{X}_{\mathbf{u}} &= (L^2(0,T; W^{1,2}(\mathbb{T}^3)), w),\\
\mathcal{X}_W &= C([0,T]; \mathfrak{U}_0),\\
\mathcal{X}_E &= (L^\infty(0,T; \mathcal{M}(\mathbb{T}^3)), w^*),\\
\mathcal{X}_\nu &= (L^\infty((0,T)\times \mathbb{T}^3; \mathcal{P}(\mathbb{R}^{13})), w^*),
\end{aligned}
$$

where β is taken from (4.207). Here the path space is interpreted exactly as in Section 4.4.3, with the only modification that the energy is defined as

$$
E_\delta(\varrho_\delta, \mathbf{u}_\delta)(\tau) = \left[\frac{1}{2}\varrho_\delta|\mathbf{u}_\delta|^2 + P_\delta(\varrho_\delta)\right](\tau) + \int_0^\tau [\varrho_\delta^\gamma + \delta\varrho_\delta^\Gamma]\varrho_\delta^\beta \,\mathrm{d}t.
$$

Let

$$
\left[\varrho_{0,\delta}, \mathbf{q}_{0,\delta}, \frac{\mathbf{q}_{0,\delta}}{\sqrt{\varrho_{0,\delta}}}, \varrho_\delta, \varrho_\delta\mathbf{u}_\delta, \mathbf{u}_\delta, W, E_\delta(\varrho_\delta,\mathbf{u}_\delta), \delta_{[\varrho_\delta,\mathbf{u}_\delta,\nabla\mathbf{u}_\delta]}\right]
$$

be the family associated to the weak martingale solution $[\varrho_\delta, \mathbf{u}_\delta, W]$ with the corresponding family of joint laws

$$
\left\{\mathcal{L}\left[\varrho_{0,\delta}, \mathbf{q}_{0,\delta}, \frac{\mathbf{q}_{0,\delta}}{\sqrt{\varrho_{0,\delta}}}, \varrho_\delta, \varrho_\delta\mathbf{u}_\delta, \mathbf{u}_\delta, W, E_\delta(\varrho_\delta,\mathbf{u}_\delta), \delta_{[\varrho_\delta,\mathbf{u}_\delta,\nabla\mathbf{u}_\delta]}\right]; \delta \in (0,1)\right\}.
$$

Similarly to Section 4.4.3, tightness of this family of joint laws on $\mathcal{X}$ follows from the estimates obtained in Section 4.5.1 and the construction of the initial data. The analogues of Proposition 4.4.6 and Proposition 4.4.7 are straightforward. The only change lies in the proof of tightness for $\{\mathcal{L}[\varrho_\delta\mathbf{u}_\delta]; \delta \in (0,1)\}$.

Proposition 4.5.4. *The set $\{\mathcal{L}[\varrho_\delta\mathbf{u}_\delta]; \delta \in (0,1)\}$ is tight on $\mathcal{X}_{\varrho\mathbf{u}}$.*

Proof. We proceed similarly as in Proposition 4.4.5 and decompose $\varrho_\delta\mathbf{u}_\delta$ into two parts, namely, $\varrho_\delta\mathbf{u}_\delta(t) = Y^\delta(t) + Z^\delta(t)$, where

$$
\begin{aligned}
Y^\delta(t) &= \mathbf{q}(0) - \int_0^t \operatorname{div}(\varrho_\delta\mathbf{u}_\delta \otimes \mathbf{u}_\delta)\,\mathrm{d}s + \int_0^t \operatorname{div}\mathbb{S}(\nabla\mathbf{u}_\delta)\,\mathrm{d}s - \int_0^t \nabla p(\varrho_\delta)\,\mathrm{d}s\\
&\quad + \int_0^t \mathbb{G}(\varrho_\delta, \varrho_\delta\mathbf{u}_\delta)\,\mathrm{d}W(s),\\
Z^\delta(t) &= -\delta\int_0^t \nabla(\varrho_\delta + \varrho_\delta^\Gamma)\,\mathrm{d}s.
\end{aligned}
$$

By the approach of Proposition 4.4.5 (where we employ (4.207) instead of (4.148)), we obtain Hölder continuity of Y^δ, namely, there exist $\kappa > 0$ and $l > 5/2$ such that

$$
\mathbb{E}\|Y^\delta\|_{C^\kappa([0,T];W^{-l,2}(\mathbb{T}^3))} \le C.
$$

Next, we show that the set of laws $\{\mathcal{L}[Z^\delta];\ \delta \in (0,1)\}$ is tight on

$$C([0,T]; W^{-1,\frac{\Gamma+\beta}{\Gamma}}(\mathbb{T}^3))$$

and the conclusion follows by the lines of Proposition 4.4.5. We have, due to (4.207), (up to a subsequence)

$$\delta(\varrho_\delta + \varrho_\delta^\Gamma) \to 0 \quad \text{in } L^{\frac{\Gamma+\beta}{\Gamma}}((0,T)\times\mathbb{T}^3)\ \mathbb{P}\text{-a.s.}$$

Hence

$$\delta\nabla(\varrho_\delta + \varrho_\delta^\Gamma) \to 0 \quad \text{in } L^{\frac{\Gamma+\beta}{\Gamma}}(0,T; W^{-1,\frac{\Gamma+\beta}{\Gamma}}(\mathbb{T}^3))\ \mathbb{P}\text{-a.s.}$$

and

$$Z^\delta \to 0 \quad \text{in } C([0,T]; W^{-1,\frac{\Gamma+\beta}{\Gamma}}(\mathbb{T}^3))\ \mathbb{P}\text{-a.s.}$$

This leads to the convergence in law

$$Z^\delta \xrightarrow{d} 0 \quad \text{on } C([0,T]; W^{-1,\frac{\Gamma+\beta}{\Gamma}}(\mathbb{T}^3))$$

and the claim follows. □

Consequently, we may apply Jakubowski's theorem (Theorem 2.7.1) as well as Corollary 2.8.3 to obtain the following.

Proposition 4.5.5. *There exists a complete probability space* $(\tilde\Omega, \tilde{\mathfrak{F}}, \tilde{\mathbb{P}})$ *with* $\mathcal{X}$*-valued Borel measurable random variables* $[\tilde\varrho_{0,\delta}, \tilde{\mathbf{q}}_{0,\delta}, \tilde{\mathbf{k}}_{0,\delta}, \tilde\varrho_\delta, \tilde{\mathbf{q}}_\delta, \tilde{\mathbf{u}}_\delta, \tilde W_\delta, \tilde E_\delta, \tilde\nu_\delta]$, $\delta \in (0,1)$, *as well as* $[\tilde\varrho_0, \tilde{\mathbf{q}}_0, \tilde{\mathbf{k}}_0, \tilde\varrho, \tilde{\mathbf{q}}, \tilde{\mathbf{u}}, \tilde W, \tilde E, \tilde\nu]$ *such that (up to a subsequence):*

(1) *for all* $\delta \in (0,1)$, *we know that* $\mathcal{L}[\tilde\varrho_{0,\delta}, \tilde{\mathbf{q}}_{0,\delta}, \tilde{\mathbf{k}}_{0,\delta}, \tilde\varrho_\delta, \tilde{\mathbf{q}}_\delta, \tilde{\mathbf{u}}_\delta, \tilde W_\delta, \tilde E_\delta, \tilde\nu_\delta]$ *and* $\mathcal{L}[\varrho_{0,\delta}, \mathbf{q}_{0,\delta}, \frac{\mathbf{q}_{0,\delta}}{\sqrt{\varrho_{0,\delta}}}, \varrho_\delta, \varrho_\delta\mathbf{u}_\delta, \mathbf{u}_\delta, W, E_\delta(\varrho_\delta, \mathbf{u}_\delta), \delta_{[\varrho_\delta,\mathbf{u}_\delta,\nabla\mathbf{u}_\delta]}]$ *coincide. In particular, we have* $\tilde{\mathbb{P}}$*-a.s.*

$$\tilde\varrho_{0,\delta} = \tilde\varrho_\delta(0), \quad \tilde{\mathbf{q}}_{0,\delta} = \tilde\varrho_\delta\tilde{\mathbf{u}}_\delta(0), \quad \tilde{\mathbf{k}}_{0,\delta} = \frac{\tilde{\mathbf{q}}_{0,\delta}}{\sqrt{\tilde\varrho_{0,\delta}}} = \frac{\tilde\varrho_\delta\tilde{\mathbf{u}}_\delta(0)}{\sqrt{\tilde\varrho_\delta(0)}},$$

$$\tilde{\mathbf{q}}_\delta = \tilde\varrho_\delta\tilde{\mathbf{u}}_\delta, \quad \tilde E_\delta = E_\delta(\tilde\varrho_\delta, \tilde{\mathbf{u}}_\delta), \quad \tilde\nu_\delta = \delta_{[\tilde\varrho_\delta,\tilde{\mathbf{u}}_\delta,\nabla\tilde{\mathbf{u}}_\delta]},$$

as well as

$$\mathbb{E}\bigg[\bigg|\int_0^T\int_{\mathbb{T}^3}[\tilde\varrho_\delta^\gamma + \delta\tilde\varrho_\delta^\Gamma]\tilde\varrho_\delta^\beta\,\mathrm{d}x\,\mathrm{d}t\bigg|^r\bigg] + \mathbb{E}\bigg[\sup_{t\in(0,T)}\bigg|\int_{\mathbb{T}^3}\Big[\frac{1}{2}\tilde\varrho_\delta|\tilde{\mathbf{u}}_\delta|^2 + P_\delta(\tilde\varrho_\delta)\Big]\,\mathrm{d}x\bigg|^r\bigg] \lesssim c_2(r), \tag{4.212}$$

with $c_2(r)$ *given by (4.195);*

(2) *the law of* $[\tilde{\varrho}_0, \tilde{\mathbf{q}}_0, \tilde{\mathbf{k}}_0, \tilde{\varrho}, \tilde{\mathbf{q}}, \tilde{\mathbf{u}}, \tilde{W}, \tilde{E}, \tilde{\nu}]$ *on* $\mathcal{X}$ *is a Radon measure;*

(3) $[\tilde{\varrho}_{0,\delta}, \tilde{\mathbf{q}}_{0,\delta}, \tilde{\mathbf{k}}_{0,\delta}, \tilde{\varrho}_\delta, \tilde{\varrho}_\delta\tilde{\mathbf{u}}_\delta, \tilde{\mathbf{u}}_\delta, \tilde{W}_\delta, \tilde{E}_\delta, \tilde{\nu}_\delta]$ *converges in the topology of* $\mathcal{X}$ $\tilde{\mathbb{P}}$*-a.s. to* $[\tilde{\varrho}_0, \tilde{\mathbf{q}}_0, \tilde{\mathbf{k}}_0, \tilde{\varrho}, \tilde{\mathbf{q}}, \tilde{\mathbf{u}}, \tilde{W}, \tilde{E}, \tilde{\nu}]$, *i.e.,*

$$\begin{aligned}
\tilde{\varrho}_{0,\delta} &\to \tilde{\varrho}_0 \quad \text{in } L^\gamma(\mathbb{T}^3),\\
\tilde{\mathbf{q}}_{0,\delta} &\to \tilde{\mathbf{q}}_0 \quad \text{in } L^1(\mathbb{T}^3),\\
\tilde{\mathbf{k}}_{0,\delta} &\to \tilde{\mathbf{k}}_0 \quad \text{in } L^2(\mathbb{T}^3),\\
\tilde{\varrho}_\delta &\to \tilde{\varrho} \quad \text{in } C_w([0,T];L^\gamma(\mathbb{T}^3)),\\
\tilde{\varrho}_\delta &\rightharpoonup \tilde{\varrho} \quad \text{in } L^{\gamma+\beta}((0,T)\times\mathbb{T}^3), \quad \text{for some } \beta > 0,\\
\tilde{\varrho}_\delta\tilde{\mathbf{u}}_\delta &\to \tilde{\mathbf{q}} \quad \text{in } C_w([0,T];L^{\frac{2\gamma}{\gamma+1}}(\mathbb{T}^3)),\\
\tilde{\mathbf{u}}_\delta &\rightharpoonup \tilde{\mathbf{u}} \quad \text{in } L^2(0,T;W^{1,2}(\mathbb{T}^3)),\\
\tilde{W}_\delta &\to \tilde{W} \quad \text{in } C([0,T];\mathfrak{U}_0),\\
E_\delta(\tilde{\varrho}_\delta, \tilde{\mathbf{u}}_\delta) &\overset{*}{\rightharpoonup} \tilde{E} \quad \text{in } L^\infty(0,T;\mathcal{M}_b(\mathbb{T}^3)),\\
\delta_{[\tilde{\varrho}_\delta, \tilde{\mathbf{u}}_\delta, \nabla\tilde{\mathbf{u}}_\delta]} &\overset{*}{\rightharpoonup} \tilde{\nu} \quad \text{in } L^\infty_{w^*}((0,T)\times\mathbb{T}^3;\mathcal{P}(\mathbb{R}^{13})),
\end{aligned} \tag{4.213}$$

as $\delta \to 0$ $\tilde{\mathbb{P}}$*-a.s.;*

(4) *for any Carathéodory function* $H = H(t,x,\varrho,\mathbf{v},\mathbf{V})$ *where* $(t,x) \in (0,T)\times\mathbb{T}^3$, $(\varrho,\mathbf{v},\mathbf{V}) \in \mathbb{R}^{13}$, *satisfying for some* $q > 0$ *the growth condition*

$$|H(t,x,\varrho,\mathbf{v},\mathbf{V})| \lesssim 1 + |\varrho|^{q_1} + |\mathbf{v}|^{q_2} + |\mathbf{V}|^{q_2},$$

uniformly in (t,x), *denote* $\overline{H(\tilde{\varrho}, \tilde{\mathbf{u}}, \nabla\tilde{\mathbf{u}})}(t,x) = \langle\tilde{\nu}_{t,x}, H\rangle$. *Then we have*

$$\begin{aligned}
H(\tilde{\varrho}_\delta, \tilde{\mathbf{u}}_\delta, \nabla\tilde{\mathbf{u}}_\delta) &\rightharpoonup \overline{H(\tilde{\varrho}, \tilde{\mathbf{u}}, \nabla\tilde{\mathbf{u}})} \quad \text{in } L^r((0,T)\times\mathbb{T}^3)\\
&\text{for all } 1 < r \leq \frac{\gamma+\beta}{q_1} \wedge \frac{2}{q_2},
\end{aligned} \tag{4.214}$$

as $\varepsilon \to 0$ $\tilde{\mathbb{P}}$*-a.s.*

As remarked in Section 4.3.2, namely, after Proposition 4.3.10, we may deduce that the filtration

$$\tilde{\mathfrak{F}}_t := \sigma\bigg(\sigma_t[\tilde{\varrho}] \cup \sigma_t[\tilde{\mathbf{u}}] \cup \bigcup_{k=1}^{\infty} \sigma_t[\tilde{W}_k]\bigg), \quad t \in [0,T],$$

is non-anticipating with respect to $\tilde{W} = \sum_{k=1}^{\infty} e_k \tilde{W}_k$, which is a cylindrical $(\tilde{\mathfrak{F}}_t)$-Wiener process.

As in Corollary 4.4.10 we obtain the following as a direct consequence of Proposition 4.5.5.

Corollary 4.5.6. *The following convergence holds true* $\tilde{\mathbb{P}}$*-a.s.:*

$$\tilde{\varrho}_\delta\tilde{\mathbf{u}}_\delta \otimes \tilde{\mathbf{u}}_\delta \rightharpoonup \tilde{\varrho}\tilde{\mathbf{u}} \otimes \tilde{\mathbf{u}} \quad \text{in } L^1(0,T;L^1(\mathbb{T}^3)). \tag{4.215}$$

Similar to Lemma 4.4.11, the continuity equation (4.188) is satisfied by $[\tilde{\varrho}, \tilde{\mathbf{u}}]$ on the new probability space.

Lemma 4.5.7. *The random distribution $[\tilde{\varrho}, \tilde{\mathbf{u}}]$ satisfies (4.188) for all $\phi \in C_c^\infty([0,T))$ and all $\psi \in C^\infty(\mathbb{T}^3)$ $\tilde{\mathbb{P}}$-a.s.*

Next, we perform the limit $\delta \to 0$ in the momentum equation (4.124). Similarly to Section 4.4.3.1, at this stage of the proof, we are not able to identify the limit in the pressure nor in the stochastic integral. This will be done below in Section 4.5.4, where we establish strong convergence of the approximate densities.

First, we apply Proposition 4.5.5, part (4), to compositions $p(\tilde{\varrho}_\delta)$ and $\tilde{\varrho}_\delta \mathbf{F}_k(\tilde{\varrho}_\delta, \tilde{\mathbf{u}}_\delta)$, $k \in \mathbb{N}$. Specifically, there exist $\overline{p(\tilde{\varrho})}$ and $\overline{\tilde{\varrho}\mathbf{F}_k(\tilde{\varrho}, \tilde{\mathbf{u}})}$, $k \in \mathbb{N}$, such that

$$p(\tilde{\varrho}_\delta) \rightharpoonup \overline{p(\tilde{\varrho})} \quad \text{in } L^q((0,T)\times\mathbb{T}^3), \tag{4.216}$$

$$\tilde{\varrho}_\delta \mathbf{F}_k(\tilde{\varrho}_\delta, \tilde{\mathbf{u}}_\delta) \rightharpoonup \overline{\tilde{\varrho}\mathbf{F}_k(\tilde{\varrho}, \tilde{\mathbf{u}})} \quad \text{in } L^q((0,T)\times\mathbb{T}^3), \tag{4.217}$$

for some $q > 1$ $\tilde{\mathbb{P}}$-a.s. Let us now define the Hilbert–Schmidt operator $\overline{\tilde{\varrho}\mathbb{F}(\tilde{\varrho}, \tilde{\mathbf{u}})}$ by

$$\overline{\tilde{\varrho}\mathbb{F}(\tilde{\varrho}, \tilde{\mathbf{u}})} e_k := \overline{\tilde{\varrho}\mathbf{F}_k(\tilde{\varrho}, \tilde{\mathbf{u}})} \quad k \in \mathbb{N}.$$

We obtain the following result.

Proposition 4.5.8. *The random distribution $[\tilde{\varrho}, \tilde{\mathbf{u}}, \tilde{W}]$ satisfies*

$$\begin{aligned} &-\int_0^T \phi \int_{\mathbb{T}^3} \tilde{\varrho}\tilde{\mathbf{u}} \cdot \boldsymbol{\varphi} \,\mathrm{d}x\,\mathrm{d}t - \phi(0)\int_{\mathbb{T}^3} \tilde{\varrho}_0 \tilde{\mathbf{u}}_0 \cdot \boldsymbol{\varphi}\,\mathrm{d}x \\ &= \int_0^T \phi \int_{\mathbb{T}^3} [\tilde{\varrho}\tilde{\mathbf{u}} \otimes \tilde{\mathbf{u}} : \nabla \boldsymbol{\varphi} + \overline{p(\tilde{\varrho})} \operatorname{div} \boldsymbol{\varphi}]\,\mathrm{d}x\,\mathrm{d}s \\ &\quad - \int_0^T \phi \int_{\mathbb{T}^3} \mathbb{S}(\nabla \tilde{\mathbf{u}}) : \nabla \boldsymbol{\varphi}\,\mathrm{d}x\,\mathrm{d}t + \int_0^T \phi \int_{\mathbb{T}^3} \overline{\tilde{\varrho}\mathbb{F}(\tilde{\varrho}, \tilde{\mathbf{u}})} \cdot \boldsymbol{\varphi}\,\mathrm{d}x\,\mathrm{d}\tilde{W}, \end{aligned} \tag{4.218}$$

for all $\phi \in C_c^\infty([0,T))$ and all $\boldsymbol{\varphi} \in C^\infty(\mathbb{T}^3)$ $\mathbb{P}$-a.s.

Proof. Similarly to Proposition 4.4.12, we apply Theorem 2.9.1 to show that the random distributions $[\tilde{\varrho}_\delta, \tilde{\mathbf{u}}_\delta, \tilde{W}_\delta]$ solve the momentum equation (4.124). In the following we will explain how to pass to the limit in pressure and stochastic integral. As a consequence of (4.216), the definition of p_δ, and Proposition 4.5.5,

$$p_\delta(\tilde{\varrho}_\delta) \rightharpoonup \overline{p(\tilde{\varrho})} \quad \text{in } L^q((0,T)\times\mathbb{T}^3), \tag{4.219}$$

for some $q > 1$ $\tilde{\mathbb{P}}$-a.s., where $\overline{p(\tilde{\varrho})}$ was given in (4.216). Thus it remains to show

$$\tilde{\varrho}_\delta \mathbf{F}_k(\tilde{\varrho}_\delta, \tilde{\mathbf{u}}_\delta) \to \overline{\tilde{\varrho}\mathbf{F}_k(\tilde{\varrho}, \tilde{\mathbf{u}})} \quad \text{in } L^2(0,T; W^{-l,2}(\mathbb{T}^3)) \tag{4.220}$$

$\tilde{\mathbb{P}}$-a.s. for any $k \in \mathbb{N}$, where $l > \frac{3}{2}$. Due to Proposition 4.5.5 and Corollary 4.5.6, the convergence (4.220) follows from

$$\tilde{\varrho}_\delta \mathbf{F}_k(\tilde{\varrho}_\delta, \tilde{\mathbf{u}}) \to \overline{\tilde{\varrho}\mathbf{F}_k(\tilde{\varrho}, \tilde{\mathbf{u}})} \quad \text{in } L^2(0,T; W^{-l,2}(\mathbb{T}^3)) \tag{4.221}$$

$\tilde{\mathbb{P}}$-a.s. For details we refer to the proof of (4.161). As a byproduct we also obtain again, similar to (4.166),

$$b(\tilde{\varrho}_\delta) \to \overline{b(\tilde{\varrho})} \quad \text{in } C_w([0,T]; L^2(\mathbb{T}^3)) \text{ as } \delta \to 0, \tag{4.222}$$

for any $b \in C^1([0,\infty))$, $b'(\varrho) = 0$ for $\varrho \to \infty$. Finally, (4.221) follows from the renormalized form of the equation of continuity as in (4.165)–(4.166). As in (4.167) we obtain

$$\begin{aligned} &\tilde{\mathbb{E}} \int_0^T \|\tilde{\varrho}_\varepsilon \mathbb{F}_\varepsilon(\tilde{\varrho}_\varepsilon, \tilde{\mathbf{u}}_\varepsilon)\|^2_{L_2(\mathfrak{U}; W_x^{-l,2})} \,\mathrm{d}t \\ &\lesssim \tilde{\mathbb{E}} \int_0^T (\tilde{\varrho}_\varepsilon)_{\mathbb{T}^3} \int_{\mathbb{T}^3} (\tilde{\varrho}_\varepsilon + \tilde{\varrho}_\varepsilon |\tilde{\mathbf{u}}_\varepsilon|^2) \,\mathrm{d}x \,\mathrm{d}t \lesssim c_1(r), \end{aligned}$$

using (4.192) and (4.193). Consequently, (4.220) implies

$$\tilde{\varrho}_\varepsilon \mathbb{F}_\varepsilon(\tilde{\varrho}_\varepsilon, \tilde{\mathbf{u}}_\varepsilon) \to \overline{\tilde{\varrho}\mathbb{F}(\tilde{\varrho}, \tilde{\mathbf{u}})} \quad \text{in } L^2(0,T; L_2(\mathfrak{U}; W^{-l,2}(\mathbb{T}^3)))$$

$\tilde{\mathbb{P}}$-a.s. Combining this with the convergence of $\tilde{W}_\delta$ from Proposition 4.5.5, we may apply Lemma 2.6.6 to pass to the limit in the stochastic integral and hence complete the proof. □

4.5.4 Limit $\delta \to 0$ – deterministic compactness method

We consider a family $[\tilde{\varrho}_\delta, \tilde{\mathbf{u}}_\delta, \tilde{W}_\delta]$ of dissipative martingale solutions of problem (4.121)–(4.122) enjoying the compactness properties stated in (4.213) and (4.214). Our ultimate goal is to show that the limit $[\tilde{\varrho}, \tilde{\mathbf{u}}, \tilde{W}]$ is a dissipative martingale solution of our target problem (4.184)–(4.185).

4.5.4.1 Compactness of the density

As in Section 4.4.3, our ultimate task is to show strong convergence of the densities $\tilde{\varrho}_\delta$. Following the deterministic variant of the proof, we introduce the *oscillation defect measure*. We set

$$\mathbf{osc}_\alpha[\tilde{\varrho}_\delta \to \tilde{\varrho}]((0,T)\times\mathbb{T}^3) = \sup_{k\geq 1}\left(\limsup_{\delta\to 0} \tilde{\mathbb{E}} \int_0^T \int_{\mathbb{T}^3} |T_k(\tilde{\varrho}_\delta) - T_k(\tilde{\varrho})|^\alpha \,\mathrm{d}x\,\mathrm{d}t \right),$$

where $\alpha \geq 1$ and T_k is a family of cut-off functions defined for $k \in \mathbb{N}$ as

$$T_k(r) = kT\left(\frac{r}{k}\right), \quad T \in C^\infty([0,\infty)), \quad T(r) = \begin{cases} r & \text{for } 0 \leq r \leq 1, \\ T''(r) \leq 0 & \text{for } r \in (1,3), \\ 2 & \text{for } r \geq 3. \end{cases}$$

In the following we will show that

$$\mathbf{osc}_{\gamma+1}[\tilde{\varrho}_\delta \to \tilde{\varrho}]((0,\tau)\times\mathbb{T}^3) \lesssim 1 \quad \text{for any } \tau > 0. \tag{4.223}$$

This is crucial for the proof of compactness of the density.

Relation (4.223) will be used to show that the limit $[\tilde{\varrho}, \tilde{\mathbf{u}}]$ satisfies the renormalized continuity equation $\tilde{\mathbb{P}}$-a.s. Note that the theory of DiPerna–Lions (see Theorem A.3.1) does not apply to the limit as the integrability of the density $\tilde{\varrho}$ is only L^γ. We proceed as follows:

- we establish a variant of the effective viscous flux identity (4.171);
- we show that $\mathbf{osc}_\alpha[\tilde{\varrho}_\delta \to \tilde{\varrho}]((0,T)\times\mathbb{T}^3) < \infty$ for $\alpha = \gamma+1$. In particular, this implies that the limit $[\tilde{\varrho}, \tilde{\mathbf{u}}]$ satisfies the renormalized equation of continuity;
- we use the compactness argument analogous to Section 4.4.3 to complete the proof of strong convergence of $(\tilde{\varrho}_\delta)_{\delta\in(0,1)}$.

We let $\delta \to 0$ in the renormalized equation of continuity (4.132), obtaining

$$\begin{aligned}\int_{\mathbb{T}^3} \overline{b(\tilde{\varrho})}\varphi(\tau)\,\mathrm{d}x + \int_0^\tau\int_{\mathbb{T}^3} \overline{(b'(\tilde{\varrho})\tilde{\varrho} - b(\tilde{\varrho}))\operatorname{div}\tilde{\mathbf{u}}}\varphi\,\mathrm{d}x\,\mathrm{d}t &= \int_{\mathbb{T}^3} b(\tilde{\varrho}_0)\varphi(0)\,\mathrm{d}x \\ &+ \int_0^\tau\int_{\mathbb{T}^3}[\overline{b(\tilde{\varrho})}\partial_t\varphi + \overline{b(\tilde{\varrho})\tilde{\mathbf{u}}}\cdot\nabla\varphi]\,\mathrm{d}x\,\mathrm{d}t,\end{aligned} \tag{4.224}$$

for any $\varphi \in C_c^\infty([0,T]\times\mathbb{T}^3)$ and any $b \in C^1([0,\infty))$, $b'(\varrho) = 0$ for $\varrho \to \infty$. Here, $\overline{b(\tilde{\varrho})}$ and $\overline{(b'(\tilde{\varrho})\tilde{\varrho} - b(\tilde{\varrho}))\operatorname{div}\tilde{\mathbf{u}}}$ denote the weak $L^1(\tilde{\Omega}\times Q)$-limits of $b(\tilde{\varrho}_\delta)$ and $(b'(\tilde{\varrho}_\delta)\tilde{\varrho}_\delta - b(\tilde{\varrho}_\delta))\operatorname{div}\tilde{\mathbf{u}}_\delta$, respectively, obtained by means of Proposition 4.5.5, part (4). Note that we may use (4.222) to get

$$\overline{b(\tilde{\varrho})\tilde{\mathbf{u}}} = \overline{b(\tilde{\varrho})}\tilde{\mathbf{u}}.$$

With (4.224) at hand, we use

$$\nabla\Delta^{-1}[\overline{b(\tilde{\varrho})} - (\overline{b(\tilde{\varrho})})_{\mathbb{T}^3}] = \Delta^{-1}[\nabla\overline{b(\tilde{\varrho})}]$$

as a test function in (4.218) (see Section 4.5.1 for details), deducing

$$\begin{aligned}&\int_0^\tau\int_{\mathbb{T}^3}\overline{p(\tilde{\varrho})[b(\tilde{\varrho})} - (\overline{b(\tilde{\varrho})})_{\mathbb{T}^3}]\,\mathrm{d}x\,\mathrm{d}t \\ &= \left[\int_{\mathbb{T}^3}\tilde{\varrho}\tilde{\mathbf{u}}\cdot\Delta^{-1}[\nabla\overline{b(\tilde{\varrho})}]\,\mathrm{d}x\right]_{t=0}^{t=\tau} - \int_0^\tau\int_{\mathbb{T}^3}\tilde{\varrho}\tilde{\mathbf{u}}\otimes\tilde{\mathbf{u}} : \nabla\Delta^{-1}\nabla\overline{b(\tilde{\varrho})}\,\mathrm{d}x\,\mathrm{d}t \\ &\quad + \int_0^\tau\int_{\mathbb{T}^3}\mathbb{S}(\nabla\tilde{\mathbf{u}}) : \nabla\Delta^{-1}\nabla\overline{b(\tilde{\varrho})}\,\mathrm{d}x\,\mathrm{d}t + \int_0^\tau\int_{\mathbb{T}^3}\tilde{\varrho}\tilde{\mathbf{u}}\cdot\nabla\Delta^{-1}\operatorname{div}[\overline{b(\tilde{\varrho})}\tilde{\mathbf{u}}]\,\mathrm{d}x\,\mathrm{d}t \\ &\quad + \int_0^\tau\int_{\mathbb{T}^3}\tilde{\varrho}\tilde{\mathbf{u}}\cdot\Delta^{-1}[\nabla\overline{(b'(\tilde{\varrho})\tilde{\varrho} - b(\tilde{\varrho}))\operatorname{div}\tilde{\mathbf{u}}}]\,\mathrm{d}x\,\mathrm{d}t \\ &\quad - \int_0^\tau\int_{\mathbb{T}^3}\overline{\tilde{\varrho}\mathbb{F}(\tilde{\varrho},\tilde{\mathbf{u}})}\cdot\Delta^{-1}\nabla\overline{b(\tilde{\varrho})}\,\mathrm{d}x\,\mathrm{d}\tilde{W}.\end{aligned} \tag{4.225}$$

Similarly, letting $\delta \to 0$ in (4.198), denoting by $\overline{p(\tilde{\varrho})b(\tilde{\varrho})}$ the weak $L^1(\tilde{\Omega} \times Q)$-limit of $p(\tilde{\varrho}_\delta)b(\tilde{\varrho}_\delta)$ (recall Proposition 4.5.5, part (4)), and using (4.221), we get

$$\begin{aligned}
\int_0^\tau \int_{\mathbb{T}^3} [\overline{p(\tilde{\varrho})b(\tilde{\varrho})} - \overline{p(\tilde{\varrho})}(\overline{b(\tilde{\varrho})})_{\mathbb{T}^3}] \,\mathrm{d}x\,\mathrm{d}t &= \left[\int_{\mathbb{T}^3} \tilde{\varrho}\tilde{\mathbf{u}} \cdot \Delta^{-1}[\nabla \overline{b(\tilde{\varrho})}] \,\mathrm{d}x\right]_{t=0}^{t=\tau} \\
&- \lim_{\delta\to 0}\left[\int_0^\tau \int_{\mathbb{T}^3} \tilde{\varrho}_\delta \tilde{\mathbf{u}}_\delta \otimes \tilde{\mathbf{u}}_\delta : \nabla\Delta^{-1}\nabla b(\tilde{\varrho}_\delta) \,\mathrm{d}x\,\mathrm{d}t \right.\\
&- \int_0^\tau \int_{\mathbb{T}^3} \mathbb{S}(\nabla \tilde{\mathbf{u}}_\delta) : \nabla\Delta^{-1}\nabla b(\tilde{\varrho}_\delta) \,\mathrm{d}x\,\mathrm{d}t \\
&\left.- \int_0^\tau \int_{\mathbb{T}^3} \tilde{\varrho}_\delta \tilde{\mathbf{u}}_\delta \cdot \nabla\Delta^{-1}\mathrm{div}\,[b(\tilde{\varrho}_\delta)\tilde{\mathbf{u}}_\delta] \,\mathrm{d}x\,\mathrm{d}t\right] \\
&+ \int_0^\tau \int_{\mathbb{T}^3} \tilde{\varrho}\tilde{\mathbf{u}} \cdot \Delta^{-1}[\nabla \overline{(b'(\tilde{\varrho})\tilde{\varrho} - b(\tilde{\varrho}))\mathrm{div}\,\tilde{\mathbf{u}}}] \,\mathrm{d}x\,\mathrm{d}t \\
&- \int_0^\tau \int_{\mathbb{T}^3} \tilde{\varrho}\mathbb{F}(\tilde{\varrho}, \tilde{\mathbf{u}}) \cdot \Delta^{-1}\nabla \overline{b(\tilde{\varrho})} \,\mathrm{d}x\,\mathrm{d}\tilde{W}.
\end{aligned} \tag{4.226}$$

Now, the crucial step is again to apply Lemma A.1.11. Consequently, we deduce

$$\begin{aligned}
&\int_0^\tau \int_{\mathbb{T}^3} \tilde{\varrho}\tilde{\mathbf{u}} \cdot \nabla\Delta^{-1}\mathrm{div}\,[\overline{b(\tilde{\varrho})}\tilde{\mathbf{u}}] \,\mathrm{d}x\,\mathrm{d}t - \int_0^\tau \int_{\mathbb{T}^3} \tilde{\varrho}\tilde{\mathbf{u}} \otimes \tilde{\mathbf{u}} : \nabla\Delta^{-1}\nabla \overline{b(\tilde{\varrho})} \,\mathrm{d}x\,\mathrm{d}t \\
&= \lim_{\delta\to 0}\left[\int_0^\tau \int_{\mathbb{T}^3} \tilde{\varrho}_\delta \tilde{\mathbf{u}}_\delta \cdot \nabla\Delta^{-1}\mathrm{div}\,[b(\tilde{\varrho}_\delta)\tilde{\mathbf{u}}_\delta] \,\mathrm{d}x\,\mathrm{d}t - \int_0^\tau \int_{\mathbb{T}^3} \tilde{\varrho}_\delta \tilde{\mathbf{u}}_\delta \otimes \tilde{\mathbf{u}}_\delta : \nabla\Delta^{-1}\nabla b(\tilde{\varrho}_\delta) \,\mathrm{d}x\right].
\end{aligned}$$

In addition, we recall that $\eta = \lambda + \frac{\mu}{3}$ and easily see

$$\begin{aligned}
\int_{\mathbb{T}^3} \mathbb{S}(\nabla \tilde{\mathbf{u}}) : \nabla\Delta^{-1}\nabla \overline{b(\tilde{\varrho})} \,\mathrm{d}x &= (\eta + \mu) \int_{\mathbb{T}^3} \overline{b(\tilde{\varrho})}\mathrm{div}\,\tilde{\mathbf{u}} \,\mathrm{d}x \\
\lim_{\delta\to 0}\left[\int_{\mathbb{T}^3} \mathbb{S}(\nabla \tilde{\mathbf{u}}_\delta) : \nabla\Delta^{-1}\nabla b(\tilde{\varrho}_\delta) \,\mathrm{d}x\right] &= (\eta + \mu) \int_{\mathbb{T}^3} \overline{b(\tilde{\varrho})\mathrm{div}\,\tilde{\mathbf{u}}} \,\mathrm{d}x.
\end{aligned}$$

Thus, comparing (4.225) and (4.226) we obtain the following form of the effective viscous flux identity:

$$\begin{aligned}
&\int_0^\tau \int_{\mathbb{T}^3} [\overline{p(\tilde{\varrho})b(\tilde{\varrho})} - \overline{b(\tilde{\varrho})\mathrm{div}\,\tilde{\mathbf{u}}}] \,\mathrm{d}x\,\mathrm{d}t \\
&= \int_0^\tau \int_{\mathbb{T}^3} [\overline{p(\tilde{\varrho})}\;\overline{b(\tilde{\varrho})} - \overline{b(\tilde{\varrho})}\mathrm{div}\,\tilde{\mathbf{u}}] \,\mathrm{d}x\,\mathrm{d}t;
\end{aligned} \tag{4.227}$$

cf. (4.171).

Now we use (4.227) to control the oscillation defect measure. The choice $b = T_k$ yields

$$\int_0^\tau \int_{\mathbb{T}^3} [\overline{p(\tilde{\varrho})T_k(\tilde{\varrho})} - \overline{p(\tilde{\varrho})}\;\overline{T_k(\tilde{\varrho})}] \,\mathrm{d}x = \int_0^\tau \int_{\mathbb{T}^3} [\overline{T_k(\tilde{\varrho})\mathrm{div}\,\tilde{\mathbf{u}}} - \overline{T_k(\tilde{\varrho})}\mathrm{div}\,\tilde{\mathbf{u}}] \,\mathrm{d}x\,\mathrm{d}t.$$

Using (4.213), the integral on the right hand side can be estimated as

$$\begin{aligned}
&\tilde{\mathbb{E}}\int_0^\tau\int_{\mathbb{T}^3}\overline{T_k(\varrho)\operatorname{div}\mathbf{u}}-\overline{T_k(\varrho)}\operatorname{div}\tilde{\mathbf{u}}\,\mathrm{d}x\,\mathrm{d}t\\
&=\lim_{\delta\to0}\tilde{\mathbb{E}}\int_0^\tau\int_{\mathbb{T}^3}(T_k(\tilde\varrho_\delta)-T_k(\varrho))\operatorname{div}\tilde{\mathbf{u}}_\delta\,\mathrm{d}x\,\mathrm{d}t\\
&\quad+\lim_{\delta\to0}\tilde{\mathbb{E}}\int_0^\tau\int_{\mathbb{T}^3}(T_k(\tilde\varrho)-T_k(\tilde\varrho_\delta))\operatorname{div}\tilde{\mathbf{u}}\,\mathrm{d}x\,\mathrm{d}t\\
&\lesssim\limsup_{\delta\to0}\|T_k(\tilde\varrho_\delta)-T_k(\tilde\varrho)\|_{L^2(\tilde\Omega\times(0,\tau)\times\mathbb{T}^3)},
\end{aligned}\tag{4.228}$$

where the constant in the last inequality is independent of k. Now, since p is convex, non-negative and non-decreasing, we have $p(a)-p(b)\geq p(a-b)$ whenever $0\leq a\leq b$. In addition, T_k concave, so we have

$$\begin{aligned}
&\int_0^\tau\int_{\mathbb{T}^3}[\overline{p(\tilde\varrho)T_k(\tilde\varrho)}-\overline{p(\tilde\varrho)}\,\overline{T_k(\tilde\varrho)}]\,\mathrm{d}x\,\mathrm{d}t\\
&\geq\int_0^\tau\int_{\mathbb{T}^3}[\overline{p(\tilde\varrho)T_k(\tilde\varrho)}-\overline{p(\tilde\varrho)}\,\overline{T_k(\tilde\varrho)}+(\overline{p(\tilde\varrho)}-p(\tilde\varrho))(\overline{T_k(\tilde\varrho)}-T_k(\tilde\varrho))]\,\mathrm{d}x\,\mathrm{d}t\\
&\geq\int_0^\tau\int_{\mathbb{T}^3}[\overline{p(\tilde\varrho)T_k(\tilde\varrho)}-p(\tilde\varrho)\overline{T_k(\tilde\varrho)}-\overline{p(\tilde\varrho)}T_k(\tilde\varrho)+p(\tilde\varrho)T_k(\tilde\varrho)]\,\mathrm{d}x\,\mathrm{d}t\\
&=\lim_{\delta\to0}\int_0^\tau\int_{\mathbb{T}^3}(p(\tilde\varrho_\delta)-p(\tilde\varrho))(T_k(\tilde\varrho_\delta)-T_k(\tilde\varrho))\,\mathrm{d}x\,\mathrm{d}t\\
&\gtrsim\limsup_{\delta\to0}\int_0^\tau\int_{\mathbb{T}^3}|T_k(\tilde\varrho_\delta)-T_k(\tilde\varrho)|^{\gamma+1}\,\mathrm{d}x\,\mathrm{d}t.
\end{aligned}\tag{4.229}$$

Thus relations (4.227)–(4.229) put together yield the desired conclusion:

$$\limsup_{\delta\to0}\mathbb{E}\int_0^\tau\int_{\mathbb{T}^3}|T_k(\tilde\varrho_\delta)-T_k(\tilde\varrho)|^{\gamma+1}\,\mathrm{d}x\,\mathrm{d}t\leq c.\tag{4.230}$$

As the estimate is independent of k, we get (4.223).

Based on (4.223) we are going to pass to the limit in the following equation:

$$\begin{aligned}
&\partial_t b(\overline{T_k(\tilde\varrho)})+\operatorname{div}(b(\overline{T_k(\tilde\varrho)})\tilde{\mathbf{u}})+(b'(\overline{T_k(\tilde\varrho)})\overline{T_k(\tilde\varrho)}-b(\overline{T_k(\tilde\varrho)}))\operatorname{div}\tilde{\mathbf{u}}\\
&\quad=-b'(\overline{T_k(\tilde\varrho)})\overline{(T_k'(\tilde\varrho)\tilde\varrho-T_k(\tilde\varrho))\operatorname{div}\tilde{\mathbf{u}}},
\end{aligned}\tag{4.231}$$

which holds in the sense of distributions thanks to (4.224). On account of (4.223) we have, for all $p\in(1,\gamma)$,

$$\begin{aligned}
\tilde{\mathbb{E}}\|\overline{T_k(\tilde\varrho)}-\tilde\varrho\|_{L^p(\tilde\Omega\times Q)}^p&\leq\liminf_{\delta\to0}\tilde{\mathbb{E}}\|T_k(\tilde\varrho_\delta)-\tilde\varrho_\delta\|_{L_t^pL_x^p}^p\\
&\leq2^p\liminf_{\delta\to0}\tilde{\mathbb{E}}\int_{[|\tilde\varrho_\delta|\geq k]}|\tilde\varrho_\delta|^p\,\mathrm{d}x\,\mathrm{d}t\\
&\leq2^pk^{p-\gamma}\liminf_{\delta\to0}\tilde{\mathbb{E}}\int_0^T\int_{\mathbb{T}^3}|\tilde\varrho_\delta|^\gamma\,\mathrm{d}x\,\mathrm{d}t\longrightarrow0,\quad k\to\infty,
\end{aligned}$$

so we have

$$\overline{T_k(\tilde{\varrho})} \to \tilde{\varrho} \quad \text{in } L^p(\tilde{\Omega} \times (0,T) \times \mathbb{T}^3). \tag{4.232}$$

In order to pass to the limit in (4.231) we have to show

$$b'(\overline{T_k(\tilde{\varrho})})\overline{(T_k'(\tilde{\varrho})\tilde{\varrho} - T_k(\tilde{\varrho}))\operatorname{div}\tilde{\mathbf{u}}} \to 0 \quad \text{in } L^1(\tilde{\Omega} \times (0,T) \times \mathbb{T}^3). \tag{4.233}$$

Recall that b has to satisfy $b'(z) = 0$ for all $z \geq M$ for some $M = M(b)$. We define

$$Q_{k,M} := \{(\omega, t, x) \in \tilde{\Omega} \times [0,T] \times \mathbb{T}^3;\ \overline{T_k(\tilde{\varrho})} \leq M\}$$

and gain

$$\begin{aligned}
&\tilde{\mathbb{E}} \int_Q |b'(\overline{T_k(\tilde{\varrho})})\overline{(T_k'(\tilde{\varrho})\tilde{\varrho} - T_k(\tilde{\varrho}))\operatorname{div}\tilde{\mathbf{u}}}| \,\mathrm{d}x\,\mathrm{d}t \\
&\quad \leq \sup_{z \leq M} |b'(z)| \tilde{\mathbb{E}} \int_0^T \int_{\mathbb{T}^3} \chi_{Q_{k,M}} |\overline{(T_k'(\tilde{\varrho})\tilde{\varrho} - T_k(\tilde{\varrho}))\operatorname{div}\tilde{\mathbf{u}}}| \,\mathrm{d}x\,\mathrm{d}t \\
&\quad \leq C \liminf_{\delta \to 0} \tilde{\mathbb{E}} \int_0^T \int_{\mathbb{T}^3} \chi_{Q_{k,M}} |(T_k'(\tilde{\varrho}_\delta)\tilde{\varrho}_\delta - T_k(\tilde{\varrho}_\delta))\operatorname{div}\tilde{\mathbf{u}}_\delta| \,\mathrm{d}x\,\mathrm{d}t \\
&\quad \leq C \sup_\delta \|\operatorname{div}\tilde{\mathbf{u}}_\delta\|_{L^2(\tilde{\Omega} \times (0,T) \times \mathbb{T}^3)} \liminf_{\delta \to 0} \|T_k'(\tilde{\varrho}_\delta)\tilde{\varrho}_\delta - T_k(\tilde{\varrho}_\delta)\|_{L^2(Q_{k,M})}.
\end{aligned}$$

It follows from interpolation that

$$\begin{aligned}
&\|T_k'(\tilde{\varrho}_\delta)\tilde{\varrho}_\delta - T_k(\tilde{\varrho}_\delta)\|_{L^2(Q_{k,M})}^2 \\
&\quad \leq \|T_k'(\tilde{\varrho}_\delta)\tilde{\varrho}_\delta - T_k(\tilde{\varrho}_\delta)\|_{L^1(\tilde{\Omega} \times Q)}^\alpha \|T_k'(\tilde{\varrho}_\delta)\tilde{\varrho}_\delta - T_k(\tilde{\varrho}_\delta)\|_{L^{\gamma+1}(Q_{k,M})}^{(1-\alpha)(\gamma+1)},
\end{aligned} \tag{4.234}$$

where $\alpha = \frac{\gamma-1}{\gamma}$. Moreover, we show similarly to the proof of (4.232) that

$$\begin{aligned}
\|T_k'(\tilde{\varrho}_\delta)\tilde{\varrho}_\delta - T_k(\tilde{\varrho}_\delta)\|_{L^1(\tilde{\Omega} \times (0,T) \times \mathbb{T}^3)} &\leq C k^{1-\gamma} \sup_\delta \tilde{\mathbb{E}} \int_0^T \int_{\mathbb{T}^3} |\tilde{\varrho}_\delta|^\gamma \,\mathrm{d}x\,\mathrm{d}t \\
&\longrightarrow 0, \quad k \to \infty,
\end{aligned} \tag{4.235}$$

so it suffices to prove

$$\sup_\delta \|T_k'(\tilde{\varrho}_\delta)\tilde{\varrho}_\delta - T_k(\tilde{\varrho}_\delta)\|_{L^{\gamma+1}(Q_{k,M})} \leq C, \tag{4.236}$$

independently of k. As $T_k'(z)z \leq T_k(z)$, we have, by the definition of $Q_{k,M}$,

$$\begin{aligned}
&\|T_k'(\tilde{\varrho}_\delta)\tilde{\varrho}_\delta - T_k(\tilde{\varrho}_\delta)\|_{L^{\gamma+1}(Q_{k,M})} \\
&\quad \leq 2(\|T_k(\tilde{\varrho}_\delta) - T_k(\tilde{\varrho})\|_{L^{\gamma+1}(\tilde{\Omega} \times Q)} + \|T_k(\tilde{\varrho})\|_{L^{\gamma+1}(Q_{k,M})}) \\
&\quad \leq 2(\|T_k(\tilde{\varrho}_\delta) - T_k(\tilde{\varrho})\|_{L^{\gamma+1}(\tilde{\Omega} \times Q)} + \|T_k(\tilde{\varrho}) - \overline{T_k(\tilde{\varrho})}\|_{L^{\gamma+1}(\tilde{\Omega} \times Q)} + \|\overline{T_k(\tilde{\varrho})}\|_{L^{\gamma+1}(Q_{k,M})}).
\end{aligned}$$

$$\leq 2(\|T_k(\tilde{\varrho}_\delta) - T_k(\tilde{\varrho})\|_{L^{\gamma+1}(\tilde{\Omega}\times Q)} + \|T_k(\tilde{\varrho}) - \overline{T_k(\tilde{\varrho})}\|_{L^{\gamma+1}(\tilde{\Omega}\times Q)}) + CM.$$

Now (4.223), (4.222) with $b = T_k$ and the weak lower semicontinuity of the $L^{\gamma+1}$-norm, imply (4.236). On the other hand (4.234)–(4.236) imply (4.233). So we pass to the limit in (4.231) and gain

$$\partial_t b(\tilde{\varrho}) + \operatorname{div}(b(\tilde{\varrho})\tilde{\mathbf{u}}) + (b'(\tilde{\varrho})\tilde{\varrho} - b(\tilde{\varrho}))\operatorname{div}\tilde{\mathbf{u}} = 0, \tag{4.237}$$

in the sense of distributions. In particular, we have

$$\int_{\mathbb{T}^3} \tilde{\varrho} L_k(\tilde{\varrho})(\tau)\,\mathrm{d}x + \int_0^\tau \int_{\mathbb{T}^3} T_k(\tilde{\varrho})\operatorname{div}\tilde{\mathbf{u}}\,\mathrm{d}x\,\mathrm{d}t = \int_{\mathbb{T}^3} \tilde{\varrho}_0 L_k(\tilde{\varrho}_0)\,\mathrm{d}x, \tag{4.238}$$

where

$$L_k(\varrho) = \begin{cases} \log(\varrho) & \text{if } \varrho < k, \\ \log(k) & \text{if } \varrho \geq k. \end{cases}$$

Similarly, it follows from (4.224) that

$$\int_{\mathbb{T}^3} \overline{\tilde{\varrho} L_k(\tilde{\varrho})}(\tau)\,\mathrm{d}x + \int_0^\tau \int_{\mathbb{T}^3} \overline{T_k(\tilde{\varrho})\operatorname{div}\tilde{\mathbf{u}}}\,\mathrm{d}x\,\mathrm{d}t = \int_{\mathbb{T}^3} \tilde{\varrho}_0 L_k(\tilde{\varrho}_0)\,\mathrm{d}x. \tag{4.239}$$

Moreover, as p is non-decreasing, we deduce from (4.227)

$$\int_0^\tau \int_{\mathbb{T}^3} \overline{T_k(\tilde{\varrho})\operatorname{div}\tilde{\mathbf{u}}}\,\mathrm{d}x\,\mathrm{d}t \geq \int_0^\tau \int_{\mathbb{T}^3} \overline{T_k(\tilde{\varrho})}\operatorname{div}\tilde{\mathbf{u}}\,\mathrm{d}x\,\mathrm{d}t. \tag{4.240}$$

Summing up (4.238)–(4.240), we get

$$\int_{\mathbb{T}^3} [\overline{\tilde{\varrho} L_k(\tilde{\varrho})} - \tilde{\varrho} L_k(\tilde{\varrho})](\tau)\,\mathrm{d}x \leq \int_0^\tau \int_{\mathbb{T}^3} (\overline{T_k(\tilde{\varrho})} - T_k(\tilde{\varrho}))\operatorname{div}\tilde{\mathbf{u}}\,\mathrm{d}x\,\mathrm{d}t. \tag{4.241}$$

The final observation is that

$$\int_0^\tau \int_{\mathbb{T}^3} (\overline{T_k(\tilde{\varrho})} - T_k(\tilde{\varrho}))\operatorname{div}\tilde{\mathbf{u}}\,\mathrm{d}x\,\mathrm{d}t \to 0 \quad \text{as } k \to \infty,$$

because of (4.230). Letting $k \to \infty$ in (4.241), we finally obtain

$$\int_{\mathbb{T}^3} [\overline{\tilde{\varrho}\log(\tilde{\varrho})} - \tilde{\varrho}\log(\tilde{\varrho})](\tau)\,\mathrm{d}x = 0 \quad \text{for any } \tau \in [0, T].$$

As the function $\tilde{\varrho} \mapsto \tilde{\varrho}\log(\tilde{\varrho})$ is strictly convex, this yields the desired conclusion:

$$\tilde{\varrho}_\delta(\tau) \to \tilde{\varrho}(\tau) \quad \text{in } L^1(\mathbb{T}^3) \text{ for any } \tau \in [0, T]. \tag{4.242}$$

Finally, we can pass to the limit in the stochastic integral from the energy inequality as in Proposition 4.4.13.

So, we can pass to the limit in the energy inequality (4.125) using (4.210) exactly as in Section 4.4.3. We have proved the main result of this Chapter, Theorem 4.0.2.

5 Local well-posedness

In view of the existence result for dissipative martingale solutions from Chapter 4, it is natural to ask whether or not one can construct solutions that are strong in the PDE sense, at least locally in time. In the present chapter, we answer this question affirmatively and show the existence of a unique local-in-time strong solution, which exists up to a positive stopping time. Recall that the notion of a stopping time associated to a filtration was introduced in Definition 2.1.26. Roughly speaking, it is a random time up to which the solution is constructed. More specifically, every trajectory of the solution has a different life span, i.e., depends on the randomness variable $\omega \in \Omega$.

The system of equations reads

$$\mathrm{d}\varrho + \operatorname{div}(\varrho \mathbf{u})\,\mathrm{d}t = 0, \tag{5.1}$$

$$\mathrm{d}(\varrho \mathbf{u}) + \left[\operatorname{div}(\varrho \mathbf{u} \otimes \mathbf{u}) + a\nabla \varrho^{\gamma}\right]\mathrm{d}t = \operatorname{div} \mathbb{S}(\nabla \mathbf{u})\,\mathrm{d}t + \mathbb{G}(\varrho, \varrho \mathbf{u})\,\mathrm{d}W. \tag{5.2}$$

The initial conditions are the following random variables:

$$\varrho(0,\cdot) = \varrho_0, \quad \mathbf{u}(0,\cdot) = \mathbf{u}_0, \tag{5.3}$$

with sufficient space regularity specified below. As before, the driving process $W = \sum_{k=1}^{\infty} e_k W_k$ is a cylindrical $(\mathfrak{F}_t)$-Wiener process in $\mathfrak{U}$, defined on some complete probability space $(\Omega, \mathfrak{F}, \mathbb{P})$, and the coefficient $\mathbb{G}$ is generally non-linear and satisfies suitable growth assumptions. More precisely, let $\mathbb{G}(\varrho, \mathbf{q}) : \mathfrak{U} \to L^2(\mathbb{T}^3)$ be defined by

$$\mathbb{G}(\varrho, \mathbf{q})e_k = \mathbf{G}_k(\cdot, \varrho(\cdot), \mathbf{q}(\cdot)),$$

where $\varrho \in L^2(\mathbb{T}^3)$, $\varrho \geq 0$, $\mathbf{q} \in L^2(\mathbb{T}^3)$. In contrast to Chapter 4, where solutions are only weak in the PDE sense, we have to strengthen the assumptions on the coefficients $\mathbf{G}_k$. We suppose that the coefficients $\mathbf{G}_k : \mathbb{T}^3 \times [0,\infty) \times \mathbb{R}^3 \to \mathbb{R}^3$ are C^s-functions that satisfy uniformly in $x \in \mathbb{T}^3$

$$\mathbf{G}_k(\cdot, 0, 0) = 0, \tag{5.4}$$

$$|\nabla^l_{x,\varrho,\mathbf{q}} \mathbf{G}_k(x, \varrho, \mathbf{q})| \leq \alpha_k, \quad \sum_{k=1}^{\infty} \alpha_k < \infty \quad \text{for all } l \in \{1, \dots, s\}, \tag{5.5}$$

with $s \in \mathbb{N}$ specified below. A typical example we have in mind is

$$\mathbf{G}_k(x, \varrho, \mathbf{q}) = \mathbf{a}_k(x)\varrho + \mathbb{A}_k(x)\mathbf{q}, \tag{5.6}$$

where $\mathbf{a}_k : \mathbb{T}^3 \to \mathbb{R}^3$ and $\mathbb{A}_k : \mathbb{T}^3 \to \mathbb{R}^{3\times 3}$ are smooth functions. However, our analysis applies to general non-linear coefficients $\mathbf{G}_k$.

As discussed in Section 3.3, the system (5.1)–(5.3) is studied in the context of solutions that are strong in both the PDE and the probabilistic sense. The main result of this chapter is the existence of a unique *maximal strong pathwise solution* to system (5.1)–(5.3). Let us first recall the definition of a local strong pathwise solution.

https://doi.org/10.1515/9783110492552-005

Definition 5.0.1 (Local strong pathwise solution). Let $(\Omega, \mathfrak{F}, (\mathfrak{F}_t)_{t\geq 0}, \mathbb{P})$ be a stochastic basis with a complete right-continuous filtration and let W be an $(\mathfrak{F}_t)$-cylindrical Wiener process. Let $(\varrho_0, \mathbf{u}_0)$ be an $\mathfrak{F}_0$-measurable random variable in the space $W^{s,2}(\mathbb{T}^3) \times W^{s,2}(\mathbb{T}^3)$ for some $s > \frac{7}{2}$. A triplet $(\varrho, \mathbf{u}, \mathfrak{t})$ is called a *local strong pathwise solution* to system (5.1)–(5.3), provided:

(1) $\mathfrak{t}$ is a $\mathbb{P}$-a.s. strictly positive $(\mathfrak{F}_t)$-stopping time;

(2) the density ϱ is a $W^{s,2}(\mathbb{T}^N)$-valued $(\mathfrak{F}_t)$-progressively measurable stochastic process such that

$$\varrho(\cdot \wedge \mathfrak{t}) > 0, \quad \varrho(\cdot \wedge \mathfrak{t}) \in C([0,T]; W^{s,2}(\mathbb{T}^3)) \quad \mathbb{P}\text{-a.s.};$$

(3) the velocity $\mathbf{u}$ is a $W^{s,2}(\mathbb{T}^3)$-valued $(\mathfrak{F}_t)$-progressively measurable stochastic process such that

$$\mathbf{u}(\cdot \wedge \mathfrak{t}) \in C([0,T]; W^{s,2}(\mathbb{T}^3)) \cap L^2(0,T; W^{s+1,2}(\mathbb{T}^3)) \quad \mathbb{P}\text{-a.s.};$$

(4) the equation of continuity

$$\varrho(t \wedge \mathfrak{t}) = \varrho_0 - \int_0^{t\wedge\mathfrak{t}} \operatorname{div}(\varrho \mathbf{u}) \, \mathrm{d}s$$

holds for all $t \in [0,T]$ $\mathbb{P}$-a.s.;

(5) the momentum equation

$$\begin{aligned}(\varrho \mathbf{u})(t \wedge \mathfrak{t}) = \varrho_0 \mathbf{u}_0 &- \int_0^{t\wedge\mathfrak{t}} \operatorname{div}(\varrho \mathbf{u} \otimes \mathbf{u}) \, \mathrm{d}s \\ &+ \int_0^{t\wedge\mathfrak{t}} \operatorname{div} \mathbb{S}(\nabla \mathbf{u}) \, \mathrm{d}s - \int_0^{t\wedge\mathfrak{t}} \nabla p(\varrho) \, \mathrm{d}s + \int_0^{t\wedge\mathfrak{t}} \mathbb{G}(\varrho, \varrho \mathbf{u}) \, \mathrm{d}W\end{aligned}$$

holds for all $t \in [0,T]$ $\mathbb{P}$-a.s.

Maximal strong pathwise solutions are then defined as follows.

Definition 5.0.2 (Maximal strong pathwise solution). Fix a stochastic basis with a cylindrical Wiener process and an initial condition exactly as in Definition 5.0.1. A quadruplet

$$(\varrho, \mathbf{u}, (\mathfrak{t}_R)_{R\in\mathbb{N}}, \mathfrak{t})$$

is called a *maximal strong pathwise solution* to system (5.1)–(5.3), provided:

(1) $\mathfrak{t}$ is a $\mathbb{P}$-a.s. strictly positive $(\mathfrak{F}_t)$-stopping time;

(2) $(\mathfrak{t}_R)_{R\in\mathbb{N}}$ is an increasing sequence of $(\mathfrak{F}_t)$-stopping times such that $\mathfrak{t}_R < \mathfrak{t}$ on the set $[\mathfrak{t} < T]$, $\lim_{R\to\infty} \mathfrak{t}_R = \mathfrak{t}$ a.s., and

$$\sup_{t\in[0,\mathfrak{t}_R]} \|\mathbf{u}(t)\|_{W^{2,\infty}_x} \geq R \quad \text{in } [\mathfrak{t} < T]; \tag{5.7}$$

(3) each triplet $(\varrho, \mathbf{u}, \mathfrak{t}_R)$, $R \in \mathbb{N}$, is a local strong pathwise solution in the sense of Definition 5.0.1.

The stopping times $\mathfrak{t}_R$ announce the stopping time $\mathfrak{t}$, which is therefore predictable. It denotes the maximal life span of the solution, which is determined by the time of explosion of the $W^{2,\infty}$-norm of the velocity field. Indeed, it can be seen from (5.7) that

$$\sup_{t\in[0,\mathfrak{t})} \|\mathbf{u}(t)\|_{W^{2,\infty}_x} = \infty \quad \text{in } [\mathfrak{t} < T].$$

Note that the announcing sequence $(\mathfrak{t}_R)$ is not unique. Therefore, uniqueness for maximal strong solutions is understood in the sense that only the solution $(\varrho, \mathbf{u})$ and its blow-up time $\mathfrak{t}$ are unique.

We then obtain the following existence and uniqueness result.

Theorem 5.0.3. *Let $s \in \mathbb{N}$ satisfy $s > \frac{9}{2}$ and let $\gamma > 1$. Let the coefficients $\mathbf{G}_k$ satisfy hypotheses* (5.4) *and* (5.5) *and let $(\varrho_0, \mathbf{u}_0)$ be an $\mathfrak{F}_0$-measurable, $W^{s,2}(\mathbb{T}^3) \times W^{s,2}(\mathbb{T}^3)$-valued random variable such that $\varrho_0 > 0$ $\mathbb{P}$-a.s. Then there exists a unique maximal strong pathwise solution $(\varrho, \mathbf{u}, (\mathfrak{t}_R)_{R\in\mathbb{N}}, \mathfrak{t})$ to problem* (5.1)–(5.3) *in the sense of Definition* 5.0.2 *with the initial condition $(\varrho_0, \mathbf{u}_0)$.*

Remark 5.0.4. Large parts of the proof of Theorem 5.0.3 also apply to more general pressure laws. In particular, the assumption $\gamma > 1$ is not needed. This restriction arises as we rewrite (5.1)–(5.3) to a symmetric hyperbolic system, meaning we formally divide the momentum equation (5.2) by ϱ; see Section 5.1.1.

Let us point out that, later on, we will choose s in order to have the embedding $W^{s,2} \hookrightarrow W^{2,\infty}$, i.e., at least $s > \frac{7}{2}$. Observe that even though one might expect that the $W^{s,2}$-norm blows up earlier than the $W^{2,\infty}$-norm, this is not true. Indeed, according to Definition 5.0.1 and Definition 5.0.2, a maximal strong pathwise solution satisfies

$$\mathbf{u}(\cdot \wedge \mathfrak{t}_R) \in C([0,T]; W^{s,2}(\mathbb{T}^3)) \quad \mathbb{P}\text{-a.s.}$$

and hence the velocity is continuous in $W^{s,2}(\mathbb{T}^3)$ on $[0, \mathfrak{t})$. Consequently, the blow-up of the $W^{s,2}$-norm coincides with the blow-up of the $W^{2,\infty}$-norm at time $\mathfrak{t}$. This fact reflects the nature of our *a priori* estimates (see Section 5.2.2): roughly speaking, a control of the $W^{2,\infty}$-norm implies a control of the $W^{s,2}$-norm and leads to continuity of trajectories in $W^{s,2}$.

Note that the results of this chapter also apply to the situation in general dimensions N. The corresponding bounds $s > \frac{9}{2}$ and $s > \frac{7}{2}$ have to be replaced by $s > \frac{N}{2} + 3$ and $s > \frac{N}{2} + 2$, respectively. The required regularity $s > \frac{N}{2} + 3$ is higher than $s > \frac{N}{2} + 2$, needed in the deterministic situation; see Matsumura–Nishida [MN79, MN83] and Valli–Zajaczkowski [VZ86]. This is due to the loss of regularity with respect to the time variable pertinent to the stochastic problems. Possibly optimal results could be achieved by working in the framework of L^p-spaces as Cho et al. [CCK04] and to adapt this approach to the stochastic setting in the spirit of Glatt-Holtz and Vicol [GHV14].

Moreover, the method used in this chapter can easily be adapted to handle the same problem on the whole space $\mathbb{R}^3$, with relevant far field conditions for ϱ, $\mathbf{u}$, say

$$\varrho \to \overline{\varrho}, \quad \mathbf{u} \to 0 \quad \text{as } |x| \to \infty.$$

On the other hand, the case when the fluid interacts with a physical boundary, for instance $\mathcal{O}$, a bounded domain with the no-slip boundary condition for $\mathbf{u}$, would require a more elaborate treatment.

In addition, most of our analysis applies to the stochastic compressible Euler system as well. Indeed, the only point where we rely on the positive viscosity μ is the proof of continuity of trajectories of a solution in $W^{s,2}$; see Section 5.2.4. It is based on the variational approach within a Gelfand triplet, which gives a very elegant proof, especially in comparison to the Euler setting where one would need to find another reasoning; cf. Glatt-Holtz and Vicol [GHV14].

Our approach can be summarized as follows. Similarly to Kim [Kim11] (see also Glatt-Holtz and Vicol [GHV14]), we use suitable cut-off operators to render all nonlinearities in the equations globally Lipschitz continuous. The resulting stochastic system may admit *global-in-time solutions*. Still, the approach proposed by Kim [Kim11] and later revised by Glatt-Holtz and Vicol [GHV14] cannot be applied in a direct fashion for the following reasons:

(1) The energy method is only applicable to *symmetric* hyperbolic systems and their viscous perturbations.
(2) In order to symmetrize (5.1)–(5.2), the density must be strictly positive, in other words, the system must be vacuum-free.
(3) For the density to remain positive at least on a short time interval, the maximum principle must be applied to the transport equation (5.1). Accordingly, equation (5.1) must be solved exactly and not by means of a finite-dimensional approximation.
(4) To avoid technical problems with non-local operators in the transport equation, the cut-off must be applied only to the velocity field.

In view of these difficulties and anticipating strict positivity of the density, we transform the problem to a symmetric hyperbolic system perturbed by partial viscosity and the stochastic driving term; see Section 5.1.1. Then cut-off operators in the spirit of Kim [Kim11] are applied to the velocity field and this system is then studied in detail in Section 5.2. We use this technique to cut the non-linear parts as well as to guarantee the non-degeneracy of the density, which leads to global-in-time strong martingale solutions to this approximate system. The main ideas of the proof are as follows.

First, we adapt a hybrid method similar to the one presented in Chapter 4: the equation of continuity is solved directly, while the momentum equation is approximated by a finite-dimensional Galerkin scheme. On this level, we are able to gain higher order uniform energy estimates by differentiating in space. Then, using the stochastic compactness method, we prove the existence of a strong martingale solution. In Section 5.2.5 we establish pathwise uniqueness and then the method of Gyöngy–Krylov (see Section 2.10) is applied to recover the convergence of the approximate solutions on the original probability space; see Section 5.2.6. The existence of a unique strong pathwise solution therefore follows.

Finally, in Section 5.3 we employ the results of the previous sections to conclude the proof of Theorem 5.0.3. This last step is in the spirit of the recent treatment of the incompressible Euler system by Glatt-Holtz and Vicol [GHV14]. However, the analysis is more involved due to the complicated structure of (5.1)–(5.3). We rely on a delicate combination of stopping time arguments that allows one to use the equivalence of (5.1)–(5.3) with the system studied in Section 5.2. As a consequence, also the corresponding existence and uniqueness result may be applied. One of the difficulties originates in the fact that we no longer assume the initial condition to be integrable in ω. Thus the *a priori* estimates from Section 5.2 are no longer valid. We present the details of the proof of uniqueness in Section 5.3.1 and the existence of a local strong pathwise solution in Section 5.3.2 and Section 5.3.3. We conclude with the existence of a maximal strong pathwise solution in Section 5.3.4.

5.1 Preliminary considerations

To begin with, let us introduce a new variable r, related to ϱ through the following formula:

$$\varrho = \varrho(r) = \left(\frac{\gamma - 1}{2a\gamma}\right)^{\frac{1}{\gamma-1}} r^{\frac{2}{\gamma-1}},$$

together with the following associated family of diffusion coefficients:

$$\mathbf{F}_k(\cdot, r, \mathbf{u}) = \frac{1}{\varrho(r)} \mathbf{G}_k(\cdot, \varrho(r), \varrho(r)\mathbf{u}),$$

similarly to Chapter 4. Note that, for the model case (5.6), this implies

$$\mathbf{F}_k(x, r, \mathbf{u}) = \mathbf{a}_k(x) + \mathbb{A}_k(x)\mathbf{u}.$$

As we are interested in *strong* solutions for which both ϱ and $\mathbf{u}$ are bounded and ϱ is bounded below away from zero, the hypothesis (5.5) implies the same property for $\mathbf{F}_k$ restricted to this range. In addition, we have

$$\sum_{k=1}^{\infty} |\mathbf{F}_k(\cdot, r, \mathbf{u})| \le c(1 + |\mathbf{u}|). \tag{5.8}$$

Also note that it suffices to assume that (5.5) holds only locally, meaning on each compact subset of $\mathbb{T}^3 \times (0,\infty) \times \mathbb{R}^3$.

Recall that, if ϱ, $\mathbf{q}$ are $(\mathfrak{F}_t)$-progressively measurable $L^2(\mathbb{T}^3)$-valued stochastic processes such that

$$\varrho \in L^2(\Omega \times [0,T]; L^2(\mathbb{T}^3)), \quad \mathbf{q} \in L^2(\Omega \times [0,T]; L^2(\mathbb{T}^3))$$

and $\mathbb{G}$ satisfies (5.4) and (5.5), the stochastic integral

$$\int_0^t \mathbb{G}(\varrho, \varrho\mathbf{u})\,\mathrm{d}W = \sum_{k=1}^{\infty} \int_0^t \mathbf{G}_k(\cdot, \varrho, \varrho\mathbf{u})\,\mathrm{d}W_k$$

is a well-defined $(\mathfrak{F}_t)$-martingale ranging in $L^2(\mathbb{T}^3)$. We refer to Section 2.3 for more details. In addition, combining Lemma 2.3.9, the hypotheses (5.4) and (5.5), the estimate (5.11), and the embedding

$$W^{s,2}(\mathbb{T}^3) \hookrightarrow C(\mathbb{T}^3), \quad s > \frac{3}{2},$$

we obtain the following estimate for the stochastic integral appearing in (5.2).

Lemma 5.1.1. *Let $\mathbf{G}_k = \mathbf{G}_k(\varrho, \mathbf{q})$ satisfy* (5.4) *and* (5.5) *for a non-negative integer s. Let $p \geq 2$, $\alpha \in [0, \frac{1}{2})$. Suppose that*

$$\varrho, \mathbf{q} \in L^{\beta p}(\Omega \times (0,T); W^{s,2}(\mathbb{T}^3)), \quad \beta = \max\{s, 1\}.$$

Then the following holds:

(1) *If $s = 0$, then*

$$t \mapsto \int_0^t \mathbb{G}(\varrho, \mathbf{q})\,\mathrm{d}W \in L^p(\Omega; W^{\alpha,p}(0,T; L^2(\mathbb{T}^3)))$$

and

$$\mathbb{E}\left[\left\|\int_0^t \mathbb{G}(\varrho, \mathbf{q})\,\mathrm{d}W\right\|^p_{W_t^{\alpha,p} L_x^2}\right] \leq c(\alpha, p)\mathbb{E}\left[\int_0^T \|[\varrho, \mathbf{q}]\|^p_{L_x^2}\,\mathrm{d}t\right].$$

(2) *If $s > \frac{3}{2}$, then*

$$t \mapsto \int_0^t \mathbb{G}(\varrho, \mathbf{q})\,\mathrm{d}W \in L^p(\Omega; W^{\alpha,p}(0,T; W^{s,2}(\mathbb{T}^3)))$$

and

$$\mathbb{E}\left[\left\|\int_0^t \mathbb{G}(\varrho, \mathbf{q})\,\mathrm{d}W\right\|^p_{W_t^{\alpha,p} W_x^{s,2}}\right] \leq c(\alpha, p)\mathbb{E}\left[\int_0^T \|[\varrho, \mathbf{q}]\|^{sp}_{W_x^{s,2}}\,\mathrm{d}t\right].$$

The following estimates are standard in the Moser type calculus and can be found, e.g., in Majda [Maj12, Proposition 2.1]. They will be very useful when dealing with higher order derivatives.

Proposition 5.1.2. (1) *For $u, v \in W^{s,2} \cap L^\infty(\mathbb{T}^N)$ and $\alpha \in \mathbb{N}_0^N$, $|\alpha| \le s$, we have*

$$\|\partial_x^\alpha(uv)\|_{L_x^2} \le c_s(\|u\|_{L_x^\infty}\|\nabla_x^s v\|_{L_x^2} + \|v\|_{L_x^\infty}\|\nabla_x^s u\|_{L_x^2}). \tag{5.9}$$

(2) *For $u \in W^{s,2}(\mathbb{T}^N)$, $\nabla_x u \in L^\infty(\mathbb{T}^N)$, $v \in W^{s-1,2} \cap L^\infty(\mathbb{T}^N)$ and $\alpha \in \mathbb{N}_0^N$, $|\alpha| \le s$, we have*

$$\|\partial_x^\alpha(uv) - u\partial_x^\alpha v\|_{L_x^2} \le c_s(\|\nabla_x u\|_{L_x^\infty}\|\nabla_x^{s-1} v\|_{L_x^2} + \|v\|_{L_x^\infty}\|\nabla_x^s u\|_{L_x^2}). \tag{5.10}$$

(3) *Let $u \in W^{s,2} \cap C(\mathbb{T}^3)$ and let F be an s-times continuously differentiable function in an open neighborhood of the compact set $G = \mathrm{range}[u]$. Then we have, for all $\alpha \in \mathbb{N}_0^N$, $|\alpha| \le s$,*

$$\|\partial_x^\alpha F(u)\|_{L_x^2} \le c_s\|\partial_u F\|_{C^{s-1}(G)}\|u\|_{L_x^\infty}^{|\alpha|-1}\|\partial_x^\alpha u\|_{L_x^2}. \tag{5.11}$$

5.1.1 Rewriting the equations as a symmetric hyperbolic-parabolic problem

It is well known in the context of compressible fluids that existence of strong solutions is intimately related to the strict positivity of the density, i.e., the non-appearance of vacuum states. Anticipating this property in the framework of strong solutions we may rewrite (5.1)–(5.2) as a hyperbolic-parabolic system for the unknowns r, $\mathbf{u}$, where r is a function of ϱ. To be more precise, as the time derivative of ϱ satisfies the deterministic equation (5.1), we have

$$\mathrm{d}(\varrho\mathbf{u}) = \mathrm{d}\varrho\,\mathbf{u} + \varrho\,\mathrm{d}\mathbf{u},$$

where, in accordance with (5.1),

$$\mathrm{d}\varrho = -\mathrm{div}\,(\varrho\mathbf{u})\,\mathrm{d}t.$$

Consequently, the momentum equation (5.2) reads

$$\varrho\,\mathrm{d}\mathbf{u} + [\varrho\mathbf{u}\cdot\nabla\mathbf{u} + a\nabla\varrho^\gamma]\,\mathrm{d}t = \mathrm{div}\,\mathbb{S}(\nabla\mathbf{u})\,\mathrm{d}t + \mathbb{G}(\varrho, \varrho\mathbf{u})\,\mathrm{d}W,$$

or, anticipating strict positivity of the mass density,

$$\mathrm{d}\mathbf{u} + \left[\mathbf{u}\cdot\nabla\mathbf{u} + a\frac{1}{\varrho}\nabla\varrho^\gamma\right]\mathrm{d}t = \frac{1}{\varrho}\mathrm{div}\,\mathbb{S}(\nabla\mathbf{u})\,\mathrm{d}t + \frac{1}{\varrho}\mathbb{G}(\varrho, \varrho\mathbf{u})\,\mathrm{d}W.$$

Next, we rewrite

$$a\frac{1}{\varrho}\nabla\varrho^\gamma = \frac{a\gamma}{\gamma-1}\nabla\varrho^{\gamma-1} = \frac{2a\gamma}{\gamma-1}\varrho^{\frac{\gamma-1}{2}}\nabla\varrho^{\frac{\gamma-1}{2}}$$

and evoke the renormalized variant of (5.1). See for instance Remark 4.4.4 and recall that we assume that ϱ is a strong solution. This leads to

$$\mathrm{d}\varrho^{\frac{\gamma-1}{2}} + \mathbf{u}\cdot\nabla\varrho^{\frac{\gamma-1}{2}}\,\mathrm{d}t + \frac{\gamma-1}{2}\varrho^{\frac{\gamma-1}{2}}\operatorname{div}\mathbf{u}\,\mathrm{d}t = 0.$$

Thus, for a new variable

$$r \equiv \sqrt{\frac{2a\gamma}{\gamma-1}}\varrho^{\frac{\gamma-1}{2}},$$

system (5.1)–(5.2) takes the form

$$\mathrm{d}r + \mathbf{u}\cdot\nabla r\,\mathrm{d}t + \frac{\gamma-1}{2}r\operatorname{div}\mathbf{u}\,\mathrm{d}t = 0, \tag{5.12}$$

$$\mathrm{d}\mathbf{u} + [\mathbf{u}\cdot\nabla\mathbf{u} + r\nabla r]\,\mathrm{d}t = D(r)\operatorname{div}\mathbb{S}(\nabla\mathbf{u})\,\mathrm{d}t + \mathbb{F}(r,\mathbf{u})\,\mathrm{d}W, \tag{5.13}$$

where

$$D(r) = \frac{1}{\varrho(r)} = \left(\frac{\gamma-1}{2a\gamma}\right)^{-\frac{1}{\gamma-1}} r^{-\frac{2}{\gamma-1}}, \quad \mathbb{F}(r,\mathbf{u}) = \frac{1}{\varrho(r)}\mathbb{G}(\varrho(r),\varrho(r)\mathbf{u}).$$

Observe that the left hand side of (5.13) corresponds to a symmetric hyperbolic system; cf. Majda [Maj12]. For such systems, higher order energy estimates can be obtained by differentiating (5.12) and (5.13) in x up to order s; cf. Gallagher [Gal00] and Majda [Maj12]. Unlike the more elaborated treatment proposed by Cho et al. [CCK04], giving rise to the optimal regularity space for the deterministic Navier–Stokes system, the energy approach avoids differentiating the equations in the time variable – a procedure that may be delicate in the stochastic setting.

5.1.2 Outline of the proof of Theorem 5.0.3

In the *deterministic* setting, system (5.12)–(5.13) can be solved via an approximation procedure. The so-obtained local-in-time strong solution exists on a maximal time interval, the length of which can be estimated in terms of the size of the initial data. However, in the *stochastic* setting it is more convenient to work with approximate solutions defined on the whole time interval $[0,T]$. To this end, we introduce suitable cut-off operators applied to the $W^{2,\infty}$-norm of the velocity field; cf. Kim [Kim11]. Specifically, we consider the approximate system in the form

$$\mathrm{d}r + \varphi_R(\|\mathbf{u}\|_{W^{2,\infty}_x})\left[\mathbf{u}\cdot\nabla r + \frac{\gamma-1}{2}r\operatorname{div}\mathbf{u}\right]\mathrm{d}t = 0, \tag{5.14}$$

$$\begin{aligned}\mathrm{d}\mathbf{u} + \varphi_R(\|\mathbf{u}\|_{W^{2,\infty}_x})[\mathbf{u}\cdot\nabla\mathbf{u} + r\nabla r]\,\mathrm{d}t &= \varphi_R(\|\mathbf{u}\|_{W^{2,\infty}_x})D(r)\operatorname{div}\mathbb{S}(\nabla\mathbf{u})\,\mathrm{d}t\\ &\quad + \varphi_R(\|\mathbf{u}\|_{W^{2,\infty}_x})\mathbb{F}(r,\mathbf{u})\,\mathrm{d}W,\end{aligned} \tag{5.15}$$

$$r(0) = r_0, \quad \mathbf{u}(0) = \mathbf{u}_0, \tag{5.16}$$

where $\varphi_R : [0,\infty) \to [0,1]$ are smooth cut-off functions satisfying

$$\varphi_R(y) = \begin{cases} 1, & 0 \le y \le R, \\ 0, & R+1 \le y. \end{cases}$$

Our aim is to solve (5.14)–(5.16) via the stochastic compactness method presented in Section 2.6. Therefore, we first construct solutions to certain approximate problems, we then establish tightness of their laws in suitable topologies, and finally we deduce the existence of a strong martingale solution to (5.14)–(5.16). That is, a solution which is strong in the PDE sense but weak in the probabilistic sense. The necessary uniform bounds are obtained through a purely hyperbolic approach by differentiating with respect to the space variable and testing the resulting expression with suitable space derivative of the unknown functions (more precisely, by applying Itô's formula, Theorem 2.4.1).

For the above mentioned reasons, the approximate densities must be positive on time intervals of positive length. Therefore, the approximation scheme must be chosen to preserve the maximum principle for (5.14). To this end, the approximate solutions to (5.14)–(5.16) will be constructed by means of a hybrid method based on:

(1) solving the deterministic equation of continuity (5.14) for a given $\mathbf{u}$, obtaining $r = r[\mathbf{u}]$;
(2) plugging $r = r[\mathbf{u}]$ in (5.15) and using a fixed point argument to get local-in-time solutions of a Galerkin approximation of (5.15);
(3) extending the Galerkin solution to $[0, T]$ by means of *a priori* bounds.

Note that the transport equation (5.14) can be solved exactly in terms of a given velocity field $\mathbf{u}$ as the cut-off operators apply to $\mathbf{u}$ only.

5.2 The approximate system

In this section we focus on the approximate system (5.14)–(5.16). More precisely, our aim is twofold. First, we establish existence of a strong martingale solution for initial data in $L^p(\Omega; W^{s,2}(\mathbb{T}^3))$ for all $1 \le p < \infty$ and some $s > \frac{7}{2}$. Second, we prove pathwise uniqueness provided $s > \frac{9}{2}$. This in turn implies existence of a unique strong pathwise solution using the method of Gyöngy–Krylov from Section 2.10.

To this end, let us introduce these two concepts of strong solution for the approximate system (5.14)–(5.16). A strong martingale solution is strong in the PDE sense but only weak in the probabilistic sense. In other words, neither the stochastic basis nor a cylindrical Wiener process can be given in advance and become a part of the solution. Accordingly, the initial condition is stated in the form of an initial law. On the other hand, a strong pathwise solution is strong in both the PDE and the probabilistic sense, that is, the stochastic elements are given in advance.

Definition 5.2.1 (Strong martingale solution). Let Λ be a Borel probability measure on $W^{s,2}(\mathbb{T}^3) \times W^{s,2}(\mathbb{T}^3)$ and let $s \in \mathbb{N}$. A multiplet

$$((\Omega, \mathfrak{F}, (\mathfrak{F}_t)_{t\geq 0}, \mathbb{P}), r, \mathbf{u}, W)$$

is called a *strong martingale solution* to the approximate system (5.14)–(5.16) with the initial law Λ, provided:

(1) $(\Omega, \mathfrak{F}, (\mathfrak{F}_t)_{t\geq 0}, \mathbb{P})$ is a stochastic basis with a complete right-continuous filtration;
(2) W is an $(\mathfrak{F}_t)$-cylindrical Wiener process;
(3) r is a $W^{s,2}(\mathbb{T}^3)$-valued $(\mathfrak{F}_t)$-progressively measurable stochastic process satisfying

$$r \in L^2(\Omega; C([0,T]; W^{s,2}(\mathbb{T}^3))), \quad r > 0 \ \mathbb{P}\text{-a.s.};$$

(4) the velocity $\mathbf{u}$ is a $W^{s,2}(\mathbb{T}^3)$-valued $(\mathfrak{F}_t)$-progressively measurable stochastic process satisfying

$$\mathbf{u} \in L^2(\Omega; C([0,T]; W^{s,2}(\mathbb{T}^3))) \cap L^2(\Omega; L^2(0,T; W^{s+1,2}(\mathbb{T}^3)));$$

(5) there exists an $\mathfrak{F}_0$-measurable random variable $[\varrho_0, \mathbf{u}_0]$ such that $\Lambda = \mathcal{L}[\varrho_0, \mathbf{u}_0]$;
(6) the equations

$$\begin{aligned}
r(t) &= r_0 - \int_0^t \varphi_R(\|\mathbf{u}\|_{W^{2,\infty}_x})\Big[\mathbf{u} \cdot \nabla r + \frac{\gamma - 1}{2} r \operatorname{div} \mathbf{u}\Big] \, \mathrm{d}s, \\
\mathbf{u}(t) &= \mathbf{u}_0 - \int_0^t \varphi_R(\|\mathbf{u}\|_{W^{2,\infty}_x})[\mathbf{u} \cdot \nabla \mathbf{u} + r \nabla r] \, \mathrm{d}s \\
&\quad + \int_0^t \varphi_R(\|\mathbf{u}\|_{W^{2,\infty}_x}) D(r) \operatorname{div} \mathbb{S}(\nabla \mathbf{u}) \, \mathrm{d}s + \int_0^t \varphi_R(\|\mathbf{u}\|_{W^{2,\infty}_x}) \mathbb{F}(r, \mathbf{u}) \, \mathrm{d}W
\end{aligned}$$

hold, for all $t \in [0,T]$ $\mathbb{P}$-a.s.

Definition 5.2.2 (Strong pathwise solution). Let $(\Omega, \mathfrak{F}, (\mathfrak{F}_t)_{t\geq 0}, \mathbb{P})$ be a given stochastic basis with a complete right-continuous filtration, let W be a given $(\mathfrak{F}_t)$-cylindrical Wiener process, and let $s \in \mathbb{N}$. Then $(r, \mathbf{u})$ is called a *strong pathwise solution* to the approximate system (5.14)–(5.16) with the initial condition $(r_0, \mathbf{u}_0)$, provided:

(1) r is a $W^{s,2}(\mathbb{T}^3)$-valued $(\mathfrak{F}_t)$-progressively measurable stochastic process satisfying

$$r \in L^2(\Omega; C([0,T]; W^{s,2}(\mathbb{T}^3))), \quad r > 0 \ \mathbb{P}\text{-a.s.};$$

(2) the velocity $\mathbf{u}$ is a $W^{s,2}(\mathbb{T}^3)$-valued $(\mathfrak{F}_t)$-progressively measurable stochastic process satisfying

$$\mathbf{u} \in L^2(\Omega; C([0,T]; W^{s,2}(\mathbb{T}^3))) \cap L^2(\Omega; L^2(0,T; W^{s+1,2}(\mathbb{T}^3)));$$

(3) the equations

$$r(t) = r_0 - \int_0^t \varphi_R(\|\mathbf{u}\|_{W_x^{2,\infty}})\Big[\mathbf{u}\cdot\nabla r + \frac{\gamma-1}{2} r \operatorname{div}\mathbf{u}\Big]\,\mathrm{d}s,$$

$$\mathbf{u}(t) = \mathbf{u}_0 - \int_0^t \varphi_R(\|\mathbf{u}\|_{W_x^{2,\infty}})[\mathbf{u}\cdot\nabla\mathbf{u} + r\nabla r]\,\mathrm{d}s$$
$$+ \int_0^t \varphi_R(\|\mathbf{u}\|_{W_x^{2,\infty}})D(r)\operatorname{div}\mathbb{S}(\nabla\mathbf{u})\,\mathrm{d}s + \int_0^t \varphi_R(\|\mathbf{u}\|_{W_x^{2,\infty}})\mathbb{F}(r,\mathbf{u})\,\mathrm{d}W$$

hold, for all $t \in [0,T]$ $\mathbb{P}$-a.s.

The main result of this section reads as follows.

Theorem 5.2.3. *Let the coefficients* $\mathbf{G}_k$ *satisfy hypotheses* (5.4) *and* (5.5) *and let*

$$(r_0, \mathbf{u}_0) \in L^p(\Omega, \mathfrak{F}_0, \mathbb{P}; W^{s,2}(\mathbb{T}^3) \times W^{s,2}(\mathbb{T}^3)),$$

for all $1 \le p < \infty$ *and some* $s \in \mathbb{N}$, *such that* $s > \frac{7}{2}$. *In addition, suppose that*

$$\|r_0\|_{W_x^{1,\infty}} < R, \quad r_0 > \frac{1}{R}, \quad \mathbb{P}\text{-a.s.}$$

for some deterministic constant $R > 0$. *Then:*

(1) *There exists a strong martingale solution to problem* (5.14)–(5.16) *in the sense of Definition* 5.2.1 *with the initial law* $\Lambda = \mathcal{L}[(r_0, \mathbf{u}_0)]$. *Moreover, there exists a deterministic constant* $\underline{r}_R > 0$ *such that*

$$r(t,\cdot) \ge \underline{r}_R > 0 \quad \mathbb{P}\text{-a.s.} \quad \text{for all } t \in [0,T]$$

and

$$\mathbb{E}\Big[\sup_{t\in[0,T]} \|(r(t),\mathbf{u}(t))\|_{W_x^{s,2}}^2 + \int_0^T \|\mathbf{u}\|_{W_x^{s+1,2}}^2\,\mathrm{d}t\Big]^p \le c(R, r_0, \mathbf{u}_0, p) < \infty, \tag{5.17}$$

for all $1 \le p < \infty$.

(2) *If* $s > \frac{9}{2}$, *then pathwise uniqueness holds true. Specifically, if* $(r^1, \mathbf{u}^1)$, $(r^2, \mathbf{u}^2)$ *are two strong solutions to* (5.14)–(5.16) *defined on the same stochastic basis with the same Wiener process W and*

$$\mathbb{P}(r_0^1 = r_0^2,\ \mathbf{u}_0^1 = \mathbf{u}_0^2) = 1,$$

then

$$\mathbb{P}(r^1(t) = r^2(t),\ \mathbf{u}^1(t) = \mathbf{u}^2(t),\ \text{for all } t \in [0,T]) = 1.$$

Consequently, there exists a unique strong pathwise solution to (5.14)–(5.15) *in the sense of Definition* 5.2.2.

The rest of this section is dedicated to the proof of Theorem 5.2.3, which is divided into several parts. First, in Section 5.2.1 we construct the approximate solutions to (5.14)–(5.16) by employing the hybrid method delineated in Section 5.1.2. Second, in Section 5.2.2 we derive higher order energy estimates which hold true uniformly in the approximation parameter m. Third, in Section 5.2.3 we perform the stochastic compactness method, as we establish tightness of the laws of the approximate solutions and apply the classical Skorokhod representation theorem, Theorem 2.6.2. This yields existence of a new complete probability space with a sequence of converging random variables. Then, in Section 5.2.4 we identify the limit with a strong martingale solution to (5.14)–(5.16). Finally, in Section 5.2.5 we provide the proof of pathwise uniqueness under the additional assumption that $s > \frac{9}{2}$. In Section 5.2.6 we employ the Gyöngy–Krylov argument to deduce the existence of a strong pathwise solution.

5.2.1 The Galerkin approximation

To begin with, observe that, for any $\mathbf{u} \in C([0,T]; W^{2,\infty}(\mathbb{T}^3))$, the transport equation (5.14) admits a classical solution $r = r[\mathbf{u}]$, uniquely determined by the initial datum r_0; cf. Theorem A.2.5. In addition, for a certain universal constant c, we have the estimates

$$
\begin{aligned}
&\frac{1}{R}\exp(-cRt) \le \exp(-cRt)\inf_{\mathbb{T}^3} r_0 \le r(t,\cdot) \le \exp(cRt)\sup_{\mathbb{T}^3} r_0 \le R\exp(cRt) \\
&|\nabla r(t,\cdot)| \le \exp(cRt)|\nabla r_0| \le R\exp(cRt), \quad t \in [0,T].
\end{aligned}
\tag{5.18}
$$

As in Chapter 4, we consider H_m, the space of trigonometric polynomials of order m, together with an orthonormal basis $(\boldsymbol{\psi}_m)_{m\in\mathbb{N}}$ and the projections $\Pi_m : L^2(\mathbb{T}^3) \to H_m$; cf. (4.6). We look for approximate solutions $\mathbf{u}_m$ of (5.15) belonging to $L^2(\Omega; C([0,T]; H_m))$, satisfying

$$
\begin{aligned}
&\mathrm{d}\langle \mathbf{u}_m, \boldsymbol{\psi}_i\rangle + \varphi_R(\|\mathbf{u}_m\|_{W^{2,\infty}_x})\langle \mathbf{u}_m \cdot \nabla \mathbf{u}_m + r[\mathbf{u}_m]\nabla r[\mathbf{u}_m], \boldsymbol{\psi}_i\rangle\,\mathrm{d}t \\
&\quad = \varphi_R(\|\mathbf{u}_m\|_{W^{2,\infty}_x})\langle D(r[\mathbf{u}_m])\operatorname{div}\mathbb{S}(\nabla\mathbf{u}_m), \boldsymbol{\psi}_i\rangle\,\mathrm{d}t \\
&\qquad + \varphi_R(\|\mathbf{u}_m\|_{W^{2,\infty}_x})\langle \mathbb{F}(r[\mathbf{u}_m], \mathbf{u}_m), \boldsymbol{\psi}_i\rangle\,\mathrm{d}W, \quad i = 1,\dots,m. \\
&\mathbf{u}_m(0) = \Pi_m \mathbf{u}_0.
\end{aligned}
\tag{5.19}
$$

As all norms on the finite-dimensional space H_m are equivalent, solutions of (5.14) and (5.19) can be obtained in a standard way by means of the Banach fixed point argument. Specifically, setting $\mathcal{B} = L^2(\Omega; C([0,T^*]; H_m))$, we have to show that the mapping

$$
\mathbf{u} \mapsto \mathcal{T}\mathbf{u} : \mathcal{B} \to \mathcal{B},
$$

$$
\langle (\mathcal{T}\mathbf{u})(\tau), \psi_i\rangle = \langle \mathbf{u}(0), \psi_i\rangle - \int_0^\tau \varphi_R(\|\mathbf{u}\|_{W^{2,\infty}_x})\langle [\mathbf{u}\cdot\nabla\mathbf{u} + r[\mathbf{u}]\nabla r[\mathbf{u}]], \boldsymbol{\psi}_i\rangle\,\mathrm{d}t
$$

$$+\int_0^\tau \varphi_R(\|\mathbf{u}\|_{W_x^{2,\infty}})\langle D(r[\mathbf{u}])\operatorname{div}\mathbb{S}(\nabla\mathbf{u}),\boldsymbol{\psi}_i\rangle\,\mathrm{d}t$$
$$+\int_0^\tau \varphi_R(\|\mathbf{u}\|_{W_x^{2,\infty}})\langle \mathbb{F}(r[\mathbf{u}],\mathbf{u}),\boldsymbol{\psi}_i\rangle\,\mathrm{d}W,\quad i=1,\dots,m, \tag{5.20}$$

is a contraction for T^* sufficiently small. The three components of $\mathscr{T}$ appearing on the right hand side of (5.20) will be denoted by $\mathscr{T}_{\mathrm{det}}^1$, $\mathscr{T}_{\mathrm{det}}^2$, and $\mathscr{T}_{\mathrm{sto}}$, respectively.

For $r_1=r[\mathbf{v}_1]$, $r_2=r[\mathbf{v}_2]$, we get

$$\begin{aligned}
&\mathrm{d}(r_1-r_2)+\mathbf{v}_1\cdot\nabla(r_1-r_2)\,\mathrm{d}t+\frac{\gamma-1}{2}\operatorname{div}\mathbf{v}_1(r_1-r_2)\,\mathrm{d}t\\
&\quad=-\nabla r_2\cdot(\mathbf{v}_1-\mathbf{v}_2)\,\mathrm{d}t-\frac{\gamma-1}{2}r_2\operatorname{div}(\mathbf{v}_1-\mathbf{v}_2)\,\mathrm{d}t,
\end{aligned}\tag{5.21}$$

where we have set

$$\mathbf{v}_1=\varphi_R(\|\mathbf{u}\|_{W_x^{2,\infty}})\mathbf{u},\quad \mathbf{v}_2=\varphi_R(\|\mathbf{v}\|_{W_x^{2,\infty}})\mathbf{v}.$$

We multiply (5.21) by r_1-r_2 and obtain, after integrating over $\mathbb{T}^3$,

$$\begin{aligned}
\mathrm{d}\int_{\mathbb{T}^3}\frac{|r_1-r_2|^2}{2}\,\mathrm{d}x&=\int_{\mathbb{T}^3}\operatorname{div}\mathbf{v}_1\frac{|r_1-r_2|^2}{2}\,\mathrm{d}x\,\mathrm{d}t-\frac{\gamma-1}{2}\int_{\mathbb{T}^3}\operatorname{div}\mathbf{v}_1|r_1-r_2|^2\,\mathrm{d}x\,\mathrm{d}t\\
&\quad-\int_{\mathbb{T}^3}\nabla r_2\cdot(\mathbf{v}_1-\mathbf{v}_2)(r_1-r_2)\,\mathrm{d}x\,\mathrm{d}t\\
&\quad-\frac{\gamma-1}{2}\int_{\mathbb{T}^3}r_2\operatorname{div}(\mathbf{v}_1-\mathbf{v}_2)(r_1-r_2)\,\mathrm{d}x\,\mathrm{d}t.
\end{aligned}$$

Consequently, we easily deduce

$$\sup_{0\le t\le T^*}\|r[\mathbf{u}]-r[\mathbf{v}]\|_{L_x^2}^2\le T^*C(m,R,T)\sup_{0\le t\le T^*}\|\mathbf{u}-\mathbf{v}\|_{H_m}^2, \tag{5.22}$$

noting that r_1, r_2 coincide at $t=0$ and that r_j, ∇r_j are bounded by a deterministic constant depending on R; recall (5.18).

As a consequence of (5.18), (5.22), and the equivalence of norms on H_m, we can show that the mapping $\mathscr{T}_{\mathrm{det}}=\mathscr{T}_{\mathrm{det}}^1+\mathscr{T}_{\mathrm{det}}^2$ satisfies the estimate

$$\|\mathscr{T}_{\mathrm{det}}\mathbf{u}-\mathscr{T}_{\mathrm{det}}\mathbf{v}\|_{\mathcal{B}}^2\le T^*C(m,R,T)\|\mathbf{u}-\mathbf{v}\|_{\mathcal{B}}^2. \tag{5.23}$$

Setting $J_R(\mathbf{w})=\varphi_{R+1}(\|\mathbf{w}\|_{W_x^{2,\infty}})\mathbf{w}$, we have, by Burkholder–Davis–Gundy's inequality,

$$\begin{aligned}
&\|\mathscr{T}_{\mathrm{sto}}\mathbf{u}-\mathscr{T}_{\mathrm{sto}}\mathbf{v}\|_{\mathcal{B}}^2\\
&\quad=\mathbb{E}\sup_{0\le t\le T^*}\left\|\int_0^t(\varphi_R(\|\mathbf{u}\|_{W_x^{2,\infty}})\mathbb{F}(r[\mathbf{u}],\mathbf{u})-\varphi(\|\mathbf{v}\|_{W_x^{2,\infty}})\mathbb{F}(r[\mathbf{v}],\mathbf{v}))\,\mathrm{d}W\right\|_{H_m}^2\\
&\quad\le\mathbb{E}\int_0^{T^*}\sum_{k=1}^\infty\|\varphi_R(\|\mathbf{u}\|_{W_x^{2,\infty}})\mathbf{F}_k(r[\mathbf{u}],J_R(\mathbf{u}))-\varphi_R(\|\mathbf{v}\|_{W_x^{2,\infty}})\mathbf{F}_k(r[\mathbf{v}],J_R(\mathbf{v}))\|_{H_m}^2\,\mathrm{d}t
\end{aligned}$$

$$\lesssim \mathbb{E}\int_0^{T^*} |\varphi_R(\|\mathbf{u}\|_{W^{2,\infty}_x}) - \varphi_R(\|\mathbf{v}\|_{W^{2,\infty}_x})|^2 \sum_{k=1}^{\infty} \|\mathbf{F}_k(r[\mathbf{u}], J_R(\mathbf{u}))\|^2_{H_m}\, dt$$
$$+ \mathbb{E}\int_0^{T^*} \varphi_R(\|\mathbf{v}\|_{W^{2,\infty}_x})^2 \sum_{k=1}^{\infty} \|\mathbf{F}_k(r[\mathbf{u}], J_R(\mathbf{u})) - \mathbf{F}_k(r[\mathbf{v}], J_R(\mathbf{v}))\|^2_{H_m}\, dt,$$

with a constant $C = C(m, R)$. Using the growth conditions for $\mathbf{F}_k$ (see (5.8)), we gain

$$\begin{aligned}
&\|\mathscr{T}_{\mathrm{sto}}\mathbf{u} - \mathscr{T}_{\mathrm{sto}}\mathbf{v}\|^2_{\mathcal{B}} \\
&\quad\lesssim \mathbb{E}\|\mathbf{u} - \mathbf{v}\|^2_{W^{2,\infty}_x} + \mathbb{E}\int_0^{T^*} \|r[\mathbf{u}] - r[\mathbf{v}]\|^2_{L^2_x}\, ds + \mathbb{E}\int_0^{T^*} \|J_R(\mathbf{u}) - J_R(\mathbf{v})\|^2_{L^2_x}\, ds \\
&\quad\lesssim \|\mathbf{u} - \mathbf{v}\|^2_{\mathcal{B}},
\end{aligned} \tag{5.24}$$

with a constant $c = T^* C(m, R, T^*)$. Note that the last step was a consequence of (5.22) and the equivalence of norms. Combining (5.23) and (5.24) shows that $\mathscr{T}$ is a contraction for a deterministic small time $T^* > 0$. A solution to (5.14)–(5.16) on the whole interval $[0, T]$ can be obtained by decomposing $[0, T]$ into small subintervals and gluing the corresponding solutions together.

5.2.2 Uniform estimates

In this subsection, we derive estimates that hold uniformly for $m \to \infty$, yielding a basis for our compactness argument presented in Section 5.2.3. At this stage, the approximate velocity field $\mathbf{u}_m$ is smooth in the x-variable; whence the corresponding solution $r_m = r[\mathbf{u}_m]$ of the transport equation (5.14) enjoys the same smoothness as the initial datum r_0.

Let α be a multi-index such that $|\alpha| \leq s$. Differentiating (5.14) in the x-variable, we obtain

$$\begin{aligned}
&\mathrm{d}\partial_x^\alpha r_m + \varphi_R(\|\mathbf{u}_m\|_{W^{2,\infty}_x})\Big[\mathbf{u}_m \cdot \nabla\partial_x^\alpha r_m + \frac{\gamma - 1}{2} r_m \operatorname{div} \partial_x^\alpha \mathbf{u}_m\Big]\, dt \\
&\quad = \varphi_R(\|\mathbf{u}_m\|_{W^{2,\infty}_x})[\mathbf{u}_m \cdot \partial_x^\alpha \nabla r_m - \partial_x^\alpha(\mathbf{u}_m \cdot \nabla r_m)]\, \mathrm{d}t \\
&\qquad + \frac{\gamma - 1}{2}\varphi_R(\|\mathbf{u}_m\|_{W^{2,\infty}_x})[r_m \partial_x^\alpha \operatorname{div} \mathbf{u}_m - \partial_x^\alpha(r_m \operatorname{div} \mathbf{u}_m)]\, dt \\
&\quad =: T_1^m\, \mathrm{d}t + T_2^m\, \mathrm{d}t.
\end{aligned} \tag{5.25}$$

Similarly, we may use the fact that the spaces H_m are invariant with respect to the spatial derivatives. In particular, we deduce

$$\begin{aligned}
&\mathrm{d}\langle \partial_x^\alpha \mathbf{u}_m, \boldsymbol{\psi}_i\rangle + \varphi_R(\|\mathbf{u}_m\|_{W^{2,\infty}_x})\langle [\mathbf{u}_m \cdot \nabla\partial_x^\alpha \mathbf{u}_m + r_m \nabla\partial_x^\alpha r_m], \boldsymbol{\psi}_i\rangle\, \mathrm{d}t \\
&\quad - \varphi_R(\|\mathbf{u}_m\|_{W^{2,\infty}_x})\langle D(r_m) \operatorname{div} \mathbb{S}(\nabla\partial_x^\alpha \mathbf{u}_m), \boldsymbol{\psi}_i\rangle\, \mathrm{d}t \\
&= \varphi_R(\|\mathbf{u}_m\|_{W^{2,\infty}_x})\langle [\mathbf{u}_m \cdot \partial_x^\alpha \nabla \mathbf{u}_m - \partial_x^\alpha(\mathbf{u}_m \cdot \nabla \mathbf{u}_m)], \boldsymbol{\psi}_i\rangle\, \mathrm{d}t
\end{aligned}$$

$$
\begin{aligned}
&+ \varphi_R(\|\mathbf{u}_m\|_{W_x^{2,\infty}})\langle[r_m\partial_x^\alpha\nabla r_m - \partial_x^\alpha(r_m\nabla r_m)], \boldsymbol{\psi}_i\rangle\,\mathrm{d}t\\
&- \varphi_R(\|\mathbf{u}_m\|_{W_x^{2,\infty}})\langle[D(r_m)\partial_x^\alpha\operatorname{div}\mathbb{S}(\nabla\mathbf{u}_m) - \partial_x^\alpha(D(r_m)\operatorname{div}\mathbb{S}(\nabla\mathbf{u}_m))], \boldsymbol{\psi}_i\rangle\,\mathrm{d}t\\
&+ \varphi_R(\|\mathbf{u}_m\|_{W_x^{2,\infty}})\langle\partial_x^\alpha\mathbb{F}(r_m,\mathbf{u}_m), \boldsymbol{\psi}_i\rangle\,\mathrm{d}W\\
=:\ & T_3^m\,\mathrm{d}t + T_4^m\,\mathrm{d}t + T_5^m\,\mathrm{d}t + \varphi_R(\|\mathbf{u}_m\|_{W_x^{2,\infty}})\langle\partial_x^\alpha\mathbb{F}(r_m,\mathbf{u}_m), \boldsymbol{\psi}_i\rangle\,\mathrm{d}W, \qquad (5.26)
\end{aligned}
$$

for $i = 1,\dots,m$. Using (5.10), we estimate the "error" terms as follows:

$$
\begin{aligned}
\|T_1^m\|_{L_x^2} &\lesssim \varphi_R(\|\mathbf{u}_m\|_{W_x^{2,\infty}})[\|\nabla\mathbf{u}_m\|_{L_x^\infty}\|\nabla^s r_m\|_{L_x^2} + \|\nabla r_m\|_{L_x^\infty}\|\nabla^s\mathbf{u}_m\|_{L_x^2}],\\
\|T_2^m\|_{L_x^2} &\lesssim \varphi_R(\|\mathbf{u}_m\|_{W_x^{2,\infty}})[\|\nabla r_m\|_{L_x^\infty}\|\nabla^s\mathbf{u}_m\|_{L_x^2} + \|\operatorname{div}\mathbf{u}_m\|_{L_x^\infty}\|\nabla^s r_m\|_{L_x^2}],\\
\|T_3^m\|_{L_x^2} &\lesssim \varphi_R(\|\mathbf{u}_m\|_{W_x^{2,\infty}})\|\nabla\mathbf{u}_m\|_{L_x^\infty}\|\nabla^s\mathbf{u}_m\|_{L_x^2},\\
\|T_4^m\|_{L_x^2} &\lesssim \|\nabla r_m\|_{L_x^\infty}\|\nabla^s r_m\|_{L_x^2},
\end{aligned}
\qquad (5.27)
$$

while

$$
\begin{aligned}
\|T_5^n\|_2 &\lesssim \varphi_R(\|\mathbf{u}_m\|_{W_x^{2,\infty}})\|\nabla D(r_m)\|_{L_x^\infty}\|\nabla^s\mathbb{S}(\nabla\mathbf{u}_m)\|_{L_x^2}\\
&\quad + \varphi_R(\|\mathbf{u}_m\|_{W_x^{2,\infty}})\|\operatorname{div}\mathbb{S}(\nabla\mathbf{u}_m)\|_{L_x^\infty}\|\nabla^s D(r_m)\|_{L_x^2}.
\end{aligned}
\qquad (5.28)
$$

Multiplying (5.25) by $\partial_x^\alpha r_m$ and integrating over $\mathbb{T}^3$, we observe

$$
\begin{aligned}
&\mathrm{d}\int_{\mathbb{T}^3}\frac{|\partial_x^\alpha r_m|^2}{2}\,\mathrm{d}x + \int_{\mathbb{T}^3}\varphi_R(\|\mathbf{u}_m\|_{W_x^{2,\infty}})\frac{\gamma-1}{2}r_m\operatorname{div}\partial_x^\alpha\mathbf{u}_m\,\partial_x^\alpha r_m\,\mathrm{d}x\,\mathrm{d}t\\
&\quad = -\int_{\mathbb{T}^3}\varphi_R(\|\mathbf{u}_m\|_{W_x^{2,\infty}})\mathbf{u}_m\cdot\nabla\partial_x^\alpha r_m\,\partial_x^\alpha r_m\,\mathrm{d}x\,\mathrm{d}t\\
&\qquad + \int_{\mathbb{T}^3}(T_1^m + T_2^m)\,\partial_x^\alpha r_m\,\mathrm{d}x\,\mathrm{d}t.
\end{aligned}
$$

Now, using (5.27) as well as

$$
\int_{\mathbb{T}^3}\mathbf{u}_m\cdot\nabla\partial_x^\alpha r\partial_x^\alpha r_m\,\mathrm{d}x = -\frac{1}{2}\int_{\mathbb{T}^3}\operatorname{div}\mathbf{u}_m|\partial_x^\alpha r_m|^2\,\mathrm{d}x,
$$

we obtain

$$
\begin{aligned}
&\|\partial_x^\alpha r_m(\tau)\|_{L_x^2}^2 + (\gamma-1)\int_0^\tau\varphi_R(\|\mathbf{u}_m\|_{W_x^{2,\infty}})\int_{\mathbb{T}^3}r_m\operatorname{div}\partial_x^\alpha\mathbf{u}_m\partial_x^\alpha r_m\,\mathrm{d}x\,\mathrm{d}t\\
&\quad\lesssim \int_0^\tau\varphi_R(\|\mathbf{u}_m\|_{W_x^{2,\infty}})(\|\mathbf{u}_m\|_{W_x^{1,\infty}}\|r_m\|_{W_x^{s,2}} + \|r_m\|_{W_x^{1,\infty}}\|\mathbf{u}_m\|_{W_x^{s,2}})\|\partial_x^\alpha r_m\|_{L_x^2}\,\mathrm{d}t\\
&\qquad + \|\partial_x^\alpha r_0\|_{L_x^2}^2,
\end{aligned}
\qquad (5.29)
$$

provided $|\alpha| \le s$.

To apply the same treatment to (5.26), we use Itô's formula for the function $f(\mathbf{C}^m) = \int_{\mathbb{T}^3}|\partial_x^\alpha\mathbf{u}_m|^2\,\mathrm{d}x$. Here $\mathbf{C}^m = (c_1^m,\dots,c_n^m)$ are the coefficients in the expansion $\mathbf{u}_m = \sum_{i=1}^m c_i^m\boldsymbol{\psi}_i$. We have

$$
\|\partial_x^\alpha\mathbf{u}_m(\tau)\|_{L_x^2}^2 + 2\int_0^\tau\varphi_R(\|\mathbf{u}_m\|_{W_x^{2,\infty}})\int_{\mathbb{T}^3}[\mathbf{u}_m\cdot\nabla\partial_x^\alpha\mathbf{u}_m + r_m\nabla\partial_x^\alpha r_m]\cdot\partial_x^\alpha\mathbf{u}_m\,\mathrm{d}x\,\mathrm{d}t
$$

$$
\begin{aligned}
&-2\int_0^\tau \varphi_R(\|\mathbf{u}_m\|_{W_x^{2,\infty}})\int_{\mathbb{T}^3} D(r_m)\operatorname{div}\mathbb{S}(\nabla\partial_x^\alpha\mathbf{u}_m)\cdot\partial_x^\alpha\mathbf{u}_m\,\mathrm{d}x\,\mathrm{d}t\\
&=\|\partial_x^\alpha\Pi_m\mathbf{u}_0\|_{L_x^2}^2+2\int_0^\tau\int_{\mathbb{T}^3}[T_3^n+T_4^n+T_5^n]\cdot\partial_x^\alpha\mathbf{u}_m\,\mathrm{d}x\,\mathrm{d}t\\
&\quad+2\int_0^\tau \varphi_R(\|\mathbf{u}_m\|_{W_x^{2,\infty}})\int_{\mathbb{T}^3}\partial_x^\alpha\mathbb{F}(r_m,\mathbf{u}_m)\cdot\partial_x^\alpha\mathbf{u}_m\,\mathrm{d}W\\
&\quad+\sum_{k=1}^\infty\int_0^\tau \varphi_R(\|\mathbf{u}_m\|_{W_x^{2,\infty}})\int_{\mathbb{T}^3}|\partial_x^\alpha\mathbf{F}_k(r_m,\mathbf{u}_m)|^2\,\mathrm{d}x\,\mathrm{d}t.
\end{aligned}
\tag{5.30}
$$

Integrating by parts yields

$$
\begin{aligned}
&\int_{\mathbb{T}^3}[\mathbf{u}_m\cdot\nabla\partial_x^\alpha\mathbf{u}_m+r_m\nabla\partial_x^\alpha r_m]\cdot\partial_x^\alpha\mathbf{u}_m\,\mathrm{d}x\\
&=-\frac{1}{2}\int_{\mathbb{T}^3}|\partial_x^\alpha\mathbf{u}_m|^2\operatorname{div}\mathbf{u}_m\,\mathrm{d}x-\int_{\mathbb{T}^3}r_m\operatorname{div}\partial_x^\alpha\mathbf{u}_m\partial_x^\alpha r_m\,\mathrm{d}x-\int_{\mathbb{T}^3}\nabla r_m\cdot\partial_x^\alpha\mathbf{u}_m\partial_x^\alpha r_m
\end{aligned}
$$

as well as

$$
\begin{aligned}
&-\int_{\mathbb{T}^3}[D(r_n)\operatorname{div}\mathbb{S}(\nabla\partial_x^\alpha\mathbf{u}_m)]\cdot\partial_x^\alpha\mathbf{u}_m\,\mathrm{d}x\\
&=\int_{\mathbb{T}^3}\nabla D(r_n)\cdot\mathbb{S}(\nabla\partial_x^\alpha\mathbf{u}_m)\cdot\partial_x^\alpha\mathbf{u}_m\,\mathrm{d}x+\int_{\mathbb{T}^3}D(r_m)\mathbb{S}(\nabla\partial_x^\alpha\mathbf{u}_m):\nabla\partial_x^\alpha\mathbf{u}_m\,\mathrm{d}x.
\end{aligned}
$$

Summing up (5.29)–(5.30) and using (5.27)–(5.28), we observe that the term containing $r_m\partial_x^\alpha r_m\operatorname{div}\partial_x^\alpha\mathbf{u}_m$ on the left hand side of (5.29) cancels out and we infer

$$
\begin{aligned}
&\|(r_m(\tau),\mathbf{u}_m(\tau))\|_{W_x^{s,2}}^2\\
&\quad+\sum_{|\alpha|\le s}\int_0^\tau\int_{\mathbb{T}^3}\varphi_R(\|\mathbf{u}_m\|_{W_x^{2,\infty}})D(r_m)\mathbb{S}(\nabla\partial_x^\alpha\mathbf{u}_m):\nabla\partial_x^\alpha\mathbf{u}_m\,\mathrm{d}x\,\mathrm{d}t\\
&\lesssim\|(r_0,\mathbf{u}_0)\|_{W_x^{s,2}}^2+\int_0^\tau\varphi_R(\|\mathbf{u}_m\|_{W_x^{2,\infty}})\|\mathbf{u}_m\|_{W_x^{1,\infty}}(\|r_m\|_{W_x^{s,2}}^2+\|\mathbf{u}_m\|_{W_x^{s,2}}^2)\,\mathrm{d}t\\
&\quad+\int_0^\tau\varphi_R(\|\mathbf{u}_m\|_{W_x^{2,\infty}})\|r_m\|_{W_x^{1,\infty}}(\|r_m\|_{W_x^{s,2}}^2+\|\mathbf{u}_n\|_{W_x^{s,2}}^2)\,\mathrm{d}t\\
&\quad+\int_0^\tau\varphi_R(\|\mathbf{u}_m\|_{W_x^{2,\infty}})\|\operatorname{div}\mathbb{S}(\nabla\mathbf{u}_m)\|_{L_x^\infty}\|D(r_m)\|_{W_x^{s,2}}\|\mathbf{u}_m\|_{W_x^{s,2}}\,\mathrm{d}t\\
&\quad+\int_0^\tau\varphi_R(\|\mathbf{u}_m\|_{W_x^{2,\infty}})\|\nabla D(r_m)\|_{L_x^\infty}\|\mathbf{u}_m\|_{W_x^{s,2}}\|\mathbb{S}(\nabla\mathbf{u}_m)\|_{W_x^{s,2}}\,\mathrm{d}t\\
&\quad+\sum_{|\alpha|\le s}\int_0^\tau\varphi_R(\|\mathbf{u}_m\|_{W_x^{2,\infty}})\int_{\mathbb{T}^3}\partial_x^\alpha\mathbb{F}(r_m,\mathbf{u}_m)\cdot\partial_x^\alpha\mathbf{u}_m\,\mathrm{d}x\,\mathrm{d}W\\
&\quad+\sum_{|\alpha|\le s}\sum_{k=1}^\infty\int_0^\tau\varphi_R(\|\mathbf{u}_m\|_{W_x^{2,\infty}})\int_{\mathbb{T}^3}|\partial_x^\alpha\mathbf{F}_k(r_m,\mathbf{u}_m)|^2\,\mathrm{d}x\,\mathrm{d}t\\
&=(I_0)+(I_1)+\cdots+(I_6).
\end{aligned}
\tag{5.31}
$$

Remark 5.2.4. Note that the above estimate depends on R only through the cut-off function φ_R. Moreover, in accordance with (5.18),

$$
\varphi_R(\|\mathbf{u}_m\|_{W_x^{2,\infty}})\|\mathbf{u}_m\|_{W_x^{1,\infty}}+\varphi_R(\|\mathbf{u}_m\|_{W_x^{2,\infty}})\|\operatorname{div}\mathbb{S}(\nabla\mathbf{u}_m)\|_{L_x^\infty}\lesssim cR,
\tag{5.32}
$$

$$\begin{aligned}&\|r_m^{-1}\|_{L^\infty_x}+\|r_m\|_{W^{1,\infty}_x}+\|D(r_m)^{-1}\|_{L^\infty_x}+\|D(r_m)\|_{W^{1,\infty}_x}\\&\quad\lesssim c(R)\exp(cRT)(\|r_0\|_{W^{1,\infty}_x}+\|r_0^{-1}\|_{W^{1,\infty}_x})\lesssim c(R)\exp(cRT).\end{aligned}\tag{5.33}$$

Also, in view of (5.11) and (5.18),

$$\|D(r_m)\|_{W^{s,2}_x}\le c(R,T)\|r_m\|_{W^{s,2}_x}.\tag{5.34}$$

In contrast to the preceding part, the following inequalities depend on R. In the following we repeatedly use the embedding $W^{s,2}(\mathbb{T}^3)\hookrightarrow W^{2,\infty}(\mathbb{T}^3)$, which follows from the assumption $s>\frac{7}{2}$. Using (5.32)–(5.33), we easily get

$$(I_1)+(I_2)+(I_3)\lesssim c(R,T)\int_0^T\|(r_m,\mathbf{u}_m)\|^2_{W^{s,2}_x}\,\mathrm{d}t.$$

As far as $(I)_4$ is concerned we have, by (5.33) and Young's inequality,

$$\begin{aligned}(I_4)&\lesssim c(R,T)\int_0^T\varphi_R(\|\mathbf{u}_m\|_{W^{2,\infty}_x})\|\mathbf{u}_m\|_{W^{s,2}_x}\|\nabla^s\mathbb{S}(\nabla\mathbf{u}_m)\|_{L^2_x}\,\mathrm{d}t\\&\lesssim c(R,T)\int_0^T\varphi_R(\|\mathbf{u}_m\|_{W^{2,\infty}_x})\|\mathbf{u}_m\|_{W^{s,2}_x}\|\sqrt{D(r_m)}\nabla^s\mathbb{S}(\nabla\mathbf{u}_m)\|_{L^2_x}\,\mathrm{d}t\\&\lesssim\kappa\int_0^T\varphi_R(\|\mathbf{u}_m\|_{W^{2,\infty}_x})\|\sqrt{D(r_m)}\nabla^s\mathbb{S}(\nabla\mathbf{u}_m)\|^2_{L^2_x}\,\mathrm{d}t\\&\quad+c(\kappa,R,T)\int_0^T\varphi_R(\|\mathbf{u}_m\|_{W^{2,\infty}_x})\|\mathbf{u}_m\|^2_{W^{s,2}_x}\,\mathrm{d}t,\end{aligned}$$

for all $\kappa>0$. Choosing κ small enough will enable us to absorb the corresponding term in the left hand side of (5.31). By (5.5) and (5.11) we have

$$\begin{aligned}(I)_6&\lesssim\int_0^T\varphi_R(\|\mathbf{u}_n\|_{W^{2,\infty}_x})\sum_{k=1}^\infty\|\mathbf{F}_k\|^2_{C^{s-1}}\|(r_m,\mathbf{u}_m)\|^{2(s-1)}_{L^\infty_x}\|(r_m,\mathbf{u}_m)\|^2_{W^{s,2}_x}\,\mathrm{d}t\\&\lesssim c(R,T)\int_0^T\|(r_m,\mathbf{u}_m)\|^2_{W^{s,2}_x}\,\mathrm{d}t.\end{aligned}$$

Here the C^{s-1}-norm of $\mathbf{F}_k$ is taken over the bounded set $\mathrm{range}[(r_m,\mathbf{u}_m)]$. The stochastic integral can be treated in the same fashion. After applying expectations, we gain

$$\begin{aligned}&\mathbb{E}\bigg[\sup_{t\in[0,T]}\bigg|\int_0^t\varphi_R(\|\mathbf{u}_m\|_{W^{2,\infty}_x})\int_{\mathbb{T}^3}\partial_x^\alpha\mathbb{F}(r_m,\mathbf{u}_m)\cdot\partial_x^\alpha\mathbf{u}_m\,\mathrm{d}x\,\mathrm{d}W\bigg|\bigg]^p\\&\quad\lesssim\mathbb{E}\bigg[\sum_{k=1}^\infty\int_0^T\varphi_R(\|\mathbf{u}_m\|_{W^{2,\infty}_x})^2\Big(\int_{\mathbb{T}^3}\partial_x^\alpha\mathbf{F}_k(r_m,\mathbf{u}_m)\cdot\partial_x^\alpha\mathbf{u}_m\,\mathrm{d}x\Big)^2\,\mathrm{d}x\,\mathrm{d}t\bigg]^{\frac{p}{2}}\\&\quad\lesssim\mathbb{E}\bigg[\int_0^T\varphi_R(\|\mathbf{u}_m\|_{W^{2,\infty}_x})^2\Big(\sum_{k=1}^\infty\|\mathbf{F}_k(r_m,\mathbf{u}_m)\|^2_{W^{s,2}_x}\Big)\|\mathbf{u}_m\|^2_{W^{s,2}_x}\,\mathrm{d}t\bigg]^{\frac{p}{2}}\\&\quad\lesssim\mathbb{E}\bigg[\int_0^T\varphi_R(\|\mathbf{u}_m\|_{W^{2,\infty}_x})^2\|(r_m,\mathbf{u}_m)\|^{2(s-1)}_{L^\infty_x}\|(r_m,\mathbf{u}_m)\|^4_{W^{s,2}_x}\,\mathrm{d}t\bigg]^{\frac{p}{2}}\end{aligned}$$

$$\lesssim c(R,T)\mathbb{E}\bigg[\sup_{t\in[0,T]}\|(r_m,\mathbf{u}_m)\|^2_{W^{s,2}_x}\int_0^T\|(r_m,\mathbf{u}_m)\|^2_{W^{s,2}_x}\,\mathrm{d}t\bigg]^{\frac{p}{2}}$$
$$\lesssim \kappa\mathbb{E}\Big[\sup_{t\in[0,T]}\|(r_m,\mathbf{u}_m)\|^{2p}_{W^{s,2}_x}\Big]+c(\kappa,R,T)\mathbb{E}\bigg[\int_0^T\|(r_m,\mathbf{u}_m)\|^2_{W^{s,2}_x}\,\mathrm{d}t\bigg]^p.$$

Here, we also took into account Burkholder–Davis–Gundy's and weighted Young's inequality. In order to obtain the final estimate, we take the supremum in time and the pth power and we apply expectations. Summarizing the previous discussion and choosing κ small enough, we obtain

$$\mathbb{E}\bigg[\sup_{t\in[0,T]}\|(r_m,\mathbf{u}_n)\|^2_{W^{s,2}_x}+\int_0^T\varphi_R(\|\mathbf{u}_m\|_{W^{2,\infty}_x})\int_{\mathbb{T}^3}D(r_m)|\nabla^{s+1}\mathbf{u}_m|^2\,\mathrm{d}x\,\mathrm{d}t\bigg]^p$$
$$\lesssim c(R,T,s)\mathbb{E}\bigg[\|(r_0,\mathbf{u}_0)\|^2_{W^{s,2}_x}+\int_0^T\|(r_m,\mathbf{u}_m)\|^2_{W^{s,2}_x}\,\mathrm{d}t+1\bigg]^p.$$

Finally, we apply Gronwall's lemma and use (5.33) to conclude

$$\mathbb{E}\bigg[\sup_{t\in[0,T]}\|(r_m,\mathbf{u}_n)\|^2_{W^{s,2}_x}+\int_0^T\varphi_R(\|\mathbf{u}_m\|_{W^{2,\infty}_x})\int_{\mathbb{T}^3}D(r_m)|\nabla^{s+1}\mathbf{u}_m|^2\,\mathrm{d}x\,\mathrm{d}t\bigg]^p$$
$$\lesssim c(R,T,s)\mathbb{E}\big[\|(r_0,\mathbf{u}_0)\|^{2p}_{W^{s,2}_x}+1\big], \tag{5.35}$$

whenever $s>\frac{7}{2}$.

5.2.3 Compactness

Now everything is in hand to set up our compactness argument leading to the existence part of Theorem 5.2.3. Let us define the path space $\mathcal{X}=\mathcal{X}_r\times\mathcal{X}_{\mathbf{u}}\times\mathcal{X}_W$. We have

$$\mathcal{X}_{\mathbf{u}}=C([0,T];W^{\beta,2}(\mathbb{T}^3)),\quad \mathcal{X}_r=C([0,T];W^{\beta,2}(\mathbb{T}^3)),\quad \mathcal{X}_W=C([0,T];\mathfrak{U}_0),$$

where $\beta<s$ (not necessarily integer) can be chosen arbitrarily close to s such that $\beta>\frac{7}{2}$. Hence we have the embedding $W^{\beta,2}(\mathbb{T}^3)\hookrightarrow W^{2,\infty}(\mathbb{T}^3)$, which is needed to pass to the limit in the cut-off operators.

We denote by $\mathcal{L}[r_m]$ and $\mathcal{L}[\mathbf{u}_m]$ the law of r_m and $\mathbf{u}_m$, respectively, on the corresponding path space. By $\mathcal{L}[W]$ we denote the law of W on $\mathcal{X}_W$ and their joint law on $\mathcal{X}$ is denoted by $\mathcal{L}[r_m,\mathbf{u}_m,W]$. To proceed, it is necessary to establish their tightness.

Proposition 5.2.5. *The set $\{\mathcal{L}[\mathbf{u}_m];\,m\in\mathbb{N}\}$ is tight on $\mathcal{X}_{\mathbf{u}}$.*

Proof. We start with a compact embedding relation. We have

$$C([0,T];W^{s,2}(\mathbb{T}^3))\cap C^\gamma([0,T];L^2(\mathbb{T}^3))\overset{c}{\hookrightarrow}C([0,T];W^{\beta,2}(\mathbb{T}^3)),\quad \gamma>0,\ \beta<s,$$

which follows directly from the abstract Arzelà–Ascoli theorem, Theorem 1.1.1.

Due to (5.35), $\mathbf{u}_m$ satisfies

$$\mathbf{u}_m(\tau) = \Pi_m \mathbf{u}_0 - \int_0^\tau \varphi_R(\|\mathbf{u}_m\|_{W_x^{2,\infty}})\Pi_m[\mathbf{u}_m \cdot \nabla \mathbf{u}_m + r_m \nabla r_m]\,\mathrm{d}t$$
$$+ \int_0^\tau \varphi_R(\|\mathbf{u}_m\|_{W_x^{2,\infty}})\Pi_m[D(r_m)\operatorname{div}\mathbb{S}(\nabla \mathbf{u}_m)]\,\mathrm{d}t$$
$$+ \int_0^\tau \varphi_R(\|\mathbf{u}_m\|_{W_x^{2,\infty}})\Pi_m \mathbb{F}(r_m, \mathbf{u}_m)\,\mathrm{d}W.$$

Now we decompose $\mathbf{u}_n$ into two parts, namely, $\mathbf{u}_m = \mathbf{Y}_m + \mathbf{Z}_m$, where

$$\mathbf{Y}_m(\tau) = \Pi_m \mathbf{u}_0 - \int_0^\tau \varphi_R(\|\mathbf{u}_m\|_{W_x^{2,\infty}})\Pi_m[\mathbf{u}_m \cdot \nabla \mathbf{u}_m + r_m \nabla r_m]\,\mathrm{d}t$$
$$+ \int_0^\tau \varphi_R(\|\mathbf{u}_m\|_{W_x^{2,\infty}})\Pi_m[D(r_m)\operatorname{div}\mathbb{S}(\nabla \mathbf{u}_m)]\,\mathrm{d}t,$$
$$\mathbf{Z}_m(\tau) = \int_0^\tau \varphi_R(\|\mathbf{u}_m\|_{W_x^{2,\infty}})\Pi_m \mathbb{F}(r_m, \mathbf{u}_m)\,\mathrm{d}W.$$

By (5.35) and the continuity of Π_m on $L^2(\mathbb{T}^3)$ from (4.6), we know, for any $\kappa \in (0,1)$, that

$$\mathbb{E}\big[\|\mathbf{Y}_n\|_{C_t^\kappa L_x^2}\big] \leq c(R).$$

On the other hand (5.35), combined with Lemma 5.1.1 (for $s = 0$) and (2.19), yields the same conclusion for $\mathbf{Z}_m$, with $0 < \kappa < 1/2$. □

Proposition 5.2.6. *The set $\{\mathcal{L}[r_m];\, n \in \mathbb{N}\}$ is tight on $\mathcal{X}_r$.*

Proof. The proof is completely analogous to Proposition 5.2.5 using the equation (5.14) for r_m and the uniform estimate (5.35). □

Since also the law $\mathcal{L}[W]$ is tight, being a Radon measure on the Polish space $\mathcal{X}_W$, we finally deduce the desired tightness of the joint laws $\mathcal{L}[r_m, \mathbf{u}_m, W]$.

Corollary 5.2.7. *The set $\{\mathcal{L}[r_m, \mathbf{u}_m, W];\, m \in \mathbb{N}\}$ is tight on $\mathcal{X}$.*

Since the path space $\mathcal{X}$ is a Polish space, we may use the classical Skorokhod representation theorem (Theorem 2.6.2). That is, passing to a weakly convergent subsequence due to Prokhorov's theorem (Theorem 2.6.1) we infer the following result.

Proposition 5.2.8. *There exists a complete probability space $(\tilde{\Omega}, \tilde{\mathfrak{F}}, \tilde{\mathbb{P}})$ with $\mathcal{X}$-valued Borel measurable random variables $(\tilde{r}_m, \tilde{\mathbf{u}}_m, \tilde{W}_m)$, $m \in \mathbb{N}$, and $(\tilde{r}, \tilde{\mathbf{u}}, \tilde{W})$ such that (up to a subsequence):*
1. *the law of $(\tilde{r}_m, \tilde{\mathbf{u}}_m, \tilde{W}_m)$ is given by $\mathcal{L}[r_m, \mathbf{u}_m, W]$, $m \in \mathbb{N}$;*
2. *the law of $(\tilde{r}, \tilde{\mathbf{u}}, \tilde{W})$ is a Radon measure;*

(3) $(\tilde{r}_m, \tilde{\mathbf{u}}_m, \tilde{W}_m)$ *converges* $\tilde{\mathbb{P}}$*-a.s. to* $(\tilde{r}, \tilde{\mathbf{u}}, \tilde{W})$ *in the topology of* $\mathcal{X}$*, i.e.,*

$$\begin{aligned} \tilde{r}_m &\to \tilde{r} \quad \text{in } C([0,T]; W^{\beta,2}(\mathbb{T}^3)), \\ \tilde{\mathbf{u}}_m &\to \tilde{\mathbf{u}}_m \quad \text{in } C([0,T]; W^{\beta,2}(\mathbb{T}^3)), \\ \tilde{W}_m &\to \tilde{W} \quad \text{in } C([0,T]; \mathfrak{U}_0), \end{aligned}$$

as $m \to \infty$ $\tilde{\mathbb{P}}$*-a.s.*

5.2.4 Identification of the limit

As the next step, we will identify the limit obtained in Proposition 5.2.8 with a strong martingale solution to (5.14)–(5.16), completing the proof of Theorem 5.2.3. We will proceed similarly to Chapter 4 (see, in particular, Section 4.1.2.4).

Since $\tilde{r}$ and $\tilde{\mathbf{u}}$ are stochastic processes with continuous trajectories (cf. Definition 2.1.11), progressive measurability with respect to their respective canonical filtrations follows from Proposition 2.1.18. Consequently, they are progressively measurable with respect to the canonical filtration generated by $[\tilde{r}, \tilde{\mathbf{u}}, \tilde{W}]$, namely,

$$\tilde{\mathfrak{F}}_t := \sigma\left(\sigma_t[\tilde{r}] \cup \sigma_t[\tilde{\mathbf{u}}] \cup \bigcup_{k=1}^{\infty} \sigma_t[\tilde{W}_k] \right), \quad t \in [0,T].$$

In view of Lemma 2.1.35, the process $\tilde{W}$ is a cylindrical Wiener processes with respect to its canonical filtration. In order to show that $\tilde{W}$ is a cylindrical Wiener process with respect to $(\tilde{\mathfrak{F}}_t)_{t\geq 0}$, we intend to apply Corollary 2.1.36. Hence we need to show that the filtration is non-anticipative with respect to $\tilde{W}$. To this end, we first recall Theorem 2.9.1 and deduce that, for every $m \in \mathbb{N}$, $\tilde{W}_m = \sum_{k=1}^{\infty} e_k \tilde{W}_{m,k}$ is a cylindrical Wiener process with respect to

$$\sigma\left(\sigma_t[\tilde{r}_m] \cup \sigma_t[\tilde{\mathbf{u}}_m] \cup \bigcup_{k=1}^{\infty} \sigma_t[\tilde{W}_{m,k}] \right), \quad t \in [0,T].$$

In other words, this filtration is non-anticipative with respect to $\tilde{W}_m$. Lemma 2.9.3, together with Proposition 5.2.8, then yields the claim.

We claim that $[\tilde{r}, \tilde{\mathbf{u}}, \tilde{W}]$ is a strong martingale solution to (5.14)–(5.16). Indeed, in order to identify (5.14), we observe that, due to Theorem 2.9.1, the stochastic process $[\tilde{r}_m, \tilde{\mathbf{u}}_m]$ solves (5.14) on the new probability space. With Proposition 5.2.8 and (5.35) at hand, we may pass to the limit and deduce that $[\tilde{r}, \tilde{\mathbf{u}}]$ solves (5.14).

Similarly, we may identify (5.15). In accordance with Theorem 2.9.1, the approximate problem (5.19) is solved by $[\tilde{r}_m, \tilde{\mathbf{u}}_m, \tilde{W}_m]$ on the new probability space. Finally, due to Proposition 5.2.8 and the uniform moment estimates from (5.35), we may pass to the limit using Lemma 2.6.6 and (5.15) follows, which completes the existence part of the proof of Theorem 5.2.3. Note that the strong continuity of $\tilde{r}$ and $\tilde{\mathbf{u}}$ in $W^{s,2}(\mathbb{T}^3)$ $\tilde{\mathbb{P}}$-a.s.

can be deduced directly from the equations. Indeed, using the variational approach, the momentum equation (5.15) is solved in the Gelfand triplet

$$W^{s+1,2}(\mathbb{T}^3) \hookrightarrow W^{s,2}(\mathbb{T}^3) \hookrightarrow W^{s-1,2}(\mathbb{T}^3).$$

The stochastic integral has continuous trajectories in $W^{s,2}(\mathbb{T}^3)$ due to the uniform estimates, Lemma 5.1.1 (part (ii)) and (2.19), while the coefficients of the deterministic part in the momentum equation belong to the space $L^2(0,T;W^{s-1,2}(\mathbb{T}^3))$ a.s. Hence Theorem 2.4.3 applies and yields the desired continuity of the velocity field $\tilde{\mathbf{u}}$. The continuity of $\tilde{r}$ then follows from the equation of continuity.

5.2.5 Pathwise uniqueness

To show pathwise uniqueness, we mimic the approach of Section 5.2.2. The difference of two solutions $[r^j, \mathbf{u}^j]$, $j = 1, 2$, satisfies

$$\begin{aligned} \mathrm{d}\partial_x^\alpha(r^1 - r^2) &= -\varphi_R(\|\mathbf{u}^1\|_{W_x^{2,\infty}})\partial_x^\alpha\Big(\mathbf{u}^1\cdot\nabla r^1 + \frac{\gamma-1}{2}r^1\operatorname{div}\mathbf{u}^1\Big)\,\mathrm{d}t \\ &\quad + \varphi_R(\|\mathbf{u}^2\|_{W_x^{2,\infty}})\partial_x^\alpha\Big(\mathbf{u}^2\cdot\nabla r^2 + \frac{\gamma-1}{2}r^2\operatorname{div}\mathbf{u}^2\Big)\,\mathrm{d}t \end{aligned} \tag{5.36}$$

and

$$\begin{aligned} \mathrm{d}\partial_x^\alpha(\mathbf{u}_1 - \mathbf{u}_2) &= -\varphi_R(\|\mathbf{u}_1\|_{W_x^{2,\infty}})\partial_x^\alpha(\mathbf{u}_1\cdot\nabla\mathbf{u}_1 + r_1\nabla r_1 - D(r_1)\operatorname{div}\mathbb{S}(\nabla\mathbf{u}_1))\,\mathrm{d}t \\ &\quad + \varphi_R(\|\mathbf{u}_2\|_{W_x^{2,\infty}})\partial_x^\alpha(\mathbf{u}_2\cdot\nabla\mathbf{u}_2 + r_2\nabla r_2 - D(r_2)\operatorname{div}\mathbb{S}(\nabla\mathbf{u}_2))\,\mathrm{d}t \\ &\quad + [\varphi_R(\|\mathbf{u}_1\|_{W_x^{2,\infty}})\partial_x^\alpha\mathbb{F}(r_1,\mathbf{u}_1) - \varphi_R(\|\mathbf{u}_2\|_{W_x^{2,\infty}})\partial_x^\alpha\mathbb{F}(r_2,\mathbf{u}_2)]\,\mathrm{d}W \end{aligned}$$

for $|\alpha| \le s'$ (later we choose $s' \in \mathbb{N}$ such that $s' \le s-1$).

Multiplying (5.36) by $\partial_x^\alpha(r^1 - r^2)$, we get

$$\begin{aligned} \frac{1}{2}\mathrm{d}|\partial_x^\alpha(r^1 - r^2)|^2 &= -\varphi_R(\|\mathbf{u}^1\|_{W_x^{2,\infty}})\partial_x^\alpha\Big(\mathbf{u}^1\cdot\nabla r^1 + \frac{\gamma-1}{2}r^1\operatorname{div}\mathbf{u}^1\Big)\partial_x^\alpha(r^1 - r^2)\,\mathrm{d}t \\ &\quad + \varphi_R(\|\mathbf{u}^2\|_{W_x^{2,\infty}})\partial_x^\alpha\Big(\mathbf{u}^2\cdot\nabla r^2 + \frac{\gamma-1}{2}r^2\operatorname{div}\mathbf{u}^2\Big)\partial_x^\alpha(r^1 - r^2)\,\mathrm{d}t. \end{aligned} \tag{5.37}$$

Similarly, using Itô's product rule, Proposition 2.4.2, we obtain

$$\begin{aligned} &\frac{1}{2}\mathrm{d}|\partial_x^\alpha(\mathbf{u}^1 - \mathbf{u}^2)|^2 \\ &\quad = -\varphi_R(\|\mathbf{u}^1\|_{W_x^{2,\infty}})\partial_x^\alpha(\mathbf{u}^1\cdot\nabla\mathbf{u}^1 + r^1\nabla r^1 - D(r^1)\operatorname{div}\mathbb{S}(\nabla\mathbf{u}^1))\cdot\partial_x^\alpha(\mathbf{u}^1 - \mathbf{u}^2)\,\mathrm{d}t \\ &\quad\quad + \varphi_R(\|\mathbf{u}^2\|_{W_x^{2,\infty}})\partial_x^\alpha(\mathbf{u}^2\cdot\nabla\mathbf{u}^2 + r^2\nabla r^2 - D(r^2)\operatorname{div}\mathbb{S}(\nabla\mathbf{u}^2))\cdot\partial_x^\alpha(\mathbf{u}^1 - \mathbf{u}^2)\,\mathrm{d}t \\ &\quad\quad + [\varphi_R(\|\mathbf{u}^1\|_{W_x^{2,\infty}})\partial_x^\alpha\mathbb{F}(r^1,\mathbf{u}^1) - \varphi_R(\|\mathbf{u}^2\|_{W_x^{2,\infty}})\partial_x^\alpha\mathbb{F}(r^2,\mathbf{u}^2)]\cdot\partial_x^\alpha(\mathbf{u}^1 - \mathbf{u}^2)\,\mathrm{d}W \end{aligned}$$

$$+\frac{1}{2}\sum_{k=1}^{\infty}[\varphi_R(\|\mathbf{u}^1\|_{W_x^{2,\infty}})\partial_x^\alpha\mathbf{F}_k(r^1,\mathbf{u}^1)-\varphi_R(\|\mathbf{u}^2\|_{W_x^{2,\infty}})\partial_x^\alpha\mathbf{F}_k(r^2,\mathbf{u}^2)]^2\,\mathrm{d}t. \tag{5.38}$$

Now observe that, by virtue of the standard embedding relation, we have

$$\begin{aligned}|\varphi_R(\|\mathbf{u}^1\|_{W_x^{2,\infty}})-\varphi_R(\|\mathbf{u}^2\|_{W_x^{2,\infty}})|&\leq c_1(R)\|\mathbf{u}^1-\mathbf{u}^2\|_{W_x^{2,\infty}}\\&\leq c_2(R)\|\mathbf{u}^1-\mathbf{u}^2\|_{W_x^{s',2}},\end{aligned}$$

as soon as $s'>\frac{7}{2}$. We sum (5.37) and (5.38), integrate over the physical space, and perform the same estimates as in Section 5.2.2. Note that the highest order terms in (5.37) read

$$\begin{aligned}&\varphi_R(\|\mathbf{u}^1\|_{W_x^{2,\infty}})\int_{\mathbb{T}^3}(\mathbf{u}^1\cdot\nabla\partial_x^\alpha r^1-\mathbf{u}^2\cdot\nabla\partial_x^\alpha r^2)\partial_x^\alpha(r^1-r^2)\,\mathrm{d}x\\&\quad+\frac{\gamma-1}{2}\varphi_R(\|\mathbf{u}^1\|_{W_x^{2,\infty}})\int_{\mathbb{T}^3}(r^1\operatorname{div}\partial_x^\alpha\mathbf{u}^1-r^2\operatorname{div}\partial_x^\alpha\mathbf{u}^2)\partial_x^\alpha(r^1-r^2)\,\mathrm{d}x\\&=\varphi_R(\|\mathbf{u}^1\|_{W_x^{2,\infty}})\int_{\mathbb{T}^3}((\mathbf{u}^1-\mathbf{u}^2)\cdot\nabla\partial_x^\alpha r^1)\partial_x^\alpha(r^1-r^2)\,\mathrm{d}x\\&\quad+\varphi_R(\|\mathbf{u}^1\|_{W_x^{2,\infty}})\int_{\mathbb{T}^3}\frac{1}{2}\operatorname{div}\mathbf{u}^2|\partial_x^\alpha(r^1-r^2)|^2\,\mathrm{d}x\\&\quad+\frac{\gamma-1}{2}\varphi_R(\|\mathbf{u}^1\|_{W_x^{2,\infty}})\int_{\mathbb{T}^3}(r^1-r^2)\operatorname{div}\partial_x^\alpha\mathbf{u}^2\partial_x^\alpha(r^1-r^2)\,\mathrm{d}x\\&\quad+\frac{\gamma-1}{2}\varphi_R(\|\mathbf{u}^1\|_{W_x^{2,\infty}})\int_{\mathbb{T}^3}r^1\operatorname{div}\partial_x^\alpha(\mathbf{u}^1-\mathbf{u}^2)\partial_x^\alpha(r^1-r^2)\,\mathrm{d}x.\end{aligned}$$

Here, the last integral

$$\varphi_R(\|\mathbf{u}^1\|_{W_x^{2,\infty}})\int_{\mathbb{T}^3}r^1\operatorname{div}(\partial_x^\alpha(\mathbf{u}^1-\mathbf{u}^2))\partial_x^\alpha(r^1-r^2)\,\mathrm{d}x$$

cancels, after integration by parts, with its counterpart in (5.38), namely,

$$\begin{aligned}&\varphi_R(\|\mathbf{u}^1\|_{W_x^{2,\infty}})\int_{\mathbb{T}^3}r^1(\partial_x^\alpha(\mathbf{u}^1-\mathbf{u}^2))\cdot\nabla\partial_x^\alpha(r^1-r^2)\,\mathrm{d}x\\&\quad=-\varphi_R(\|\mathbf{u}^1\|_{W_x^{2,\infty}})\int_{\mathbb{T}^3}\nabla r^1\cdot(\partial_x^\alpha(\mathbf{u}^1-\mathbf{u}^2))\partial_x^\alpha(r^1-r^2)\,\mathrm{d}x\\&\quad\quad-\varphi_R(\|\mathbf{u}^1\|_{W_x^{2,\infty}})\int_{\mathbb{T}^3}r^1\operatorname{div}(\partial_x^\alpha(\mathbf{u}^1-\mathbf{u}^2))\partial_x^\alpha(r^1-r^2)\,\mathrm{d}x.\end{aligned}$$

If $s'>\frac{7}{2}$, we deduce, exactly as in Subsection 5.2.2,

$$\begin{aligned}&\mathrm{d}(\|r^1-r^2\|_{W_x^{s',2}}^2+\|\mathbf{u}^1-\mathbf{u}^2\|_{W_x^{s',2}}^2)\\&\quad\leq G(t)(\|r^1-r^2\|_{W_x^{s',2}}^2+\|\mathbf{u}^1-\mathbf{u}^2\|_{W_x^{s',2}}^2)\,\mathrm{d}t\\&\quad\quad+\sum_{|\alpha|\leq s'}[\varphi_R(\|\mathbf{u}^1\|_{W_x^{2,\infty}})\partial_x^\alpha\mathbb{F}(r^1,\mathbf{u}^1)-\varphi_R(\|\mathbf{u}^2\|_{W_x^{2,\infty}})\partial_x^\alpha\mathbb{F}(r^2,\mathbf{u}^2)]\cdot\partial_x^\alpha(\mathbf{u}^1-\mathbf{u}^2)\,\mathrm{d}W,\end{aligned}$$

where we set

$$G(t) = c(R)\left(1 + \sum_{j=1}^{2}(\|r^j(t)\|^2_{W_x^{s'+1,2}} + \|\mathbf{u}^j(t)\|^2_{W_x^{s'+2,2}})\right).$$

We observe that, if $s \geq s' + 1$, the *a priori* estimates from Section 5.2.2 imply in particular that $G \in L^1(0,T)$ a.s. Applying Itô's formula to the product, Proposition 2.4.2, we obtain

$$\begin{aligned}
&\mathrm{d}[\mathrm{e}^{-\int_0^t G(\sigma)\,\mathrm{d}\sigma}(\|r^1 - r^2\|^2_{W_x^{s',2}} + \|\mathbf{u}^1 - \mathbf{u}^2\|^2_{W_x^{s',2}})] \\
&\quad = -G(t)\mathrm{e}^{-\int_0^t G(\sigma)\,\mathrm{d}\sigma}(\|r^1 - r^2\|^2_{W_x^{s',2}} + \|\mathbf{u}^1 - \mathbf{u}^2\|^2_{W_x^{s',2}})\,\mathrm{d}t \\
&\qquad + \mathrm{e}^{-\int_0^t G(\sigma)\,\mathrm{d}\sigma}\mathrm{d}(\|r^1 - r^2\|^2_{W_x^{s',2}} + \|\mathbf{u}^1 - \mathbf{u}^2\|^2_{W_x^{s',2}}) \\
&\quad \leq \sum_{|\alpha|\leq s'} \mathrm{e}^{-\int_0^t G(\sigma)\,\mathrm{d}\sigma}\varphi_R(\|\mathbf{u}^1\|_{W_x^{2,\infty}})\partial_x^\alpha \mathbb{F}(r^1,\mathbf{u}^1)\cdot\partial_x^\alpha(\mathbf{u}^1 - \mathbf{u}^2)\,\mathrm{d}W \\
&\qquad - \sum_{|\alpha|\leq s'} \mathrm{e}^{-\int_0^t G(\sigma)\,\mathrm{d}\sigma}\varphi_R(\|\mathbf{u}^2\|_{W_x^{2,\infty}})\partial_x^\alpha \mathbb{F}(r^2,\mathbf{u}^2)\cdot\partial_x^\alpha(\mathbf{u}^1 - \mathbf{u}^2)\,\mathrm{d}W.
\end{aligned}$$

Integrating over $[0,t]$ and taking expectation we observe that the stochastic integral vanishes due to the assumptions on r, $\mathbf{u}$ in Definition 5.2.1. Consequently, we infer

$$\mathbb{E}[\mathrm{e}^{-\int_0^t G(\sigma)\,\mathrm{d}\sigma}(\|r^1(t) - r^2(t)\|^2_{W_x^{s',2}} + \|\mathbf{u}^1(t) - \mathbf{u}^2(t)\|^2_{W_x^{s',2}})] = 0,$$

whenever

$$\mathbb{E}[\|r_0^1 - r_0^2\|^2_{W_x^{s',2}} + \|\mathbf{u}_0^1 - \mathbf{u}_0^2\|^2_{W_x^{s',2}}] = 0.$$

Since

$$\mathrm{e}^{-\int_0^t G(\sigma)\,\mathrm{d}\sigma} > 0 \quad \mathbb{P}\text{-a.s.}$$

and the trajectories of r^i, $\mathbf{u}^i$, $i = 1,2$, are continuous in $W^{s',2}(\mathbb{T}^3)$ a.s., the pathwise uniqueness from Theorem 5.2.3 follows.

5.2.6 Existence of a strong pathwise approximate solution

In order to complete the proof of Theorem 5.2.3, we make use of the Gyöngy–Krylov characterization of convergence in probability; see Lemma 2.10.1. It applies to situations when pathwise uniqueness and existence of a martingale solution are valid and allows one to establish existence of a pathwise solution.

We start with regular initial data corresponding to $s > \frac{9}{2}$ required for pathwise uniqueness of strong solutions to the approximate problem (5.14)–(5.16). Going back to the construction of approximate solutions, we consider the joint law

$$\mathcal{L}[r_m, \mathbf{u}_m, r_n, \mathbf{u}_n, W] \quad \text{on the space } \mathcal{X}^J = \mathcal{X}_r \times \mathcal{X}_\mathbf{u} \times \mathcal{X}_r \times \mathcal{X}_\mathbf{u} \times \mathcal{X}_W,$$

where r_m, $\mathbf{u}_m$, r_n, $\mathbf{u}_n$ are the Galerkin solutions. The following result follows easily from the arguments of Section 5.2.3.

Proposition 5.2.9. *The set* $\{\mathcal{L}[r_m, \mathbf{u}_m, r_n, \mathbf{u}_n, W];\, m, n \in \mathbb{N}\}$ *is tight on* $\mathcal{X}^J$.

Let us take any subsequence $[r_{m_k}, \mathbf{u}_{m_k}, r_{n_k}, \mathbf{u}_{n_k}, W]_{k\in\mathbb{N}}$. By the Skorokhod representation theorem (Theorem 2.6.2), we infer (for a further subsequence but without loss of generality we keep the same notation) the existence of a complete probability space $(\bar{\Omega}, \bar{\mathfrak{F}}, \bar{\mathbb{P}})$ with a sequence of random variables

$$[\hat{r}_{n_k}, \hat{\mathbf{u}}_{n_k}, \check{r}_{m_k}, \check{\mathbf{u}}_{m_k}, \bar{W}_k], \quad k \in \mathbb{N},$$

and a random variable $[\hat{r}, \hat{\mathbf{u}}, \check{r}, \check{\mathbf{u}}, \bar{W}]$ such that

$$[\hat{r}_{n_k}, \hat{\mathbf{u}}_{n_k}, \check{r}_{m_k}, \check{\mathbf{u}}_{m_k}, \bar{W}_k] \to [\hat{r}, \hat{\mathbf{u}}, \check{r}, \check{\mathbf{u}}, \bar{W}] \quad \text{in } \mathcal{X}^J \tag{5.39}$$

$\bar{\mathbb{P}}$-a.s. and

$$\mathcal{L}[\hat{r}_{n_k}, \hat{\mathbf{u}}_{n_k}, \check{r}_{m_k}, \check{\mathbf{u}}_{m_k}, \bar{W}_k] = \mathcal{L}[r_{m_k}, \mathbf{u}_{m_k}, r_{n_k}, \mathbf{u}_{n_k}, W]$$

on $\mathcal{X}^J$. Observe that, in particular, $\mathcal{L}[r_{m_k}, \mathbf{u}_{m_k}, r_{n_k}, \mathbf{u}_{n_k}, W]$ converges weakly to the measure

$$\mathcal{L}[\hat{r}, \hat{\mathbf{u}}, \check{r}, \check{\mathbf{u}}, \bar{W}].$$

As the next step, we recall the technique established in Section 5.2.4. Analogously, it can be applied to both

$$[\hat{r}_{n_k}, \hat{\mathbf{u}}_{n_k}, \bar{W}_k], \quad [\hat{r}, \hat{\mathbf{u}}, \bar{W}]$$

and

$$[\check{r}_{m_k}, \check{\mathbf{u}}_{m_k}, \bar{W}_k], \quad [\check{r}, \check{\mathbf{u}}, \bar{W}]$$

in order to show that $[\hat{r}, \hat{\mathbf{u}}, \bar{W}]$ and $[\check{r}, \check{\mathbf{u}}, \bar{W}]$ are strong martingale solutions to the approximate system (5.14)–(5.15). Finally, since $r_{n_k}(0) = r_{m_k}(0) = r_0$, it follows that

$$\bar{\mathbb{P}}(\hat{r}(0) = \check{r}(0)) = 1.$$

Since $\mathbf{u}_{n_k}(0) = \Pi_{n_k}\mathbf{u}_0$, $\mathbf{u}_{m_k}(0) = \Pi_{m_k}\mathbf{u}_0$, we obtain, for every $\ell \le n_k \wedge m_k$,

$$\bar{\mathbb{P}}(\Pi_\ell \hat{\mathbf{u}}_{n_k}(0) = \Pi_\ell \check{\mathbf{u}}_{m_k}(0)) = \mathbb{P}(\Pi_\ell \mathbf{u}_{n_k}(0) = \Pi_\ell \mathbf{u}_{m_k}(0)) = 1.$$

This leads to

$$\bar{\mathbb{P}}(\hat{\mathbf{u}}(0) = \check{\mathbf{u}}(0)) = 1,$$

using (4.6) and (5.39). Hence, in accordance with the pathwise uniqueness established in Theorem 5.2.3, we get the desired conclusion

$$\mathcal{L}[\hat{r},\hat{\mathbf{u}},\check{r},\check{\mathbf{u}}]([r_1,\mathbf{u}_1,r_2,\mathbf{u}_2];\ [r_1,\mathbf{u}_1]=[r_2,\mathbf{u}_2]) = \bar{\mathbb{P}}([\hat{r},\hat{\mathbf{u}}]=[\check{r},\check{\mathbf{u}}]) = 1.$$

Thus, everything is in hand to apply Lemma 2.10.1, which implies that the original sequence $[r_m,\mathbf{u}_m]$, defined on the initial probability space $(\Omega,\mathfrak{F},\mathbb{P})$, converges in probability in the topology of $\mathcal{X}_r\times\mathcal{X}_{\mathbf{u}}$ to a random variable $[r,\mathbf{u}]$. Without loss of generality, we assume that the convergence is almost sure (recall (5.39)) and, again by the method from Section 5.2.4, we finally deduce that the limit is the unique strong pathwise solution to the approximate problem (5.14)–(5.16). Let us denote this solution by $[r_R,\mathbf{u}_R]$.

5.3 Proof of Theorem 5.0.3

Throughout the remainder of the chapter, we go back to the original problem (5.1)–(5.3) and prove Theorem 5.0.3. Our approach relies on the equivalence between (5.1)–(5.2) and (5.12)–(5.13), which is valid, provided the density remains strictly positive; cf. Section 5.1.1. In addition, introducing suitable stopping times allows us to work with (5.14)–(5.15) instead of (5.12)–(5.13) and therefore we may apply the results of the previous section, namely, Theorem 5.2.3. Nevertheless, there is an additional difficulty that originates in the fact that the initial condition is not assumed to be integrable in ω and the initial density is not bounded from below by a positive constant. Consequently, the *a priori* estimates from Section 5.2.2 are no longer valid and the initial condition has to be truncated for Theorem 5.2.3 to be applicable. For this reason, the proof of uniqueness as well as existence of a local strong pathwise solution is divided into two steps. First, we consider an additional assumption on the initial data so that Theorem 5.2.3 applies. Second, we remove this hypothesis.

5.3.1 Uniqueness

Let us first take the following additional assumption:

$$\begin{aligned} &\varrho_0 \in L^\infty(\Omega;\mathfrak{F}_0,\mathbb{P},W^{s,2}(\mathbb{T}^3)), \quad \varrho_0 > \underline{\varrho} > 0\ \mathbb{P}\text{-a.s.},\\ &\mathbf{u}_0 \in L^\infty(\Omega;\mathfrak{F}_0,\mathbb{P},W^{s,2}(\mathbb{T}^3)), \end{aligned} \tag{5.40}$$

for some deterministic constant $\underline{\varrho} > 0$. In this case, the pathwise uniqueness of (5.1)–(5.3) is a simple consequence of the pathwise uniqueness for (5.14)–(5.15), proved in Theorem 5.2.3. To be more precise, let $[\varrho^i,\mathbf{u}^i,(\mathfrak{t}^i_R),\mathfrak{t}^i]$, $i=1,2$, be two maximal strong pathwise solutions to (5.1)–(5.3), starting from $[\varrho_0,\mathbf{u}_0]$, satisfying (5.40). Then

$$\left[r^i := \sqrt{\frac{2a\gamma}{\gamma-1}}(\varrho^i)^{\frac{\gamma-1}{2}},\mathbf{u}^i\right],\quad i=1,2,$$

both solve (5.14)–(5.15) up to the stopping time $\mathfrak{t}_R^1 \wedge \mathfrak{t}_R^2$ and their initial conditions coincide. Besides, the *a priori* estimates from Section 5.2.2 as well as the pathwise uniqueness from Section 5.2.5 applies up to the stopping time $\mathfrak{t}_R^1 \wedge \mathfrak{t}_R^2$ and we deduce

$$\mathbb{P}([\varrho^1, \mathbf{u}^1](t \wedge \mathfrak{t}_R^1 \wedge \mathfrak{t}_R^2) = [\varrho^2, \mathbf{u}^2](t \wedge \mathfrak{t}_R^1 \wedge \mathfrak{t}_R^2), \text{ for all } t \in [0,T]) = 1.$$

Sending $R \to \infty$ implies by dominated convergence

$$\mathbb{P}([\varrho^1, \mathbf{u}^1](t \wedge \mathfrak{t}^1 \wedge \mathfrak{t}^2) = [\varrho^2, \mathbf{u}^2](t \wedge \mathfrak{t}^1 \wedge \mathfrak{t}^2), \text{ for all } t \in [0,T]) = 1.$$

As a consequence, the two solutions coincide up to the stopping time $\mathfrak{t}^1 \wedge \mathfrak{t}^2$ and due to maximality of $\mathfrak{t}^1$ as well as $\mathfrak{t}^2$, it necessarily follows that $\mathfrak{t}^1 = \mathfrak{t}^2$ a.s. This completes the proof of uniqueness under the additional assumption (5.40).

Now, assume that $[\varrho_0, \mathbf{u}_0]$ only satisfies the hypotheses of Theorem 5.0.3. We define for $K > 0$ the set

$$\Omega_K = \left\{ \omega \in \Omega \,\middle|\, \|\mathbf{u}_0(\omega)\|_{W_x^{s,2}} < K, \ \|r_0(\omega)\|_{W_x^{s,2}} < K, \ \inf_{\mathbb{T}^3} r_0(\omega) > \frac{1}{K} \right\}$$

and observe that $\Omega = \bigcup_{K \in \mathbb{R}} \Omega_K$. Therefore, since Ω_K is $\mathfrak{F}_0$-measurable, the *a priori* estimates from Section 5.2.2 can be employed on Ω_K. In fact we replace $\mathbb{E}[\cdot]$ in (5.35) by $\mathbb{E}[\mathbf{1}_{\Omega_K} \cdot]$ and obtain

$$\mathbb{E}\left[\mathbf{1}_{\Omega_K} \left(\sup_{t \in [0, T \wedge \mathfrak{t}_R^i]} \|(r^i(t), \mathbf{u}^i(t))\|_{W_x^{s,2}}^2 + \int_0^{T \wedge \mathfrak{t}_R^i} \|\mathbf{u}^i(t)\|_{W_x^{s+1,2}}^2 \, \mathrm{d}t \right)^p \right] \lesssim c(R, T, s, K). \tag{5.41}$$

Accordingly, the method of pathwise uniqueness from Section 5.2.5 can be applied to Ω_K, which yields

$$\mathbb{P}(\mathbf{1}_{\Omega_K} [\varrho^1, \mathbf{u}^1](t \wedge \mathfrak{t}_R^1 \wedge \mathfrak{t}_R^2) = \mathbf{1}_{\Omega_K} [\varrho^2, \mathbf{u}^2](t \wedge \mathfrak{t}_R^1 \wedge \mathfrak{t}_R^2), \text{ for all } t \in [0,T]) = 1$$

and since $\mathbf{1}_{\Omega_K} \to \mathbf{1}_\Omega$, $\mathfrak{t}_R^i \to \mathfrak{t}^i$, $i = 1, 2$, a.s., we may send $R, K \to \infty$ and apply the dominated convergence theorem to deduce

$$\mathbb{P}([\varrho^1, \mathbf{u}^1](t \wedge \mathfrak{t}^1 \wedge \mathfrak{t}^2) = [\varrho^2, \mathbf{u}^2](t \wedge \mathfrak{t}^1 \wedge \mathfrak{t}^2), \text{ for all } t \in [0,T]) = 1.$$

The uniqueness part of Theorem 5.0.3 is thus complete.

5.3.2 Existence of a local strong solution for bounded initial data

Finally, we are ready to go back to our original problem (5.1)–(5.3) and establish the existence of a local strong pathwise solution up to an a.s. strictly positive stopping time. Let us first take the additional assumption (5.40) which will be removed later.

Having constructed strong solutions for the approximate problem (5.14)–(5.16) in Section 5.2.6, which we denote by $[r_R, \mathbf{u}_R]$, we define

$$\tau_R = \inf\{t \in [0,T] \mid \|\mathbf{u}_R(t)\|_{W^{2,\infty}_x} \geq R\},$$

with the convention $\inf \emptyset = T$. Since $\mathbf{u}_R$ has continuous trajectories in $W^{s,2}(\mathbb{T}^3)$, which is embedded into $W^{2,\infty}(\mathbb{T}^3)$, τ_R is a well-defined stopping time. Moreover, due to (5.40), the stopping time τ_R is a.s. positive, provided R is chosen large enough. Next, we recall, as stated in Theorem 5.2.3,

$$r_R \geq \underline{r}_R > 0 \quad \text{for a.e. } (\omega, t, x),$$

for some deterministic constant $\underline{r}_R$. Consequently, the density given by

$$\varrho_R := \left(\frac{\gamma - 1}{2a\gamma}\right)^{\frac{1}{\gamma-1}} r_R^{\frac{2}{\gamma-1}} \tag{5.42}$$

remains uniformly positive as well. Therefore, the unique solution $[r_R, \mathbf{u}_R]$ of the approximate system (5.14)–(5.15) with the initial condition

$$\left(r_0 := \sqrt{\frac{2a\gamma}{\gamma-1}} \varrho_0^{\frac{\gamma-1}{2}}, \mathbf{u}_0\right)$$

generates the local strong pathwise solution

$$\left(\varrho_R := \left(\frac{\gamma - 1}{2a\gamma}\right)^{\frac{1}{\gamma-1}} r_R^{\frac{2}{\gamma-1}}, \mathbf{u}_R, \tau_R\right)$$

to the original problem (5.1)–(5.3) with the initial condition $[\varrho_0, \mathbf{u}_0]$.

5.3.3 Existence of a local strong solution for general initial data

In order to relax the additional assumption upon the initial datum (5.40), consider again a solution $[r_R, \mathbf{u}_R]$ of the approximate problem (5.14)–(5.15). Now we consider the following stopping time:

$$\begin{aligned}
\tau_K &= \tau_K^1 \wedge \tau_K^2 \wedge \tau_K^3, \\
\tau_K^1 &= \inf\{t \in [0,T] \mid \|\mathbf{u}_R(t)\|_{W^{s,2}_x} \geq K\}, \\
\tau_K^2 &= \inf\{t \in [0,T] \mid \|r_R(t)\|_{W^{s,2}_x} \geq K\}, \\
\tau_K^3 &= \inf\left\{t \in [0,T] \;\middle|\; \inf_{\mathbb{T}^3} r_R(t) \leq \frac{1}{K}\right\},
\end{aligned}$$

with $K = K(R) \to \infty$ as $R \to \infty$, and

$$K(R) < R \min\left\{1, \frac{1}{c_{1,\infty}}, \frac{1}{c_{2,\infty}}\right\},$$

where $c_{1,\infty}$, $c_{2,\infty}$ are the constants in the embedding inequalities

$$\|r\|_{W_x^{1,\infty}} \leq c_{1,\infty}\|r\|_{W_x^{s,2}}, \quad \|\mathbf{u}\|_{W_x^{2,\infty}} \leq c_{2,\infty}\|\mathbf{u}\|_{W_x^{s,2}}.$$

Recall that $s > \frac{7}{2}$. The stopping time τ_K is chosen in such a way that, on $[0, \tau_K)$,

$$\sup_{t\in[0,\tau_K]} \|\mathbf{u}_R(t)\|_{W_x^{2,\infty}} < R, \quad \sup_{t\in[0,\tau_K]} \|r_R(t)\|_{W_x^{1,\infty}} < R, \quad \inf_{t\in[0,\tau_K]} \inf_{\mathbb{T}^3} r_R(t) > \frac{1}{R},$$

$\mathbb{P}$-a.s. Next we observe that Theorem 5.2.3 can be used to construct solutions with the stopping time τ_K for general initial data as in Theorem 5.0.3. Indeed, let $[r_0, \mathbf{u}_0]$ be an $\mathfrak{F}_0$-measurable random variable taking values in $W^{s,2}(\mathbb{T}^3) \times W^{s,2}(\mathbb{T}^3)$ such that $r_0 > 0$ $\mathbb{P}$-a.s. and define the set

$$U_{K(R)} = \left\{[r, \mathbf{u}] \in W^{s,2}(\mathbb{T}^3) \times W^{s,2}(\mathbb{T}^3) \;\middle|\; \|r\|_{W_x^{s,2}} < K,\ \|\mathbf{u}\|_{W_x^{s,2}} < K,\ r > \frac{1}{K}\right\}.$$

Theorem 5.2.3 then provides a unique solution $[r_M, \mathbf{u}_M]$ to (5.14)–(5.15) with $R = M$ and with the initial condition $[r_0, \mathbf{u}_0]\mathbf{1}_{[r_0,\mathbf{u}_0]\in\{U_{K(M)}\setminus\bigcup_{J=1}^{M-1} U_{K(J)}\}}$. It also solves the original system (5.12)–(5.13) up to the stopping time $\tau_{K(M)}$. Next, we find that

$$[r, \mathbf{u}] = \sum_{M=1}^{\infty} [r_M, \mathbf{u}_M]\mathbf{1}_{[r_0,\mathbf{u}_0]\in\{U_{K(M)}\setminus\bigcup_{J=1}^{M-1} U_{K(J)}\}} \tag{5.43}$$

solves the same problem with the initial data $[r_0, \mathbf{u}_0]$ up to the a.s. strictly positive stopping time

$$\tau = \sum_{M=1}^{\infty} \tau_{K(M)}\mathbf{1}_{[r_0,\mathbf{u}_0]\in\{U_{K(M)}\setminus\bigcup_{J=1}^{M-1} U_{K(J)}\}}.$$

Note in particular that $[r, \mathbf{u}]$ has a.s. continuous trajectories in $W^{s,2}(\mathbb{T}^3) \times W^{s,2}(\mathbb{T}^3)$ and the velocity also belongs to $L^2(0, T; W^{s+1,2}(\mathbb{T}^3))$ $\mathbb{P}$-a.s. Indeed, there exists a disjoint collection of sets $\Omega_M \subset \Omega$, $M \in \mathbb{N}$, satisfying $\bigcup_M \Omega_M = \Omega$ such that $[r, \mathbf{u}](\omega) = [r_M, \mathbf{u}_M](\omega)$ for a.e. $\omega \in \Omega_M$. Due to Theorem 5.2.3, the trajectories of $[r_M, \mathbf{u}_M]$ are a.s. continuous in $W^{s,2}(\mathbb{T}^3) \times W^{s,2}(\mathbb{T}^3)$. On the other hand, we lose the integrability in ω as the initial condition is only assumed to be in $W^{s,2}(\mathbb{T}^3) \times W^{s,2}(\mathbb{T}^3)$ a.s. and no integrability in ω is assumed. In particular, the estimate (5.17) is no longer valid for the solution (5.43).

To conclude, after the straightforward transformation to the original variables $[\varrho, \mathbf{u}]$ (recall (5.42)), we obtain the existence of a local strong pathwise solution to problem (5.1)–(5.3) with a strictly positive stopping time τ.

5.3.4 Existence of a maximal strong solution

In order to extend the solution $[\varrho, \mathbf{u}]$ to a maximal time of existence $\mathfrak{t}$, let $\mathcal{T}$ denote the set of all possible a.s. strictly positive stopping times corresponding to the solution starting from the initial datum $[\varrho_0, \mathbf{u}_0]$. According to the above proof, this set is non-empty. Moreover, it is closed with respect to finite minimum and finite maximum operations. More precisely,

$$\sigma_1, \sigma_2 \in \mathcal{T} \quad \Rightarrow \quad \sigma_1 \vee \sigma_2 \in \mathcal{T}$$

and

$$\sigma_1, \sigma_2 \in \mathcal{T} \quad \Rightarrow \quad \sigma_1 \wedge \sigma_2 \in \mathcal{T}$$

for any stopping times σ_1, σ_2. Let $\mathfrak{t} = \operatorname{ess\,sup}_{\sigma \in \mathcal{T}} \sigma$. Then we may choose an increasing sequence $(\sigma_M) \subset \mathcal{T}$ such that $\lim_{M\to\infty} \sigma_M = \mathfrak{t}$ a.s. Let $[\varrho_M, \mathbf{u}_M]$ be the corresponding sequence of solutions on $[0, \sigma_M]$. Due to uniqueness, this sequence defines a solution $[\varrho, \mathbf{u}]$ on $\bigcup_M [0, \sigma_M]$ by setting $[\varrho, \mathbf{u}] := [\varrho_M, \mathbf{u}_M]$ on $[0, \sigma_M]$. For each $R \in \mathbb{N}$, we now define

$$\tau_R = \mathfrak{t} \wedge \inf\{t \in [0,T] \mid \|\mathbf{u}(t)\|_{W^{2,\infty}_x} \geq R\}.$$

Then $(\varrho, \mathbf{u})$ is a solution on $[0, \sigma_M \wedge \tau_R]$ and sending $M \to \infty$ we obtain $(\varrho, \mathbf{u})$ is a solution on $[0, \tau_R]$. Note that τ_R is not a.s. strictly positive unless $\|\mathbf{u}_0\|_{W^{2,\infty}_x} < R$. Nevertheless, since $\mathbf{u}_0 \in W^{s,2}(\mathbb{T}^3)$ a.s. we may deduce that, for almost every ω, there exists $R = R(\omega)$ such that $\mathfrak{t}_{R(\omega)}(\omega) > 0$. To guarantee the strict positivity, we combine the two sequences of stopping times (σ_R) and (τ_R) and define $\mathfrak{t}_R = \sigma_R \vee \tau_R$. Then each triplet $(\varrho, \mathbf{u}, \mathfrak{t}_R)$, $R \in \mathbb{N}$, is a local strong pathwise solution with an a.s. strictly positive stopping time.

Next, we observe that, by repeating the construction of a local strong pathwise solution, a solution on $[0, \mathfrak{t}_R]$ can be extended to a solution on $[0, \mathfrak{t}_R + \sigma]$ for a $\mathbb{P}$-a.s. strictly positive stopping time σ. Indeed, with the method from Section 5.3.3 we can construct a new solution starting from $[\varrho(\mathfrak{t}_R), \mathbf{u}(\mathfrak{t}_R)]$ as the initial condition as follows. Define a stochastic basis

$$(\tilde{\Omega}, \tilde{\mathfrak{F}}, (\tilde{\mathfrak{F}}_t)_{t\geq 0}, \tilde{\mathbb{P}}) := (\Omega, \mathfrak{F}, (\mathfrak{F}_{\mathfrak{t}_R + t})_{t\geq 0}, \mathbb{P}) \tag{5.44}$$

together with a cylindrical $(\tilde{\mathfrak{F}}_t)$-Wiener process

$$\tilde{W} = (\tilde{W}_k)_{k\in\mathbb{N}}, \quad \tilde{W}_k(t) := W_k(\mathfrak{t}_R + t) - W_k(\mathfrak{t}_R), \quad k \in \mathbb{N}. \tag{5.45}$$

Due to the proof in Section 5.3.3, the datum $[\varrho(\mathfrak{t}_R), \mathbf{u}(\mathfrak{t}_R)]$ satisfies the assumptions on the initial condition. Hence we obtain $(\tilde{\varrho}, \tilde{\mathbf{u}}, \sigma)$, which is a local strong pathwise solution to (5.1)–(5.3) relative to the cylindrical Wiener process (5.45) on the stochastic basis (5.44). We stress that the stopping time σ is $\mathbb{P}$-a.s. strictly positive. Since solutions are unique (cf. Section 5.3.1), setting

$$(\bar{\varrho}, \bar{\mathbf{u}})(t) := \begin{cases} (\varrho, \mathbf{u})(t) & \text{if } t \leq \mathfrak{t}_R, \\ (\tilde{\varrho}, \tilde{\mathbf{u}})(t) & \text{if } \mathfrak{t}_R < t \leq \mathfrak{t}_R + \sigma \end{cases}$$

yields a local strong pathwise solution to (5.1)–(5.3), defined up to the stopping time $\mathfrak{t}_R + \sigma > \mathfrak{t}_R$.

Thus, in order to show that $\mathfrak{t}_R < \mathfrak{t}$ on $[\mathfrak{t} < T]$, assume for a contradiction that $\mathbb{P}(\mathfrak{t}_R = \mathfrak{t} < T) > 0$. Then we have $\mathfrak{t}_R + \sigma \in \mathcal{T}$ and hence $\mathbb{P}(\mathfrak{t} < \mathfrak{t}_R + \sigma) > 0$, which contradicts the maximality of $\mathfrak{t}$. Consequently, $(\mathfrak{t}_R)$ is an increasing sequence of stopping times converging to $\mathfrak{t}$. Moreover, on the set $[\mathfrak{t} < T]$ we have

$$\sup_{t \in [0, \mathfrak{t}_R]} \|\mathbf{u}(t)\|_{W_x^{2,\infty}} \geq R.$$

Thus, the existence part of Theorem 5.0.3 is complete.

6 Relative energy inequality and weak–strong uniqueness

The concept of *weak solution* was introduced in mathematical fluid mechanics to handle the unsurmountable difficulties related to the hypothetical or effective possibility of singularities experienced by solutions of the corresponding systems of PDEs. However, as shown in the seminal work of DeLellis–Székelyhidi [DLS10], the so far well-accepted criteria, derived from the underlying physical principles as the second law of thermodynamics, are not sufficient to guarantee the expected well-posedness of the associated initial and/or boundary value problems in the class of weak solutions. The approach based on *relative entropy/energy*, introduced by Dafermos [Daf79], has become an important and rather versatile tool whenever a weak solution is expected to be, or at least to approach, a smooth one; see Leger–Vasseur [LV11], Mellet–Vasseur [MV08], Masmoudi [Mas01], and Saint-Raymond [SR09] for various applications. In particular, the problem of weak–strong uniqueness for the compressible Navier–Stokes and the Navier–Stokes–Fourier system were addressed by Germain [Ger11] and finally solved in [FN12, FNS11]. Our main goal is to derive a relative energy inequality for the system

$$\mathrm{d}\varrho + \operatorname{div}(\varrho\mathbf{u})\,\mathrm{d}t = 0, \tag{6.1}$$

$$\mathrm{d}(\varrho\mathbf{u}) + \operatorname{div}(\varrho\mathbf{u}\otimes\mathbf{u})\,\mathrm{d}t + \nabla p(\varrho)\,\mathrm{d}t = \operatorname{div}\mathbb{S}(\nabla\mathbf{u})\,\mathrm{d}t + \mathbb{G}(\varrho,\varrho\mathbf{u})\,\mathrm{d}W, \tag{6.2}$$

with $p(\varrho) = a\varrho^\gamma$, $a > 0$, analogous to that obtained in the deterministic case in [FNS11]. For the sake of simplicity, we restrict ourselves to the physically relevant case $N = 3$, seeing that our arguments can be easily adapted for $N = 1, 2$.

We proceed in several steps. First, let us recall that under suitable assumptions on the initial law, existence of global solutions to (6.1)–(6.2) was established in Chapter 4, namely in Theorem 4.0.2. The corresponding notion of solution was the so-called dissipative martingale solution, defined as follows.

Definition 6.0.1 (Dissipative martingale solution). Let $\Lambda = \Lambda(\varrho,\mathbf{q})$ be a Borel probability measure on $L^1(\mathbb{T}^3)\times L^1(\mathbb{T}^3)$ such that

$$\Lambda\{\varrho \geq 0\} = 1, \quad \int_{L^1_x\times L^1_x}\left|\int_{\mathbb{T}^3}\left[\frac{|\mathbf{q}|^2}{\varrho} + P(\varrho)\right]\mathrm{d}x\right|^r \mathrm{d}\Lambda(\varrho,\mathbf{q}) < \infty,$$

where the pressure potential is given by

$$P(\varrho) = a\varrho\int_1^\varrho z^{\gamma-2}\,\mathrm{d}z$$

and $r \geq 1$. The quantity $((\Omega,\mathfrak{F},(\mathfrak{F}_t)_{t\geq 0},\mathbb{P}),\varrho,\mathbf{u},W)$ is called a *dissipative martingale solution* to (6.1)–(6.2) with the initial law Λ if:

https://doi.org/10.1515/9783110492552-006

(1) $(\Omega, \mathfrak{F}, (\mathfrak{F}_t)_{t\geq 0}, \mathbb{P})$ is a stochastic basis with a complete right-continuous filtration;
(2) W is a cylindrical $(\mathfrak{F}_t)$-Wiener process;
(3) the density ϱ and the velocity $\mathbf{u}$ are random distributions adapted to $(\mathfrak{F}_t)_{t\geq 0}$, $\varrho \geq 0$ $\mathbb{P}$-a.s.;
(4) there exists an $\mathfrak{F}_0$-measurable random variable $[\varrho_0, \mathbf{u}_0]$ such that $\Lambda = \mathcal{L}[\varrho_0, \varrho_0\mathbf{u}_0]$;
(5) the equation of continuity

$$-\int_0^T \partial_t\phi \int_{\mathbb{T}^3} \varrho\psi \,\mathrm{d}x\,\mathrm{d}t = \phi(0)\int_{\mathbb{T}^3} \varrho_0\psi\,\mathrm{d}x + \int_0^T \phi \int_{\mathbb{T}^3} \varrho\mathbf{u}\cdot\nabla\psi\,\mathrm{d}x\,\mathrm{d}t$$

holds for all $\phi \in C_c^\infty([0,T))$ and all $\psi \in C^\infty(\mathbb{T}^3)$ $\mathbb{P}$-a.s.;
(6) the momentum equation

$$\begin{aligned}
&-\int_0^T \partial_t\phi \int_{\mathbb{T}^3} \varrho\mathbf{u}\cdot\boldsymbol{\varphi}\,\mathrm{d}x\,\mathrm{d}t - \phi(0)\int_{\mathbb{T}^3} \varrho_0\mathbf{u}_0\cdot\boldsymbol{\varphi}\,\mathrm{d}x \\
&= \int_0^T \phi \int_{\mathbb{T}^3} [\varrho\mathbf{u}\otimes\mathbf{u} : \nabla\boldsymbol{\varphi} + p(\varrho)\mathrm{div}\,\boldsymbol{\varphi}]\,\mathrm{d}x\,\mathrm{d}t - \int_0^T \phi \int_{\mathbb{T}^3} \mathbb{S}(\nabla\mathbf{u}) : \nabla\boldsymbol{\varphi}\,\mathrm{d}x\,\mathrm{d}t \\
&\quad + \sum_{k=1}^{\infty}\int_0^T \phi \int_{\mathbb{T}^3} \mathbf{G}_k(\varrho,\varrho\mathbf{u})\cdot\boldsymbol{\varphi}\,\mathrm{d}x\,\mathrm{d}W_k
\end{aligned}$$

holds for all $\phi \in C_c^\infty([0,T))$ and all $\boldsymbol{\varphi} \in C^\infty(\mathbb{T}^3)$ $\mathbb{P}$-a.s.;
(7) the energy inequality

$$\begin{aligned}
&-\int_0^T \partial_t\phi \int_{\mathbb{T}^3} \left[\frac{1}{2}\varrho|\mathbf{u}|^2 + P(\varrho)\right]\mathrm{d}x\,\mathrm{d}t + \int_0^T \phi\int_{\mathbb{T}^3} \mathbb{S}(\nabla\mathbf{u}) : \nabla\mathbf{u}\,\mathrm{d}x\,\mathrm{d}t \\
&\leq \phi(0)\int_{\mathbb{T}^3}\left[\frac{1}{2}\varrho_0|\mathbf{u}_0|^2 + P(\varrho_0)\right]\mathrm{d}x + \frac{1}{2}\sum_{k=1}^{\infty}\int_0^T \phi\int_{\mathbb{T}^3} \varrho^{-1}|\mathbf{G}_k(\varrho,\varrho\mathbf{u})|^2\,\mathrm{d}x\,\mathrm{d}t \\
&\quad + \sum_{k=1}^{\infty}\int_0^T \phi\int_{\mathbb{T}^3} \mathbf{G}_k(\varrho,\varrho\mathbf{u})\cdot\mathbf{u}\,\mathrm{d}x\,\mathrm{d}W_k
\end{aligned} \tag{6.3}$$

holds for all $\phi \in C_c^\infty([0,T))$, $\phi \geq 0$, $\mathbb{P}$-a.s.;
(8) if $b \in C^1(\mathbb{R})$ such that $b'(z) = 0$ for all $z \geq M_b$, then, for all $\phi \in C_c^\infty([0,T))$ and all $\psi \in C^\infty(\mathbb{T}^3)$, we have $\mathbb{P}$-a.s.

$$\begin{aligned}
-\int_0^T \partial_t\phi \int_{\mathbb{T}^3} b(\varrho)\psi\,\mathrm{d}x\,\mathrm{d}t &= \phi(0)\int_{\mathbb{T}^3} b(\varrho_0)\psi\,\mathrm{d}x + \int_0^T\phi\int_{\mathbb{T}^3} b(\varrho)\mathbf{u}\cdot\nabla\psi\,\mathrm{d}x\,\mathrm{d}t \\
&\quad - \int_0^T \phi\int_{\mathbb{T}^3} (b'(\varrho)\varrho - b(\varrho))\,\mathrm{div}\,\mathbf{u}\,\psi\,\mathrm{d}x\,\mathrm{d}t.
\end{aligned}$$

We will see below that the key towards the relative energy inequality is the energy inequality (6.3). Recall that the latter one is not automatically satisfied by weak solutions, so it has to be included in the definition. In order to measure the distance between a dissipative martingale solution $[\varrho, \mathbf{u}]$ of system (6.1)–(6.2) and a pair of arbitrary (smooth) processes $[r, \mathbf{U}]$, we introduce the *relative energy* functional

$$\mathcal{E}(\varrho,\mathbf{u}|r,\mathbf{U}) = \int_{\mathbb{T}^3}\left[\frac{1}{2}\varrho|\mathbf{u}-\mathbf{U}|^2 + P(\varrho) - P'(r)(\varrho - r) - P(r)\right]\mathrm{d}x. \tag{6.4}$$

Note that $\mathcal{E} \geq 0$ since P is a convex function. In view of future applications, it is convenient if the behavior of the test functions $[r, \mathbf{U}]$ mimics that of $[\varrho, \mathbf{u}]$. Accordingly, we require r and $\mathbf{U}$ to be stochastic processes adapted to $(\mathfrak{F}_t)_{t\geq 0}$ such that

$$\mathrm{d}r = \mathrm{d}^{\mathrm{d}} r \, \mathrm{d}t + \mathrm{d}^{\mathrm{s}} r \, \mathrm{d}W, \quad \mathrm{d}\mathbf{U} = \mathrm{d}^{\mathrm{d}} \mathbf{U} \, \mathrm{d}t + \mathrm{d}^{\mathrm{s}} \mathbf{U} \, \mathrm{d}W. \tag{6.5}$$

We assume that $\mathrm{d}^{\mathrm{d}} r, \mathrm{d}^{\mathrm{d}} \mathbf{U}$ are functions of (ω, t, x), sufficiently integrable and regular with respect to the space variable, and $\mathrm{d}^{\mathrm{s}} r, \mathrm{d}^{\mathrm{s}} \mathbf{U}$ belong to $L_2(\mathfrak{U}; L^2(\mathbb{T}^3))$ a.e. in (ω, t), and have appropriate integrability and space regularity. Under these circumstances, the *relative energy inequality* reads

$$\begin{aligned} &- \int_0^T \partial_t \phi \, \mathcal{E}(\varrho, \mathbf{u} | r, \mathbf{U}) \, \mathrm{d}t + \int_0^T \phi \int_{\mathbb{T}^3} (\mathbb{S}(\nabla \mathbf{u}) - \mathbb{S}(\nabla \mathbf{U})) : \nabla(\mathbf{u} - \mathbf{U}) \, \mathrm{d}x \, \mathrm{d}t \\ &\leq \phi(0) \mathcal{E}(\varrho, \mathbf{u} | r, \mathbf{U})(0) + \int_0^T \phi \mathcal{R}(\varrho, \mathbf{u} | r, \mathbf{U}) \, \mathrm{d}t + \int_0^T \phi \, \mathrm{d}M_{\mathrm{RE}}, \end{aligned} \tag{6.6}$$

for all $\phi \in C_c^\infty([0, T))$, $\phi \geq 0$, $\mathbb{P}$-a.s. Here, M_{RE} is a real-valued square integrable martingale given by

$$\begin{aligned} M_{\mathrm{RE}}(t) = &\int_0^t \int_{\mathbb{T}^3} (\mathbf{u} - \mathbf{U}) \cdot \mathbb{G}(\varrho, \varrho \mathbf{u}) \, \mathrm{d}x \, \mathrm{d}W - \int_0^t \int_{\mathbb{T}^3} \varrho(\mathbf{u} - \mathbf{U}) \cdot \mathrm{d}^{\mathrm{s}} \mathbf{U} \, \mathrm{d}x \, \mathrm{d}W \\ &+ \int_0^t \int_{\mathbb{T}^3} (p'(r) - \varrho P''(r)) \, \mathrm{d}^{\mathrm{s}} r \, \mathrm{d}x \, \mathrm{d}W. \end{aligned} \tag{6.7}$$

The remainder term reads

$$\begin{aligned} \mathcal{R}(\varrho, \mathbf{u} | r, \mathbf{U}) = &\int_{\mathbb{T}^3} \mathbb{S}(\nabla \mathbf{U}) : (\nabla \mathbf{U} - \nabla \mathbf{u}) \, \mathrm{d}x + \int_{\mathbb{T}^3} \varrho (\mathrm{d}^{\mathrm{d}} \mathbf{U} + \mathbf{u} \cdot \nabla \mathbf{U})(\mathbf{U} - \mathbf{u}) \, \mathrm{d}x \\ &+ \int_{\mathbb{T}^3} ((r - \varrho) P''(r) \, \mathrm{d}^{\mathrm{d}} r + \nabla P'(r)(r \mathbf{U} - \varrho \mathbf{u})) \, \mathrm{d}x - \int_{\mathbb{T}^3} \operatorname{div} \mathbf{U} (p(\varrho) - p(r)) \, \mathrm{d}x \\ &+ \frac{1}{2} \sum_{k=1}^\infty \int_{\mathbb{T}^3} \varrho \left| \frac{\mathbf{G}_k(\varrho, \varrho \mathbf{u})}{\varrho} - \mathrm{d}^{\mathrm{s}} \mathbf{U}(e_k) \right|^2 \mathrm{d}x - \frac{1}{2} \sum_{k=1}^\infty \int_{\mathbb{T}^3} \varrho P'''(r) |\mathrm{d}^{\mathrm{s}} r(e_k)|^2 \, \mathrm{d}x \\ &+ \frac{1}{2} \sum_{k=1}^\infty \int_{\mathbb{T}^3} p''(r) |\mathrm{d}^{\mathrm{s}} r(e_k)|^2 \, \mathrm{d}x. \end{aligned} \tag{6.8}$$

We will show that the relative energy inequality holds true for dissipative martingale solutions to (6.1)–(6.2). Recall that existence of such solutions defined on some probability space was given in Chapter 4. The proof of (6.6) is presented in Section 6.1. The main ingredients are the energy inequality (6.3) and a careful application of Itô's stochastic calculus.

As a corollary of the relative energy inequality we obtain a weak–strong uniqueness property (pathwise and in law) for the stochastic Navier–Stokes system (6.1)–(6.2), established in Section 6.2. In particular, we prove a Yamada–Watanabe type result that says, roughly speaking, that pathwise weak–strong uniqueness implies weak–strong uniqueness in law; see Theorem 6.2.3.

Remark 6.0.2. A dissipative martingale solution satisfying the energy inequality in the differential form (6.3) may be seen as an analogue of the *a.s. super-martingale* solution, introduced by Flandoli–Romito [FR08] and further developed by Debussche–Romito [DR14] in the context of the incompressible Navier–Stokes system.

It follows from (6.3) that the limits

$$\operatorname{ess}\lim_{\tau\to s+}\int_{\mathbb{T}^3}\left[\frac{1}{2}\varrho|\mathbf{u}|^2+P(\varrho)\right](\tau)\,\mathrm{d}x,\quad \operatorname{ess}\lim_{\tau\to t-}\int_{\mathbb{T}^3}\left[\frac{1}{2}\varrho|\mathbf{u}|^2+P(\varrho)\right](\tau)\,\mathrm{d}x$$

exist $\mathbb{P}$-a.s. for any $0\le s\le t\le T$,

$$\lim_{\tau\to 0+}\int_{\mathbb{T}^3}\left[\frac{1}{2}\varrho|\mathbf{u}|^2+P(\varrho)\right](\tau)\,\mathrm{d}x=\int_{\mathbb{T}^3}\left[\frac{1}{2}\varrho|\mathbf{u}|^2+P(\varrho)\right](0)\,\mathrm{d}x,$$

and

$$\begin{aligned}&\left[\int_{\mathbb{T}^3}\left[\frac{1}{2}\varrho|\mathbf{u}|^2+P(\varrho)\right](\tau)\,\mathrm{d}x\right]_{\tau\to s+}^{\tau\to t-}+\int_s^t\int_{\mathbb{T}^3}\mathbb{S}(\nabla\mathbf{u}):\nabla\mathbf{u}\,\mathrm{d}x\,\mathrm{d}t\\&\quad\le\frac{1}{2}\int_s^t\int_{\mathbb{T}^3}\sum_{k=1}^{\infty}\varrho^{-1}|\mathbf{G}_k(\varrho,\varrho\mathbf{u})|^2\,\mathrm{d}x\,\mathrm{d}t+\int_s^t\psi\int_{\mathbb{T}^3}\mathbb{G}(\varrho,\varrho\mathbf{u})\cdot\mathbf{u}\,\mathrm{d}x\,\mathrm{d}W\end{aligned}\tag{6.9}$$

$\mathbb{P}$-a.s. Finally, in view of the weak lower semi-continuity of convex functionals,

$$\operatorname{ess}\lim_{\tau\to t-}\int_{\mathbb{T}^3}\left[\frac{1}{2}\varrho|\mathbf{u}|^2+P(\varrho)\right](\tau)\,\mathrm{d}x\ge\int_{\mathbb{T}^3}\left[\frac{1}{2}\varrho|\mathbf{u}|^2+P(\varrho)\right](t)\,\mathrm{d}x,$$

for any $t\in[0,T)$ $\mathbb{P}$-a.s. Similar observations hold for the relative energy inequality (6.6) that can be rewritten as

$$\begin{aligned}&\mathcal{E}(\varrho,\mathbf{u}|r,\mathbf{U})(t)+\int_s^t\int_{\mathbb{T}^3}(\mathbb{S}(\nabla_x\mathbf{u})-\mathbb{S}(\nabla\mathbf{U})):(\nabla\mathbf{u}-\nabla\mathbf{U})\,\mathrm{d}x\,\mathrm{d}r\\&\quad\le\mathcal{E}(\varrho,\mathbf{u}|r,\mathbf{U})(s^+)+M_{\mathrm{RE}}(t)-M_{\mathrm{RE}}(s)+\int_s^t\mathcal{R}(\varrho,\mathbf{u}|r,\mathbf{U})\,\mathrm{d}r,\end{aligned}\tag{6.10}$$

for any $0\le s\le t\le T$ $\mathbb{P}$-a.s., with

$$\mathcal{E}(\varrho,\mathbf{u}|r,\mathbf{U})(0^+)=\mathcal{E}(\varrho,\mathbf{u}|r,\mathbf{U})(0).$$

6.1 Relative energy inequality

The main result of this chapter is the following theorem. It states that any dissipative martingale solution, the existence of which is guaranteed by Theorem 4.0.2, satisfies the relative energy inequality.

Theorem 6.1.1. *Let*

$$((\Omega,\mathfrak{F},(\mathfrak{F}_t)_{t\ge0},\mathbb{P}),\varrho,\mathbf{u},W)$$

be a dissipative martingale solution of problem (6.1)–(6.2) *in the sense of Definition* 6.0.1. *Suppose that functions* r, $\mathbf{U}$ *are stochastic processes adapted to* $(\mathfrak{F}_t)_{t\geq 0}$,

$$
\begin{aligned}
&r \in C([0,T]; W^{1,q}(\mathbb{T}^3)), \quad \mathbf{U} \in C([0,T]; W^{1,q}(\mathbb{T}^3)) \quad \mathbb{P}\text{-a.s. for all } 1 \leq q < \infty,\\
&\mathbb{E}\Big[\sup_{t\in[0,T]} \|r\|^2_{W^{1,q}_x}\Big]^q + \mathbb{E}\Big[\sup_{t\in[0,T]} \|\mathbf{U}\|^2_{W^{1,q}_x}\Big]^q \leq c(q), \\
&0 < \underline{r} \leq r(t,x) \leq \overline{r} \quad \mathbb{P}\text{-a.s.}
\end{aligned}
\tag{6.11}
$$

Moreover, r, $\mathbf{U}$ *satisfy* (6.5), *where*

$$
\begin{aligned}
&\mathrm{d}^{\mathrm{d}} r, \mathrm{d}^{\mathrm{d}}\mathbf{U} \in L^q(\Omega; L^q(0,T; W^{1,q}(\mathbb{T}^3))),\\
&\mathrm{d}^{\mathrm{s}} r, \mathrm{d}^{\mathrm{s}}\mathbf{U} \in L^2(\Omega; L^2(0,T; L_2(\mathfrak{U}; L^2(\mathbb{T}^3)))),\\
&\Big(\sum_{k=1}^{\infty} |\mathrm{d}^{\mathrm{s}} r(e_k)|^q\Big)^{\frac{1}{q}} \in L^q(\Omega; L^q(0,T; L^q(\mathbb{T}^3))),\\
&\Big(\sum_{k=1}^{\infty} |\mathrm{d}^{\mathrm{s}} \mathbf{U}(e_k)|^q\Big)^{\frac{1}{q}} \in L^q(\Omega; L^q(0,T; L^q(\mathbb{T}^3)))
\end{aligned}
$$

for all $1 \leq q < \infty$. *Then the relative energy inequality* (6.6) *holds for all* $\phi \in C^\infty_c([0,T))$, $\phi \geq 0$, $\mathbb{P}$*-a.s., where the martingale* M_{RE} *is given in* (6.7) *and the remainder term is given in* (6.8). *In particular, the relative energy inequality* (6.10) *holds.*

Remark 6.1.2. Hypothesis (6.11) seems rather restrictive and even unrealistic in view of the expected properties of random processes. On the other hand, it is necessary to handle the compositions of the non-linearities, in particular the pressure $p = p(r)$. Note that (6.11) can always be achieved replacing r by $\tilde{r}$, where

$$
\tilde{r}(t) = r(t \wedge \tau_{\underline{r},\overline{r}}),
$$

where $\tau_{\underline{r},\overline{r}}$ is the stopping time given by

$$
\tau_{\underline{r},\overline{r}} = \inf\Big\{t \in [0,T] : \inf_{\mathbb{T}^3} r(t,\cdot) < \underline{r} \text{ or } \sup_{\mathbb{T}^3} r(t,\cdot) > \overline{r}\Big\}.
$$

Remark 6.1.3. For the sake of simplicity, we prove Theorem 6.1.1 in the natural three-dimensional setting. However, the same result holds in the one-dimensional and two-dimensional settings.

Proof of Theorem 6.1.1. We start by writing

$$
\begin{aligned}
\mathcal{E}(\varrho, \mathbf{u} | r, \mathbf{U}) = &\int_{\mathbb{T}^3} \Big[\frac{1}{2}\varrho|\mathbf{u}|^2 + P(\varrho)\Big]\,\mathrm{d}x - \int_{\mathbb{T}^3} \varrho\mathbf{u}\cdot\mathbf{U}\,\mathrm{d}x\\
&+ \int_{\mathbb{T}^3} \frac{1}{2}\varrho|\mathbf{U}|^2\,\mathrm{d}x - \int_{\mathbb{T}^3} \varrho P'(r)\,\mathrm{d}x + \int_{\mathbb{T}^3} [P'(r)r - P(r)]\,\mathrm{d}x.
\end{aligned}
$$

As the time evolution of the first integral is governed by the energy inequality (6.3), it remains to compute the time differentials of the remaining terms with the help of Theorem A.4.1.

Step 1: To compute $\mathrm{d}\int_{\mathbb{T}^3} \varrho\mathbf{u}\cdot\mathbf{U}\,\mathrm{d}x$, we recall that $s = \varrho\mathbf{u}$ satisfies hypotheses (A.10)–(A.12) with $m = 1$ and some $1 < q < \infty$. Applying Theorem A.4.1 we obtain

$$\begin{aligned}\mathrm{d}\Big(\int_{\mathbb{T}^3} \varrho\mathbf{u}\cdot\mathbf{U}\,\mathrm{d}x\Big) &= \Big(\int_{\mathbb{T}^3} \varrho(\mathbf{u}\cdot\mathrm{d}^{\mathrm{d}}\mathbf{U} + \mathbf{u}\cdot\nabla\mathbf{U}\cdot\mathbf{u})\,\mathrm{d}x\Big)\mathrm{d}t \\ &\quad + \Big(\int_{\mathbb{T}^3} \operatorname{div}\mathbf{U}p(\varrho) - \mathbb{S}(\nabla\mathbf{u}):\nabla\mathbf{U}\,\mathrm{d}x\Big)\mathrm{d}t \\ &\quad + \sum_{k=1}^{\infty}\int_{\mathbb{T}^3} \mathrm{d}^{\mathrm{s}}\mathbf{U}(e_k)\cdot\mathbf{G}_k(\varrho,\varrho\mathbf{u})\,\mathrm{d}x\,\mathrm{d}t + \mathrm{d}M_1,\end{aligned} \tag{6.12}$$

where

$$M_1(t) = \int_0^t\int_{\mathbb{T}^3} \mathbf{U}\cdot\mathbb{G}(\varrho,\varrho\mathbf{u})\,\mathrm{d}x\,\mathrm{d}W + \int_0^t\int_{\mathbb{T}^3} \varrho\mathbf{u}\cdot\mathrm{d}^{\mathrm{s}}\mathbf{U}\,\mathrm{d}x\,\mathrm{d}W$$

is a square integrable martingale.

Step 2: Similarly, we compute

$$\begin{aligned}\mathrm{d}\Big(\int_{\mathbb{T}^3} \frac{1}{2}\varrho|\mathbf{U}|^2\,\mathrm{d}x\Big) &= \int_{\mathbb{T}^3} \varrho\mathbf{u}\cdot\nabla\mathbf{U}\cdot\mathbf{U}\,\mathrm{d}x\,\mathrm{d}t + \int_{\mathbb{T}^3} \varrho\mathbf{U}\cdot\mathrm{d}^{\mathrm{d}}\mathbf{U}\,\mathrm{d}x\,\mathrm{d}t \\ &\quad + \frac{1}{2}\sum_{k=1}^{\infty}\int_{\mathbb{T}^3} \varrho|\mathrm{d}^{\mathrm{s}}\mathbf{U}(e_k)|^2\,\mathrm{d}x\,\mathrm{d}t + \mathrm{d}M_2, \\ M_2 &= \int_0^t\int_{\mathbb{T}^3} \varrho\mathbf{U}\cdot\mathrm{d}^{\mathrm{s}}\mathbf{U}\,\mathrm{d}x\,\mathrm{d}W,\end{aligned} \tag{6.13}$$

$$\begin{aligned}\mathrm{d}\Big(\int_{\mathbb{T}^3} [P'(r)r - P(r)]\,\mathrm{d}x\Big) &= \int_{\mathbb{T}^3} p'(r)\,\mathrm{d}^{\mathrm{d}}r\,\mathrm{d}x\,\mathrm{d}t \\ &\quad + \frac{1}{2}\sum_{k=1}^{\infty}\int_{\mathbb{T}^3} p''(r)|\mathrm{d}^{\mathrm{s}}r(e_k)|^2\,\mathrm{d}x\,\mathrm{d}t + \mathrm{d}M_3, \\ M_3 &= \int_0^t\int_{\mathbb{T}^3} p'(r)\,\mathrm{d}^{\mathrm{s}}r\,\mathrm{d}x\,\mathrm{d}W,\end{aligned} \tag{6.14}$$

and, finally,

$$\begin{aligned}\mathrm{d}\Big(\int_{\mathbb{T}^3} \varrho P'(r)\,\mathrm{d}x\Big) &= \int_{\mathbb{T}^3} \varrho\nabla P'(r)\cdot\mathbf{u}\,\mathrm{d}x\,\mathrm{d}t + \int_{\mathbb{T}^3} \varrho P''(r)\,\mathrm{d}^{\mathrm{d}}r\,\mathrm{d}x\,\mathrm{d}t \\ &\quad + \frac{1}{2}\sum_{k=1}^{\infty}\int_{\mathbb{T}^3} \varrho P'''(r)|\mathrm{d}^{\mathrm{s}}r(e_k)|^2\,\mathrm{d}x\,\mathrm{d}t + \mathrm{d}M_4, \\ M_4(t) &= \int_0^t\int_{\mathbb{T}^3} \varrho P''(r)\,\mathrm{d}^{\mathrm{s}}r\,\mathrm{d}x\,\mathrm{d}W.\end{aligned} \tag{6.15}$$

From the equations (6.3), (6.12)–(6.15) we obtain the following formula for the martingale M_{RE} appearing in (6.6):

$$M_{\mathrm{RE}}(t) = \int_0^t \int_{\mathbb{T}^3} (\mathbf{u} - \mathbf{U}) \cdot \mathbb{G}(\varrho, \varrho\mathbf{u}) \,\mathrm{d}x \,\mathrm{d}W - \int_0^t \int_{\mathbb{T}^3} \varrho(\mathbf{u} - \mathbf{U}) \cdot \mathrm{d}^s\mathbf{U} \,\mathrm{d}x \,\mathrm{d}W$$
$$+ \int_0^t \int_{\mathbb{T}^3} (p'(r) - \varrho P''(r)) \,\mathrm{d}^s r \,\mathrm{d}x \,\mathrm{d}W.$$

Step 3: Now, we can derive a differential form of (6.12)–(6.15) similar to (6.3) by applying Theorem A.4.1 to the product with a test function ψ. Summing up the resulting expressions and adding the sum to (6.3), we obtain (6.6). We have proved Theorem 6.1.1. □

6.2 Weak–strong uniqueness

As the first application of Theorem 6.1.1, we present a weak–strong uniqueness result. Here, weak solutions have to be understood in the sense of Definition 6.0.1, whereas strong solutions solve (6.1)–(6.2) in the sense of Definition 5.0.1, which we recall in the following for the convenience of the reader.

Definition 6.2.1 (Local strong pathwise solution). Let $(\Omega, \mathfrak{F}, (\mathfrak{F}_t)_{t\geq 0}, \mathbb{P})$ be a stochastic basis with a complete right-continuous filtration and let W be an $(\mathfrak{F}_t)$-cylindrical Wiener process. Let $(\varrho_0, \mathbf{u}_0)$ be an $\mathfrak{F}_0$-measurable random variable in the space $W^{s,2}(\mathbb{T}^3) \times W^{s,2}(\mathbb{T}^3)$ for some $s > \frac{7}{2}$. A triplet $(\varrho, \mathbf{u}, \mathfrak{t})$ is called a *local strong pathwise solution* to system (6.1)–(6.2), provided:

(1) $\mathfrak{t}$ is a $\mathbb{P}$-a.s. strictly positive $(\mathfrak{F}_t)$-stopping time;

(2) the density ϱ is a $W^{s,2}(\mathbb{T}^3)$-valued $(\mathfrak{F}_t)$-progressively measurable stochastic process such that

$$\varrho(\cdot \wedge \mathfrak{t}) > 0, \quad \varrho(\cdot \wedge \mathfrak{t}) \in C([0,T]; W^{s,2}(\mathbb{T}^3)) \quad \mathbb{P}\text{-a.s.};$$

(3) the velocity $\mathbf{u}$ is a $W^{s,2}(\mathbb{T}^3)$-valued $(\mathfrak{F}_t)$-progressively measurable stochastic process such that

$$\mathbf{u}(\cdot \wedge \mathfrak{t}) \in C([0,T]; W^{s,2}(\mathbb{T}^3)) \cap L^2(0,T; W^{s+1,2}(\mathbb{T}^3)) \quad \mathbb{P}\text{-a.s.};$$

(4) the equation of continuity

$$\varrho(t \wedge \mathfrak{t}) = \varrho_0 - \int_0^{t\wedge\mathfrak{t}} \operatorname{div}(\varrho\mathbf{u}) \,\mathrm{d}s$$

holds for all $t \in [0,T]$ $\mathbb{P}$-a.s.;

(5) the momentum equation

$$(\varrho\mathbf{u})(t\wedge \mathfrak{t}) = \varrho_0\mathbf{u}_0 - \int_0^{t\wedge\mathfrak{t}} \operatorname{div}(\varrho\mathbf{u}\otimes\mathbf{u})\,\mathrm{d}s + \int_0^{t\wedge\mathfrak{t}} \operatorname{div}\mathbb{S}(\nabla\mathbf{u})\,\mathrm{d}s - \int_0^{t\wedge\mathfrak{t}} \nabla p(\varrho)\,\mathrm{d}s + \int_0^{t\wedge\mathfrak{t}} \mathbb{G}(\varrho,\varrho\mathbf{u})\,\mathrm{d}W$$

holds for all $t\in[0,T]$ $\mathbb{P}$-a.s.

6.2.1 Pathwise weak–strong uniqueness

We claim the following pathwise variant of the weak–strong uniqueness principle.

Theorem 6.2.2. *The pathwise weak–strong uniqueness holds true for the system* (6.1)–(6.2) *in the following sense. Let* $[(\Omega,\mathfrak{F},(\mathfrak{F}_t),\mathbb{P}),\varrho,\mathbf{u},W]$ *be a dissipative martingale solution to system* (6.1)–(6.2) *in the sense of Definition* 6.0.1 *and let* $(\tilde\varrho,\tilde{\mathbf{u}},\mathfrak{t})$ *be a local strong pathwise solution to* (6.1)–(6.2) *in the sense of Definition* 6.2.1 *with* $s\geq 4$, *defined on the same stochastic basis with the same Wiener process and with the initial data*

$$\tilde\varrho(0,\cdot)=\varrho(0,\cdot),\quad \tilde\varrho(0,\cdot)\tilde{\mathbf{u}}(0,\cdot)=(\varrho\mathbf{u})(0,\cdot)\quad \mathbb{P}\text{-a.s.},$$
$$\varrho(0,\cdot)\geq\underline{\varrho}>0\quad \mathbb{P}\text{-a.s.} \tag{6.16}$$

Then $\varrho(\cdot\wedge\mathfrak{t})=\tilde\varrho(\cdot\wedge\mathfrak{t})$ *and* $\varrho\mathbf{u}(\cdot\wedge\mathfrak{t})=\tilde\varrho\tilde{\mathbf{u}}(\cdot\wedge\mathfrak{t})$ *a.s.*

Proof of Theorem 6.2.2. **Step 1:** We start by introducing a stopping time

$$\tau_L=\inf\{t\in(0,T)\mid \|\tilde{\mathbf{u}}(s,\cdot)\|_{W^{4,2}_x}>L\}.$$

As $(\tilde\varrho,\tilde{\mathbf{u}})$ is a strong solution,

$$\mathbb{P}\Big[\lim_{L\to\infty}\tau_L=\mathfrak{t}\Big]=1,$$

whence it suffices to show the result for a fixed L.

Step 2: Given $L>0$, we obtain, as a direct consequence of the embedding relation $W^{2,2}(\mathbb{T}^3)\hookrightarrow C(\mathbb{T}^3)$,

$$\sup_{t\in[0,\tau_L]}\|\nabla^2\tilde{\mathbf{u}}(t,\cdot)\|_{L^\infty_x}\leq c(L). \tag{6.17}$$

Moreover, since $\tilde\varrho$ satisfies the equation of continuity on the time interval $[0,\mathfrak{t}]$ and hypothesis (6.16) holds,

$$0<\underline{\varrho}_L\leq\tilde\varrho(t\wedge\mathfrak{t})\leq\overline{\varrho}_L\quad \text{for } t\in[0,\tau_L], \tag{6.18}$$

for some deterministic constants $\underline{\varrho}_L, \overline{\varrho}_L$, cf. Remark A.2.6. Next, it is easy to check that, for any $\delta > 0$ small enough,

$$P(\varrho) - P'(r)(\varrho - r) - P(r) \geq c(\delta) \begin{cases} |\varrho - r|^2 & \text{if } \delta < r, \varrho < \delta^{-1}, \\ 1 + \varrho^\gamma & \text{if } \delta < r < \delta^{-1},\ \varrho \notin [\delta/2, 2\delta]. \end{cases} \tag{6.19}$$

This motivates the following definition. For

$$\Phi_L \in C_0^\infty(0,\infty), \quad 0 \leq \Phi_L \leq 1, \quad \Phi_L(r) = 1 \quad \text{for all } r \in [\underline{\varrho}_L/2, 2\overline{\varrho}_L],$$

we introduce

$$[h]_{\mathrm{ess}} = \Phi_L(\varrho)h, \quad [h]_{\mathrm{res}} = h - \Phi_L(\varrho)h \quad \text{for any } h \in L^1(\Omega \times (0,T) \times \mathbb{T}^3).$$

It follows from (6.19) that

$$\mathcal{E}(\varrho, \mathbf{u}|\tilde{\varrho}, \tilde{\mathbf{u}}) \geq c(L)\big[\|[\mathbf{u} - \tilde{\mathbf{u}}]_{\mathrm{ess}}\|_{L_x^2}^2 + \|[\varrho - \tilde{\varrho}]_{\mathrm{ess}}\|_{L_x^2}^2\big] \tag{6.20}$$

and, similarly,

$$\mathcal{E}(\varrho, \mathbf{u}|\tilde{\varrho}, \tilde{\mathbf{u}}) \geq c(L)\big[\|\sqrt{\varrho}[\mathbf{u} - \tilde{\mathbf{u}}]_{\mathrm{res}}\|_{L_x^2}^2 + \|[1 + \varrho^\gamma]_{\mathrm{res}}\|_{L_x^1}\big], \tag{6.21}$$

whenever $t \in [0, \tau_L]$.

Step 3: Our goal now is to apply the relative energy inequality (6.6) to $r = \tilde{\varrho}$, $\mathbf{U} = \tilde{\mathbf{u}}$ on the time interval $[0, \tau_L \wedge \mathrm{t}]$. To this end, we compute

$$\mathrm{d}\tilde{\mathbf{u}} = \mathrm{d}\Big(\frac{\tilde{\varrho}\tilde{\mathbf{u}}}{\tilde{\varrho}}\Big) = \frac{1}{\tilde{\varrho}}\mathrm{d}(\tilde{\varrho}\tilde{\mathbf{u}}) - \frac{\partial_t \tilde{\varrho}}{\tilde{\varrho}}\tilde{\mathbf{u}}\,\mathrm{d}t.$$

Hence we deduce from (6.6)

$$\begin{aligned} &\mathcal{E}(\varrho, \mathbf{u}|\tilde{\varrho}, \tilde{\mathbf{u}})(t \wedge \tau_L \wedge \mathrm{t}) + \int_0^{t\wedge\tau_L\wedge \mathrm{t}} \int_{\mathbb{T}^3} (\mathbb{S}(\nabla \mathbf{u}) - \mathbb{S}(\nabla \tilde{\mathbf{u}})) : (\nabla \mathbf{u} - \nabla \tilde{\mathbf{u}})\,\mathrm{d}x\,\mathrm{d}s \\ &\quad \leq M_{\mathrm{RE}}(t \wedge \tau_L \wedge \mathrm{t}) - M_{\mathrm{RE}}(0) + \int_0^{t\wedge\tau_L\wedge\mathrm{t}} \mathcal{R}(\varrho, \mathbf{u}|\tilde{\varrho}, \tilde{\mathbf{u}})\,\mathrm{d}s, \end{aligned} \tag{6.22}$$

with

$$\begin{aligned} \mathcal{R}(\varrho, \mathbf{u}|\tilde{\varrho}, \tilde{\mathbf{u}}) = &\int_{\mathbb{T}^3} \mathbb{S}(\nabla\tilde{\mathbf{u}}) : (\nabla\tilde{\mathbf{u}} - \nabla\mathbf{u})\,\mathrm{d}x \\ &- \int_{\mathbb{T}^3} \frac{\varrho}{\tilde{\varrho}}(\partial_t \tilde{\varrho}\tilde{\mathbf{u}} + \operatorname{div}(\tilde{\varrho}\tilde{\mathbf{u}} \otimes \tilde{\mathbf{u}})) \cdot (\tilde{\mathbf{u}} - \mathbf{u})\,\mathrm{d}x \\ &+ \int_{\mathbb{T}^3} \varrho\mathbf{u} \cdot \nabla\tilde{\mathbf{u}}(\tilde{\mathbf{u}} - \mathbf{u})\,\mathrm{d}x + \int_{\mathbb{T}^3} \frac{\varrho}{\tilde{\varrho}}(\operatorname{div}\mathbb{S}(\nabla\tilde{\mathbf{u}}) - \nabla p(\tilde{\varrho})) \cdot (\tilde{\mathbf{u}} - \mathbf{u})\,\mathrm{d}x \\ &+ \int_{\mathbb{T}^3} ((\tilde{\varrho} - \varrho)P''(\tilde{\varrho})\partial_t\tilde{\varrho} + \nabla P'(\tilde{\varrho})(\tilde{\varrho}\tilde{\mathbf{u}} - \varrho\mathbf{u}))\,\mathrm{d}x - \int_{\mathbb{T}^3} \operatorname{div}\tilde{\mathbf{u}}(p(\varrho) - p(\tilde{\varrho}))\,\mathrm{d}x \end{aligned}$$

$$+\frac{1}{2}\sum_{k=1}^{\infty}\int_{\mathbb{T}^3}\varrho\left|\frac{\mathbf{G}_k(\varrho,\varrho\mathbf{u})}{\varrho}-\frac{\mathbf{G}_k(\tilde{\varrho},\tilde{\varrho}\tilde{\mathbf{u}})}{\tilde{\varrho}}\right|^2\mathrm{d}x.$$

This can be rewritten

$$\begin{aligned}
\mathcal{R}(\varrho,\mathbf{u}|\tilde{\varrho},\tilde{\mathbf{u}}) &= \int_{\mathbb{T}^3}\frac{1}{\tilde{\varrho}}(\varrho-\tilde{\varrho})\operatorname{div}\mathbb{S}(\nabla\tilde{\mathbf{u}})\cdot(\tilde{\mathbf{u}}-\mathbf{u})\,\mathrm{d}x\\
&\quad+\int_{\mathbb{T}^3}\varrho(\mathbf{u}-\tilde{\mathbf{u}})\cdot\nabla\tilde{\mathbf{u}}\cdot(\tilde{\mathbf{u}}-\mathbf{u})\,\mathrm{d}x-\int_{\mathbb{T}^3}\frac{\varrho}{\tilde{\varrho}}\nabla p(\tilde{\varrho})\cdot(\tilde{\mathbf{u}}-\mathbf{u})\,\mathrm{d}x\\
&\quad+\int_{\mathbb{T}^3}((\tilde{\varrho}-\varrho)P''(\tilde{\varrho})\partial_t\tilde{\varrho}+\nabla P'(\tilde{\varrho})(\tilde{\varrho}\tilde{\mathbf{u}}-\varrho\mathbf{u}))\,\mathrm{d}x-\int_{\mathbb{T}^3}\operatorname{div}\tilde{\mathbf{u}}(p(\varrho)-p(\tilde{\varrho}))\,\mathrm{d}x\\
&\quad+\frac{1}{2}\sum_{k=1}^{\infty}\int_{\mathbb{T}^3}\varrho\left|\frac{\mathbf{G}_k(\varrho,\varrho\mathbf{u})}{\varrho}-\frac{\mathbf{G}_k(\tilde{\varrho},\tilde{\varrho}\tilde{\mathbf{u}})}{\tilde{\varrho}}\right|^2\mathrm{d}x\\
&= \int_{\mathbb{T}^3}\frac{1}{\tilde{\varrho}}(\varrho-\tilde{\varrho})\operatorname{div}\mathbb{S}(\nabla\tilde{\mathbf{u}})\cdot(\tilde{\mathbf{u}}-\mathbf{u})\,\mathrm{d}x+\int_{\mathbb{T}^3}\varrho(\mathbf{u}-\tilde{\mathbf{u}})\cdot\nabla\tilde{\mathbf{u}}\cdot(\tilde{\mathbf{u}}-\mathbf{u})\,\mathrm{d}x\\
&\quad-\int_{\mathbb{T}^3}\operatorname{div}\tilde{\mathbf{u}}(p(\varrho)-p'(\tilde{\varrho})(\varrho-\tilde{\varrho})-p(\tilde{\varrho}))\,\mathrm{d}x\\
&\quad+\frac{1}{2}\sum_{k=1}^{\infty}\int_{\mathbb{T}^3}\varrho\left|\frac{\mathbf{G}_k(\varrho,\varrho\mathbf{u})}{\varrho}-\frac{\mathbf{G}_k(\tilde{\varrho},\tilde{\varrho}\tilde{\mathbf{u}})}{\tilde{\varrho}}\right|^2\mathrm{d}x\\
&= \mathcal{T}_1+\mathcal{T}_2+\mathcal{T}_3+\mathcal{T}_4.
\end{aligned}\tag{6.23}$$

The goal is to estimate the terms $\mathcal{T}_1,\dots,\mathcal{T}_4$ and to absorb them in the left hand side of (6.22) via Gronwall's lemma. The first three terms can be estimated similarly to the deterministic case; see [FNS11, Section 4.1]. Using (6.17) and (6.18) we estimate

$$\begin{aligned}
\mathcal{T}_1 &\le \underline{\varrho}_L^{-1}\sup_{t\in[0,\tau_L]}\|\nabla_x^2\tilde{\mathbf{u}}(t,\cdot)\|_{L_x^\infty}\int_{\mathbb{T}^3}|\varrho-\tilde{\varrho}||\tilde{\mathbf{u}}-\mathbf{u}|\,\mathrm{d}x\\
&\le c(L)\int_{\mathbb{T}^3}\Phi_L(\varrho)^2|\varrho-\tilde{\varrho}||\tilde{\mathbf{u}}-\mathbf{u}|\,\mathrm{d}x+\int_{\mathbb{T}^3}(1-\Phi_L(\varrho)^2)|\varrho-\tilde{\varrho}||\tilde{\mathbf{u}}-\mathbf{u}|\,\mathrm{d}x\\
&=: c(L)\mathcal{T}_1^1+c(L)\mathcal{T}_1^2.
\end{aligned}$$

Using (6.17), (6.18), (6.19), and (6.20), we obtain

$$\begin{aligned}
\mathcal{T}_1^1 &\le c(L)\left(\|\Phi_L(\varrho)(\mathbf{u}-\tilde{\mathbf{u}})\|_{L_x^2}^2+\|\Phi_L(\varrho)(\varrho-\tilde{\varrho})\|_{L_x^2}^2\right)\\
&= c(L)\left(\|[\mathbf{u}-\tilde{\mathbf{u}}]_{\mathrm{ess}}\|_{L_x^2}^2+\|[\varrho-\tilde{\varrho}]_{\mathrm{ess}}\|_{L_x^2}^2\right)\\
&\le c(L)\mathcal{E}(\varrho,\mathbf{u}|\tilde{\varrho},\tilde{\mathbf{u}})
\end{aligned}$$

and similarly, by (6.19),

$$\begin{aligned}
\mathcal{T}_1^2 &\le c(L)\int_{\mathbb{T}^3}(1-\Phi_L(\varrho))^2\varrho|\tilde{\mathbf{u}}-\mathbf{u}|\,\mathrm{d}x\\
&\le c(L)\left(\|(1-\Phi_L(\varrho))\sqrt{\varrho}(\mathbf{u}-\tilde{\mathbf{u}})\|_{L_x^2}^2+\|(1-\Phi_L(\varrho))\sqrt{\varrho}\|_{L_x^2}^2\right)\\
&\le c(L)\left(\|(1-\Phi_L(\varrho))\sqrt{\varrho}(\mathbf{u}-\tilde{\mathbf{u}})\|_{L_x^2}^2+\|(1-\Phi_L(\varrho))(1+\varrho^\gamma)\|_{L_x^1}\right)
\end{aligned}$$

$$
\begin{aligned}
&= c(L)\left(\|\sqrt{\varrho}[\mathbf{u}-\tilde{\mathbf{u}}]_{\mathrm{res}}\|_{L_x^2}^2 + \|[1+\varrho^\gamma]_{\mathrm{res}}\|_{L_x^1}\right)\\
&\leq c(L)\mathcal{E}(\varrho,\mathbf{u}|\tilde{\varrho},\tilde{\mathbf{u}}).
\end{aligned}
$$

By (6.17) we easily find

$$
\mathscr{T}_2 \leq \sup_{t\in[0,\tau_L]} \|\nabla_x\tilde{\mathbf{u}}(t,\cdot)\|_{L_x^\infty}\int_{\mathbb{T}^3}\varrho|\mathbf{u}-\tilde{\mathbf{u}}|^2\,\mathrm{d}x \leq c(L)\mathcal{E}(\varrho,\mathbf{u}|\tilde{\varrho},\tilde{\mathbf{u}}).
$$

Finally, we have

$$
\begin{aligned}
\mathscr{T}_3 &\leq \sup_{t\in[0,\tau_L]} \|\nabla_x\tilde{\mathbf{u}}(t,\cdot)\|_{L_x^\infty}\int_{\mathbb{T}^3}(P(\varrho)-P'(\tilde{\varrho})(\varrho-\tilde{\varrho})-P(\tilde{\varrho}))\,\mathrm{d}x\\
&\leq c(L)\mathcal{E}(\varrho,\mathbf{u}|\tilde{\varrho},\tilde{\mathbf{u}}).
\end{aligned}
$$

We conclude

$$
\mathscr{T}_1+\mathscr{T}_2+\mathscr{T}_3 \leq c(L)\mathcal{E}(\varrho,\mathbf{u}|\tilde{\varrho},\tilde{\mathbf{u}}). \tag{6.24}
$$

Now we estimate the part arising from the correction term and decompose

$$
\begin{aligned}
\mathscr{T}_4 &= \frac{1}{2}\sum_{k=1}^{\infty}\int_{\mathbb{T}^3}\chi_{\varrho\leq\frac{\tilde{\varrho}}{2}}\varrho\left|\frac{\mathbf{G}_k(\varrho,\varrho\mathbf{u})}{\varrho}-\frac{\mathbf{G}_k(\tilde{\varrho},\tilde{\varrho}\tilde{\mathbf{u}})}{\tilde{\varrho}}\right|^2\mathrm{d}x\\
&\quad+\frac{1}{2}\sum_{k=1}^{\infty}\int_{\mathbb{T}^3}\chi_{\frac{\tilde{\varrho}}{2}\leq\varrho\leq2\tilde{\varrho}}\varrho\left|\frac{\mathbf{G}_k(\varrho,\varrho\mathbf{u})}{\varrho}-\frac{\mathbf{G}_k(\tilde{\varrho},\tilde{\varrho}\tilde{\mathbf{u}})}{\tilde{\varrho}}\right|^2\mathrm{d}x\\
&\quad+\frac{1}{2}\sum_{k=1}^{\infty}\int_{\mathbb{T}^3}\chi_{\varrho\geq2\tilde{\varrho}}\varrho\left|\frac{\mathbf{G}_k(\varrho,\varrho\mathbf{u})}{\varrho}-\frac{\mathbf{G}_k(\tilde{\varrho},\tilde{\varrho}\tilde{\mathbf{u}})}{\tilde{\varrho}}\right|^2\mathrm{d}x\\
&= \mathscr{T}_4^1+\mathscr{T}_4^2+\mathscr{T}_4^3.
\end{aligned}
$$

Using (5.4), (5.5), (6.18), and (6.19), we obtain

$$
\begin{aligned}
\mathscr{T}_4^1 &\leq c(L)\int_{\mathbb{T}^3}\chi_{\varrho\leq\frac{\tilde{\varrho}}{2}}(1+\varrho|\mathbf{u}|^2+\varrho|\tilde{\mathbf{u}}|^2)\,\mathrm{d}x\\
&\leq c(L)\int_{\mathbb{T}^3}\chi_{\varrho\leq\frac{\tilde{\varrho}}{2}}\,\mathrm{d}x + c(L)\,\mathbb{E}\int_{\mathbb{T}^3}\varrho|\mathbf{u}-\tilde{\mathbf{u}}|^2\,\mathrm{d}x\\
&\leq c(L)\int_{\mathbb{T}^3}\chi_{\varrho\leq\frac{\tilde{\varrho}}{2}}(P(\varrho)-P'(\tilde{\varrho})(\varrho-\tilde{\varrho})-P(\tilde{\varrho}))\,\mathrm{d}x + c(L)\int_{\mathbb{T}^3}\varrho|\mathbf{u}-\tilde{\mathbf{u}}|^2\,\mathrm{d}x\\
&\leq c(L)\,\mathcal{E}(\varrho,\mathbf{u}|\tilde{\varrho},\tilde{\mathbf{u}}).
\end{aligned}
$$

Similarly, we obtain by (6.18) and (6.19) and the mean-value theorem

$$
\begin{aligned}
\mathscr{T}_4^2 &\leq \frac{1}{2}\sum_{k=1}^{\infty}\int_{\mathbb{T}^3}\chi_{\frac{\tilde{\varrho}}{2}\leq\varrho\leq2\tilde{\varrho}}\varrho\left|\frac{\mathbf{G}_k(\varrho,\varrho\mathbf{u})}{\varrho}-\frac{\mathbf{G}_k(\tilde{\varrho},\varrho\mathbf{u})}{\tilde{\varrho}}\right|^2\mathrm{d}x\\
&\quad+\frac{1}{2}\sum_{k=1}^{\infty}\int_{\mathbb{T}^3}\chi_{\frac{\tilde{\varrho}}{2}\leq\varrho\leq2\tilde{\varrho}}\varrho\left|\frac{\mathbf{G}_k(\tilde{\varrho},\varrho\mathbf{u})}{\varrho}-\frac{\mathbf{G}_k(\tilde{\varrho},\tilde{\varrho}\tilde{\mathbf{u}})}{\tilde{\varrho}}\right|^2\mathrm{d}x
\end{aligned}
$$

$$
\begin{aligned}
&\le c(L) \int_{\mathbb{T}^3} \chi_{\frac{\tilde\varrho}{2} \le \varrho \le 2\tilde\varrho} (|\varrho - \tilde\varrho|^2 (1 + |\varrho \mathbf{u}|^2) + |\varrho \mathbf{u} - \tilde\varrho \tilde{\mathbf{u}}|^2) \, \mathrm{d}x \\
&\le c(L) \int_{\mathbb{T}^3} \chi_{\frac{\tilde\varrho}{2} \le \varrho \le 2\tilde\varrho} (|\varrho - \tilde\varrho|^2 (1 + |\tilde{\mathbf{u}}|^2) + |\varrho (\mathbf{u} - \tilde{\mathbf{u}})|^2) \, \mathrm{d}x \\
&\le c(L) \int_{\mathbb{T}^3} \chi_{\frac{\tilde\varrho}{2} \le \varrho \le 2\tilde\varrho} |\varrho - \tilde\varrho|^2 \, \mathrm{d}x + \int_{\mathbb{T}^3} \varrho |\mathbf{u} - \tilde{\mathbf{u}}|^2 \, \mathrm{d}x \\
&\le c(L) \int_{\mathbb{T}^3} (P(\varrho) - P'(\tilde\varrho)(\varrho - \tilde\varrho) - P(\tilde\varrho)) \, \mathrm{d}x + \mathcal{E}([\varrho, \mathbf{u}] | [\tilde\varrho, \tilde{\mathbf{u}}]) \\
&\le c(L) \, \mathcal{E}(\varrho, \mathbf{u} | \tilde\varrho, \tilde{\mathbf{u}}).
\end{aligned}
$$

Finally, (6.19) yields

$$
\begin{aligned}
\mathcal{T}_4^3 &\le c(L) \int_{\mathbb{T}^3} \chi_{\varrho \ge 2\tilde\varrho} (\varrho + \varrho |\mathbf{u}|^2 + \varrho |\tilde{\mathbf{u}}|^2) \, \mathrm{d}x \\
&\le c(L) \int_{\mathbb{T}^3} \chi_{\varrho \ge 2\tilde\varrho} (\varrho + \varrho |\mathbf{u} - \tilde{\mathbf{u}}|^2 + \varrho |\tilde{\mathbf{u}}|^2) \, \mathrm{d}x \\
&\le c(L) \int_{\mathbb{T}^3} \chi_{\varrho \ge 2\tilde\varrho} (\varrho^\gamma (1 + |\tilde{\mathbf{u}}|^2) + \varrho |\mathbf{u} - \tilde{\mathbf{u}}|^2) \, \mathrm{d}x \\
&\le c(L) \int_{\mathbb{T}^3} (P(\varrho) - P'(\tilde\varrho)(\varrho - r) - P(r)) \, \mathrm{d}x + \mathcal{E}(\varrho, \mathbf{u} | \tilde\varrho, \tilde{\mathbf{u}}) \\
&\le c(L) \, \mathcal{E}(\varrho, \mathbf{u} | \tilde\varrho, \tilde{\mathbf{u}}).
\end{aligned}
$$

Plugging everything together, we deduce

$$
\mathcal{E}(\varrho, \mathbf{u} | \tilde\varrho, \tilde{\mathbf{u}})(t \wedge \tau_L \wedge \mathfrak{t}) \le M_{\mathrm{RE}}(t \wedge \tau_L \wedge \mathfrak{t}) - M_{\mathrm{RE}}(0) + c(L) \int_0^{t \wedge \tau_L \wedge \mathfrak{t}} \mathcal{E}(\varrho, \mathbf{u} | \tilde\varrho, \tilde{\mathbf{u}}) \, \mathrm{d}t.
$$

Taking expectation and applying Gronwall's lemma, the claim follows. □

6.2.2 Weak–strong uniqueness in law

Strictly speaking, the strong solution and the dissipative martingale solutions of problem (6.1)–(6.2) may not be defined on the same probability space and with the same Wiener process W. However, as a consequence of Theorem 6.2.2, we also obtain weak–strong uniqueness in law.

Theorem 6.2.3. *The weak–strong uniqueness in law holds true for the system* (6.1)–(6.2) *in the following sense. If*

$$
[(\Omega^1, \mathfrak{F}^1, (\mathfrak{F}_t^1)_{t \ge 0}, \mathbb{P}^1), \varrho^1, \mathbf{u}^1, W^1]
$$

is a dissipative martingale solution to system (6.1)–(6.2) *in the sense of Definition* 6.0.1 *and*

$$
[\varrho^2, \mathbf{u}^2, \mathfrak{t}]
$$

is a local strong pathwise solution to (6.1)–(6.2) *in the sense of Definition* 6.2.1 *defined on a stochastic basis* $(\Omega^2, \mathfrak{F}^2, (\mathfrak{F}_t^2)_{t\geq 0}, \mathbb{P}^2)$ *with Wiener process* W^2 *such that*

$$\Lambda = \mathcal{L}[\varrho^1(0), \varrho^1\mathbf{u}^1(0)] = \mathcal{L}[\varrho^2(0), \varrho^2\mathbf{u}^2(0)],$$

then

$$\mathcal{L}[\varrho^1(\cdot \wedge t), \varrho^1\mathbf{u}^1(\cdot \wedge t)] = \mathcal{L}[\varrho^2(\cdot \wedge t), \varrho^2\mathbf{u}^2(\cdot \wedge t)]. \tag{6.25}$$

Proof. The proof is based on ideas of the classical result of Yamada–Watanabe for SDEs as presented for instance by Karatzas–Shreve [KS91, Proposition 3.20]. However, we need to face several substantial difficulties that originate in the complicated structure of system (6.1)–(6.2).

Let $R^1 := \varrho^1 - \varrho^1(0)$, $R^2 := \varrho^2 - \varrho^2(0)$, $\mathbf{Q}^1 := \varrho^1\mathbf{u}^1 - (\varrho^1\mathbf{u}^1)(0)$, $\mathbf{Q}^2 := \varrho^2\mathbf{u}^2 - (\varrho^2\mathbf{u}^2)(0)$. Let M^1 be the real-valued martingale from the energy inequality (6.3) of the dissipative solution $(\varrho^1, \mathbf{u}^1)$ and let $M^2 \equiv 0$. Set

$$\begin{aligned}\Theta := {} & L^\gamma(\mathbb{T}^3) \times L^{\frac{2\gamma}{\gamma+1}}(\mathbb{T}^3) \times C([0,T]; \mathfrak{U}_0) \times C([0,T]) \\ & \times C_w([0,T]; L^\gamma(\mathbb{T}^3)) \times C_w([0,T]; L^{\frac{2\gamma}{\gamma+1}}(\mathbb{T}^3)) \times L^2(0,T; W^{1,2}(\mathbb{T}^3)).\end{aligned}$$

We denote by $\theta = (r_0, \mathbf{q}_0, w, m, r, \mathbf{q}, \mathbf{v})$ a generic element of Θ. Let $\mathfrak{B}_T(\Theta)$ denote the σ-field on Θ, given by

$$\begin{aligned}\mathfrak{B}_T(\Theta) := {} & \mathfrak{B}(L^\gamma(\mathbb{T}^3)) \otimes \mathfrak{B}(L^{\frac{2\gamma}{\gamma+1}}(\mathbb{T}^3)) \otimes \mathfrak{B}(C([0,T]; \mathfrak{U}_0)) \otimes \mathfrak{B}(C([0,T])) \\ & \otimes \mathfrak{B}_T(C_w([0,T]; L^\gamma(\mathbb{T}^3))) \otimes \mathfrak{B}_T(C_w([0,T]; L^{\frac{2\gamma}{\gamma+1}}(\mathbb{T}^3))) \otimes \mathfrak{B}(L^2(0,T; W^{1,2}(\mathbb{T}^3))),\end{aligned}$$

where for a separable Banach space X we denote by $\mathfrak{B}(X)$ its Borel σ-field and by $\mathfrak{B}_T(C_w([0,T];X))$ the σ-field generated by the mappings

$$C_w([0,T];X) \to X, \quad h \mapsto h(s), \quad s \in [0,T].$$

The discussion by Brzeźniak et al. [BOS16, Section 3] shows that

$$(C_w([0,T];X), \mathfrak{B}_T(C_w([0,T];X)))$$

is a Radon space, i.e., every probability measure on $(C_w([0,T];X), \mathfrak{B}_T(C_w([0,T];X)))$ is Radon. Since the same is true for any Polish space equipped with the Borel σ-field and since the topological product of a countable collection of Radon spaces is a Radon space, we deduce that $(\Theta, \mathfrak{B}_T(\Theta))$ is a Radon space. As discussed by Leão et al. [LJFR99, Theorem 3.2], every Radon space enjoys the regular conditional probability property. Namely, if P is a probability measure on $(\Theta, \mathfrak{B}_T(\Theta))$, $(E, \mathcal{E})$ is a measurable space, and

$$\mathfrak{T} : (\Theta, \mathfrak{B}_T(\Theta), P) \to (E, \mathcal{E})$$

is a measurable mapping, then there exists a regular conditional probability with respect to $\mathfrak{T}$, that is, there exists a function $K : E \times \mathfrak{B}_T(\Theta) \to [0,1]$, called a transition probability, such that:

(i) $K(x,\cdot)$ is a probability measure on $\mathfrak{B}_T(\Theta)$, for all $x \in E$;

(ii) $K(\cdot,A)$ is a measurable function on $(E,\mathcal{E})$, for all $A \in \mathfrak{B}_T(\Theta)$;

(iii) for all $A \in \mathfrak{B}_T(\Theta)$ and all $B \in \mathcal{E}$, we have

$$P(A \cap \mathfrak{T}^{-1}(B)) = \int_B K(x,A)\,(\mathfrak{T}_* P)(\mathrm{d}x),$$

where $\mathfrak{T}_* P$ denotes the pushforward measure on $(E,\mathcal{E})$.

Let $j \in \{1,2\}$, let μ^j denote the joint law of

$$(\varrho^j(0), (\varrho^j \mathbf{u}^j)(0), W^j, M^j, R^j, \mathbf{Q}^j, \mathbf{u}^j)$$

on Θ, and let $\mathbb{P}^W$ be the Wiener measure on $C([0,T];\mathfrak{U}_0)$ which also coincides with the projection to w of μ^j. The law of $(r_0, \mathbf{q}_0)$ is Λ and the law of $(r_0, \mathbf{q}_0, w)$ is the product measure $\Lambda \otimes \mathbb{P}^W$ since $(\varrho^j(0), (\varrho^j \mathbf{u}^j)(0))$ is $\mathfrak{F}_0^j$-measurable and W^j is independent of $\mathfrak{F}_0^j$. Furthermore,

$$\mu^j[(r(0), \mathbf{q}(0)) = 0] = 1.$$

Now, everything is in hand to bring the two solutions $(\varrho^1, \mathbf{u}^1, W^1)$ and $(\varrho^2, \mathbf{u}^2, W^2)$ to the same probability space while preserving their joint laws. To this end, we recall that on $(\Theta, \mathfrak{B}_T(\Theta), \mu^j)$ there exists a regular conditional probability with respect to $(r_0, \mathbf{q}_0, w)$, denoted by K^j. Besides, since Θ is a product space and $(r_0, \mathbf{q}_0, w)$ is the projection to the first three coordinates, we regard K^j as a function on

$$\begin{aligned}
&[L^\gamma(\mathbb{T}^3) \times L^{\frac{2\gamma}{\gamma+1}}(\mathbb{T}^3) \times C([0,T];\mathfrak{U}_0)] \\
&\quad \times [\mathfrak{B}(C[0,T]) \otimes \mathfrak{B}_T(C_w([0,T];L^\gamma(\mathbb{T}^3))) \\
&\quad \otimes \mathfrak{B}_T(C_w([0,T];L^{\frac{2\gamma}{\gamma+1}}(\mathbb{T}^3))) \otimes \mathfrak{B}(L^2(0,T;W^{1,2}(\mathbb{T}^3)))].
\end{aligned}$$

Property (iii) above rewrites as follows. Let

$$A_1 \in \mathfrak{B}(L^\gamma(\mathbb{T}^3)) \otimes \mathfrak{B}(L^{\frac{2\gamma}{\gamma+1}}(\mathbb{T}^3)) \otimes \mathfrak{B}(C([0,T];\mathfrak{U}_0))$$

and

$$\begin{aligned}
A_2 \in\ &\mathfrak{B}(C([0,T])) \otimes \mathfrak{B}_T(C_w([0,T];L^\gamma(\mathbb{T}^3))) \\
&\otimes \mathfrak{B}_T(C_w([0,T];L^{\frac{2\gamma}{\gamma+1}}(\mathbb{T}^3))) \otimes \mathfrak{B}(L^2(0,T;W^{1,2}(\mathbb{T}^3))).
\end{aligned}$$

Then

$$\mu^j[A_1 \times A_2] = \int_{A_1} K^j(r_0, \mathbf{q}_0, w, A_2)\Lambda(\mathrm{d}(r_0, \mathbf{q}_0))\mathbb{P}^W(\mathrm{d}w). \tag{6.26}$$

Finally, we define

$$\Omega := \Theta \times C([0,T]) \times C_w([0,T];L^\gamma(\mathbb{T}^3)) \times C_w([0,T];L^{\frac{2\gamma}{\gamma+1}}(\mathbb{T}^3)) \times L^2(0,T;W^{1,2}(\mathbb{T}^3))$$

and denote by $\mathfrak{F}$ the σ-field on Ω given as the completion of

$$\begin{aligned}&\mathfrak{B}_T(\Theta) \otimes \mathfrak{B}(C[0,T]) \otimes \mathfrak{B}_T(C_w([0,T];L^\gamma(\mathbb{T}^3)))\\ &\quad \otimes \mathfrak{B}_T(C_w([0,T];L^{\frac{2\gamma}{\gamma+1}}(\mathbb{T}^3))) \otimes \mathfrak{B}(L^2(0,T;W^{1,2}(\mathbb{T}^3))),\end{aligned}$$

with respect to the probability measure

$$\begin{aligned}\mathbb{P}(\mathrm{d}\omega) :=\, & K^1(r_0,\mathbf{q}_0,w,\mathrm{d}(m_1,r_1,\mathbf{q}_1,\mathbf{v}_1))\\ &\times K^2(r_0,\mathbf{q}_0,w,\mathrm{d}(m_2,r_2,\mathbf{q}_2,\mathbf{v}_2))\Lambda(\mathrm{d}(r_0,\mathbf{q}_0))\mathbb{P}^W(\mathrm{d}w).\end{aligned} \tag{6.27}$$

Here we have denoted by $\omega = (r_0,\mathbf{q}_0,w,m_1,r_1,\mathbf{q}_1,\mathbf{v}_1,m_2,r_2,\mathbf{q}_2,\mathbf{v}_2)$ a canonical element of Ω. In order to endow $(\Omega,\mathfrak{F},\mathbb{P})$ with a filtration that satisfies the usual conditions, we take

$$\begin{aligned}&\mathfrak{G}_t := \sigma((r_0,\mathbf{q}_0,w(s),m_1(s),r_1(s),\mathbf{q}_1(s),\mathbf{v}_1(s),m_2(s),r_2(s),\mathbf{q}_2(s),\mathbf{v}_2(s));\, 0\le s\le t),\\ &\tilde{\mathfrak{G}}_t := \sigma(\mathfrak{G}_t \cup \{N;\, \mathbb{P}(N)=0\}), \quad \mathfrak{F}_t := \bigcap_{\varepsilon\in(0,T-t)} \tilde{\mathfrak{G}}_{t+\varepsilon}, \quad t\in[0,T).\end{aligned}$$

Then, from (6.27) and (6.26), it follows that

$$\begin{aligned}&\mathbb{P}[\omega\in\Omega;\, (r_0,\mathbf{q}_0,w,m_j,r_j,\mathbf{q}_j,\mathbf{v}_j)\in A_1\times A_2]\\ &\quad = \int_{A_1\times A_2} K^j(r_0,\mathbf{q}_0,w,\mathrm{d}(m_j,r_j,\mathbf{q}_j,\mathbf{v}_j))\Lambda(\mathrm{d}(r_0,\mathbf{q}_0))\mathbb{P}^W(\mathrm{d}w)\\ &\quad = \int_{A_1} K^j(r_0,\mathbf{q}_0,w,A_2)\Lambda(\mathrm{d}(r_0,\mathbf{q}_0))\mathbb{P}^W(\mathrm{d}w)\\ &\quad = \mu^j[A_1\times A_2]\\ &\quad = \mathbb{P}^j[(\varrho^j(0),(\varrho^j\mathbf{u}^j)(0),W^j,M^j,R^j,\mathbf{Q}^j,\mathbf{u}^j)\in A_1\times A_2].\end{aligned}$$

Hence the law of $(r_0,\mathbf{q}_0,w,m_j,r_j,\mathbf{q}_j,\mathbf{v}_j)$ under $\mathbb{P}$ coincides with the law of

$$(\varrho^j(0),(\varrho^j\mathbf{u}^j)(0),W^j,M^j,R^j,\mathbf{Q}^j,\mathbf{u}^j)$$

under $\mathbb{P}^j$. As a consequence, the law of $(r_0+r_j,\mathbf{q}_0+\mathbf{q}_j,\mathbf{v}_j,w,m_j)$ under $\mathbb{P}$ coincides with the law of $(\varrho^j,\varrho^j\mathbf{u}^j,\mathbf{u}^j,W^j,M^j)$ under $\mathbb{P}^j$. In particular, w is an $(\mathfrak{F}_t)$-cylindrical Wiener process.

To summarize, we have defined a stochastic basis $(\Omega,\mathfrak{F},(\mathfrak{F}_t)_{t\ge0},\mathbb{P})$ with random variables $(r_0+r_j,\mathbf{q}_0+\mathbf{q}_j,\mathbf{v}_j,w)$ that have the same law as the original solutions $(\varrho^j,\varrho^j\mathbf{u}^j,\mathbf{u}^j,W^j)$, $j=1,2$. As a consequence,

$$\mathbb{P}[\mathbf{q}_0+\mathbf{q}_j = (r_0+r_j)\mathbf{v}_j] = 1$$

and $(r_0 + r_j, \mathbf{q}_0 + \mathbf{q}_j, \mathbf{v}_j, w)$ solves (6.1)–(6.2) in the weak sense. This can be verified by Theorem 2.9.1; see also the proof of Propositions 4.3.14 and 4.4.12. Besides, since the law of $(\varrho^2, \mathbf{u}^2)$ is actually supported on a space of functions with higher regularity (see Definition 5.0.1) and $\varrho^2 > 0$, we deduce that $(r_0 + r_2, \mathbf{v}_2, w)$ is a strong solution to (6.1)–(6.2).

By the same reasoning as in Remark 6.0.2, we obtain the following version of the energy inequality (6.3), which holds true for all $0 \le s \le t \le T$, $\mathbb{P}^1$-a.s.:

$$
\begin{aligned}
&\int_{\mathbb{T}^3} \Big[\frac{1}{2}\varrho^1|\mathbf{u}^1|^2 + P(\varrho^1)\Big](t)\,\mathrm{d}x + \int_s^t \int_{\mathbb{T}^3} \mathbb{S}(\nabla_x \mathbf{u}^1) : \nabla_x \mathbf{u}^1 \,\mathrm{d}x\,\mathrm{d}r \\
&\le \int_{\mathbb{T}^3} \Big[\frac{|(\varrho^1\mathbf{u}^1)(s^+)|^2}{2\varrho^1(s^+)} + P(\varrho^1(s^+))\Big]\,\mathrm{d}x + \frac{1}{2}\int_s^t \Big(\int_{\mathbb{T}^3} \sum_{k=1}^{\infty} \frac{|\mathbf{G}_k(\varrho^1, \varrho^1\mathbf{u}^1)|^2}{\varrho^1}\,\mathrm{d}x\Big)\,\mathrm{d}r \\
&\quad + M^1(t) - M^1(s).
\end{aligned}
$$

Hence the equality of joint laws of $(r_0 + r_1, \mathbf{q}_0 + \mathbf{q}_1, \mathbf{v}_1, m_1)$ and $(\varrho^1, \varrho^1\mathbf{u}^1, \mathbf{u}^1, M^1)$ implies the corresponding inequality satisfied by $(r_0 + r_1, \mathbf{q}_0 + \mathbf{q}_1, \mathbf{v}_1, m_1)$. Since, in view of Remark 6.0.2, this is exactly the version of (6.3) that is used in the proof of pathwise weak–strong uniqueness, Theorem 6.2.2 applies and yields

$$
\mathbb{P}[r_0 + r_1 = r_0 + r_2,\ \mathbf{q}_0 + \mathbf{q}_1 = \mathbf{q}_0 + \mathbf{q}_2] = 1
$$

or, equivalently,

$$
\mathbb{P}\big[\omega = (r_0, \mathbf{q}_0, \mathbf{w}, m_1, r_1, \mathbf{q}_1, \mathbf{v}_1, m_2, r_2, \mathbf{q}_2, \mathbf{v}_2) \in \Omega;\ r_1 = r_2,\ \mathbf{q}_1 = \mathbf{q}_2\big] = 1.
$$

Hence, for all $A \in \mathfrak{B}_T(C_w([0,T]; L^\gamma(\mathbb{T}^3))) \otimes \mathfrak{B}_T(C_w([0,T]; L^{\frac{2\gamma}{\gamma+1}}(\mathbb{T}^3)))$,

$$
\begin{aligned}
\mathbb{P}^1\big[(\varrho^1, \varrho^1\mathbf{u}^1) \in A\big] &= \mathbb{P}\big[\omega \in \Omega;\ (r_0 + r_1, \mathbf{q}_0 + \mathbf{q}_1) \in A\big] \\
&= \mathbb{P}\big[\omega \in \Omega;\ (r_0 + r_2, \mathbf{q}_0 + \mathbf{q}_2) \in A\big] \\
&= \mathbb{P}^2\big[(\varrho^2, \varrho^2\mathbf{u}^2) \in A\big]
\end{aligned}
$$

and (6.25) follows. □

Part III: **Applications**

7 Stationary solutions

The main goal of this chapter is to show the existence of stationary solutions to

$$\mathrm{d}\varrho + \operatorname{div}(\varrho\mathbf{u})\,\mathrm{d}t = 0, \tag{7.1}$$

$$\mathrm{d}(\varrho\mathbf{u}) + \operatorname{div}(\varrho\mathbf{u}\otimes\mathbf{u})\,\mathrm{d}t + \nabla p(\varrho)\,\mathrm{d}t = \operatorname{div}\mathbb{S}(\nabla\mathbf{u})\,\mathrm{d}t + \mathbb{G}(\varrho,\varrho\mathbf{u})\,\mathrm{d}W, \tag{7.2}$$

where $p(\varrho) = a\varrho^{\gamma}$, $a > 0$, in the framework of dissipative martingale solutions constructed in Chapter 4. To this end, we use a direct method based on a multi-layer approximation scheme similar to Chapter 4. Nevertheless, the uniform estimates necessary for the existence theory are in general not suitable to study the long-time behavior of the system. They are based on the application of Gronwall's lemma and therefore grow exponentially with the final time T. Hence, the major challenge is to derive new estimates which are uniform with respect to all the approximation parameters as well as in T. This is the heart of the construction. Let us point out that the standard methods used for the incompressible system, as for instance by Flandoli–Gątarek [FG95] and Flandoli–Romito [FR08], are not applicable in the compressible case. Indeed, system (7.1)–(7.2) is of mixed hyperbolic-parabolic type and the dissipation term does not contain the density. Consequently, the forcing terms on the right hand side of the energy balance cannot be absorbed in the dissipative term appearing on the left hand side in an obvious and straightforward manner.

Furthermore, it does not seem to be possible to find universal estimates that would be uniform in all the parameters R, m, ε, δ as well as in T. Instead, during each approximation step we develop new estimates which are then used for the particular passage to the limit at hand. More precisely, at the starting level, that is for fixed parameters $R, m \in \mathbb{N}$, $\varepsilon, \delta > 0$, we show existence, uniqueness, and continuous dependence on the initial condition. Thus, the resulting system is Markovian and the transition semigroup is Feller. Consequently, the existence of invariant measures can be shown with the help of the standard Krylov–Bogoliubov method in the infinite-dimensional setting (see Section 2.12). This generates a family of approximate stationary solutions. Note that we lose uniqueness already after the first passage to the limit (in R). Hence, the usual Krylov–Bogoliubov approach cannot be employed anymore and even the concept of invariant measure becomes ambiguous. To overcome this problem, we construct stationary solutions on the next level as limits of the corresponding approximate stationary solutions from the previous level.

At each approximation step, there are essentially three necessary estimates: for the energy, the velocity, and the pressure. At the deepest level, we are able to obtain the first two estimates uniformly in R, m, but the third one depends on all the parameters R, m, ε, δ and is therefore not suitable for any limit procedure. The key observation is that these estimates may be significantly improved if we take stationarity into account. Therefore, working directly with stationary solutions given by the Krylov–Bogoliubov

https://doi.org/10.1515/9783110492552-007

method, we derive an estimate for the energy as well as the velocity, which is uniform in all the approximation parameters. The estimate for the pressure is more delicate and has to be reproved at each level by applying a suitable test function to (7.1)–(7.2). The proof is then concluded by performing the limit for vanishing approximation parameters based on a combination of deterministic and probabilistic tools, similarly to Chapter 4.

It is remarkable that our result holds for the same range of the adiabatic exponent $\gamma > 3/2$ as in the nowadays available existence theory. Note that the relevant deterministic problem, namely the existence of bounded absorbing sets and attractors, requires a rather inconvenient technical restriction $\gamma > 5/3$; see [Fei00, FP01]. Indeed, consider the iconic example of the driving force $\varrho\mathbf{f}(x)\,\mathrm{d}W$ in (7.2). If we replace it by the deterministic forcing $\varrho\mathbf{f}(x)\,\mathrm{d}t$, then, to the best of our knowledge, it is not known if the global-in-time weak solutions remain uniformly bounded for $t \to \infty$ for γ in the physically relevant range $1 \leq \gamma \leq 5/3$. On the other hand, the stochastic forcing $\varrho\mathbf{f}(x)\,\mathrm{d}W$ gives rise to stationary solutions for any $\gamma > 3/2$, as shown in Theorem 7.0.3. The reason is the cancellation of certain terms in the energy balance due to stochastic averaging. We therefore observe a kind of regularizing effect due to the presence of noise. Note, however, that the growth conditions imposed on the diffusion coefficients $\mathbb{G}(\varrho, \varrho\mathbf{u})$ (see (7.9) below) appearing in the driving term are more restrictive than in Chapter 4.

In comparison to the result of Chapter 4, the existence of stationary solutions requires somewhat stronger assumptions on the model. As above, we consider the periodic boundary conditions, where the physical domain may be identified with the flat torus

$$\mathbb{T}^3 \equiv \left([-1,1]|_{\{-1,1\}}\right)^3.$$

However, our method leans essentially on the dissipative effect of the viscosity, represented by the viscous stress $\mathbb{S}$ in (3.2). In particular, it is convenient to keep a kind of Korn–Poincaré inequality in force. Following the idea of Ebin [Ebi83], we consider the physically relevant *complete-slip* conditions

$$\mathbf{u} \cdot \mathbf{n}|_{\partial\mathcal{O}} = 0, \quad [\mathbb{S}(\nabla\mathbf{u}) \cdot \mathbf{n}] \times \mathbf{n}|_{\partial\mathcal{O}} = 0, \tag{7.3}$$

imposed on the boundary of the cube

$$\mathcal{O} = [0,1]^3.$$

The crucial observation is that the constraint (7.3) is automatically satisfied by *periodic* functions ϱ, $\mathbf{u}$ defined on torus $\mathbb{T}^3$ and belonging to the symmetry class

$$\begin{aligned} \varrho(t,-x) &= \varrho(t,x) \quad x \in \mathbb{T}^3, \\ u^i(t,\cdot,-x_i,\cdot) &= -u^i(t,\cdot,x_i,\cdot) \quad i = 1,2,3, \\ u^i(t,\cdot,-x_j,\cdot) &= u^i(t,\cdot,x_j,\cdot) \quad i \neq j,\ i,j = 1,2,3; \end{aligned} \tag{7.4}$$

cf. [Ebi83]. In such a way, we may eliminate the problems connected to the presence of a physical boundary by considering periodic functions defined on $\mathbb{T}^3$ and belonging, in addition, to the symmetry class (7.4). Note that, for $\mathbf{u}$ in the class (7.4), we have the Korn–Poincaré inequality

$$\int_{\mathbb{T}^3} \mathbb{S}(\nabla \mathbf{u}) : \nabla \mathbf{u} \, \mathrm{d}x \geq c_{\mathrm{KP}} \|\mathbf{u}\|^2_{W^{1,2}_x}; \tag{7.5}$$

see Remark A.1.9.

In order to complement the above, we prescribe the total mass

$$\int_{\mathbb{T}^3} \varrho(t,x) \, \mathrm{d}x = M_0, \quad t \in [0,\infty), \tag{7.6}$$

where $M_0 > 0$ is a deterministic constant. The assumption that M_0 is deterministic is taken for simplicity, in order to avoid unnecessary technicalities. A more general case of random M_0 satisfying

$$\underline{m} \leq M_0 \leq \overline{m} \quad \mathbb{P}\text{-a.s.} \tag{7.7}$$

for some deterministic constants $\underline{m}, \overline{m} \in (0,\infty)$ can also be considered. In that case, one would prescribe the law of M_0, so that (7.7) holds.

As before, the stochastic integral in (3.2) is understood in the following sense:

$$\mathbb{G}(\varrho, \varrho\mathbf{u}) \, \mathrm{d}W = \sum_{k=1}^{\infty} \mathbf{G}_k(x, \varrho, \varrho\mathbf{u}) \, \mathrm{d}W_k.$$

In agreement with (7.4), we suppose that the coefficients $\mathbf{G}_k(x, \varrho, \mathbf{q})$ satisfy

$$\begin{aligned} G^i_k(\cdot, -x_i, \cdot, -q^i, \cdot) &= -G^i_k(\cdot, x_i, \cdot, q^i, \cdot), \quad i = 1,2,3, \\ G^i_k(\cdot, -x_j, \cdot, -q^j, \cdot) &= G^i_k(\cdot, x_j, \cdot, q^j, \cdot), \quad i \neq j, \; i,j = 1,2,3. \end{aligned} \tag{7.8}$$

The reason for (7.8) is to keep the spatially periodic solutions in the symmetry class (7.4) as long as the initial data belong to (7.4) $\mathbb{P}$-a.s. More specifically, under the hypothesis (7.8) all non-linearities in (7.1) and (7.2) map the class of functions satisfying (7.4) into itself. Accordingly, a direct inspection of the existence proof reveals that the solutions ϱ, $\mathbf{u}$ constructed in the proof of Theorem 4.0.2 will remain in (7.4) for any time as long as the initial data ϱ_0, $\mathbf{u}_0$ belong to (7.4).

As discussed in Section 3.5, our approach to the construction of stationary solutions relies on the concept of dissipative martingale solution introduced in Section 3.4. However, the problem of finding stationary solutions to (7.1)–(7.2) is very different from the evolutionary Cauchy problem. Indeed, the initial law becomes irrelevant and the global-in-time bounds cannot be controlled by the initial condition. Hence, we give a definition of a dissipative martingale solution which is adapted to our purposes.

Definition 7.0.1 (Dissipative martingale solution). The quantity

$$((\Omega, \mathfrak{F}, (\mathfrak{F}_t)_{t\geq 0}, \mathbb{P}), \varrho, \mathbf{u}, W)$$

is called a *dissipative martingale solution* to (7.1)–(7.2) if:

(1) $(\Omega, \mathfrak{F}, (\mathfrak{F}_t)_{t\geq 0}, \mathbb{P})$ is a stochastic basis with a complete right-continuous filtration;

(2) W is a cylindrical $(\mathfrak{F}_t)$-Wiener process;

(3) the density ϱ and the velocity $\mathbf{u}$ are random distributions adapted to $(\mathfrak{F}_t)_{t\geq 0}$, $\varrho \geq 0$ $\mathbb{P}$-a.s.;

(4) the equation of continuity

$$-\int_0^\infty \partial_t \phi \int_{\mathbb{T}^3} \varrho \psi \,\mathrm{d}x\,\mathrm{d}t = \int_0^\infty \phi \int_{\mathbb{T}^3} \varrho \mathbf{u} \cdot \nabla \psi \,\mathrm{d}x\,\mathrm{d}t$$

holds for all $\phi \in C_c^\infty((0,\infty))$ and all $\psi \in C^\infty(\mathbb{T}^3)$ $\mathbb{P}$-a.s.;

(5) the momentum equation

$$\begin{aligned} &-\int_0^\infty \partial_t \phi \int_{\mathbb{T}^3} \varrho \mathbf{u} \cdot \boldsymbol{\varphi} \,\mathrm{d}x\,\mathrm{d}t \\ &= \int_0^\infty \phi \int_{\mathbb{T}^3} [\varrho \mathbf{u} \otimes \mathbf{u} : \nabla \boldsymbol{\varphi} + p(\varrho) \operatorname{div} \boldsymbol{\varphi}] \,\mathrm{d}x\,\mathrm{d}t - \int_0^\infty \phi \int_{\mathbb{T}^3} \mathbb{S}(\nabla \mathbf{u}) : \nabla \boldsymbol{\varphi} \,\mathrm{d}x\,\mathrm{d}t \\ &\quad + \sum_{k=1}^\infty \int_0^\infty \phi \int_{\mathbb{T}^3} \mathbf{G}_k(\varrho, \varrho \mathbf{u}) \cdot \boldsymbol{\varphi} \,\mathrm{d}x\,\mathrm{d}W_k \end{aligned}$$

holds for all $\phi \in C_c^\infty((0,\infty))$ and all $\boldsymbol{\varphi} \in C^\infty(\mathbb{T}^N)$ $\mathbb{P}$-a.s.;

(6) the energy inequality

$$\begin{aligned} &-\int_0^\infty \partial_t \phi \int_{\mathbb{T}^3} \Big[\frac{1}{2} \varrho |\mathbf{u}|^2 + P(\varrho)\Big] \,\mathrm{d}x\,\mathrm{d}t + \int_0^\infty \phi \int_{\mathbb{T}^3} \mathbb{S}(\nabla \mathbf{u}) : \nabla \mathbf{u} \,\mathrm{d}x\,\mathrm{d}t \\ &\leq \frac{1}{2} \sum_{k=1}^\infty \int_0^\infty \phi \int_{\mathbb{T}^3} \varrho^{-1} |\mathbf{G}_k(\varrho, \varrho \mathbf{u})|^2 \,\mathrm{d}x\,\mathrm{d}t + \sum_{k=1}^\infty \int_0^\infty \phi \int_{\mathbb{T}^3} \mathbf{G}_k(\varrho, \varrho \mathbf{u}) \cdot \mathbf{u} \,\mathrm{d}x\,\mathrm{d}W_k \end{aligned}$$

holds for all $\phi \in C_c^\infty((0,\infty))$, $\phi \geq 0$, $\mathbb{P}$-a.s., with the pressure potential given by

$$P(\varrho) = \varrho \int_1^\varrho \frac{p(z)}{z^2} \,\mathrm{d}z;$$

(7) if $b \in C^1(\mathbb{R})$ such that $b'(z) = 0$ for all $z \geq M_b$, then, for all $\phi \in C_c^\infty((0,\infty))$ and all $\psi \in C^\infty(\mathbb{T}^3)$, we have $\mathbb{P}$-a.s.

$$\begin{aligned} -\int_0^\infty \partial_t \phi \int_{\mathbb{T}^3} b(\varrho) \psi \,\mathrm{d}x\,\mathrm{d}t &= \int_0^\infty \phi \int_{\mathbb{T}^3} b(\varrho) \mathbf{u} \cdot \nabla \psi \,\mathrm{d}x\,\mathrm{d}t \\ &\quad - \int_0^\infty \phi \int_{\mathbb{T}^3} (b'(\varrho)\varrho - b(\varrho)) \operatorname{div} \mathbf{u}\, \psi \,\mathrm{d}x\,\mathrm{d}t. \end{aligned}$$

Note that, as opposed to Definition 3.4.1, the initial state does not play any role and we consider test functions in time that are compactly supported in $(0,\infty)$.

In accordance with the available *a priori* bounds provided by the energy estimates, a suitable state space for $[\varrho, \varrho\mathbf{u}]$ is

$$\varrho \in L^{\gamma}(\mathbb{T}^3), \quad \varrho\mathbf{u} \in L^{\frac{2\gamma}{\gamma+1}}(\mathbb{T}^3),$$

where γ is the adiabatic exponent in the state equation (7.2). Accordingly, we consider initial laws Λ defined on the Borel σ-algebra of the product space $L^{\gamma}(\mathbb{T}^3) \times L^{\frac{2\gamma}{\gamma+1}}(\mathbb{T}^3)$. In addition, we say that a dissipative martingale solution $[\varrho, \mathbf{u}]$ satisfies the complete-slip boundary conditions (7.3), if $[\varrho(t,\cdot), \varrho\mathbf{u}(t,\cdot)]$ belong to the symmetry class (7.4) for any $t \in [0,T]$ $\mathbb{P}$-a.s.

Finally, everything is in hand to define the notion of a stationary solution to (7.1)–(7.2).

Definition 7.0.2. A dissipative martingale solution $[\varrho, \mathbf{u}, W]$ to (7.1)–(7.2) in the sense of Definition 7.0.1 is called *stationary*, provided the joint law of the time shift $[\mathcal{S}_\tau\varrho, \mathcal{S}_\tau\mathbf{u}, \mathcal{S}_\tau W - W(\tau)]$ on

$$L^1_{\rm loc}(0,\infty; L^{\gamma}(\mathbb{T}^3)) \times L^1_{\rm loc}(0,\infty; W^{1,2}(\mathbb{T}^3)) \times C_{\rm loc}([0,\infty); \mathfrak{U}_0)$$

is independent of $\tau \geq 0$.

Our result then reads as follows.

Theorem 7.0.3. *Let $M_0 \in (0,\infty)$ be given and let $\gamma > \frac{3}{2}$. Suppose that the diffusion coefficients $\mathbf{G}_k$ belong to the symmetry class* (7.8) *and there exist functions $\mathbf{F}_k \in C^1(\mathbb{T}^3 \times (0,\infty) \times \mathbb{R}^3)$ and constants $f_k \geq 0$, $k \in \mathbb{N}$, such that*

$$\begin{aligned} \mathbf{G}_k(x,\varrho,\mathbf{q}) &= \varrho\,\mathbf{F}_k(x,\varrho,\mathbf{u}), \\ |\nabla_{\mathbf{u}}\mathbf{F}_k(x,\varrho,\mathbf{u})| + |\mathbf{F}_k(x,\varrho,\mathbf{u})| &\leq f_k, \quad \sum_{k=1}^{\infty} f_k^2 = F < \infty. \end{aligned} \tag{7.9}$$

Then problem (7.1)–(7.2), (7.3), *and* (7.6) *admits a stationary martingale solution $[\varrho, \mathbf{u}, W]$ in the sense of Definition* 7.0.2, *satisfying the complete-slip boundary conditions* (7.3).

Note that, if for instance $\mathbf{G}_k(x,\varrho,0) = 0$ for all $x \in \mathbb{T}^3$, $\varrho \in [0,\infty)$, $k \in \mathbb{N}$, (7.1)–(7.2) admits a trivial stationary solution, namely, $\mathbf{u} \equiv 0$ and $\varrho \equiv$ const. Nevertheless, Theorem 7.0.3 applies to more general diffusion coefficients $\mathbf{G}_k$ where such trivial solutions do not exist.

We point out that the moments in (3.27)–(3.29) for stationary solutions can only be estimated up to a certain order $p > 1$. This is in contrast to the existence theory from Chapter 4, where arbitrarily large moments are finite, provided the initial data posses sufficient integrability.

The rest of this chapter is devoted to the proof of Theorem 7.0.3. In Section 7.1, we introduce the basic finite-dimensional approximation and construct a family of approximate solutions adapting the standard Krylov–Bogoliubov method. In Section 7.2, we develop global-in-time estimates for stationary solutions and pass to the limit $R \to \infty$ and $m \to \infty$. Section 7.3 is devoted to the vanishing viscosity limit, i.e., $\varepsilon \to 0$. Finally, in Section 7.4, we perform the limit for vanishing artificial pressure, i.e., $\delta \to 0$, obtaining the desired stationary solution to (7.1)–(7.2).

7.1 Basic finite-dimensional approximation

In this section, we introduce the zero-level approximate system to (7.1)–(7.2) and study its long-time behavior for suitable initial data belonging to the symmetry class (7.4). More precisely, based on an energy estimate, Proposition 7.1.1, and bounds for the density, Lemma 7.1.2, we apply the Krylov–Bogoliubov method to deduce the existence of an invariant measure.

We stress that, in accordance with hypothesis (7.8), the solutions can be constructed to be spatially periodic solutions, i.e., they belong to the symmetry class (7.4), as long as the initial data belong to the same class (7.4). We always tacitly assume this fact without specifying it explicitly in the future.

Let

$$H_m = \left\{ v = \sum_{\mathbf{m}, \max_{j=1,2,3} |m_j| \le m} [a_{\mathbf{m}} \cos(\pi \mathbf{m} \cdot x) + b_{\mathbf{m}} \sin(\pi \mathbf{m} \cdot x)] \mid a_{\mathbf{m}}, b_{\mathbf{m}} \in \mathbb{R} \right\}^3$$

be the space of trigonometric polynomials of order m, endowed with the Hilbert structure of the Lebesgue space $L^2(\mathbb{T}^3)$ and let $\|\cdot\|_{H_m}$ denote the corresponding norm. Let

$$\Pi_m : L^2(\mathbb{T}^3) \to H_m$$

be the associated L^2-orthogonal projection. Recall that the following holds:

$$\|\Pi_m v\|_{L^p_x} \le c_p \|v\|_{L^p_x} \quad \forall v \in L^p(\mathbb{T}^3) \tag{7.10}$$

and

$$\Pi_m v \to v \quad \text{in } L^p(\mathbb{T}^3),$$

for any $p \in (1, \infty)$; cf. Grafakos [Gra08, Chapter 3].

7.1.1 Approximate field equations

Fix $R \in \mathbb{N}$, $m \in \mathbb{N}$, $\varepsilon > 0$, $\delta > 0$ and let $\Gamma > \max\{\frac{9}{2}, \gamma\}$. The approximate solutions $\varrho = \varrho_m$, $\mathbf{u} = \mathbf{u}_m$, $\mathbf{u}_m(t) \in H_m$ for any t are constructed to satisfy the following system of equations:

$$
\begin{aligned}
&\mathrm{d}\varrho + \operatorname{div}(\varrho[\mathbf{u}]_R)\,\mathrm{d}t = \varepsilon\Delta\varrho\,\mathrm{d}t - 2\varepsilon\varrho\,\mathrm{d}t + \chi\Big(\frac{1}{M_0}\int_{\mathbb{T}^3}\varrho\,\mathrm{d}x\Big)\,\mathrm{d}t,\\
&\mathrm{d}\int_{\mathbb{T}^3}\varrho\mathbf{u}\cdot\boldsymbol{\varphi}\,\mathrm{d}x - \int_{\mathbb{T}^3}\varrho[\mathbf{u}]_R\otimes\mathbf{u}:\nabla\boldsymbol{\varphi}\,\mathrm{d}x\,\mathrm{d}t - \int_{\mathbb{T}^3}a\varrho^\gamma\chi(\|\mathbf{u}\|_{H_m}-R)\operatorname{div}\boldsymbol{\varphi}\,\mathrm{d}x\,\mathrm{d}t\\
&\quad= -\int_{\mathbb{T}^3}\mathbb{S}(\nabla\mathbf{u}):\nabla\boldsymbol{\varphi}\,\mathrm{d}x\,\mathrm{d}t + \sum_{k=1}^{\infty}\int_{\mathbb{T}^3}\varrho\Pi_m\mathbf{F}_k(\varrho,\mathbf{u})\cdot\boldsymbol{\varphi}\,\mathrm{d}x\,\mathrm{d}W_k\\
&\qquad+\varepsilon\int_{\mathbb{T}^3}\varrho\mathbf{u}\cdot\Delta\boldsymbol{\varphi}\,\mathrm{d}x\,\mathrm{d}t - \varepsilon\int_{\mathbb{T}^3}\varrho\mathbf{u}\cdot\boldsymbol{\varphi}\,\mathrm{d}x\,\mathrm{d}t + \delta\int_{\mathbb{T}^3}\varrho^\Gamma\chi(\|u\|_{H_m}-R)\operatorname{div}\boldsymbol{\varphi}\,\mathrm{d}x\,\mathrm{d}t,
\end{aligned}
\tag{7.11}
$$

for any test function $\varphi \in H_m$, where

$$[\mathbf{u}]_R = \chi(\|\mathbf{u}\|_{H_m} - R)\mathbf{u},$$

with

$$\chi \in C^\infty(\mathbb{R}),\quad \chi = \begin{cases} 1 & \text{on } (-\infty, 0],\\ \text{a decreasing function on } (0,1), \\ 0 & \text{on } [1,\infty).\end{cases}$$

Similar to before, we use the following notation for the corresponding energy:

$$E_\delta(\varrho,\varrho\mathbf{u}) := \frac{1}{2}\frac{|\varrho\mathbf{u}|^2}{\varrho} + \frac{a}{\gamma-1}\varrho^\gamma + \frac{\delta}{\Gamma-1}\varrho^\Gamma.$$

Note that the basic approximate system (7.11) is not the same as the one from Chapter 4, namely, (4.14)–(4.15). To be more precise, in order to obtain global-in-time estimates we are forced to include two more "stabilizing" terms in the continuity equation and to modify the momentum equation accordingly. Nevertheless, similarly to Section 4.1, it can be shown that problem (7.11) admits a unique strong pathwise solution for any $\mathfrak{F}_0$-measurable initial data $[\varrho_0, (\varrho\mathbf{u})_0]$ satisfying, for some $\nu > 0$,

$$
\begin{aligned}
&\varrho_0 \in C^{2+\nu}(\mathbb{T}^3),\ 0 < \underline{\varrho} < \varrho_0 < \overline{\varrho},\quad (\varrho\mathbf{u})_0 \in C^2(\mathbb{T}^3)\quad \mathbb{P}\text{-a.s.},\\
&\mathbb{E}\Big[\Big(\int_{\mathbb{T}^3}E_\delta(\varrho_0,(\varrho\mathbf{u})_0)\,\mathrm{d}x\Big)^n\Big] \le c(n)\quad \text{for all } 1 \le n < \infty,
\end{aligned}
\tag{7.12}
$$

where $\underline{\varrho}$, $\overline{\varrho}$ are deterministic constants and where the associated initial value of $\mathbf{u}$ is uniquely determined by

$$\mathbf{u}_0 \in H_m,\quad \int_{\mathbb{T}^3}\varrho_0\mathbf{u}_0\cdot\boldsymbol{\varphi}\,\mathrm{d}x = \int_{\mathbb{T}^3}(\varrho\mathbf{u})_0\cdot\boldsymbol{\varphi}\,\mathrm{d}x\quad \text{for all } \boldsymbol{\varphi} \in H_m.$$

7.1.2 Basic energy estimates

The energy estimates obtained from the corresponding energy balance similar to (4.55) are not well-suited for the construction of stationary solutions. Indeed, the application of Gronwall's lemma leads to an in time exponentially growing right hand side. In this Section we derive improved energy estimates which overcome this problem and hold true uniformly in t. However, it is important to note that, at this stage of the proof, we are not able to obtain estimates independent of all the approximation parameters, namely, the following bounds blow up as $\varepsilon \to 0$. The necessary uniform estimates for the passage to the limit in ε will be derived directly for stationary solutions in Section 7.3.

Proposition 7.1.1. *Let* $(\varrho, \mathbf{u})$ *be a solution to* (7.11) *starting from*

$$\varrho_0 = 1, \quad (\varrho \mathbf{u})_0 = \mathbf{u}_0 = 0. \tag{7.13}$$

Then the following bounds hold true:

$$\mathbb{E}\Big[\Big(\int_{\mathbb{T}^3} E_\delta(\varrho, \varrho\mathbf{u})(\tau, x)\,\mathrm{d}x\Big)^n\Big] \leq c(n, \varepsilon, F), \quad n \in \mathbb{N}, \tag{7.14}$$

$$\frac{1}{T}\mathbb{E}\Big[\int_0^T \Big(\|\mathbf{u}\|^2_{W^{1,2}_x} + \frac{2a\varepsilon}{\gamma}\|\nabla \varrho^{\gamma/2}\|^2_{L^2_x} + \frac{2\delta\varepsilon}{\Gamma}\|\nabla \varrho^{\Gamma/2}\|^2_{L^2_x}\Big)\,\mathrm{d}t\Big] \leq c(\varepsilon, F). \tag{7.15}$$

Proof. As in Proposition 4.1.14, we apply Itô's chain rule to (7.11) to deduce the basic energy balance. We have

$$\begin{aligned}
&\mathrm{d}\int_{\mathbb{T}^3} E_\delta(\varrho, \varrho\mathbf{u})\,\mathrm{d}x + 2\varepsilon \int_{\mathbb{T}^3}\Big[\frac{1}{2}\varrho|\mathbf{u}|^2 + \frac{a\gamma}{\gamma-1}\varrho^\gamma + \frac{\delta\Gamma}{\Gamma-1}\varrho^\Gamma\Big]\,\mathrm{d}x\,\mathrm{d}t \\
&\quad + \int_{\mathbb{T}^3} \mathbb{S}(\nabla\mathbf{u}) : \nabla\mathbf{u}\,\mathrm{d}x\,\mathrm{d}t + \varepsilon\int_{\mathbb{T}^3}\varrho|\nabla\mathbf{u}|^2\,\mathrm{d}x\,\mathrm{d}t + \varepsilon\int_{\mathbb{T}^3}(a\gamma\varrho^{\gamma-2} + \delta\varrho^{\Gamma-2})|\nabla\varrho|^2\,\mathrm{d}x\,\mathrm{d}t \\
&\quad + \varepsilon\int_{\mathbb{T}^3}\frac{1}{2}\chi\Big(\frac{1}{M_0}\int_{\mathbb{T}^3}\varrho\,\mathrm{d}x\Big)|\mathbf{u}|^2\,\mathrm{d}x\,\mathrm{d}t \\
&= \sum_{k=1}^\infty \int_{\mathbb{T}^3}\varrho\Pi_m\mathbf{F}_k(\varrho, \mathbf{u})\cdot\mathbf{u}\,\mathrm{d}x\,\mathrm{d}W_k + \frac{1}{2}\sum_{k=1}^\infty\int_{\mathbb{T}^3}\frac{1}{\varrho}|\varrho\Pi_m\mathbf{F}_k(\varrho, \mathbf{u})|^2\,\mathrm{d}x\,\mathrm{d}t \\
&\quad + \chi\Big(\frac{1}{M_0}\int_{\mathbb{T}^3}\varrho\,\mathrm{d}x\Big)\int_{\mathbb{T}^3}\Big(\frac{a\gamma}{\gamma-1}\varrho^{\gamma-1} + \frac{\delta\Gamma}{\Gamma-1}\varrho^{\Gamma-1}\Big)\,\mathrm{d}x\,\mathrm{d}t.
\end{aligned} \tag{7.16}$$

In view of hypothesis (7.9) and the continuity of Π_m (7.10), we have

$$\begin{aligned}
\sum_{k=1}^\infty \int_{\mathbb{T}^3}\frac{1}{\varrho}|\varrho\Pi_m\mathbf{F}_k(\varrho, \mathbf{u})|^2\,\mathrm{d}x &\leq c\|\varrho\|_{L^\gamma_x}\sum_{k=1}^\infty\|\mathbf{F}_k(\varrho, \mathbf{u})\|^2_{L^{2\gamma'}_x} \\
&\leq c\|\varrho\|_{L^\gamma_x}\sum_{k=1}^\infty\|\mathbf{F}_k(\varrho, \mathbf{u})\|^2_{L^\infty_x} \leq c(F)\|\varrho\|_{L^\gamma_x},
\end{aligned} \tag{7.17}$$

where $\frac{1}{\gamma} + \frac{1}{\gamma'} = 1$. Remark that the function $\hat{\varrho} = \int_{\mathbb{T}^3}\varrho\,\mathrm{d}x$ satisfies the deterministic ODE

$$\frac{\mathrm{d}}{\mathrm{d}t}\hat{\varrho} = -2\varepsilon\hat{\varrho} + \chi\Big(\frac{\hat{\varrho}}{M_0}\Big). \tag{7.18}$$

In particular, the function $\hat{\varrho}$ is bounded by a constant that depends solely on the initial mass M_0. Taking expectation in (7.16) leads to

$$\begin{aligned}
&\frac{\mathrm{d}}{dt}\mathbb{E}\Big[\int_{\mathbb{T}^3} E_\delta(\varrho,\varrho\mathbf{u})\,\mathrm{d}x\Big] + 2\varepsilon\mathbb{E}\Big[\int_{\mathbb{T}^3}\Big[\frac{1}{2}\varrho|\mathbf{u}|^2 + \frac{a\gamma}{\gamma-1}\varrho^\gamma + \frac{\delta\Gamma}{\Gamma-1}\varrho^\Gamma\Big]\,\mathrm{d}x\Big]\\
&\quad + \mathbb{E}\Big[\int_{\mathbb{T}^3}\mathbb{S}(\nabla\mathbf{u}):\nabla\mathbf{u}\,\mathrm{d}x\Big] + \varepsilon\mathbb{E}\Big[\int_{\mathbb{T}^3}(a\gamma\varrho^{\gamma-2} + \delta\Gamma\varrho^{\Gamma-2})|\nabla\varrho|^2\,\mathrm{d}x\Big]\\
&\quad\le c(F)\mathbb{E}\|\varrho\|_{L^\gamma_x} + \mathbb{E}\Big[\chi\Big(\frac{1}{M_0}\int_{\mathbb{T}^3}\varrho\,\mathrm{d}x\Big)\int_{\mathbb{T}^3}\Big(\frac{a\gamma}{\gamma-1}\varrho^{\gamma-1} + \frac{\delta\Gamma}{\Gamma-1}\varrho^{\Gamma-1}\Big)\,\mathrm{d}x\Big].
\end{aligned}\tag{7.19}$$

Now, we observe that both terms on the right hand side can be estimated by the weighted Young inequality. Indeed, for any $\kappa > 0$, we have

$$c(F)\mathbb{E}\|\varrho\|_{L^\gamma_x} \le \frac{\varepsilon}{2}\mathbb{E}\Big[\int_{\mathbb{T}^3}\frac{a\gamma}{\gamma-1}\varrho^\gamma\,\mathrm{d}x\Big] + c(\varepsilon)$$

as well as

$$\begin{aligned}
&\mathbb{E}\Big[\chi\Big(\frac{1}{M_0}\int_{\mathbb{T}^3}\varrho\,\mathrm{d}x\Big)\int_{\mathbb{T}^3}\Big(\frac{a\gamma}{\gamma-1}\varrho^{\gamma-1} + \frac{\delta\Gamma}{\Gamma-1}\varrho^{\Gamma-1}\Big)\,\mathrm{d}x\Big]\\
&\quad\le\frac{\varepsilon}{2}\mathbb{E}\Big[\int_{\mathbb{T}^3}\Big[\frac{a\gamma}{\gamma-1}\varrho^\gamma + \frac{\delta\Gamma}{\Gamma-1}\varrho^\Gamma\Big]\,\mathrm{d}x\Big] + c(\varepsilon),
\end{aligned}$$

using $\chi \le 1$. So, both κ-terms can be absorbed in the left hand side of (7.19).

This readily implies (7.14) for $n = 1$ with an ε-dependent constant on the right hand side that blows up as $\varepsilon \to 0$. In addition, keeping (7.13) in mind and applying the Korn–Poincaré inequality (7.5), we deduce the estimate for the ergodic averages (7.15).

As the next step, we apply the Itô formula to (7.16) to obtain, for $n \in \mathbb{N}$,

$$\begin{aligned}
&\mathrm{d}\Big(\int_{\mathbb{T}^3} E_\delta(\varrho,\varrho\mathbf{u})\,\mathrm{d}x\Big)^n + 2\varepsilon n\Big(\int_{\mathbb{T}^3}\Big[\frac{1}{2}\varrho|\mathbf{u}|^2 + \frac{a\gamma}{\gamma-1}\varrho^\gamma + \frac{\delta\Gamma}{\Gamma-1}\varrho^\Gamma\Big]\,\mathrm{d}x\Big)^n\,\mathrm{d}t\\
&\quad + n\Big(\int_{\mathbb{T}^3} E_\delta(\varrho,\varrho\mathbf{u})\,\mathrm{d}x\Big)^{n-1}\Big[\int_{\mathbb{T}^3}\mathbb{S}(\nabla\mathbf{u}):\nabla\mathbf{u}\,\mathrm{d}x\,\mathrm{d}t + \varepsilon\int_{\mathbb{T}^3}\varrho|\nabla\mathbf{u}|^2\,\mathrm{d}x\,\mathrm{d}t\\
&\quad + \varepsilon\int_{\mathbb{T}^3}(a\gamma\varrho^{\gamma-2} + \delta\varrho^{\Gamma-2})|\nabla\varrho|^2\,\mathrm{d}x\,\mathrm{d}t + \varepsilon\int_{\mathbb{T}^3}\frac{1}{2}\chi\Big(\frac{1}{M_0}\int_{\mathbb{T}^3}\varrho\,\mathrm{d}x\Big)|\mathbf{u}|^2\,\mathrm{d}x\,\mathrm{d}t\Big]\\
&= n\sum_{k=1}^\infty\Big(\int_{\mathbb{T}^3} E_\delta(\varrho,\varrho\mathbf{u})\,\mathrm{d}x\Big)^{n-1}\int_{\mathbb{T}^3}\varrho\Pi_m\mathbf{F}_k(\varrho,\mathbf{u})\cdot\mathbf{u}\,\mathrm{d}x\,\mathrm{d}W_k\\
&\quad + n\Big(\int_{\mathbb{T}^3} E_\delta(\varrho,\varrho\mathbf{u})\,\mathrm{d}x\Big)^{n-1}\Big[\frac{1}{2}\sum_{k=1}^\infty\int_{\mathbb{T}^3}\frac{1}{\varrho}|\varrho\Pi_m\mathbf{F}_k(\varrho,\mathbf{u})|^2\,\mathrm{d}x\,\mathrm{d}t\\
&\quad + \chi\Big(\frac{1}{M_0}\int_{\mathbb{T}^3}\varrho\,\mathrm{d}x\Big)\int_{\mathbb{T}^3}\Big(\frac{a\gamma}{\gamma-1}\varrho^{\gamma-1} + \frac{\delta\Gamma}{\Gamma-1}\varrho^{\Gamma-1}\Big)\,\mathrm{d}x\,\mathrm{d}t\Big]\\
&\quad + \frac{n(n-1)}{2}\Big(\int_{\mathbb{T}^3} E_\delta(\varrho,\varrho\mathbf{u})\,\mathrm{d}x\Big)^{n-2}\sum_{k=1}^\infty\Big(\int_{\mathbb{T}^3}\varrho\Pi_m\mathbf{F}_k(\varrho,\mathbf{u})\cdot\mathbf{u}\,\mathrm{d}x\Big)^2\,\mathrm{d}t =: \mathcal{K}.
\end{aligned}\tag{7.20}$$

By virtue of (7.9) and the continuity of Π_m (7.10), we have

$$\begin{aligned}\sum_{k=1}^{\infty}\Big(\int_{\mathbb{T}^3}\varrho\Pi_m\mathbf{F}_k(\varrho,\mathbf{u})\cdot\mathbf{u}\,\mathrm{d}x\Big)^2 &\le \sum_{k=1}^{\infty}\|\sqrt{\varrho}\,\Pi_m\mathbf{F}_k(\varrho,\mathbf{u})\|_{L^2_x}^2\|\sqrt{\varrho}\mathbf{u}\|_{L^2_x}^2\\ &\le c\sum_{k=1}^{\infty}\|\varrho\|_{L^\gamma_x}\|\Pi_m\mathbf{F}_k(\varrho,\mathbf{u})\|_{L^{2\gamma'}_x}^2\|\sqrt{\varrho}\mathbf{u}\|_{L^2_x}^2\\ &\le c\sum_{k=1}^{\infty}\|\varrho\|_{L^\gamma_x}\|\mathbf{F}_k(\varrho,\mathbf{u})\|_{L^{2\gamma'}_x}^2\|\sqrt{\varrho}\mathbf{u}\|_{L^2_x}^2\\ &\le c(F)\|\varrho\|_{L^\gamma_x}\|\sqrt{\varrho}\mathbf{u}\|_{L^2_x}^2\\ &\le c(F)\|\varrho\|_{L^\gamma_x}\int_{\mathbb{T}^3}E_\delta(\varrho,\varrho\mathbf{u})\,\mathrm{d}x. \end{aligned}\tag{7.21}$$

Therefore, passing to expectations, the right hand side of (7.20) may be estimated by

$$\begin{aligned}\mathbb{E}\mathcal{K} \le n\,\mathbb{E}\Big[\Big(\int_{\mathbb{T}^3}E_\delta(\varrho,\varrho\mathbf{u})\,\mathrm{d}x\Big)^{n-1}\int_{\mathbb{T}^3}\Big(\frac{a\gamma}{\gamma-1}\varrho^{\gamma-1}+\frac{\delta\Gamma}{\Gamma-1}\varrho^{\Gamma-1}\Big)\,\mathrm{d}x\Big]\,\mathrm{d}t\\ +c(n,F)\mathbb{E}\Big[\Big(\int_{\mathbb{T}^3}E_\delta(\varrho,\varrho\mathbf{u})\,\mathrm{d}x\Big)^{n-1}\|\varrho\|_{L^\gamma_x}\Big]\,\mathrm{d}t,\end{aligned}\tag{7.22}$$

using also (7.17). Now, after a repeated application of the weighted Young inequality, we obtain

$$\begin{aligned}\mathbb{E}\mathcal{K} &\le \kappa\,\mathbb{E}\Big[\Big(\int_{\mathbb{T}^3}E_\delta(\varrho,\varrho\mathbf{u})\,\mathrm{d}x\Big)^{n-1}\int_{\mathbb{T}^3}\Big(\frac{a\gamma}{\gamma-1}\varrho^{\gamma}+\frac{\delta\Gamma}{\Gamma-1}\varrho^{\Gamma}\Big)\,\mathrm{d}x\Big]\,\mathrm{d}t\\ &\quad+\kappa\,\mathbb{E}\Big[\Big(\int_{\mathbb{T}^3}E_\delta(\varrho,\varrho\mathbf{u})\,\mathrm{d}x\Big)^{n-1}\int_{\mathbb{T}^3}\frac{a\gamma}{\gamma-1}\varrho^{\gamma}\,\mathrm{d}x\Big]\,\mathrm{d}t\\ &\quad+c(\kappa)\mathbb{E}\Big[\Big(\int_{\mathbb{T}^3}E_\delta(\varrho,\varrho\mathbf{u})\,\mathrm{d}x\Big)^{n-1}\Big]\le 3\kappa\mathbb{E}\Big[\Big(\int_{\mathbb{T}^3}E_\delta(\varrho,\varrho\mathbf{u})\,\mathrm{d}x\Big)^{n}\Big]+c(\kappa),\end{aligned}$$

for all $\kappa > 0$. Choosing κ small enough (depending on ε), both terms in (7.22) can be absorbed in the second term on the left hand side of (7.20), yielding a constant that blows up as $\varepsilon \to 0$. Hence we may infer (7.14) for any solution of (7.11) starting from regular initial data (7.12). □

7.1.3 Regularity of the density

Making use of the additional damping terms in the first equation in (7.11), we are able to show strong statements about the regularity of the solution depending on the parameters.

Lemma 7.1.2. *Let $\mathbf{u} \in C([0,\infty); H_m)$. Let ϱ be a classical solution to*

$$\partial_t\varrho + \operatorname{div}(\varrho[\mathbf{u}]_R) = \varepsilon\Delta\varrho - 2\varepsilon\varrho + \chi\Big(\frac{1}{M_0}\int_{\mathbb{T}^3}\varrho\,\mathrm{d}x\Big),\tag{7.23}$$

with $\varrho(0) \in C^{2+\nu}(\mathbb{T}^3)$ such that $\varrho(0) > 0$ and $\int_{\mathbb{T}^3}\varrho(0)\,\mathrm{d}x \le \overline{m}$. Then:

(a) *We have*

$$\|\varrho(\tau,\cdot)\|_{W^{k,p}_x} \le c(\overline{m},k,p,m,R,\varepsilon) \quad \forall\tau \ge 1, \tag{7.24}$$

for all $k \in \mathbb{N}$ *and* $p \in (1,\infty)$.

(b) *There exists a deterministic constant* $\underline{\varrho} = \underline{\varrho}(\overline{m},m,R,\varepsilon) > 0$ *such that*

$$\varrho(\tau,\cdot) \ge \underline{\varrho} \quad \forall\tau \ge 1. \tag{7.25}$$

In particular, the constants are independent of $\mathbf{u}$.

Proof. We start with equation (7.18) for the density averages that is independent of $\mathbf{u}$. Since (7.18) is a first order deterministic ODE, an easy observation shows

$$\hat{\varrho}(t) \to M_\varepsilon \quad \text{as } t \to \infty, \tag{7.26}$$

where $M_\varepsilon > 0$ is the unique solution to the equation $2\varepsilon M_\varepsilon = \chi(\frac{M_\varepsilon}{M_0})$. The convergence above is uniform in the sense that, for every $\kappa > 0$, there is $T = T(\overline{m},\varepsilon,\kappa)$ deterministic such that $|\hat{\varrho}(t) - M_\varepsilon| < \kappa$ for all $t \ge T$.

The next step is to show that ϱ is uniformly bounded from below as claimed in (b). Returning to the equation of continuity, we have

$$\partial_t\varrho - \varepsilon\Delta\varrho + \nabla\varrho\cdot[\mathbf{u}]_R = -(2\varepsilon + \operatorname{div}[\mathbf{u}]_R)\varrho + \chi\Big(\frac{1}{M_0}\hat{\varrho}\Big).$$

Seeing that

$$|\operatorname{div}[\mathbf{u}]_R| \le D(R,m)$$

for some constant $D(R,m)$, we use the comparison principle (see Protter–Weinberger [PW67]) to deduce

$$\varrho(t,\cdot) \ge \underline{\varrho}(t),$$

where $\underline{\varrho}$ solves the equation

$$\frac{\mathrm{d}\underline{\varrho}}{\mathrm{d}t} = -\underline{\varrho}(2\varepsilon + D(R,m)) + \chi\Big(\frac{1}{M_0}\hat{\varrho}\Big), \quad 0 < \underline{\varrho}(0) \le \inf_{\mathbb{T}^3}\varrho(0). \tag{7.27}$$

In accordance with (7.26), we have

$$\chi\Big(\frac{1}{M_0}\hat{\varrho}(t)\Big) \to \chi\Big(\frac{M_\varepsilon}{M_0}\Big) = 2\varepsilon M_\varepsilon > 0 \quad \text{as } t \to \infty.$$

Since any solution to (7.27) is asymptotically stabilized towards this equilibrium, we conclude that $\hat{\varrho}(t) > 0$ for any $t > 0$,

$$\underline{\varrho}(t) \to \frac{\chi(\frac{M_\varepsilon}{M_0})}{2\varepsilon + D(R,m)} \quad \text{as } t \to \infty,$$

and finally (7.25) follows.

Now we are going to prove part (a). First, note that (7.26) implies

$$\hat{\varrho}(t) = \|\varrho(t)\|_{L^1_x} \leq c(\overline{m}). \tag{7.28}$$

We apply maximal regularity theory (see Theorem A.2.2) to equation (7.23) to obtain

$$\begin{aligned}
&\|\partial_t \varrho\|_{L^2(T,T+1;W_x^{-2,q})} + \|\Delta\varrho\|_{L^2(T,T+1;W_x^{-2,q})} \\
&\quad \leq c\Big(\|\varrho(T)\|_{W_x^{-1,q}} + \|\operatorname{div}(\varrho[\mathbf{u}]_R)\|_{L^2(T,T+1;W_x^{-2,q})} \\
&\quad + \Big\|\chi\Big(\frac{1}{M_0}\int_{\mathbb{T}^3} \varrho \,\mathrm{d}x\Big)\Big\|_{L^2(T,T+1;W_x^{-2,q})}\Big),
\end{aligned}$$

where q is chosen such that $1 < q < 3/2$. Since $L^1(\mathbb{T}^3) \hookrightarrow W^{-1,q}(\mathbb{T}^3)$, using (7.28) we have

$$\begin{aligned}
&\|\partial_t \varrho\|_{L^2(T,T+1;W_x^{-2,q})} + \|\varrho\|_{L^2(T,T+1;L_x^q)} \\
&\quad \leq c\,(\|\varrho(T)\|_{W_x^{-1,q}} + \|\varrho\|_{L^2(T,T+1;W_x^{-1,q})} + 1) \\
&\quad \leq c\,(\|\varrho(T)\|_{L_x^1} + \|\varrho\|_{L^2(T,T+1;L_x^1)} + 1) \\
&\quad \leq c(\|\varrho\|_{L^\infty(T,T+1;L_x^1)} + 1) \leq c,
\end{aligned}$$

where c depends on R and ε but is independent of T. Consequently, there is $\tau = \tau(T) \in [T, T+1]$ such that $\varrho(\tau)$ is bounded in $L^q(\mathbb{T}^3)$ independently of T. A similar argument as above shows

$$\begin{aligned}
&\|\partial_t \varrho\|_{L^2(\tau,\tau+1;W_x^{-1,q})} + \|\varrho\|_{L^2(T,T+1;W_x^{1,q})} \\
&\quad \leq c\,(\|\varrho(\tau)\|_{L_x^q} + \|\varrho\|_{L^2(T,T+1;L_x^q)} + 1) \leq c.
\end{aligned}$$

So we have

$$\varrho \in L^2(T, T+1; W^{1,q}(\mathbb{T}^3)),$$

with a bound independent of T. Now, we can bootstrap the argument to obtain the claim. □

7.1.4 Approximate invariant measures

With estimates (7.14), (7.15), and (7.24) at hand, we are ready to apply the method of Krylov–Bogoliubov (see Section 2.12) to construct an invariant measure for system (7.11) with fixed parameters R, m, ε, and δ. For $\underline{r} > 0$, we define the set

$$\mathcal{R} = \mathcal{R}_{\underline{r}} = \{(r, \mathbf{v}) \in C^{2+\nu}(\mathbb{T}^3) \times H_m;\ \underline{r}^{-1} \leq r \leq \underline{r},\ \|\nabla r\|_{L_x^\infty} \leq \underline{r}\},$$

which will be the state space for solutions to (7.11). By $C_b(\mathcal{R})$ we denote the space of bounded continuous functions on $\mathcal{R}$.

First of all, we recall that the approximate system (7.11) can be solved using the same method as in Section 4.1. In what follows, for an $\mathfrak{F}_s$-measurable $\mathcal{R}$-valued random variable η, we denote by $\mathbf{U}^{\eta}_{s,t} = (\varrho^{\eta}_{s,t}, \mathbf{u}^{\eta}_{s,t})$ the solution of (7.11) at time t starting at time s from the initial condition η. If $s = 0$, then we write simply $\mathbf{U}^{\eta}_t$. We obtain the following result.

Theorem 7.1.3. *There is $\underline{r}$ large enough such that the following holds. Let $0 \le s < t$ be given. Let η be an $\mathfrak{F}_s$-measurable $\mathcal{R}$-valued initial condition. Then there exists $\mathbf{U}^{\eta}_s = (\varrho^{\eta}_s, \mathbf{u}^{\eta}_s) \in L^2(\Omega; C([s,t]; \mathcal{R}))$, which is the unique strong pathwise solution to (7.11) starting from η at time s. In addition, if η_1, η_2 are two such initial conditions, there is $\beta \in (0,2)$ such that*

$$\mathbb{E}\|\mathbf{U}^{\eta_1}_{s,t} - \mathbf{U}^{\eta_2}_{s,t}\|^2_{\mathcal{R}} \le C(t-s, R, m, \varepsilon, \delta)\, \mathbb{E}\|\eta_1 - \eta_2\|^{\beta}_{\mathcal{R}}. \tag{7.29}$$

Proof. The existence of the unique strong pathwise solution follows by the arguments of Section 4.1. In addition, by means of Lemma 7.1.2, the solution belongs to $L^2(\Omega; C([s,t]; \mathcal{R}))$ if we choose $\underline{r}$ large enough. Following the ideas of Proposition 4.1.9 (see, in particular, Corollary 4.1.10), we obtain

$$\mathbb{E}\|\mathbf{u}^{\eta_1}_{s,t} - \mathbf{u}^{\eta_2}_{s,t}\|^2_{H_m} \le \mathbb{E} \sup_{s\le\sigma\le t} \|\mathbf{u}^{\eta_1}_{s,\sigma} - \mathbf{u}^{\eta_2}_{s,\sigma}\|^2_{H_m} \le C(t-s, R, m, \varepsilon, \delta)\, \mathbb{E}\|\eta_1 - \eta_2\|^2_{\mathcal{R}}.$$

Moreover, Lemma A.2.7 implies

$$\sup_{s\le\sigma\le t} \|\varrho^{\eta_1}_{s,\sigma} - \varrho^{\eta_2}_{s,\sigma}\|_{W^{1,2}_x} \le C(t-s, R, m, \varepsilon, \delta) \sup_{s\le\sigma\le t} \|\mathbf{u}^{\eta_1}_{s,\sigma} - \mathbf{u}^{\eta_2}_{s,\sigma}\|_{H_m}$$

$\mathbb{P}$-a.s. and hence

$$\mathbb{E}\|\varrho^{\eta_1}_{s,t} - \varrho^{\eta_2}_{s,t}\|^{\beta}_{W^{1,2}_x} \le C(t-s, R, m, \varepsilon, \delta)\, \mathbb{E}\|\eta_1 - \eta_2\|^{\beta}_{\mathcal{R}},$$

for any $\beta > 0$. In order to obtain the final estimate, we choose $l \in \mathbb{N}$ such that $W^{l,2}(\mathbb{T}^3) \hookrightarrow C^{2+\nu}(\mathbb{T}^3)$ and interpolate $W^{l,2}(\mathbb{T}^3)$ between $W^{l+1,2}(\mathbb{T}^3)$ and $W^{1,2}(\mathbb{T}^3)$. Using Lemma 7.1.2, this implies, for some $\beta \in (0,2)$,

$$\begin{aligned}
\mathbb{E}\|\varrho^{\eta_1}_{s,t} - \varrho^{\eta_2}_{s,t}\|^2_{C^{2+\nu}_x} &\le c\,\mathbb{E}\|\varrho^{\eta_1}_{s,t} - \varrho^{\eta_2}_{s,t}\|^2_{W^{l,2}_x} \\
&\le c\,\mathbb{E}\|\varrho^{\eta_1}_{s,t} - \varrho^{\eta_2}_{s,t}\|^{\beta}_{W^{1,2}_x} \|\varrho^{\eta_1}_{s,t} - \varrho^{\eta_2}_{s,t}\|^{2-\beta}_{W^{l+1,2}_x} \\
&\le C(t-s, R, m, \varepsilon, \delta)\, \mathbb{E}\|\eta_1 - \eta_2\|^{\beta}_{\mathcal{R}}.
\end{aligned}$$ □

Let us now define the operators P_t by

$$(P_t\varphi)(\eta) := \mathbb{E}[\varphi(\mathbf{U}^{\eta}_t)] \quad \varphi \in C_b(\mathcal{R}).$$

Corollary 7.1.4. *Equation* (7.11) *defines a Feller–Markov process, that is,* $P_t : C_b(\mathcal{R}) \to C_b(\mathcal{R})$ *and*

$$\mathbb{E}[\varphi(\mathbf{U}^\eta_{t+s})|\mathfrak{F}_t] = (P_s\varphi)(\mathbf{U}^\eta_t) \quad \forall \varphi \in C_b(\mathcal{R}),\ \forall \eta \in H,\ \forall t,s > 0. \tag{7.30}$$

Besides, the semi-group property $P_{t+s} = P_t \circ P_s$ *holds true.*

Proof. The Feller property $P_t : C_b(\mathcal{R}) \to C_b(\mathcal{R})$ is an immediate consequence of (7.29) and the dominated convergence theorem.

In order to establish the Markov property (7.30), we shall prove

$$\mathbb{E}[\varphi(\mathbf{U}^\eta_{t+s})Z] = \mathbb{E}[(P_s\varphi)(\mathbf{U}^\eta_t)Z],$$

for all $\mathfrak{F}_t$-measurable random variables $Z \in L^1(\Omega)$. By uniqueness,

$$\mathbf{U}^\eta_{t+s} = \mathbf{U}^{\mathbf{U}^\eta_t}_{t,t+s} \quad \mathbb{P}\text{-a.s.}$$

It is therefore sufficient to show that

$$\mathbb{E}[\varphi(\mathbf{U}^{\mathbf{V}}_{t,t+s})Z] = \mathbb{E}[(P_s\varphi)(\mathbf{V})Z]$$

holds true for every $\mathfrak{F}_t$-measurable random variable $\mathbf{V}$. By approximation (we use dominated convergence and the fact that $\mathbf{V}_n \to \mathbf{V}$ in $\mathcal{R}$ implies $P_t\varphi(\mathbf{V}_n) \to P_t\varphi(\mathbf{V})$ in $\mathbb{R}$), it suffices to prove it for random variables $\mathbf{V} = \sum_{i=1}^k \mathbf{V}^i \mathbf{1}_{A^i}$, where $\mathbf{V}^i \in \mathcal{R}$ are deterministic and $(A^i) \subset \mathfrak{F}_t$ is a collection of disjoint sets such that $\bigcup_i A^i = \Omega$. Consequently, it suffices to prove it for every deterministic $\mathbf{V} \in E$. Now, the random variable $\mathbf{U}^{\mathbf{V}}_{t,t+s}$ depends only on the increments of the Brownian motion between t and $t+s$ and hence it is independent of $\mathfrak{F}_t$. Therefore

$$\mathbb{E}[\varphi(\mathbf{U}^{\mathbf{V}}_{t,t+s})Z] = \mathbb{E}[\varphi(\mathbf{U}^{\mathbf{V}}_{t,t+s})]\mathbb{E}[Z].$$

Since $\mathbf{U}^{\mathbf{V}}_{t,t+s}$ has the same law as $\mathbf{U}^{\mathbf{V}}_s$ by uniqueness, we have

$$\mathbb{E}[\varphi(\mathbf{U}^{\mathbf{V}}_{t,t+s})Z] = \mathbb{E}[\varphi(\mathbf{U}^{\mathbf{V}}_s)]\mathbb{E}[Z] = P_s\varphi(\mathbf{V})\mathbb{E}[Z] = \mathbb{E}[P_s\varphi(\mathbf{V})Z]$$

and the proof of (7.30) is complete.

Taking expectation in (7.30), we get on the one hand

$$\mathbb{E}[\mathbb{E}[\varphi(\mathbf{U}^\eta_{t+s})|\mathfrak{F}_t]] = \mathbb{E}[\varphi(\mathbf{U}^\eta_{t+s})] = (P_{t+s}\varphi)(\eta)$$

and on the other hand

$$\mathbb{E}[(P_s\varphi)(\mathbf{U}^\eta_t)] = (P_t(P_s\varphi))(\eta).$$

Thus the semi-group property follows. □

For an $\mathfrak{F}_0$-measurable random variable $\eta \in \mathcal{R}$, let $\mu_{t,\eta}$ denote the law of $\mathbf{U}_t^\eta$. If the law of η is μ, then it follows from the definition of the operator P_t that $\mu_{t,\eta} = P_t^* \mu$. For the application of the Krylov–Bogoliubov method (cf. Corollary 2.12.4), we shall prove the following result.

Proposition 7.1.5. *Let the initial condition be given by (7.13), that is, $\eta \equiv (1,0) \in \mathcal{R}$. Then the set of laws*

$$\left\{ \frac{1}{T} \int_0^T \mu_{s,\eta} \, \mathrm{d}s;\ T > 0 \right\}$$

is tight on $\mathcal{R}$.

Proof. Recall that $\mu_{s,\eta}$ are laws on the space $\mathcal{R}$. In particular, the second component is finite-dimensional whereas the first one is not. Let $\mu_{s,\eta}^{\varrho}$ and $\mu_{s,\eta}^{\mathbf{u}}$ denote the marginals of $\mu_{s,\eta}$ corresponding, respectively, to the first and second component of the solution. That is, $\mu_{s,\eta}^{\varrho}$ is the law of ϱ_s^η on $C^{2+\nu}(\mathbb{T}^3)$ and $\mu_{s,\eta}^{\mathbf{u}}$ is the law of $\mathbf{u}_s^\eta$ on H_m. Then it suffices to establish tightness of both following sets separately:

$$\left\{ \frac{1}{T} \int_0^T \mu_{s,\eta}^{\mathbf{u}} \, \mathrm{d}s;\ T > 0 \right\}, \quad \left\{ \frac{1}{T} \int_0^T \mu_{s,\eta}^{\varrho} \, \mathrm{d}s;\ T > 0 \right\}. \tag{7.31}$$

As a consequence of (7.15) and the equivalence of norms on H_m, we have

$$\frac{1}{T} \mathbb{E} \left[\int_0^T \|\mathbf{u}_t^\eta\|_{H_m}^2 \, \mathrm{d}t \right] \le c(m, \varepsilon, F).$$

Consequently, for compact sets

$$B_L := \{\mathbf{v} \in H_m;\ \|\mathbf{v}\|_{H_m} \le L\} \subset H_m,$$

by means of Chebyshev's inequality, we obtain

$$\frac{1}{T} \int_0^T \mu_{s,\eta}^{\mathbf{u}}(B_L^c) \, \mathrm{d}s = \frac{1}{T} \int_0^T \mathbb{P}(\|\mathbf{u}_s^\eta\|_{H_m} > L) \, \mathrm{d}s \le \frac{1}{L^2} \frac{1}{T} \mathbb{E} \left[\int_0^T \|\mathbf{u}_t^\eta\|_{H_m}^2 \, \mathrm{d}t \right],$$

which in turn implies tightness of the first set in (7.31). In order to establish tightness in the second component, we define

$$B_L := \{ r \in W^{k,p}(\mathbb{T}^3);\ \|r\|_{W_x^{k,p}} \le L \}.$$

For $p \in (1, \infty)$ and $k \in \mathbb{N}$ sufficiently large, this is a compact set in $C^{2+\nu}(\mathbb{T}^3)$. Hence, making use of (7.24), we have

$$\frac{1}{T} \int_0^T \mu_{s,\eta}^{\varrho}(B_L^c) \, \mathrm{d}s = \frac{1}{T} \int_0^T \mathbb{P}(\|\varrho_s^\eta\|_{W_x^{k,p}} > L) \, \mathrm{d}s \le \frac{1}{L} \sup_{t \ge 0} \mathbb{E} \|\varrho_t^\eta\|_{W_x^{k,p}}$$

and the desired tightness follows. □

Finally, the Krylov–Bogoliubov theorem applies and yields the following.

Corollary 7.1.6. *Fix $R, m \in \mathbb{N}$, ε, $\delta > 0$. Then there exists an invariant measure $\mathcal{L}_{\varrho,\mathbf{u}}$ for the dynamics, given by* (7.11). *In addition,*

$$\mathcal{L}_{\varrho,\mathbf{u}}[r \geq \underline{\varrho}] = 1, \quad \mathcal{L}_{\varrho,\mathbf{u}}\left[\int_{\mathbb{T}^3} r \,\mathrm{d}x = M_\varepsilon\right] = 1.$$

Proof. The existence part follows from Corollary 2.12.4. The second part of the statement is then a consequence of Lemma 7.1.2, namely, (7.25) and (7.26). □

7.2 First limit procedures: $R \to \infty$, $m \to \infty$

The existence of an invariant measure for the zero-level approximate problem (7.11) implies the existence of a stationary solution $[\varrho_R, \mathbf{u}_R]$. Our ultimate goal is to perform successively the limits for $R \to \infty$, $m \to \infty$, $\varepsilon \to 0$, and finally $\delta \to 0$. Even though this may look like a straightforward modification of the arguments in Chapter 4, there are several new aspects that must be handled. First of all, the uniform bounds used in Chapter 4 are controlled by the initial data. This is not the case for the stationary solution for which the "initial value" is not *a priori* known and the necessary estimates must be deduced from the energy balance (7.20) using the fact that the solution possesses the same law at any time. Moreover, the estimates derived in the previous section, that is, Proposition 7.1.1 and Lemma 7.1.2, do not hold independently of the approximation parameters R, m, ε, δ and are therefore not suitable for the limit procedure. In addition, since the point values of the density are not compact, the proof of the strong convergence of the approximate densities based on continuity of the effective viscous flux must be modified.

Let $[\varrho_R, \mathbf{u}_R]$ be a solution of the approximate problem (7.11), whose law at every time t is given by the invariant measure $\mathcal{L}_{\varrho_R,\mathbf{u}_R}$, constructed in Corollary 7.1.6. As the first step, we show a new uniform bound for $[\varrho_R, \mathbf{u}_R]$ that can be deduced from the energy balance (7.20). Note that at this stage, the estimate still blows up as $\varepsilon \to 0$.

Proposition 7.2.1. *Let $[\varrho_R, \mathbf{u}_R]$ be a stationary solution to* (7.11) *given by the invariant measure from Corollary* 7.1.6. *Then we have, for all $n \in \mathbb{N}$ and a.e. $t \in (0, \infty)$,*

$$\begin{aligned} &\mathbb{E}\left[\left(\int_{\mathbb{T}^3}\left[\frac{1}{2}\varrho_R|\mathbf{u}_R|^2 + \frac{a\gamma}{\gamma-1}\varrho_R^\gamma + \frac{\delta\Gamma}{\Gamma-1}\varrho_R^\Gamma\right]\mathrm{d}x\right)^n\right] \leq c(n, F, \varepsilon), \\ &\mathbb{E}\left[\|\mathbf{u}_R\|^2_{W^{1,2}_x}\right] \leq c(F, \varepsilon). \end{aligned} \tag{7.32}$$

Proof. After taking expectations in (7.20), we observe, due to stationarity of $[\varrho_R, \mathbf{u}_R]$, that the first term is constant in time, thus its time derivative vanishes. This is a consequence of Corollary 2.11.9. By the same reasoning we may ultimately omit the time integrals in all the remaining expressions. Then we apply (7.17) and (7.21) to estimate the terms coming from the stochastic integral and obtain

$$
\begin{aligned}
&\varepsilon\,\mathbb{E}\Big(\int_{\mathbb{T}^3}\Big[\frac{1}{2}\varrho_R|\mathbf{u}_R|^2+\frac{a\gamma}{\gamma-1}\varrho_R^\gamma+\frac{\delta\Gamma}{\Gamma-1}\varrho_R^\Gamma\Big]\,\mathrm{d}x\Big)^n\\
&\quad+\mathbb{E}\Big[\Big(\int_{\mathbb{T}^3}E_\delta(\varrho_R,\varrho_R\mathbf{u}_R)\,\mathrm{d}x\Big)^{n-1}\int_{\mathbb{T}^3}\mathbb{S}(\nabla\mathbf{u}_R):\nabla\mathbf{u}_R\,\mathrm{d}x\Big]\\
&\le c(n,F)\,\mathbb{E}\Big[\Big(\int_{\mathbb{T}^3}E_\delta(\varrho_R,\varrho_R\mathbf{u}_R)\,\mathrm{d}x\Big)^{n-1}\int_{\mathbb{T}^3}\Big(\frac{a\gamma}{\gamma-1}\varrho_R^{\gamma-1}+\frac{\delta\Gamma}{\Gamma-1}\varrho_R^{\Gamma-1}\Big)\,\mathrm{d}x\Big]\\
&\quad+c(n,F)\mathbb{E}\Big[\Big(\int_{\mathbb{T}^3}E_\delta(\varrho_R,\varrho_R\mathbf{u}_R)\,\mathrm{d}x\Big)^{n-1}\|\varrho_R\|_{L_x^\gamma}\Big].
\end{aligned}
$$

The application of the weighted Young inequality (similar to the estimates after (7.22)) allows one to absorb both terms on the right hand side into the first term on the left hand side and (7.32) follows. The estimate of the velocity is then obtained from the Korn–Poincaré inequality (7.5) by taking $n = 1$. □

Proposition 7.2.2. *Let $[\varrho_R,\mathbf{u}_R]$ be a stationary solution to (7.11) whose law is given by the invariant measure from Corollary 7.1.6. Then we have, for all $n \in \mathbb{N}$, a.e. $T \in (0,\infty)$, and $\tau > 0$,*

$$
\begin{aligned}
&\mathbb{E}\Big(\sup_{t\in[T,T+\tau]}\int_{\mathbb{T}^3}E_\delta(\varrho_R,\varrho_R\mathbf{u}_R)\,\mathrm{d}x\Big)^n\\
&\quad+2\varepsilon\,\mathbb{E}\Big(\int_T^{T+\tau}\int_{\mathbb{T}^3}\Big[\frac{1}{2}\varrho_R|\mathbf{u}_R|^2+\frac{a\gamma}{\gamma-1}\varrho_R^\gamma+\frac{\delta\Gamma}{\Gamma-1}\varrho_R^\Gamma\Big]\,\mathrm{d}x\,\mathrm{d}t\Big)^n\\
&\quad+\mathbb{E}\Big(\int_T^{T+\tau}\|\mathbf{u}_R\|_{W_x^{1,2}}^2\,\mathrm{d}t\Big)^n+\varepsilon\,\mathbb{E}\Big(\int_T^{T+\tau}\int_{\mathbb{T}^3}(a\gamma\varrho_R^{\gamma-2}+\delta\varrho_R^{\Gamma-2})|\nabla\varrho_R|^2\,\mathrm{d}x\,\mathrm{d}t\Big)^n\\
&\le c(n,F,\varepsilon,\tau),
\end{aligned}
\tag{7.33}
$$

where the constant on the right hand side does not depend on T.

Proof. Taking the nth power and expectation in the energy balance (7.16), we estimate the corresponding stochastic integral using Burkholder–Davis–Gundy's inequality and (7.21) as follows:

$$
\begin{aligned}
&\mathbb{E}\Big(\sup_{t\in[T,T+\tau]}\Big|\sum_{k=1}^\infty\int_0^t\int_{\mathbb{T}^3}\varrho_R\,\Pi_m\mathbf{F}_k(\varrho_R,\mathbf{u}_R)\cdot\mathbf{u}_R\,\mathrm{d}x\,\mathrm{d}W_k\Big|^n\Big)\\
&\le c(n)\mathbb{E}\Big(\int_T^{T+\tau}\sum_{k=1}^\infty\Big(\int_{\mathbb{T}^3}\varrho_R\Pi_m\mathbf{F}_k(\varrho_R,\mathbf{u}_R)\cdot\mathbf{u}_R\,\mathrm{d}x\Big)^2\mathrm{d}t\Big)^{\frac{n}{2}}\\
&\le c(n,F)\mathbb{E}\Big(\int_T^{T+\tau}\|\varrho_R\|_{L_x^\gamma}\int_{\mathbb{T}^3}\varrho_R|\mathbf{u}_R|^2\,\mathrm{d}x\,\mathrm{d}t\Big)^{\frac{n}{2}}.
\end{aligned}
$$

The second term on the right hand side of (7.16) is estimated using (7.17) and the dissipative term uses the Korn–Poincaré inequality (7.5). Therefore, we deduce

$$
\begin{aligned}
&\mathbb{E}\Big(\sup_{t\in[T,T+\tau]}\int_{\mathbb{T}^3} E_\delta(\varrho_R,\varrho_R\mathbf{u}_R)\,\mathrm{d}x\Big)^n \\
&\quad + 2\varepsilon\,\mathbb{E}\Big(\int_T^{T+\tau}\int_{\mathbb{T}^3}\Big[\frac{1}{2}\varrho_R|\mathbf{u}_R|^2+\frac{a\gamma}{\gamma-1}\varrho_R^\gamma+\frac{\delta\Gamma}{\Gamma-1}\varrho_R^\Gamma\Big]\,\mathrm{d}x\,\mathrm{d}t\Big)^n \\
&\quad + \mathbb{E}\Big(\int_T^{T+\tau}\|\mathbf{u}_R\|_{W^{1,2}_x}\,\mathrm{d}t\Big)^n + \varepsilon\,\mathbb{E}\Big(\int_T^{T+\tau}\int_{\mathbb{T}^3}(a\gamma\varrho_R^{\gamma-2}+\delta\varrho_R^{\Gamma-2})|\nabla\varrho_R|^2\,\mathrm{d}x\,\mathrm{d}t\Big)^n \\
&\le c(n)\,\mathbb{E}\Big(\int_{\mathbb{T}^3} E_\delta(\varrho_R,\varrho_R\mathbf{u}_R)(T)\,\mathrm{d}x\Big)^n \\
&\quad + c(n,F)\,\mathbb{E}\Big(\int_T^{T+\tau}\|\varrho_R\|_{L^\gamma_x}\int_{\mathbb{T}^3}\varrho_R|\mathbf{u}_R|^2\,\mathrm{d}x\,\mathrm{d}t\Big)^{\frac{n}{2}} + c(n,F)\,\mathbb{E}\Big(\int_T^{T+\tau}\|\varrho_R\|_{L^\gamma_x}\,\mathrm{d}t\Big)^n \\
&\quad + c(n)\,\mathbb{E}\Big(\int_T^{T+\tau}\int_{\mathbb{T}^3}\Big[\frac{a\gamma}{\gamma-1}\varrho_R^{\gamma-1}+\frac{\delta\Gamma}{\Gamma-1}\varrho_R^{\Gamma-1}\Big]\,\mathrm{d}x\,\mathrm{d}t\Big)^n.
\end{aligned}
\tag{7.34}
$$

Due to (7.32), the first term on the right hand side can be estimated by a constant $c(n,F,\varepsilon)$. The third term on the right hand side can be estimated by Young's inequality as follows:

$$
\begin{aligned}
&\mathbb{E}\Big(\int_T^{T+\tau}\|\varrho_R\|_{L^\gamma_x}\,\mathrm{d}t\Big)^n \\
&\le \frac{\varepsilon}{2}\,\mathbb{E}\Big(\int_T^{T+\tau}\int_{\mathbb{T}^3}\Big[\frac{1}{2}\varrho_R|\mathbf{u}_R|^2+\frac{a\gamma}{\gamma-1}\varrho_R^\gamma+\frac{\delta\Gamma}{\Gamma-1}\varrho_R^\Gamma\Big]\,\mathrm{d}x\,\mathrm{d}t\Big)^n + c(n,\varepsilon,\tau).
\end{aligned}
\tag{7.35}
$$

Subsequently it can be absorbed into the second term on the left hand side of (7.34). A similar approach applies to the last term on the right hand side of (7.34). For the remaining term we write

$$
\begin{aligned}
&\mathbb{E}\Big(\int_T^{T+\tau}\|\varrho\|_{L^\gamma_x}\int_{\mathbb{T}^3}\varrho_R|\mathbf{u}_R|^2\,\mathrm{d}x\,\mathrm{d}t\Big)^{\frac{n}{2}} \\
&\le \mathbb{E}\Big(\sup_{t\in[T,T+\tau]}\int_{\mathbb{T}^3}\frac{1}{2}\varrho_R|\mathbf{u}_R|^2\,\mathrm{d}x\int_T^{T+\tau}\|\varrho_R\|_{L^\gamma_x}\,\mathrm{d}t\Big)^{\frac{n}{2}} \\
&\le \kappa\,\mathbb{E}\Big(\sup_{t\in[T,T+\tau]}\int_{\mathbb{T}^3}\frac{1}{2}\varrho_R|\mathbf{u}_R|^2\,\mathrm{d}x\Big)^n + c(\kappa)\,\mathbb{E}\Big(\int_T^{T+\tau}\|\varrho_R\|_{L^\gamma_x}\,\mathrm{d}t\Big)^n,
\end{aligned}
$$

where the last term can be again estimated as in (7.35). Choosing κ sufficiently small yields the claim. □

In view of the uniform bounds provided by Proposition 7.2.2, for fixed $\varepsilon,\delta>0$, the asymptotic limits for $R\to\infty$ and $m\to\infty$ can be carried over exactly as for the initial value problem in Section 4.2. In the limit, we obtain the following approximate system:

- **Regularized equation of continuity.** We have

$$-\int_0^\infty \partial_t \phi \int_{\mathbb{T}^3} \varrho \psi \,\mathrm{d}x\,\mathrm{d}t = \int_0^\infty \phi \int_{\mathbb{T}^3} \varrho \mathbf{u} \cdot \nabla \psi \,\mathrm{d}x\,\mathrm{d}t - \varepsilon \int_0^\infty \phi \int_{\mathbb{T}^3} [\nabla \varrho \cdot \nabla \psi + 2\varrho \psi] \,\mathrm{d}x\,\mathrm{d}t$$
$$+ 2\varepsilon \int_0^\infty \phi \int_{\mathbb{T}^3} M_\varepsilon \psi \,\mathrm{d}x\,\mathrm{d}t, \tag{7.36}$$

for all $\phi \in C_c^\infty((0,\infty))$ and all $\psi \in C^\infty(\mathbb{T}^3)$ $\mathbb{P}$-a.s.

- **Regularized momentum equation.** We have

$$\begin{aligned}
-&\int_0^\infty \partial_t \phi \int_{\mathbb{T}^3} \varrho \mathbf{u} \cdot \boldsymbol{\varphi} \,\mathrm{d}x\,\mathrm{d}t \\
&= \int_0^\infty \phi \int_{\mathbb{T}^3} \varrho \mathbf{u} \otimes \mathbf{u} : \nabla \boldsymbol{\varphi} \,\mathrm{d}x\,\mathrm{d}t + \int_0^\infty \phi \int_{\mathbb{T}^3} (a\varrho^\gamma + \delta \varrho^\Gamma) \mathrm{div}\, \boldsymbol{\varphi} \,\mathrm{d}x\,\mathrm{d}t \\
&\quad - \int_0^\infty \phi \int_{\mathbb{T}^3} \mathbb{S}(\nabla \mathbf{u}) : \nabla \boldsymbol{\varphi} \,\mathrm{d}x\,\mathrm{d}t + \varepsilon \int_0^\infty \phi \int_{\mathbb{T}^3} \varrho \mathbf{u} \cdot \Delta \boldsymbol{\varphi} \,\mathrm{d}x\,\mathrm{d}t \\
&\quad - \varepsilon \int_0^\infty \phi \int_{\mathbb{T}^3} \varrho \mathbf{u} \cdot \boldsymbol{\varphi} \,\mathrm{d}x\,\mathrm{d}t + \int_0^\infty \phi \int_{\mathbb{T}^3} \mathbb{G}(\varrho, \varrho \mathbf{u}) \cdot \boldsymbol{\varphi} \,\mathrm{d}x\,\mathrm{d}W,
\end{aligned} \tag{7.37}$$

for all $\phi \in C_c^\infty((0,\infty))$ and all $\boldsymbol{\varphi} \in C^\infty(\mathbb{T}^3)$ $\mathbb{P}$-a.s.

To summarize, we deduce the following.

Proposition 7.2.3. *Let $\varepsilon, \delta > 0$ be given. Then there exists a stationary weak martingale solution $[\varrho_\varepsilon, \mathbf{u}_\varepsilon]$ to (7.36)–(7.37). In addition, for $n \in \mathbb{N}$ and every $\phi \in C_c^\infty((0,\infty))$, $\phi \geq 0$, the following generalized energy inequality holds true:*

$$\begin{aligned}
-&\int_0^\infty \partial_t \phi \Big(\int_{\mathbb{T}^3} E_\delta(\varrho_\varepsilon, \varrho_\varepsilon \mathbf{u}_\varepsilon) \,\mathrm{d}x \Big)^n \mathrm{d}t \\
&+ 2\varepsilon n \int_0^\infty \phi \Big(\int_{\mathbb{T}^3} \Big[\frac{1}{2} \varrho_\varepsilon |\mathbf{u}_\varepsilon|^2 + \frac{a\gamma}{\gamma - 1} \varrho_\varepsilon^\gamma + \frac{\delta \Gamma}{\Gamma - 1} \varrho_\varepsilon^\Gamma \Big] \mathrm{d}x \Big)^n \mathrm{d}t \\
&+ n \int_0^\infty \phi \Big(\int_{\mathbb{T}^3} E_\delta(\varrho_\varepsilon, \varrho_\varepsilon \mathbf{u}_\varepsilon) \,\mathrm{d}x \Big)^{n-1} \int_{\mathbb{T}^3} \mathbb{S}(\nabla \mathbf{u}_\varepsilon) : \nabla \mathbf{u}_\varepsilon \,\mathrm{d}x\,\mathrm{d}t \\
\leq{}& n \sum_{k=1}^\infty \int_0^\infty \phi \Big(\int_{\mathbb{T}^3} E_\delta(\varrho_\varepsilon, \varrho_\varepsilon \mathbf{u}_\varepsilon) \,\mathrm{d}x \Big)^{n-1} \int_{\mathbb{T}^3} \varrho_\varepsilon \mathbf{F}_k(\varrho_\varepsilon, \mathbf{u}_\varepsilon) \cdot \mathbf{u}_\varepsilon \,\mathrm{d}x\,\mathrm{d}W_k \\
&+ \frac{n}{2} \int_0^\infty \phi \Big(\int_{\mathbb{T}^3} E_\delta(\varrho_\varepsilon, \varrho_\varepsilon \mathbf{u}_\varepsilon) \,\mathrm{d}x \Big)^{n-1} \sum_{k=1}^\infty \int_{\mathbb{T}^3} \frac{1}{\varrho_\varepsilon} |\varrho_\varepsilon \mathbf{F}_k(\varrho_\varepsilon, \mathbf{u}_\varepsilon)|^2 \,\mathrm{d}x\,\mathrm{d}t \\
&+ n \int_0^\infty \phi \Big(\int_{\mathbb{T}^3} E_\delta(\varrho_\varepsilon, \varrho_\varepsilon \mathbf{u}_\varepsilon) \,\mathrm{d}x \Big)^{n-1} \chi\Big(\frac{M_\varepsilon}{M_0} \Big) \int_{\mathbb{T}^3} \Big[\frac{a\gamma}{\gamma - 1} \varrho_\varepsilon^{\gamma - 1} + \frac{\delta \Gamma}{\Gamma - 1} \varrho_\varepsilon^{\Gamma - 1} \Big] \mathrm{d}x\,\mathrm{d}t \\
&+ \frac{n(n-1)}{2} \int_0^\infty \phi \Big(\int_{\mathbb{T}^3} E_\delta(\varrho_\varepsilon, \varrho_\varepsilon \mathbf{u}_\varepsilon) \,\mathrm{d}x \Big)^{n-2} \sum_{k=1}^\infty \Big(\int_{\mathbb{T}^3} \varrho_\varepsilon \mathbf{F}_k(\varrho_\varepsilon, \mathbf{u}_\varepsilon) \cdot \mathbf{u}_\varepsilon \,\mathrm{d}x \Big)^2 \mathrm{d}t.
\end{aligned} \tag{7.38}$$

Proof. The proof follows the lines of Sections 4.2 and 4.3. The first passage to the limit as $R \to \infty$ relies on a stopping time argument from Section 4.2.2, whereas the limit $m \to \infty$ employs the stochastic compactness method based on the Jakubowski–Skorokhod representation theorem (Theorem 2.7.1) as in Section 4.3.2. We point out that all the necessary estimates in Section 4.2 and 4.3 come from the energy balance, which is controlled by the initial condition. In the present construction, the bound for the initial energy is replaced by the estimate (7.32), which holds true due to stationarity. Apart from that, the only difference from Section 4.3 is that we have to deal with path spaces containing an unbounded time interval, that is,

$$L^q_{\mathrm{loc}}([0,\infty);X), \quad (L^q_{\mathrm{loc}}([0,\infty);X), w), \quad C_{\mathrm{loc}}([0,\infty);(X,w)),$$

where $q \in (1,\infty)$ and X is a reflexive separable Banach space. Recall that $L^q_{\mathrm{loc}}([0,\infty);X)$ is a separable metric space given by

$$(f,g) \mapsto \sum_{M\in\mathbb{N}} 2^{-M}(\|f-g\|_{L^q(0,M;X)} \wedge 1)$$

and a set $\mathcal{K} \subset L^q_{\mathrm{loc}}([0,\infty);X)$ is compact, provided the set

$$\mathcal{K}_M := \{f|_{[0,M]}; f \in \mathcal{K}\} \subset L^q(0,M;X)$$

is compact for every $M \in \mathbb{N}$. On the other hand, the remaining two spaces are generally non-metrizable locally convex topological vector spaces, generated by the semi-norms

$$f \mapsto \int_0^M \langle f(t), g(t)\rangle_X \,\mathrm{d}t, \quad M \in \mathbb{N},\ g \in L^{q'}(0,\infty;X^*),\ \frac{1}{q}+\frac{1}{q'}=1$$

and

$$f \mapsto \sup_{t\in[0,M]} \langle f(t), g\rangle_X, \quad M \in \mathbb{N},\ g \in X^*,$$

respectively. As above, a set $\mathcal{K}$ is compact, provided its restriction to each interval $[0,M]$ is compact in $(L^q(0,M;X),w)$ and $C([0,M];(X,w))$, respectively. Furthermore, it can be seen that these topological spaces belong to the class of sub-Polish spaces (see Definition 2.1.3), where the Jakubowski–Skorokhod theorem applies. Indeed, in these spaces there exists a countable family of continuous functions that separate points. The proof of tightness of the corresponding laws in the current setting is therefore reduced to exactly the same method as in Section 4.3.2. Note that the key was the uniform energy bound from (4.63), which is replaced by (7.33). Consequently, the passage to the limit follows the lines of Section 4.3.2. In addition, Lemma 2.11.4 and Lemma 2.11.5 show that this limit procedure preserves stationarity and hence the limit solution is stationary. Finally, we obtain (7.38) by passing to the limit in (7.20), which holds on the new probability space according to Theorem 2.9.1. The passage to the limit in the stochastic integral can be justified as in Proposition 4.3.15. The situation here is even easier since the assumptions on the noise are more restrictive. □

Remark 7.2.4. Note that, for $n = 1$, the generalized energy inequality (7.38) corresponds to the usual energy inequality (4.134). The higher order version for $n \in \mathbb{N}$ is new and employed in order to obtain an analogue of Proposition 7.2.2, suitable for the subsequent limit procedures $\varepsilon \to 0$ and $\delta \to 0$ in Section 7.3 and Section 7.4.

7.3 Vanishing viscosity limit

Our goal in this section is to perform the passage to the limit as $\varepsilon \to 0$. This represents the most critical and delicate part of our construction. Remark that, after completing this limit procedure, we have already proved existence of stationary solutions to the stochastic Navier–Stokes system for compressible fluids – under an additional assumption upon the adiabatic exponent γ. The last passage to the limit presented in Section 7.4 is then devoted to the weakening of this additional assumption.

We point out that the key results needed for the previous limit procedure in Section 7.2, namely, Proposition 7.2.1 and consequently Proposition 7.2.2, depend on ε. Furthermore, it turns out that the global-in-time energy estimates uniform in ε and δ are very delicate. On the contrary, in the existence proof in Chapter 4, the basic energy estimate (4.63) established on the first approximation level held true uniformly in all the approximation parameters. Consequently, no further manipulations with the energy inequality were needed. This brought significant technical simplifications in comparison to the present construction of stationary solutions. To be more precise, this is due to the fact that, already after the passage to the limit $N \to \infty$, the energy balance is violated and has to be replaced by an inequality. In other words, one cannot justify the application of Itô's formula anymore and it is necessary to establish a more general version of the energy inequality; cf. (7.38).

Recall from Section 4.4 that, in addition to the usual energy estimate (4.134), a higher integrability of the density (4.148) was necessary in order to justify the compactness argument. Nevertheless, as in the deterministic setting, it was not possible to obtain strong convergence of the approximate densities directly. Consequently, the identification of the limit proceeded in two steps. First, the passage to the limit in the approximate system was done but the expressions with non-linear dependence on the density could not be identified. Second, a stochastic analogue of the effective viscous flux method originally due to Lions [Lio98] allowed one to prove strong convergence of the densities and hence to complete the proof.

Let us begin with an estimate for the velocity.

Proposition 7.3.1. *Let $[\varrho_\varepsilon, \mathbf{u}_\varepsilon]$ be the stationary solution to (7.36)–(7.37) constructed in Proposition 7.2.3. Then, for a.e. $t \in (0,\infty)$,*

$$\mathbb{E}\big[\|\mathbf{u}_\varepsilon\|^2_{W^{1,2}_x}\big] \le c(F, M_0), \tag{7.39}$$

$$\mathbb{E}\bigg[\int_{\mathbb{T}^3} E_\delta(\varrho_\varepsilon, \varrho_\varepsilon \mathbf{u}_\varepsilon)\,\mathrm{d}x \|\mathbf{u}_\varepsilon\|^2_{W^{1,2}_x}\bigg] \le c(F, M_0)\,\mathbb{E}\bigg[\int_{\mathbb{T}^3} E_\delta(\varrho_\varrho, \varrho_\varepsilon \mathbf{u}_\varepsilon)\,\mathrm{d}x\bigg] + c(M_0). \tag{7.40}$$

Proof. Taking expectation in the energy inequality (7.38) we observe that, due to stationarity of $[\varrho_\varepsilon, \mathbf{u}_\varepsilon]$, the first term is constant in time, thus its time derivative vanishes. We recall that $M_\varepsilon \le c(M_0)$ and, using (7.9), we estimate

$$\sum_{k=1}^{\infty} \int_{\mathbb{T}^3} \frac{1}{\varrho_\varepsilon} |\varrho_\varepsilon \mathbf{F}_k(\varrho_\varepsilon, \mathbf{u}_\varepsilon)|^2 \,\mathrm{d}x \le c(F) \int_{\mathbb{T}^3} \varrho_\varepsilon \,\mathrm{d}x \le c(F, M_0) \tag{7.41}$$

and

$$\begin{aligned} \sum_{k=1}^{\infty} \Big(\int_{\mathbb{T}^3} \varrho_\varepsilon \mathbf{F}_k(\varrho_\varepsilon, \mathbf{u}_\varepsilon) \cdot \mathbf{u}_\varepsilon \,\mathrm{d}x \Big)^2 &\le \sum_{k=1}^{\infty} \| \sqrt{\varrho_\varepsilon} \mathbf{F}_k(\varrho_\varepsilon, \mathbf{u}_\varepsilon) \|_{L^2_x}^2 \| \sqrt{\varrho_\varepsilon} \mathbf{u}_\varepsilon \|_{L^2_x}^2 \\ &\le c(F, M_0) \int_{\mathbb{T}^3} \frac{1}{2} \varrho_\varepsilon |\mathbf{u}_\varepsilon|^2 \,\mathrm{d}x \le c(F, M_0) \int_{\mathbb{T}^3} E_\delta(\varrho_\varepsilon, \varrho_\varepsilon \mathbf{u}_\varepsilon) \,\mathrm{d}x, \end{aligned} \tag{7.42}$$

which leads to

$$\begin{aligned} &2\varepsilon\, \mathbb{E}\Big[\Big(\int_{\mathbb{T}^3} \Big[\frac{1}{2} \varrho_\varepsilon |\mathbf{u}_\varepsilon|^2 + \frac{a\gamma}{\gamma - 1} \varrho_\varepsilon^\gamma + \frac{\delta\Gamma}{\Gamma - 1} \varrho_\varepsilon^\Gamma \Big] \,\mathrm{d}x \Big)^n \Big] \\ &\quad + \mathbb{E}\Big[\Big(\int_{\mathbb{T}^3} E_\delta(\varrho_\varepsilon, \varrho_\varepsilon \mathbf{u}_\varepsilon) \,\mathrm{d}x \Big)^{n-1} \int_{\mathbb{T}^3} \mathbb{S}(\nabla \mathbf{u}_\varepsilon) : \nabla \mathbf{u}_\varepsilon \,\mathrm{d}x \Big] \\ &\le c(n, F, M_0)\, \mathbb{E}\Big[\Big(\int_{\mathbb{T}^3} E_\delta(\varrho_\varepsilon, \varrho_\varepsilon \mathbf{u}_\varepsilon) \,\mathrm{d}x \Big)^{n-1} \Big] \\ &\quad + n\, \mathbb{E}\Big[\Big(\int_{\mathbb{T}^3} E_\delta(\varrho_\varepsilon, \varrho_\varepsilon \mathbf{u}_\varepsilon) \,\mathrm{d}x \Big)^{n-1} \chi\Big(\frac{M_\varepsilon}{M_0} \Big) \int_{\mathbb{T}^3} \Big(\frac{a\gamma}{\gamma - 1} \varrho_\varepsilon^{\gamma - 1} + \frac{\delta\Gamma}{\Gamma - 1} \varrho_\varepsilon^{\Gamma - 1} \Big) \,\mathrm{d}x \Big]. \end{aligned} \tag{7.43}$$

Moreover, it follows from Corollary 7.1.6 that

$$\chi\Big(\frac{M_\varepsilon}{M_0} \Big) = 2\varepsilon M_\varepsilon.$$

Hence, setting $n = 1$ and applying the Korn–Poincaré inequality (7.5) yields

$$\begin{aligned} &2\varepsilon\, \mathbb{E}\Big[\int_{\mathbb{T}^3} \Big[\frac{1}{2} \varrho_\varepsilon |\mathbf{u}_\varepsilon|^2 + \frac{a\gamma}{\gamma - 1} \varrho_\varepsilon^\gamma + \frac{\delta\Gamma}{\Gamma - 1} \varrho_\varepsilon^\Gamma \Big] \,\mathrm{d}x \Big] + \mathbb{E}\big[\|\mathbf{u}_\varepsilon\|_{W^{1,2}_x}^2 \big] \\ &\quad \le c(F, M_0) + \varepsilon c(M_0)\, \mathbb{E}\Big[\int_{\mathbb{T}^3} \Big(\frac{a\gamma}{\gamma - 1} \varrho_\varepsilon^{\gamma - 1} + \frac{\delta\Gamma}{\Gamma - 1} \varrho_\varepsilon^{\Gamma - 1} \Big) \,\mathrm{d}x \Big]. \end{aligned}$$

Since the second term on the right hand side in (7.43) can be absorbed in the first term on the left hand side (cf. the estimates after (7.19)), the bound (7.39) follows.

Setting $n = 2$ in (7.43), we obtain

$$\begin{aligned} &2\varepsilon\, \mathbb{E}\Big[\Big(\int_{\mathbb{T}^3} \Big[\frac{1}{2} \varrho_\varepsilon |\mathbf{u}_\varepsilon|^2 + \frac{a\gamma}{\gamma - 1} \varrho_\varepsilon^\gamma + \frac{\delta\Gamma}{\Gamma - 1} \varrho_\varepsilon^\Gamma \Big] \,\mathrm{d}x \Big)^2 \Big] \\ &\quad + \mathbb{E}\Big[\int_{\mathbb{T}^3} E_\delta(\varrho_\varepsilon, \varrho_\varepsilon \mathbf{u}_\varepsilon) \,\mathrm{d}x \, \|\mathbf{u}_\varepsilon\|_{W^{1,2}_x}^2 \Big] \\ &\le c(F, M_0)\, \mathbb{E}\Big[\int_{\mathbb{T}^3} E_\delta(\varrho_\varepsilon, \varrho_\varepsilon \mathbf{u}_\varepsilon) \,\mathrm{d}x \Big] \\ &\quad + \varepsilon c(M_0)\, \mathbb{E}\Big[\Big(\int_{\mathbb{T}^3} E_\delta(\varrho_\varepsilon, \varrho_\varepsilon \mathbf{u}_\varepsilon) \,\mathrm{d}x \Big) \int_{\mathbb{T}^3} \Big(\frac{a\gamma}{\gamma - 1} \varrho_\varepsilon^{\gamma - 1} + \frac{\delta\Gamma}{\Gamma - 1} \varrho_\varepsilon^{\Gamma - 1} \Big) \,\mathrm{d}x \Big] \\ &=: I_1 + I_2. \end{aligned}$$

Here we argue again with the weighted Young inequality. More precisely, we estimate the second term on the right hand side as

$$
\begin{aligned}
I_2 &\leq \frac{\varepsilon}{2}\mathbb{E}\bigg[\bigg(\int_{\mathbb{T}^3}\bigg[\frac{1}{2}\varrho_\varepsilon|\mathbf{u}_\varepsilon|^2 + \frac{a\gamma}{\gamma-1}\varrho_\varepsilon^\gamma + \frac{\delta\Gamma}{\Gamma-1}\varrho_\varepsilon^\Gamma\bigg]\,\mathrm{d}x\bigg)^2\bigg] \\
&\quad + \varepsilon c(M_0)\mathbb{E}\bigg[\bigg(\int_{\mathbb{T}^3}\bigg[\frac{a\gamma}{\gamma-1}\varrho_\varepsilon^{\gamma-1} + \frac{\delta\Gamma}{\Gamma-1}\varrho_\varepsilon^{\Gamma-1}\bigg]\,\mathrm{d}x\bigg)^2\bigg] \\
&\leq \varepsilon\mathbb{E}\bigg[\bigg(\int_{\mathbb{T}^3}\bigg[\frac{1}{2}\varrho_\varepsilon|\mathbf{u}_\varepsilon|^2 + \frac{a\gamma}{\gamma-1}\varrho_\varepsilon^\gamma + \frac{\delta\Gamma}{\Gamma-1}\varrho_\varepsilon^\Gamma\bigg]\,\mathrm{d}x\bigg)^2\bigg] + \varepsilon c(M_0).
\end{aligned}
$$

Thus (7.40) follows. □

We point out that the corresponding bound for the energy which can be obtained from (7.43), i.e.,

$$
\varepsilon\mathbb{E}\bigg[\int_{\mathbb{T}^3}\bigg[\frac{1}{2}\varrho_\varepsilon|\mathbf{u}_\varepsilon|^2 + \frac{a\gamma}{\gamma-1}\varrho_\varepsilon^\gamma + \frac{\delta\Gamma}{\Gamma-1}\varrho_\varepsilon^\Gamma\bigg]\,\mathrm{d}x\bigg] \leq c(F, M_0),
$$

still depends on ε and is therefore not suitable for the passage to the limit $\varepsilon \to 0$. As the next step, we derive an improved estimate for the energy as well as for the pressure.

Proposition 7.3.2. *Let $[\varrho_\varepsilon, \mathbf{u}_\varepsilon]$ be the stationary solution to (7.36)–(7.37) constructed in Proposition 7.2.3. Then the following uniform bound holds true for a.e. $t \in (0,\infty)$:*

$$
\mathbb{E}\bigg[\int_{\mathbb{T}^3}\bigg[a\varrho_\varepsilon^{\gamma+1} + \delta\varrho_\varepsilon^{\Gamma+1} + \frac{1}{3}\varrho_\varepsilon^2|\mathbf{u}_\varepsilon|^2\bigg]\,\mathrm{d}x\bigg] \leq c(\delta, F, M_0). \tag{7.44}
$$

In addition, if $s \in (1, \frac{\Gamma+1}{\Gamma-1} \wedge \frac{2(\gamma+1)}{\gamma+2}]$, then, for a.e. $T > 0$ and $\tau > 0$,

$$
\mathbb{E}\bigg[\bigg(\sup_{t\in[T,T+\tau]}\int_{\mathbb{T}^3} E_\delta(\varrho_\varepsilon, \varrho_\varepsilon\mathbf{u}_\varepsilon)\,\mathrm{d}x\bigg)^s\bigg] + \mathbb{E}\bigg[\bigg(\int_T^{T+\tau}\|\mathbf{u}_\varepsilon\|_{W_x^{1,2}}^2\,\mathrm{d}t\bigg)^s\bigg] \leq c(\tau, \delta, M_0, F, s), \tag{7.45}
$$

where the constant is independent of T.

Proof. Similarly to Section 4.4.2, our goal is to use $\nabla\Delta^{-1}[\varrho_\varepsilon - (\varrho_\varepsilon)_{\mathbb{T}^3}]$ as a test function in the momentum equation. Here Δ is the periodic Laplacian and the mean value satisfies $(\varrho_\varepsilon)_{\mathbb{T}^3} = |\mathbb{T}^3|^{-1}M_\varepsilon$. We apply Itô's formula to the functional

$$
f(\varrho, \mathbf{q}) = \int_{\mathbb{T}^3}\mathbf{q}\cdot\nabla\Delta^{-1}[\varrho - (\varrho)_{\mathbb{T}^3}]\,\mathrm{d}x = \int_{\mathbb{T}^3}\mathbf{q}\cdot\Delta^{-1}\nabla\varrho\,\mathrm{d}x.
$$

This step can be made rigorous by mollifying the equation. We obtain formally, from (7.37),

$$
\begin{aligned}
&\int_{\mathbb{T}^3}(a\varrho_\varepsilon^{\gamma+1} + \delta\varrho_\varepsilon^{\Gamma+1})\,\mathrm{d}x\,\mathrm{d}t \\
&= \mathrm{d}\int_{\mathbb{T}^3}\varrho_\varepsilon\mathbf{u}_\varepsilon\cdot\Delta^{-1}[\nabla\varrho_\varepsilon]\,\mathrm{d}x - \int_{\mathbb{T}^3}\varrho_\varepsilon\mathbf{u}_\varepsilon\otimes\mathbf{u}_\varepsilon : \nabla\Delta^{-1}\nabla\varrho_\varepsilon\,\mathrm{d}x\,\mathrm{d}t
\end{aligned}
$$

$$+ \int_{\mathbb{T}^3} \mathbb{S}(\nabla \mathbf{u}_\varepsilon) : \nabla \Delta^{-1} \nabla \varrho_\varepsilon \, dx \, dt - \varepsilon \int_{\mathbb{T}^3} \varrho_\varepsilon \mathbf{u}_\varepsilon \cdot \nabla \varrho_\varepsilon \, dx \, dt$$
$$+ \varepsilon \int_{\mathbb{T}^3} \varrho_\varepsilon \mathbf{u}_\varepsilon \cdot \Delta^{-1} \nabla \varrho_\varepsilon \, dx \, dt - \int_{\mathbb{T}^3} \varrho_\varepsilon \mathbb{F}(\varrho_\varepsilon, \mathbf{u}_\varepsilon) \cdot \Delta^{-1} \nabla \varrho_\varepsilon \, dx \, dW$$
$$- \int_{\mathbb{T}^3} \varrho_\varepsilon \mathbf{u}_\varepsilon \cdot \nabla \Delta^{-1} (\partial_t \varrho_\varepsilon) \, dx \, dt.$$

After a rather tedious but straightforward manipulation, we deduce, from (7.36),

$$\begin{aligned}
&\int_T^{T+1} \int_{\mathbb{T}^3} (a\varrho_\varepsilon^{\gamma+1} + \delta \varrho_\varepsilon^{\Gamma+1}) \, dx \, dt + \int_T^{T+1} \int_{\mathbb{T}^3} \frac{1}{3} \varrho_\varepsilon^2 |\mathbf{u}_\varepsilon|^2 \, dx \, dt \\
&= \frac{M_\varepsilon}{|\mathbb{T}^3|} \int_T^{T+1} \int_{\mathbb{T}^3} (a\varrho_\varepsilon^{\gamma} + \delta \varrho_\varepsilon^{\Gamma}) \, dx \, dt + \frac{1}{3} \frac{M_\varepsilon}{|\mathbb{T}^3|} \int_T^{T+1} \int_{\mathbb{T}^3} \varrho_\varepsilon |\mathbf{u}_\varepsilon|^2 \, dx \, dt \\
&+ \int_T^{T+1} \int_{\mathbb{T}^3} (\mu + \eta) \operatorname{div} \mathbf{u}_\varepsilon \, \varrho_\varepsilon \, dx \, dt \\
&+ 2\varepsilon \int_T^{T+1} \int_{\mathbb{T}^3} \varrho_\varepsilon \mathbf{u}_\varepsilon \cdot \nabla \Delta^{-1} \left[\varrho_\varepsilon - \frac{M_\varepsilon}{|\mathbb{T}^3|} \right] dx \, dt + \varepsilon \int_T^{T+1} \int_{\mathbb{T}^3} \varrho_\varepsilon^2 \operatorname{div} \mathbf{u}_\varepsilon \, dx \, dt \\
&- \int_T^{T+1} \int_{\mathbb{T}^3} \left(\varrho_\varepsilon \mathbf{u}_\varepsilon \otimes \mathbf{u}_\varepsilon - \frac{1}{3} \varrho_\varepsilon |\mathbf{u}_\varepsilon|^2 \mathbb{I} \right) : \nabla \Delta^{-1} \nabla \varrho_\varepsilon \, dx \, dt \\
&+ \left[\int_{\mathbb{T}^3} \varrho_\varepsilon \mathbf{u}_\varepsilon \cdot \nabla \Delta^{-1} \left[\varrho_\varepsilon - \frac{M_\varepsilon}{|\mathbb{T}^3|} \right] dx \right]_{t=T}^{t=T+1} \\
&+ \int_T^{T+1} \int_{\mathbb{T}^3} \varrho_\varepsilon \mathbf{u}_\varepsilon \cdot \nabla \Delta^{-1} [\operatorname{div} (\varrho_\varepsilon \mathbf{u}_\varepsilon)] \, dx \, dt \\
&- \sum_{k=1}^{\infty} \int_T^{T+1} \int_{\mathbb{T}^3} \mathbf{G}_k (\varrho_\varepsilon, \varrho_\varepsilon \mathbf{u}_\varepsilon) \cdot \nabla \Delta^{-1} \left[\varrho_\varepsilon - \frac{M_\varepsilon}{|\mathbb{T}^3|} \right] dx \, dW_k,
\end{aligned} \tag{7.46}$$

where we denote again $\eta = \mu + \frac{1}{3}\lambda$. Note that in the above the second term on the left hand side, the second term on the right hand side, and the second summand on the fifth line were added artificially and they cancel out. Passing to expectations in (7.46) and keeping in mind that the processes are stationary, we deduce

$$\begin{aligned}
&\mathbb{E}\left[\int_{\mathbb{T}^3} \left[a\varrho_\varepsilon^{\gamma+1} + \delta \varrho_\varepsilon^{\Gamma+1} + \frac{1}{3} \varrho_\varepsilon^2 |\mathbf{u}_\varepsilon|^2 \right] dx \right] \\
&\leq c(M_0) \, \mathbb{E}\left[\int_{\mathbb{T}^3} E_\delta (\varrho_\varepsilon, \varrho_\varepsilon \mathbf{u}_\varepsilon) \, dx \right] - \mathbb{E}\left[\int_{\mathbb{T}^3} \left(\varrho_\varepsilon \mathbf{u}_\varepsilon \otimes \mathbf{u}_\varepsilon - \frac{1}{3} \varrho_\varepsilon |\mathbf{u}_\varepsilon|^2 \mathbb{I} \right) : \nabla \Delta^{-1} \nabla \varrho_\varepsilon \, dx \right] \\
&\quad + \mathbb{E}\left[\int_{\mathbb{T}^3} (\mu + \eta) \operatorname{div} \mathbf{u}_\varepsilon \, \varrho_\varepsilon \, dx \, dt \right] + \mathbb{E}\left[\int_T^{T+1} \int_{\mathbb{T}^3} \varrho_\varepsilon \mathbf{u}_\varepsilon \cdot \nabla \Delta^{-1} \operatorname{div} (\varrho_\varepsilon \mathbf{u}_\varepsilon) \, dx \right] \\
&\quad + 2\varepsilon \, \mathbb{E}\left[\int_{\mathbb{T}^3} \varrho_\varepsilon \mathbf{u}_\varepsilon \cdot \nabla \Delta^{-1} \left[\varrho_\varepsilon - \frac{M_\varepsilon}{|\mathbb{T}^3|} \right] dx \right] + \varepsilon \, \mathbb{E}\left[\int_{\mathbb{T}^3} \varrho_\varepsilon^2 \operatorname{div} \mathbf{u}_\varepsilon \, dx \right] \\
&=: (I) + (II) + (III) + (IV) + (V) + (VI).
\end{aligned}$$

Now, we estimate each term separately. By Young's inequality we obtain, for every $\kappa > 0$,

$$(I) \leq \kappa \, \mathbb{E}\left[\int_{\mathbb{T}^3} \left(\frac{1}{3} \varrho_\varepsilon^2 |\mathbf{u}_\varepsilon|^2 + a\varrho_\varepsilon^{\gamma+1} + \delta \varrho_\varepsilon^{\Gamma+1} \right) dx \right] + c(\kappa, M_0) \, \mathbb{E}\left[\int_{\mathbb{T}^3} (|\mathbf{u}_\varepsilon|^2 + 1) \, dx \right]$$

$$\leq \kappa\,\mathbb{E}\Big[\int_{\mathbb{T}^3}\Big(\frac{1}{3}\varrho_\varepsilon^2|\mathbf{u}_\varepsilon|^2 + a\varrho_\varepsilon^{\gamma+1} + \delta\varrho_\varepsilon^{\Gamma+1}\Big)\,\mathrm{d}x\Big] + c(\kappa,F,M_0),$$

using the uniform bound (7.39). In order to control the remaining integrals on the right hand side, we first use Hölder's and Sobolev's inequalities to obtain

$$\begin{aligned}(II) &\leq c\,\mathbb{E}[\|\sqrt{\varrho_\varepsilon}\mathbf{u}_\varepsilon\|_{L_x^2}\|\mathbf{u}_\varepsilon\|_{L_x^6}\|\sqrt{\varrho_\varepsilon}\,\nabla\Delta^{-1}\nabla\varrho_\varepsilon\|_{L_x^3}]\\ &\leq c(\mathbb{E}[\|\sqrt{\varrho_\varepsilon}\mathbf{u}_\varepsilon\|_{L_x^2}^2\|\mathbf{u}_\varepsilon\|_{W_x^{1,2}}^2] + \mathbb{E}[\|\sqrt{\varrho_\varepsilon}\,\nabla\Delta^{-1}\nabla\varrho_\varepsilon\|_{L_x^3}^2]).\end{aligned}$$

Furthermore, since $\Gamma \geq 9/2$, we have

$$\begin{aligned}\mathbb{E}[\|\varrho_\varepsilon{}^{1/2}\nabla\Delta^{-1}\nabla\varrho_\varepsilon\|_{L_x^3}^2] &\leq \mathbb{E}[\|\varrho_\varepsilon\|_{L_x^{\frac{9}{2}}}\|\nabla\Delta^{-1}\nabla\varrho_\varepsilon\|_{L_x^{\frac{9}{2}}}^2]\\ &\leq c\,\mathbb{E}[\|\varrho_\varepsilon\|_{L_x^{\frac{9}{2}}}^3]\,\mathrm{d}t \leq c\,\mathbb{E}[\|\varrho_\varepsilon\|_{L_x^\Gamma}^3] \leq \kappa\delta\mathbb{E}[\|\varrho_\varepsilon\|_{L_x^\Gamma}^\Gamma] + c(\kappa,\delta).\end{aligned}$$

Note that we also used the continuity of $\nabla\Delta^{-1}\nabla$ and Young's inequality for arbitrary $\kappa > 0$. Similarly, we estimate

$$\begin{aligned}(IV) &\leq \mathbb{E}[\|\mathbf{u}_\varepsilon\|_{L_x^6}\|\varrho_\varepsilon\|_{L_x^3}\|\nabla\Delta^{-1}\mathrm{div}\,(\varrho_\varepsilon\mathbf{u}_\varepsilon)\|_{L_x^2}] \leq c\,\mathbb{E}[\|\mathbf{u}_\varepsilon\|_{L_x^6}\|\varrho_\varepsilon\|_{L_x^3}\|\varrho_\varepsilon\mathbf{u}_\varepsilon\|_{L_x^2}]\\ &\leq c\,\mathbb{E}[\|\mathbf{u}_\varepsilon\|_{W_x^{1,2}}^2\|\varrho_\varepsilon\|_{L_x^3}^2] \leq c\,\mathbb{E}[\|\mathbf{u}_\varepsilon\|_{W_x^{1,2}}^2\|\varrho_\varepsilon\|_{L_x^\Gamma}^\Gamma] + c(F,M_0),\end{aligned}$$

using (7.39). We also have

$$\begin{aligned}(III) &\leq \kappa\delta\,\mathbb{E}[\|\varrho_\varepsilon\|_{L_x^2}^2] + c(\kappa,\delta)\mathbb{E}[\|\nabla\mathbf{u}_\varepsilon\|_{L_x^2}^2]\\ &\leq \kappa\delta\,\mathbb{E}[\|\varrho_\varepsilon\|_{L_x^\Gamma}^\Gamma] + c(\kappa,\delta,F,M_0)\end{aligned}$$

as well as

$$(VI) \leq \kappa\delta\,\mathbb{E}\|\varrho_\varepsilon\|_{L_x^\Gamma}^\Gamma + c(\kappa,\delta,F,M_0).$$

Finally, continuity of $\nabla\Delta^{-1}$ and (7.39) implies

$$\begin{aligned}(V) &\leq \kappa\delta\Big(\mathbb{E}\|\varrho_\varepsilon\|_{L_x^4}^4 + \mathbb{E}\Big\|\nabla\Delta^{-1}\Big[\varrho_\varepsilon - \frac{M_\varepsilon}{|\mathbb{T}^3|}\Big]\Big\|_{L_x^4}^4\Big) + c(\kappa,\delta)\|\mathbf{u}_\varepsilon\|_{L_x^2}^2\\ &\leq c\kappa\delta\,\mathbb{E}\|\varrho_\varepsilon\|_{L_x^\Gamma}^\Gamma + c(\kappa,\delta,F,M_0).\end{aligned}$$

Summing up the inequalities above, choosing κ small enough, and using stationarity, we obtain

$$\begin{aligned}&\mathbb{E}\Big[\int_{\mathbb{T}^3}\Big[a\varrho_\varepsilon^{\gamma+1} + \delta\varrho_\varepsilon^{\Gamma+1} + \frac{1}{3}\varrho_\varepsilon^2|\mathbf{u}_\varepsilon|^2\Big]\,\mathrm{d}x\Big]\\ &\quad\leq \mathbb{E}\Big[\int_{\mathbb{T}^3} E_\delta(\varrho_\varepsilon,\varrho_\varrho\mathbf{u}_\varepsilon)\,\mathrm{d}x\,\|\mathbf{u}_\varepsilon\|_{W_x^{1,2}}^2\Big] + c(\delta,F,M_0).\end{aligned}$$

Thus, due to (7.40) and Young's inequality, we conclude that the stationary solution $[\varrho_\varepsilon, \mathbf{u}_\varepsilon]$ admits the uniform bound (7.44) as well as

$$\mathbb{E}\left[\left(1 + \int_{\mathbb{T}^3} E_\delta(\varrho_\varepsilon, \varrho_\varepsilon \mathbf{u}_\varepsilon)\,\mathrm{d}x\right)\|\mathbf{u}_\varepsilon\|^2_{W^{1,2}_x}\right] \le c(\delta, F, M_0). \tag{7.47}$$

Finally, let us show (7.45). To this end, we may go back to the energy inequality (7.38) for $n = 1$, obtaining

$$\begin{aligned}
&\mathbb{E}\left[\left(\sup_{t\in[T,T+\tau]} \int_{\mathbb{T}^3} E_\delta(\varrho_\varepsilon, \varrho_\varepsilon \mathbf{u}_\varepsilon)\,\mathrm{d}x\right)^s\right] + \mathbb{E}\left[\left(\int_T^{T+\tau} \|\mathbf{u}_\varepsilon\|^2_{W^{1,2}_x}\,\mathrm{d}t\right)^s\right] \\
&\le c(s)\,\mathbb{E}\left[\sup_{t\in[T,T+\tau]}\left|\sum_{k=1}^\infty \int_T^t \int_{\mathbb{T}^3} \varrho_\varepsilon \mathbf{F}_k(\varrho_\varepsilon, \mathbf{u}_\varepsilon)\cdot \mathbf{u}_\varepsilon\,\mathrm{d}x\,\mathrm{d}W_k\right|^s\right] \\
&\quad + c(s)\,\mathbb{E}\left[\sup_{t\in[T,T+\tau]}\left|\sum_{k=1}^\infty \int_T^t \int_{\mathbb{T}^3} \frac{1}{\varrho_\varepsilon}|\varrho_\varepsilon \mathbf{F}_k(\varrho_\varepsilon, \mathbf{u}_\varepsilon)|^2\,\mathrm{d}x\,\mathrm{d}r\right|^s\right] \\
&\quad + c(s)\,\mathbb{E}\left[\sup_{t\in[T,T+\tau]}\left|\int_T^t \int_{\mathbb{T}^3} \varrho_\varepsilon^{\gamma-1} + \varrho_\varepsilon^{\Gamma-1}\,\mathrm{d}x\,\mathrm{d}r\right|^s\right].
\end{aligned}$$

The first term on the right hand side is estimated using the Burkholder–Davis–Gundy inequality and (7.42); the second term using (7.41). We deduce

$$\begin{aligned}
&\mathbb{E}\left[\left(\sup_{t\in[T,T+\tau]} \int_{\mathbb{T}^3} E_\delta(\varrho_\varepsilon, \varrho_\varrho \mathbf{u}_\varepsilon)\,\mathrm{d}x\right)^s\right] \le c(s,\tau, M_0, F) \\
&\times\left(1 + \mathbb{E}\left[\left(\int_T^{T+\tau}\int_{\mathbb{T}^3} \varrho_\varepsilon|\mathbf{u}_\varepsilon|^2\,\mathrm{d}x\,\mathrm{d}r\right)^{\frac{s}{2}}\right] + \mathbb{E}\left[\left|\int_T^{T+\tau}\int_{\mathbb{T}^3} \varrho_\varepsilon^{\gamma-1} + \varrho_\varepsilon^{\Gamma-1}\,\mathrm{d}x\,\mathrm{d}r\right|^s\right]\right).
\end{aligned}$$

Now, by Hölder's and Sobolev's inequalities, stationarity, (7.44) and (7.47), for $s \in (1,2)$,

$$\begin{aligned}
&\mathbb{E}\left(\int_T^{T+\tau}\int_{\mathbb{T}^3} \varrho_\varepsilon|\mathbf{u}_\varepsilon|^2\,\mathrm{d}x\,\mathrm{d}r\right)^{\frac{s}{2}} \le c(\tau,s)\,\mathbb{E}\left(\int_T^{T+\tau} (\|\sqrt{\varrho_\varepsilon}\mathbf{u}_\varepsilon\|_{L^2_x}\|\mathbf{u}_\varepsilon\|_{L^6_x})^s \|\sqrt{\varrho_\varepsilon}\|^s_{L^3_x}\,\mathrm{d}t\right)^{\frac{1}{2}} \\
&\le c(\tau,s)\,\mathbb{E}\left(\int_T^{T+\tau} \|\sqrt{\varrho_\varepsilon}\mathbf{u}_\varepsilon\|^2_{L^2_x}\|\mathbf{u}_\varepsilon\|^2_{L^6_x} + \|\sqrt{\varrho_\varepsilon}\|^{\frac{2s}{2-s}}_{L^3_x}\,\mathrm{d}t\right)^{\frac{1}{2}} \\
&\le c(\tau,s)(\mathbb{E}[\|\sqrt{\varrho_\varepsilon}\mathbf{u}_\varepsilon\|^2_{L^2_x}\|\mathbf{u}_\varepsilon\|^2_{W^{1,2}_x}] + \mathbb{E}\|\varrho_\varepsilon\|^{\frac{s}{2-s}}_{L^\gamma_x}) \\
&\le c(\tau,s,\delta,F,M_0)\left(1 + \mathbb{E}\left(\int_{\mathbb{T}^3} \varrho_\varepsilon^{\gamma+1}\,\mathrm{d}x\right)^{\frac{s}{(\gamma+1)(2-s)}}\right) \\
&\le c(\tau,s,\delta,F,M_0)\left(1 + \mathbb{E}\int_{\mathbb{T}^3} \varrho_\varepsilon^{\gamma+1}\,\mathrm{d}x\right) \le c(\tau,s,\delta,F,M_0),
\end{aligned} \tag{7.48}$$

provided $s \le \frac{2(\gamma+1)}{\gamma+2}$. Similarly,

$$\left(\int_{\mathbb{T}^3} \varrho_\varepsilon^{\gamma-1} + \varrho_\varepsilon^{\Gamma-1}\,\mathrm{d}x\right)^s \le c\left(1 + \int_{\mathbb{T}^3} (\varrho_\varepsilon^{\gamma+1} + \varrho_\varepsilon^{\Gamma+1})\,\mathrm{d}x\right),$$

provided $s \le \frac{\Gamma+1}{\Gamma-1}$. Consequently, (7.45) follows due to (7.44). □

With Proposition 7.3.1 and Proposition 7.3.2 at hand, we are able to follow the compactness argument of Section 4.4. To be more precise, as $\varepsilon \to 0$ we aim at constructing stationary solutions to the following system:

– **Equation of continuity.** We have

$$-\int_0^\infty \partial_t \phi \int_{\mathbb{T}^3} \varrho \psi \,\mathrm{d}x\,\mathrm{d}t = \int_0^\infty \phi \int_{\mathbb{T}^3} \varrho \mathbf{u} \cdot \nabla \psi \,\mathrm{d}x\,\mathrm{d}t, \tag{7.49}$$

for all $\phi \in C_c^\infty((0,\infty))$ and all $\psi \in C^\infty(\mathbb{T}^3)$ $\mathbb{P}$-a.s.

– **Regularized momentum equation.** We have

$$\begin{aligned} -\int_0^\infty \partial_t \phi \int_{\mathbb{T}^3} \varrho \mathbf{u} \cdot \boldsymbol{\varphi} \,\mathrm{d}x\,\mathrm{d}t &\\ = \int_0^\infty \phi \int_{\mathbb{T}^3} \varrho \mathbf{u} \otimes \mathbf{u} : \nabla \boldsymbol{\varphi} \,\mathrm{d}x\,\mathrm{d}t + \int_0^\infty \phi \int_{\mathbb{T}^3} (a\varrho^\gamma + \delta \varrho^\Gamma) \operatorname{div} \boldsymbol{\varphi} \,\mathrm{d}x\,\mathrm{d}t &\\ - \int_0^\infty \phi \int_{\mathbb{T}^3} \mathbb{S}(\nabla \mathbf{u}) : \nabla \boldsymbol{\varphi} \,\mathrm{d}x\,\mathrm{d}t + \int_0^\infty \phi \int_{\mathbb{T}^3} \mathbb{G}(\varrho, \varrho\mathbf{u}) \cdot \boldsymbol{\varphi} \,\mathrm{d}x\,\mathrm{d}W, & \end{aligned} \tag{7.50}$$

for all $\phi \in C_c^\infty((0,\infty))$ and all $\boldsymbol{\varphi} \in C^\infty(\mathbb{T}^3)$ $\mathbb{P}$-a.s.

Note that, unlike the energy estimate in Chapter 4, the bound (7.45) only gives limited moment estimates, i.e., s cannot be arbitrarily large. Nevertheless, (7.45) is sufficient to perform the passage to the limit. We also point out that the assumption (7.9) on the noise coefficients is actually stronger than the one in Chapter 4 and consequently the convergence of the stochastic integral is more straightforward.

We deduce the following.

Proposition 7.3.3. *Let $\delta > 0$ be given. Then there exists a stationary solution $[\varrho_\delta, \mathbf{u}_\delta]$ to (7.49)–(7.50). Moreover, we have the estimates*

$$\mathbb{E}\big[\|\mathbf{u}_\delta(t)\|_{W_x^{1,2}}^2\big] \le c(F, M_0) \tag{7.51}$$

and

$$\begin{aligned} \mathbb{E}\Big[\int_{\mathbb{T}^3} E_\delta(\varrho_\delta, \varrho_\delta\mathbf{u}_\delta)\,\mathrm{d}x \int_{\mathbb{T}^3} \|\mathbf{u}_\delta\|_{W_x^{1,2}}^2 \,\mathrm{d}x\Big] \le c(F, M_0)\mathbb{E}\Big[\int_{\mathbb{T}^3} E_\delta(\varrho_\delta, \varrho_\delta\mathbf{u}_\delta)\,\mathrm{d}x\Big] &\\ + c(M_0), & \end{aligned} \tag{7.52}$$

for a.e. $t \in (0,\infty)$. In addition, the equation of continuity (7.49) holds true in the renormalized sense and for all $\phi \in C_c^\infty((0,\infty))$, $\phi \ge 0$, the following energy inequality holds true:

$$\begin{aligned} -\int_0^\infty \partial_t \phi \Big(\int_{\mathbb{T}^3} E_\delta(\varrho_\delta, \varrho_\delta\mathbf{u}_\delta)\,\mathrm{d}x \Big)\mathrm{d}t + \int_0^\infty \phi \int_{\mathbb{T}^3} \mathbb{S}(\nabla \mathbf{u}_\delta) : \nabla \mathbf{u}_\delta \,\mathrm{d}x\,\mathrm{d}t &\\ \le \sum_{k=1}^\infty \int_0^\infty \phi \int_{\mathbb{T}^3} \varrho_\delta \mathbf{F}_k(\varrho_\delta, \mathbf{u}_\delta) \cdot \mathbf{u}_\delta \,\mathrm{d}x\,\mathrm{d}W_k &\\ + \frac{1}{2} \int_0^\infty \phi \sum_{k=1}^\infty \int_{\mathbb{T}^3} \frac{1}{\varrho_\delta} |\varrho_\delta \mathbf{F}_k(\varrho_\delta, \mathbf{u}_\delta)|^2 \,\mathrm{d}x\,\mathrm{d}t. & \end{aligned} \tag{7.53}$$

Proof. First, we proceed as in Section 4.4.3 and establish the necessary tightness of the joint law of $[\varrho_\varepsilon, \varrho_\varepsilon \mathbf{u}_\varepsilon, \mathbf{u}_\varepsilon, W]$ as well as the other necessary quantities. The only difference is that the corresponding path spaces are replaced by their local-in-time analogues, as discussed in the proof of Proposition 7.2.3. Consequently, the Jakubowski–Skorokhod theorem applies and we obtain a new family of martingale solutions, still denoted by $[\varrho_\varepsilon, \varrho_\varepsilon \mathbf{u}_\varepsilon, \mathbf{u}_\varepsilon, W_\varepsilon]$, obeying the same laws and converging alsmot surely with respect to a new basis, still denoted by $(\Omega, \mathfrak{F}, (\mathfrak{F}_t)_{t\geq 0}, \mathbb{P})$. In addition, the limit satisfies

$$-\int_0^\infty \partial_t \phi \int_{\mathbb{T}^3} \varrho \psi \,\mathrm{d}x\,\mathrm{d}t = \int_0^\infty \phi \int_{\mathbb{T}^3} \varrho \mathbf{u} \cdot \nabla \psi \,\mathrm{d}x\,\mathrm{d}t, \qquad \int_{\mathbb{T}^3} \varrho \,\mathrm{d}x = M_0, \tag{7.54}$$

for all $\phi \in C_c^\infty((0,\infty))$ and all $\psi \in C^\infty(\mathbb{T}^3)$ $\mathbb{P}$-a.s., and

$$\begin{aligned} &-\int_0^\infty \partial_t \phi \int_{\mathbb{T}^3} \varrho \mathbf{u} \cdot \boldsymbol{\varphi} \,\mathrm{d}x\,\mathrm{d}t \\ &= \int_0^\infty \phi \int_{\mathbb{T}^3} \varrho \mathbf{u} \otimes \mathbf{u} : \nabla \boldsymbol{\varphi} \,\mathrm{d}x\,\mathrm{d}t + \int_0^\infty \phi \int_{\mathbb{T}^3} (a\overline{\varrho^\gamma} + \delta \overline{\varrho^\Gamma}) \operatorname{div} \boldsymbol{\varphi} \,\mathrm{d}x\,\mathrm{d}t \\ &\quad - \int_0^\infty \phi \int_{\mathbb{T}^3} \mathbb{S}(\nabla \mathbf{u}) : \nabla \boldsymbol{\varphi} \,\mathrm{d}x\,\mathrm{d}t + \int_0^\infty \phi \int_{\mathbb{T}^3} \overline{\mathbb{G}(\varrho, \varrho \mathbf{u})} \cdot \boldsymbol{\varphi} \,\mathrm{d}x\,\mathrm{d}W, \end{aligned} \tag{7.55}$$

for all $\phi \in C_c^\infty((0,\infty))$ and all $\boldsymbol{\varphi} \in C^\infty(\mathbb{T}^3)$ $\mathbb{P}$-a.s. Here, the bars denote the corresponding weak limits with respect to t, x. In addition, ϱ satisfies the renormalized equation of continuity, that is,

$$\partial_t b(\varrho) + \operatorname{div}(b(\varrho)\mathbf{u}) + (b'(\varrho)\varrho - b(\varrho)) \operatorname{div} \mathbf{u} = 0, \tag{7.56}$$

in the sense of distribution on $(0,\infty) \times \mathbb{T}^3$ for every $b \in C^1([0,\infty))$, with $b'(z) = 0$ for $z \geq M_b$ for some constant $M_b > 0$. However, as discussed in [FNP01, Remark 1.1], the assumption on b' can be weakened to

$$|b'(z)z| \leq c(z^\theta + z^{\frac{\gamma}{2}}) \quad \text{for all } z > 0 \text{ and some } \theta \in \left(0, \frac{\gamma}{2}\right).$$

This in particular includes the function $b(z) = z \log z$ employed below.

In order to complete the proof, it suffices to show strong convergence of the densities as in Section 4.4.3. More specifically, we prove that

$$\limsup_{\varepsilon \to 0} \mathbb{E}[\|\varrho_\varepsilon - \varrho\|_{L_x^{\Gamma+1}}^{\Gamma+1}] \leq \limsup_{\varepsilon \to 0} \mathbb{E}\left[\int_{\mathbb{T}^3} (\varrho_\varepsilon^{\Gamma+1} - \overline{\varrho^\Gamma}\varrho) \,\mathrm{d}x\right] \leq 0 \tag{7.57}$$

holds true for any $t > 0$. Note that the first inequality follows from the algebraic inequality

$$(A - B)^{\Gamma+1} = (A - B)^\Gamma (A - B) \leq (A^\Gamma - B^\Gamma)(A - B) \quad \text{whenever } A, B \geq 0.$$

In order to see the rightmost inequality in (7.57), we use the method of Lions [Lio98] based on regularity of the effective viscous flux. More specifically, mimicking the technique from the proof of Proposition 7.3.2, we derive from (7.36)–(7.37) the following identity:

$$\begin{aligned}
&\int_T^{T+1}\int_{\mathbb{T}^3}(a\varrho_\varepsilon^{\gamma+1}+\delta\varrho_\varepsilon^{\Gamma+1})\,\mathrm{d}x\,\mathrm{d}t=\frac{M_\varepsilon}{|\mathbb{T}^3|}\int_T^{T+1}\int_{\mathbb{T}^3}(a\varrho_\varepsilon^{\gamma}+\delta\varrho_\varepsilon^{\Gamma})\,\mathrm{d}x\,\mathrm{d}t\\
&\quad+\int_T^{T+1}\int_{\mathbb{T}^3}(\varrho_\varepsilon\mathbf{u}_\varepsilon\cdot\nabla\Delta^{-1}\mathrm{div}\,(\varrho_\varepsilon\mathbf{u}_\varepsilon)-\varrho_\varepsilon\mathbf{u}_\varepsilon\otimes\mathbf{u}_\varepsilon:\nabla\Delta^{-1}\nabla\varrho_\varepsilon)\,\mathrm{d}x\,\mathrm{d}t\\
&\quad+\int_T^{T+1}\int_{\mathbb{T}^3}(\mu+\eta)\mathrm{div}\,\mathbf{u}_\varepsilon\,\varrho_\varepsilon\,\mathrm{d}x\,\mathrm{d}t\\
&\quad+2\varepsilon\int_T^{T+1}\int_{\mathbb{T}^3}\varrho_\varepsilon\mathbf{u}_\varepsilon\cdot\nabla\Delta^{-1}\Big[\varrho_\varepsilon-\frac{M_\varepsilon}{|\mathbb{T}^3|}\Big]\,\mathrm{d}x\,\mathrm{d}t+\varepsilon\int_T^{T+1}\int_{\mathbb{T}^3}\varrho_\varepsilon^2\mathrm{div}\,\mathbf{u}_\varepsilon\,\mathrm{d}x\,\mathrm{d}t\\
&\quad+\Big[\int_{\mathbb{T}^3}\varrho\mathbf{u}\cdot\nabla\Delta^{-1}\Big[\varrho_\varepsilon-\frac{M_\varepsilon}{|\mathbb{T}^3|}\Big]\,\mathrm{d}x\Big]_{t=T}^{t=T+1}\\
&\quad-\sum_{k=1}^{\infty}\int_T^{T+1}\int_{\mathbb{T}^3}\mathbf{G}_k(\varrho_\varepsilon,\varrho_\varepsilon\mathbf{u}_\varepsilon)\cdot\nabla\Delta^{-1}\Big[\varrho_\varepsilon-\frac{M_\varepsilon}{|\mathbb{T}^3|}\Big]\,\mathrm{d}x\,\mathrm{d}W_k^\varepsilon.
\end{aligned}\tag{7.58}$$

In addition, since ϱ_ε satisfies the equation of continuity in the strong sense, we have

$$\begin{aligned}
\mathrm{d}(\varrho_\varepsilon\log\varrho_\varepsilon)&=-\mathrm{div}\,(\varrho_\varepsilon\log\varrho_\varepsilon\,\mathbf{u}_\varepsilon)-\varrho_\varepsilon\,\mathrm{div}\,\mathbf{u}_\varepsilon+\varepsilon\Delta(\varrho_\varepsilon\log\varrho_\varepsilon)-\varepsilon\frac{|\nabla\varrho_\varepsilon|^2}{\varrho_\varepsilon}\\
&\quad-2\varepsilon(\varrho_\varepsilon\log\varrho_\varepsilon+\varrho_\varepsilon)+2\varepsilon M_\varepsilon(\log\varrho_\varepsilon+1)\\
&\le-\mathrm{div}\,(\varrho_\varepsilon\log\varrho_\varepsilon\,\mathbf{u}_\varepsilon)-\varrho_\varepsilon\,\mathrm{div}\,\mathbf{u}_\varepsilon+\varepsilon\Delta(\varrho_\varepsilon\log\varrho_\varepsilon)+2\varepsilon c(M_0)(\varrho_\varepsilon+1).
\end{aligned}$$

In the above we applied the estimate

$$-2\varepsilon(\varrho_\varepsilon\log\varrho_\varepsilon+\varrho_\varepsilon)+2\varepsilon M_\varepsilon(\log\varrho_\varepsilon+1)\le\varepsilon c(M_0)(\varrho_\varepsilon+1),$$

which follows since $-\varrho_\varepsilon\log\varrho_\varepsilon$ is bounded from above by a constant and $\log\varrho_\varepsilon$ is bounded by ϱ_ε. Inserting this into (7.58) implies

$$\begin{aligned}
&\int_T^{T+1}\int_{\mathbb{T}^3}(a\varrho_\varepsilon^{\gamma+1}+\delta\varrho_\varepsilon^{\Gamma+1})\,\mathrm{d}x\,\mathrm{d}t\le M_\varepsilon\int_T^{T+1}\int_{\mathbb{T}^3}(a\varrho_\varepsilon^{\gamma}+\delta\varrho_\varepsilon^{\Gamma})\,\mathrm{d}x\,\mathrm{d}t\\
&\quad+\int_T^{T+1}\int_{\mathbb{T}^3}(\varrho_\varepsilon\mathbf{u}_\varepsilon\cdot\nabla\Delta^{-1}\mathrm{div}\,(\varrho_\varepsilon\mathbf{u}_\varepsilon)-\varrho_\varepsilon\mathbf{u}_\varepsilon\otimes\mathbf{u}_\varepsilon:\nabla\Delta_x^{-1}\nabla\varrho_\varepsilon)\,\mathrm{d}x\,\mathrm{d}t\\
&\quad-(\mu+\eta)\Big[\int_{\mathbb{T}^3}\varrho_\varepsilon\log\varrho_\varepsilon\,\mathrm{d}x\Big]_{t=T}^{t=T+1}+\varepsilon\, c(M_0)\int_T^{T+1}\int_{\mathbb{T}^3}(\varrho_\varepsilon+1)\,\mathrm{d}x\,\mathrm{d}t\\
&\quad+2\varepsilon\int_T^{T+1}\int_{\mathbb{T}^3}\varrho_\varepsilon\mathbf{u}_\varepsilon\cdot\nabla\Delta^{-1}\Big[\varrho_\varepsilon-\frac{M_\varepsilon}{|\mathbb{T}^3|}\Big]\,\mathrm{d}x\,\mathrm{d}t+\varepsilon\int_T^{T+1}\int_{\mathbb{T}^3}\varrho_\varepsilon^2\mathrm{div}\,\mathbf{u}_\varepsilon\,\mathrm{d}x\,\mathrm{d}t\\
&\quad+\Big[\int_{\mathbb{T}^3}\varrho_\varepsilon\mathbf{u}_\varepsilon\cdot\nabla\Delta_x^{-1}\Big[\varrho_\varepsilon-\frac{M_\varepsilon}{|\mathbb{T}^3|}\Big]\,\mathrm{d}x\Big]_{t=T}^{t=T+1}\\
&\quad-\sum_{k=1}^{\infty}\int_T^{T+1}\int_{\mathbb{T}^3}\varrho\mathbf{F}_k(\varrho_\varepsilon,\varrho_\varepsilon\mathbf{u}_\varepsilon)\cdot\nabla\Delta^{-1}\Big[\varrho_\varepsilon-\frac{M_\varepsilon}{|\mathbb{T}^3|}\Big]\,\mathrm{d}x\,\mathrm{d}W_k^\varepsilon.
\end{aligned}\tag{7.59}$$

Similarly, as the limit density ϱ also satisfies the renormalized equation of continuity (7.56), cf. Theorem A.3.1, we choose $b(z) = z\log z$ and deduce that

$$\mathrm{d}(\varrho\log\varrho) = -\mathrm{div}\,(\varrho\log\varrho\,\mathbf{u})\,\mathrm{d}t - \varrho\,\mathrm{div}\,\mathbf{u}\,\mathrm{d}t$$

holds true in the sense of distributions. Therefore, we obtain from the limit equations (7.54) and (7.55)

$$\begin{aligned}\int_T^{T+1}\int_{\mathbb{T}^3}(a\overline{\varrho^\gamma}+\delta\overline{\varrho^\Gamma})\varrho\,\mathrm{d}x\,\mathrm{d}t &= \frac{M_0}{|\mathbb{T}^3|}\int_T^{T+1}\int_{\mathbb{T}^3}(a\overline{\varrho^\gamma}+\delta\overline{\varrho^\Gamma})\,\mathrm{d}x\,\mathrm{d}t\\ &+\int_T^{T+1}\int_{\mathbb{T}^3}(\varrho\mathbf{u}\cdot\nabla\Delta^{-1}\mathrm{div}\,(\varrho\mathbf{u})-\varrho\mathbf{u}\otimes\mathbf{u}:\nabla\Delta^{-1}\nabla\varrho)\,\mathrm{d}x\,\mathrm{d}t\\ &-(\mu+\eta)\Big[\int_{\mathbb{T}^3}\varrho\log\varrho\,\mathrm{d}x\Big]_{t=T}^{t=T+1}\\ &+\Big[\int_{\mathbb{T}^3}\varrho\mathbf{u}\cdot\nabla\Delta^{-1}\Big[\varrho-\frac{M_0}{|\mathbb{T}^3|}\Big]\,\mathrm{d}x\Big]_{t=T}^{t=T+1}\\ &-\sum_{k=1}^{\infty}\int_T^{T+1}\int_{\mathbb{T}^3}\overline{\varrho\mathbf{F}_k(\varrho,\varrho\mathbf{u})}\cdot\nabla\Delta^{-1}\Big[\varrho-\frac{M_0}{|\mathbb{T}^3|}\Big]\,\mathrm{d}x\,\mathrm{d}W_k.\end{aligned}$$

Thus passing to expectations and using the fact that the processes are stationary, we get

$$\begin{aligned}\mathbb{E}\Big[\int_T^{T+1}\int_{\mathbb{T}^3}(a\varrho_\varepsilon^{\gamma+1}+\delta\varrho_\varepsilon^\Gamma)\varrho_\varepsilon\,\mathrm{d}x\,\mathrm{d}t\Big] &\le \frac{M_\varepsilon}{|\mathbb{T}^3|}\mathbb{E}\Big[\int_T^{T+1}\int_{\mathbb{T}^3}(a\varrho_\varepsilon^\gamma+\delta\varrho_\varepsilon^\Gamma)\,\mathrm{d}x\,\mathrm{d}t\Big]\\ &+\mathbb{E}\Big[\int_T^{T+1}\int_{\mathbb{T}^3}(\varrho_\varepsilon\mathbf{u}_\varepsilon\cdot\nabla\Delta^{-1}\mathrm{div}\,(\varrho_\varepsilon\mathbf{u}_\varepsilon)-\varrho_\varepsilon\mathbf{u}_\varepsilon\otimes\mathbf{u}_\varepsilon:\nabla\Delta^{-1}\nabla\varrho_\varepsilon)\,\mathrm{d}x\,\mathrm{d}t\Big]\\ &+2\varepsilon\,\mathbb{E}\Big[\int_{\mathbb{T}^3}\varrho_\varepsilon\mathbf{u}_\varepsilon\cdot\nabla\Delta^{-1}\Big[\varrho_\varepsilon-\frac{M_\varepsilon}{|\mathbb{T}^3|}\Big]\,\mathrm{d}x\Big]+\varepsilon\,\mathbb{E}\Big[\int_{\mathbb{T}^3}\varrho_\varepsilon^2\mathrm{div}\,\mathbf{u}_\varepsilon\,\mathrm{d}x\Big]+\varepsilon\,c(M_0).\end{aligned}\tag{7.60}$$

Note that the inequality uses $\int_{\mathbb{T}^3}\varrho_\varepsilon\,\mathrm{d}x = M_\varepsilon \le c(M_0)$ for all $\varepsilon > 0$. Similarly, we obtain

$$\begin{aligned}\mathbb{E}\Big[\int_T^{T+1}\int_{\mathbb{T}^3}(a\overline{\varrho^\gamma}+\delta\overline{\varrho^\Gamma})\varrho\,\mathrm{d}x\,\mathrm{d}t\Big] &= M_0\,\mathbb{E}\Big[\int_T^{T+1}\int_{\mathbb{T}^3}(a\overline{\varrho^\gamma}+\delta\overline{\varrho^\Gamma})\,\mathrm{d}x\,\mathrm{d}t\Big]\\ &+\mathbb{E}\Big[\int_T^{T+1}\int_{\mathbb{T}^3}(\varrho\mathbf{u}\cdot\nabla\Delta^{-1}\mathrm{div}\,(\varrho\mathbf{u})-\varrho\mathbf{u}\otimes\mathbf{u}:\nabla\Delta^{-1}\nabla\varrho)\,\mathrm{d}x\,\mathrm{d}t\Big].\end{aligned}\tag{7.61}$$

Note that the ε-terms in (7.60) vanish due to Proposition 7.3.2 and we have $M_\varepsilon \to M_0$ as $\varepsilon \to 0$. Consequently, the desired conclusion (7.57) follows as soon as we observe

$$\begin{aligned}&\lim_{\varepsilon\to0}\mathbb{E}\Big[\int_T^{T+1}\int_{\mathbb{T}^3}(\varrho_\varepsilon\mathbf{u}_\varepsilon\cdot\nabla\Delta^{-1}\mathrm{div}\,(\varrho_\varepsilon\mathbf{u}_\varepsilon)-\varrho_\varepsilon\mathbf{u}_\varepsilon\otimes\mathbf{u}_\varepsilon:\nabla\Delta^{-1}\nabla\varrho_\varepsilon)\,\mathrm{d}x\,\mathrm{d}t\Big]\\ &\quad=\mathbb{E}\Big[\int_T^{T+1}\int_{\mathbb{T}^3}(\varrho\mathbf{u}\cdot\nabla\Delta^{-1}\mathrm{div}\,(\varrho\mathbf{u})-\varrho\mathbf{u}\otimes\mathbf{u}:\nabla\Delta^{-1}\nabla\varrho)\,\mathrm{d}x\,\mathrm{d}t\Big].\end{aligned}\tag{7.62}$$

In fact, (7.62) in combination with (7.60) and (7.61) implies

$$\limsup_{\varepsilon\to 0}\mathbb{E}\left[\int_T^{T+1}\int_{\mathbb{T}^3}(a\varrho_\varepsilon^\gamma+\delta\varrho_\varepsilon^\Gamma)\varrho_\varepsilon\,\mathrm{d}x\,\mathrm{d}t\right]\leq\mathbb{E}\left[\int_T^{T+1}\int_{\mathbb{T}^3}(a\overline{\varrho^\gamma}+\delta\overline{\varrho^\Gamma})\varrho\,\mathrm{d}x\,\mathrm{d}t\right],$$

which shows strong convergence of ϱ_ε by monotonicity arguments. Relation (7.62) can be established by compensated compactness arguments (Lemma A.1.11 applied $\mathbb{P}$-a.s.) if we show that the expressions under expectations are $\mathbb{P}$-equi-integrable. Considering the two summands separately and using continuity of $\nabla\Delta^{-1}\nabla$ and Sobolev's embedding, we have

$$\begin{aligned}\left|\int_{\mathbb{T}^3}\varrho_\varepsilon\mathbf{u}_\varepsilon\cdot\nabla\Delta^{-1}\mathrm{div}\,(\varrho_\varepsilon\mathbf{u}_\varepsilon)\,\mathrm{d}x\right| &\leq c\|\sqrt{\varrho_\varepsilon}\mathbf{u}_\varepsilon\|_{L_x^2}\|\sqrt{\varrho_\varepsilon}\|_{L_x^{2\Gamma}}\|\nabla\Delta^{-1}\nabla\varrho_\varepsilon\mathbf{u}_\varepsilon\|_{L_x^{\frac{2\Gamma}{\Gamma-1}}}\\ &\leq c\|\sqrt{\varrho_\varepsilon}\mathbf{u}_\varepsilon\|_{L_x^2}\|\sqrt{\varrho_\varepsilon}\|_{L_x^{2\Gamma}}\|\varrho_\varepsilon\mathbf{u}_\varepsilon\|_{L_x^{\frac{2\Gamma}{\Gamma-1}}}\\ &\leq c\|\sqrt{\varrho_\varepsilon}\mathbf{u}_\varepsilon\|_{L_x^2}\|\sqrt{\varrho_\varepsilon}\|_{L_x^{2\Gamma}}\|\mathbf{u}_\varepsilon\|_{L_x^6}\|\varrho_\varepsilon\|_{L_x^\Gamma}\\ &\leq c\|\sqrt{\varrho_\varepsilon}\mathbf{u}_\varepsilon\|_{L_x^2}\|\sqrt{\varrho_\varepsilon}\|_{L_x^{2\Gamma}}\|\mathbf{u}_\varepsilon\|_{W_x^{1,2}}\|\varrho_\varepsilon\|_{L_x^\Gamma},\end{aligned}$$

as $\Gamma\geq\frac{9}{2}$. Similarly, we have

$$\left|\int_{\mathbb{T}^3}\varrho_\varepsilon\mathbf{u}_\varepsilon\otimes\mathbf{u}_\varepsilon:\nabla\Delta^{-1}\nabla\varrho_\varepsilon\,\mathrm{d}x\right|\leq c\|\sqrt{\varrho_\varepsilon}\mathbf{u}_\varepsilon\|_{L_x^2}\|\mathbf{u}\|_{W_x^{1,2}}\|\sqrt{\varrho_\varepsilon}\|_{L_x^{2\Gamma}}\|\varrho_\varepsilon\|_{L_x^\Gamma}.$$

Here, in accordance with (7.47),

$$\mathbb{E}[\|\sqrt{\varrho_\varepsilon}\mathbf{u}_\varepsilon\|_{L_x^2}^2\|\mathbf{u}\|_{W_x^{1,2}}^2]\leq c(\delta,F,M_0),$$

while, by virtue of (7.44),

$$\|\sqrt{\varrho_\varepsilon}\|_{L_x^{2\Gamma}}\|\varrho_\varepsilon\|_{L_x^\Gamma}=\|\varrho_\varepsilon\|_{L_x^\Gamma}^{\frac{3}{2}}\in L^q(\Omega),\quad q=\frac{2\Gamma}{3}>2.$$

We have shown the claimed $\mathbb{P}$-equi-integrable equation (7.57), whence strong convergence of ϱ_ε. Consequently, as in Section 4.4.3, we may identify the non-linear terms in (7.55) and hence $[\varrho,\mathbf{u}]$ is a weak martingale solution to (7.49)–(7.50). Stationarity then follows by Lemma 2.11.4 and Lemma 2.11.5. The estimates (7.51) and (7.52) are obtained by weak lower semi-continuity from (7.39) and (7.40), respectively, since the constants were uniform in ε. The same arguments give the energy inequality and hence (7.53). Note that the passage to the limit in the stochastic integral can be justified, for instance with the help of Lemma 2.6.6. □

Remark 7.3.4. It is important to note that there is an essential difference between the strong convergence of the density in the existence theory in Chapter 4 and the above proof. More specifically, the existence theory requires compactness of the initial data which is not available in the present setting. Instead, the fact that the solution is stationary must be used.

7.4 Vanishing artificial pressure limit

As the final step of the proof of our main result, Theorem 7.0.3, it remains to perform the last limit procedure, that is $\delta \to 0$. Recall that, according to Proposition 7.3.3, the stationary solutions constructed in the previous section already satisfy the uniform bounds (7.51) and (7.52). Nevertheless, the pressure estimate as well as the estimate for the energy and velocity from Proposition 7.3.2 all blow up as δ vanishes. Therefore, in order to apply the compactness argument from Section 4.5, it is necessary to improve these estimates. The rest of the construction then proceeds exactly as in Section 4.5.

Proposition 7.4.1. *Let $[\varrho_\delta, \mathbf{u}_\delta]$ be the stationary solution to (7.49)–(7.50), constructed in Proposition 7.3.3. Then the following uniform bound holds true for some $\alpha > 0$ and a.e. $t \in (0,\infty)$:*

$$\mathbb{E}\left[\int_{\mathbb{T}^3} [a\varrho_\delta^{\gamma+\alpha} + \delta\varrho_\delta^{\Gamma+\alpha} + \varrho_\delta^{1+\alpha}|\mathbf{u}_\delta|^2]\,\mathrm{d}x\right] \le c(F, M_0). \tag{7.63}$$

In addition, for some $s > 1$ and for a.e. $T > 0$ and $\tau > 0$,

$$\mathbb{E}\left[\left(\sup_{t\in[T,T+\tau]}\int_{\mathbb{T}^3} E_\delta(\varrho_\delta, \varrho_\delta\mathbf{u}_\delta)\,\mathrm{d}x\right)^s\right] + \mathbb{E}\left[\left(\int_T^{T+\tau} \|\mathbf{u}_\delta\|^2_{W^{1,2}_x}\,\mathrm{d}t\right)^s\right] \le c(\tau, M_0, F, s), \tag{7.64}$$

where the constant is independent of T.

Proof. As far as the pressure estimates are concerned, we use the test function

$$\nabla\Delta^{-1}[\varrho^\alpha - (\varrho^\alpha)_{\mathbb{T}^3}], \quad \alpha > 0.$$

We obtain after a rather tedious but straightforward manipulation the following analogue of (7.46):

$$\begin{aligned}
&\int_T^{T+1}\int_{\mathbb{T}^3} (a\varrho_\delta^{\gamma+\alpha} + \delta\varrho_\delta^{\Gamma+\alpha})\,\mathrm{d}x\,\mathrm{d}t + \int_T^{T+1}\int_{\mathbb{T}^3} \frac{1}{3}\varrho_\delta^{1+\alpha}|\mathbf{u}|^2\,\mathrm{d}x\,\mathrm{d}t \\
&= \int_T^{T+1}\left(\int_{\mathbb{T}^3} (a\varrho_\delta^\gamma + \delta\varrho_\delta^\Gamma)\,\mathrm{d}x\,(\varrho_\delta^\alpha)_{\mathbb{T}^3}\right)\mathrm{d}t + \frac{1}{3}\int_T^{T+1}\left(\int_{\mathbb{T}^3} \varrho_\delta|\mathbf{u}_\delta|^2\,\mathrm{d}x\,\mathrm{d}x\,(\varrho_\delta^\alpha)_{\mathbb{T}^3}\right)\mathrm{d}t \\
&\quad + \int_T^{T+1}\int_{\mathbb{T}^3} (\mu+\eta)\operatorname{div}\mathbf{u}_\delta\,\varrho_\delta^\alpha\,\mathrm{d}x\,\mathrm{d}t \\
&\quad - \int_T^{T+1}\int_{\mathbb{T}^3} \left(\varrho_\delta\mathbf{u}_\delta\otimes\mathbf{u}_\delta - \frac{1}{3}\varrho_\delta|\mathbf{u}_\delta|^2\mathbb{I}\right) : \nabla\Delta^{-1}\nabla[\varrho_\delta^\alpha]\,\mathrm{d}x\,\mathrm{d}t \\
&\quad + \left[\int_{\mathbb{T}^3} \varrho_\delta\mathbf{u}_\delta\cdot\nabla\Delta^{-1}[\varrho_\delta^\alpha - (\varrho_\delta^\alpha)_{\mathbb{T}^3}]\,\mathrm{d}x\right]_{t=T}^{t=T+1} \\
&\quad - \int_T^{T+1}\int_{\mathbb{T}^3} \varrho_\delta\mathbf{u}_\delta\cdot\nabla\Delta^{-1}[\mathrm{d}(\varrho_\delta^\alpha)]\,\mathrm{d}x \\
&\quad - \sum_{k=1}^\infty\int_T^{T+1}\int_{\mathbb{T}^3} \mathbf{G}_k(\varrho_\delta, \varrho_\delta\mathbf{u}_\delta)\cdot\nabla\Delta^{-1}[\varrho_\delta^\alpha - (\varrho_\delta^\alpha)_{\mathbb{T}^3}]\,\mathrm{d}x\,\mathrm{d}W_k.
\end{aligned} \tag{7.65}$$

Next, we evoke the renormalized equation of continuity (7.56)

$$\mathrm{d}\varrho_\delta^\alpha + \operatorname{div}(\varrho_\delta^\alpha \mathbf{u}_\delta)\,\mathrm{d}t + (\alpha - 1)\varrho_\delta^\alpha \operatorname{div}\mathbf{u}_\delta\,\mathrm{d}t = 0,$$

deducing from (7.65)

$$\begin{aligned}
&\int_T^{T+1}\int_{\mathbb{T}^3} (a\varrho_\delta^{\gamma+\alpha} + \delta\varrho_\delta^{\Gamma+\alpha})\,\mathrm{d}x\,\mathrm{d}t + \int_T^{T+1}\int_{\mathbb{T}^3} \frac{1}{3}\varrho_\delta^{1+\alpha}|\mathbf{u}_\delta|^2\,\mathrm{d}x\,\mathrm{d}t \\
&= \int_T^{T+1}\Big(\int_{\mathbb{T}^3} (a\varrho_\delta^\gamma + \delta\varrho^\Gamma)\,\mathrm{d}x\,(\varrho_\delta^\alpha)_{\mathbb{T}^3}\Big)\mathrm{d}t + \frac{1}{3}\int_T^{T+1}\Big(\int_{\mathbb{T}^3} \varrho_\delta|\mathbf{u}_\delta|^2\,\mathrm{d}x\,(\varrho_\delta^\alpha)_{\mathbb{T}^3}\Big)\mathrm{d}t \\
&\quad + \int_T^{T+1}\int_{\mathbb{T}^3} (\mu + \eta)\operatorname{div}\mathbf{u}_\delta\,\varrho_\delta^\alpha\,\mathrm{d}x\,\mathrm{d}t \\
&\quad - \int_T^{T+1}\int_{\mathbb{T}^3} \Big(\varrho_\delta\mathbf{u}_\delta \otimes \mathbf{u}_\delta - \frac{1}{3}\varrho|\mathbf{u}_\delta|^2\mathbb{I}\Big) : \nabla\Delta^{-1}\nabla[\varrho_\delta^\alpha]\,\mathrm{d}x\,\mathrm{d}t \\
&\quad + \Big[\int_{\mathbb{T}^3} \varrho_\delta\mathbf{u}_\delta \cdot \nabla\Delta^{-1}[\varrho_\delta^\alpha - (\varrho_\delta^\alpha)_{\mathbb{T}^3}]\,\mathrm{d}x\Big]_{t=T}^{t=T+1} \\
&\quad + \int_T^{T+1}\int_{\mathbb{T}^3} \varrho_\delta\mathbf{u}_\delta \cdot \nabla\Delta^{-1}[\operatorname{div}(\varrho_\delta^\alpha\mathbf{u}_\delta) + (\alpha - 1)\varrho_\delta^\alpha\operatorname{div}\mathbf{u}_\delta]\,\mathrm{d}x\,\mathrm{d}t \\
&\quad - \sum_{k=1}^\infty \int_T^{T+1}\int_{\mathbb{T}^3} \mathbf{G}_k(\varrho_\delta, \varrho_\delta\mathbf{u}_\delta) \cdot \nabla\Delta^{-1}[\varrho_\delta^\alpha - (\varrho_\delta^\alpha)_{\mathbb{T}^3}]\,\mathrm{d}x\,\mathrm{d}W_k.
\end{aligned} \tag{7.66}$$

Before proceeding, we make the assumption that $0 < \alpha < 1/3$, which implies in particular

$$|(\varrho_\delta^\alpha)_{\mathbb{T}^3}| \le c(M_0), \quad \|\nabla\Delta^{-1}[\varrho_\delta^\alpha - (\varrho_\delta^\alpha)_{\mathbb{T}^3}]\|_{L_x^\infty} \le c(M_0),$$

using Hölder's inequality, Sobolev's embedding, and continuity of $\nabla\Delta^{-1}\nabla$. Passing to expectations in (7.66) and keeping in mind that the processes are stationary, we deduce

$$\begin{aligned}
&\mathbb{E}\Big[\int_{\mathbb{T}^3} [a\varrho_\delta^{\gamma+\alpha} + \delta\varrho_\delta^{\Gamma+\alpha} + \varrho_\delta^{1+\alpha}|\mathbf{u}_\delta|^2]\,\mathrm{d}x\Big] \\
&\le c(M_0)\Big(\mathbb{E}\Big[\int_{\mathbb{T}^3} E_\delta(\varrho_\delta, \varrho_\delta\mathbf{u}_\delta)\,\mathrm{d}x\Big] + 1\Big) + \mathbb{E}\Big[\int_{\mathbb{T}^3} (\mu + \eta)\operatorname{div}\mathbf{u}_\delta\,\varrho_\delta^\alpha\,\mathrm{d}x\Big] \\
&\quad + \mathbb{E}\Big[\int_{\mathbb{T}^3} \Big(\varrho_\delta\mathbf{u}_\delta \otimes \mathbf{u}_\delta - \frac{1}{3}\varrho_\delta|\mathbf{u}_\delta|^2\mathbb{I}\Big) : \nabla\Delta^{-1}\nabla[\varrho_\delta^\alpha]\,\mathrm{d}x\Big] \\
&\quad + \mathbb{E}\Big[\int_{\mathbb{T}^3} \varrho_\delta\mathbf{u}_\delta \cdot \nabla\Delta^{-1}[\operatorname{div}(\varrho_\delta^\alpha\mathbf{u}_\delta) + (\alpha - 1)\varrho_\delta^\alpha\operatorname{div}\mathbf{u}_\delta]\,\mathrm{d}x\Big].
\end{aligned} \tag{7.67}$$

Moreover, using the uniform bound (7.51) we further reduce (7.67) to

$$\begin{aligned}
&\mathbb{E}\Big[\int_{\mathbb{T}^3} [a\varrho_\delta^{\gamma+\alpha} + \delta\varrho_\delta^{\Gamma+\alpha} + \varrho_\delta^{1+\alpha}|\mathbf{u}_\delta|^2]\,\mathrm{d}x\Big] \\
&\le \mathbb{E}\Big[\int_{\mathbb{T}^3} \Big(\varrho_\delta\mathbf{u}_\delta \otimes \mathbf{u}_\delta - \frac{1}{3}\varrho_\delta|\mathbf{u}_\delta|^2\mathbb{I}\Big) : \nabla\Delta^{-1}\nabla[\varrho_\delta^\alpha]\,\mathrm{d}x\Big] \\
&\quad + \mathbb{E}\Big[\int_{\mathbb{T}^3} \varrho_\delta\mathbf{u}_\delta \cdot \nabla\Delta^{-1}[\operatorname{div}(\varrho_\delta^\alpha\mathbf{u}_\delta) + (\alpha - 1)\varrho_\delta^\alpha\operatorname{div}\mathbf{u}_\delta]\,\mathrm{d}x\Big] + c(F, M_0).
\end{aligned}$$

Note that we have applied Young's inequality to the first and second term on the right hand side of (7.67) in order to absorb the arising terms eventually. To control the remaining integrals on the right hand side, we first use Hölder's inequality to obtain

$$\begin{aligned}&\left|\mathbb{E}\left[\int_{\mathbb{T}^3}\left(\varrho_\delta\mathbf{u}_\delta\otimes\mathbf{u}_\delta-\frac{1}{3}\varrho_\delta|\mathbf{u}_\delta|^2\mathbb{I}\right):\nabla\Delta^{-1}\nabla[\varrho_\delta^\alpha]\,\mathrm{d}x\right]\right|\\&\quad\leq c\,\mathbb{E}[\|\sqrt{\varrho_\delta}\mathbf{u}_\delta\|_{L_x^2}\|\mathbf{u}_\delta\|_{L_x^6}\|\sqrt{\varrho_\delta}\nabla\Delta^{-1}\nabla[\varrho_\delta^\alpha]\|_{L_x^3}]\\&\quad\leq c(\mathbb{E}[\|\sqrt{\varrho}\mathbf{u}_\delta\|_{L_x^2}^2\|\mathbf{u}_\delta\|_{W_x^{1,2}}^2]+\mathbb{E}[\|\sqrt{\varrho_\delta}\nabla\Delta^{-1}\nabla[\varrho_\delta^\alpha]\|_{L_x^3}^2]).\end{aligned}$$

Furthermore, we have

$$\begin{aligned}&\|\sqrt{\varrho_\delta}\nabla\Delta^{-1}\nabla[\varrho_\delta^\alpha]\|_{L_x^3}^2\leq\|\sqrt{\varrho_\delta}\|_{L_x^{2\gamma}}^2\|\nabla\Delta^{-1}\nabla[\varrho_\delta^\alpha]\|_{L_x^q}^2,\\&\frac{1}{2\gamma}+\frac{1}{q}=\frac{1}{3},\quad\gamma>\frac{3}{2}.\end{aligned}$$

Now, we choose $\alpha>0$ so small that $\alpha q\leq 1$, to conclude

$$\|\sqrt{\varrho_\delta}\operatorname{div}\nabla\Delta^{-1}\nabla[\varrho_\delta^\alpha]\|_{L_x^3}^2\leq c(M_0)\|\varrho_\delta\|_{L_x^\gamma}.$$

Similarly, we estimate

$$\left|\int_{\mathbb{T}^3}\varrho_\delta\mathbf{u}_\delta\cdot\nabla\Delta^{-1}\operatorname{div}[\varrho_\delta^\alpha\mathbf{u}_\delta]\,\mathrm{d}x\right|\leq\|\sqrt{\varrho_\delta}\mathbf{u}_\delta\|_{L_x^2}\|\sqrt{\varrho_\delta}\|_{L_x^{2\gamma}}\|\nabla\Delta^{-1}\nabla[\varrho_\delta^\alpha\mathbf{u}_\delta]\|_{L_x^q},$$

where

$$\frac{1}{2}+\frac{1}{2\gamma}+\frac{1}{q}=1,\quad\text{in particular } q<6 \text{ if } \gamma>\frac{3}{2},$$

and where

$$\|\nabla\Delta^{-1}\nabla[\varrho_\delta^\alpha\mathbf{u}]\|_{L_x^q}\leq\|\varrho_\delta^\alpha\mathbf{u}_\delta\|_{L_x^q}\leq\|\mathbf{u}_\delta\|_{L_x^6}\|\varrho_\delta^\alpha\|_{L_x^s},\quad\frac{1}{6}+\frac{1}{s}=\frac{1}{q}.$$

Taking $\alpha s\leq 1$ we get, similarly to the above,

$$\left|\int_{\mathbb{T}^3}\varrho_\delta\mathbf{u}_\delta\cdot\nabla\Delta^{-1}\operatorname{div}[\varrho_\delta^\alpha\mathbf{u}_\delta]\,\mathrm{d}x\right|\leq c(M_0)(\|\sqrt{\varrho_\delta}\mathbf{u}_\delta\|_{L_x^2}^2\|\mathbf{u}_\delta\|_{W_x^{1,2}}^2+\|\varrho_\delta\|_{L_x^\gamma}).$$

Finally, we have

$$\begin{aligned}\left|\int_{\mathbb{T}^3}\varrho_\delta\mathbf{u}_\delta\cdot\nabla\Delta^{-1}[\varrho_\delta^\alpha\operatorname{div}\mathbf{u}_\delta]\,\mathrm{d}x\right|&\leq\|\sqrt{\varrho_\delta}\|_{L_x^{2\gamma}}\|\sqrt{\varrho_\delta}\mathbf{u}_\delta\|_{L_x^2}\|\nabla\Delta_x^{-1}[\varrho_\delta^\alpha\operatorname{div}\mathbf{u}_\delta]\|_{L_x^q}\\&\leq\frac{1}{2}(\|\varrho_\delta\|_{L_x^\gamma}+\|\sqrt{\varrho_\delta}\mathbf{u}\|_{L_x^2}^2\|\nabla\Delta_x^{-1}[\varrho_\delta^\alpha\operatorname{div}\mathbf{u}_\delta]\|_{L_x^q}^2),\end{aligned}$$

where

$$\frac{1}{2\gamma}+\frac{1}{2}+\frac{1}{q}=1,\quad q<6\text{ if }\gamma>\frac{3}{2}.$$

As the $\nabla\Delta^{-1}$-operator gains one derivative, we get, by means of the standard Sobolev embedding,

$$\|\nabla\Delta^{-1}[\varrho_\delta^\alpha \operatorname{div}\mathbf{u}_\delta]\|_{L_x^q} \le \|\varrho_\delta^\alpha \operatorname{div}\mathbf{u}_\delta\|_{L_x^r}, \quad r<2.$$

Thus, similarly to the previous steps, we conclude

$$\left|\int_{\mathbb{T}^3} \varrho_\delta \mathbf{u}_\delta \cdot \nabla\Delta^{-1}[\varrho_\delta^\alpha \operatorname{div}\mathbf{u}_\delta]\,\mathrm{d}x\right| \le c(M_0)(\|\sqrt{\varrho_\delta}\mathbf{u}_\delta\|_{L_x^2}^2 \|\mathbf{u}_\delta\|_{W_x^{1,2}}^2 + \|\varrho_\delta\|_{L_x^\gamma}).$$

Summing up the above estimates, we obtain

$$\mathbb{E}\left[\int_{\mathbb{T}^3}\left[a\varrho_\delta^{\gamma+\alpha} + \delta\varrho_\delta^{\Gamma+\alpha} + \frac{1}{3}\varrho_\delta^{1+\alpha}|\mathbf{u}_\delta|^2\right]\mathrm{d}x\right] \le \mathbb{E}\left[\int_{\mathbb{T}^3} E_\delta(\varrho_\delta,\varrho_\delta\mathbf{u}_\delta)\,\mathrm{d}x \int_{\mathbb{T}^3}\|\mathbf{u}_\delta\|_{W_x^{1,2}}^2\,\mathrm{d}x\right] + c(F,M_0),$$

where we have absorbed the term $\|\varrho_\delta\|_{L_x^\gamma}$ in the left hand side. We close the estimates by evoking (7.52) and absorbing the $E_\delta(\varrho_\delta,\varrho_\delta\mathbf{u}_\delta)$-term in the left hand side. Thus we may conclude that any global-in-time stationary solutions admit the uniform bound (7.63) as well as

$$\mathbb{E}\left[\left(1+\int_{\mathbb{T}^3} E_\delta(\varrho_\delta,\varrho_\delta\mathbf{u}_\delta)\,\mathrm{d}x\right)\|\mathbf{u}_\delta\|_{W_x^{1,2}}^2\right] \le c(F,M_0). \tag{7.68}$$

Finally, we claim that

$$\mathbb{E}\left[\left(\int_{\mathbb{T}^3} E_\delta(\varrho_\delta,\varrho_\delta\mathbf{u}_\delta)\,\mathrm{d}x\right)^s\right] \le c$$

for a certain $s=s(\alpha)>1$. Indeed the ϱ_δ-dependent terms can be estimated directly by (7.63) while, by Hölder's inequality and Sobolev's embedding,

$$\begin{aligned}\left(\int_{\mathbb{T}^3}\varrho_\delta|\mathbf{u}_\delta|^2\,\mathrm{d}x\right)^s &\le (\|\sqrt{\varrho_\delta}\mathbf{u}_\delta\|_{L_x^2}\|\mathbf{u}_\delta\|_{L_x^6})^s\|\sqrt{\varrho_\delta}\|_{L_x^3}^s \\ &\le c\left[\|\sqrt{\varrho_\delta}\mathbf{u}_\delta\|_{L_x^2}^2\|\mathbf{u}_\delta\|_{W_x^{1,2}}^2 + \left(\int_{\mathbb{T}^3}\varrho_\delta^\gamma\,\mathrm{d}x\right)^{\frac{1}{\gamma(2-s)}}\right].\end{aligned}$$

Note that we also took into account $\gamma>\frac{3}{2}$. This can be estimated by (7.68), provided $s<2-\frac{1}{\gamma}$. The term with ϱ_δ^γ and $\delta\varrho_\delta^\Gamma$ is estimated by Jensen's inequality. Now we go back to the energy inequality (7.53). Using (7.41), we obtain, after taking the power s and the supremum in time and expectation,

$$\begin{aligned}&\mathbb{E}\left[\left(\sup_{t\in[T,T+\tau]}\int_{\mathbb{T}^3} E_\delta(\varrho_\delta,\varrho_\delta\mathbf{u}_\delta)\,\mathrm{d}x\right)^s\right] + \mathbb{E}\left[\left(\int_T^{T+\tau}\|\mathbf{u}_\delta\|_{W_x^{1,2}}^2\,\mathrm{d}t\right)^s\right] \\ &\le c(F,M_0) + \mathbb{E}\left[\sup_{t\in[T,T+\tau]}\left|\sum_{k=1}^\infty\int_T^t\int_{\mathbb{T}^3}\mathbf{G}_k(\varrho_\delta,\varrho_\delta\mathbf{u}_\delta)\cdot\mathbf{u}_\delta\,\mathrm{d}x\,\mathrm{d}W_k\right|^s\right].\end{aligned}$$

Note that all terms are well-defined by (7.68). Here, the second term on the right hand side is controlled by (7.42) and the Burkholder–Davis–Gundy inequality, similarly to (7.48), as follows:

$$\mathbb{E}\left[\sup_{t\in[T,T+\tau]}\left|\sum_{k=1}^{\infty}\int_T^t\int_{\mathbb{T}^3}\mathbf{G}_k(\varrho_\delta,\varrho_\delta\mathbf{u}_\delta)\cdot\mathbf{u}_\delta\,\mathrm{d}x\,\mathrm{d}W_k\right|^s\right]$$
$$\leq c(s,M_0)\,\mathbb{E}\left[\int_T^{T+\tau}\left(\int_{\mathbb{T}^3}\frac{1}{2}\varrho_\delta|\mathbf{u}_\delta|^2\,\mathrm{d}x\right)^{\frac{s}{2}}\mathrm{d}t\right],$$

which can be again estimated by (7.63). We therefore conclude that (7.64) holds true for a.e. $T>0$, where the constant depends on τ but is independent of T. □

Finally, everything is in hand to complete the proof of Theorem 7.0.3.

Proof of Theorem 7.0.3. We follow the lines of Section 4.5. In view of Proposition 7.4.1, we are able to apply the Jakubowski–Skorokhod representation theorem and obtain convergence of $[\varrho_\delta,\mathbf{u}_\delta]$ (in fact, we obtain a new family of martingale solutions defined on a new probability space but keep the original notation for simplicity) to a stationary weak martingale solution of

$$-\int_0^\infty\partial_t\phi\int_{\mathbb{T}^3}\varrho\psi\,\mathrm{d}x\,\mathrm{d}t=\int_0^\infty\phi\int_{\mathbb{T}^3}\varrho\mathbf{u}\cdot\nabla\psi\,\mathrm{d}x\,\mathrm{d}t,\qquad\int_{\mathbb{T}^3}\varrho\,\mathrm{d}x=M_0,$$

for all $\phi\in C_c^\infty((0,\infty))$ and all $\psi\in C^\infty(\mathbb{T}^3)$ $\mathbb{P}$-a.s., and

$$\begin{aligned}-\int_0^\infty&\partial_t\phi\int_{\mathbb{T}^3}\varrho\mathbf{u}\cdot\boldsymbol{\varphi}\,\mathrm{d}x\,\mathrm{d}t\\&=\int_0^\infty\phi\int_{\mathbb{T}^3}\varrho\mathbf{u}\otimes\mathbf{u}:\nabla\boldsymbol{\varphi}\,\mathrm{d}x\,\mathrm{d}t+\int_0^\infty\phi\int_{\mathbb{T}^3}a\overline{\varrho^\gamma}\operatorname{div}\boldsymbol{\varphi}\,\mathrm{d}x\,\mathrm{d}t\\&\quad-\int_0^\infty\phi\int_{\mathbb{T}^3}\mathbb{S}(\nabla\mathbf{u}):\nabla\boldsymbol{\varphi}\,\mathrm{d}x\,\mathrm{d}t+\int_0^\infty\phi\int_{\mathbb{T}^3}\overline{\mathbb{G}(\varrho,\varrho\mathbf{u})}\cdot\boldsymbol{\varphi}\,\mathrm{d}x\,\mathrm{d}W,\end{aligned}$$

for all $\psi\in C_c^\infty((0,\infty))$ and all $\boldsymbol{\varphi}\in C^\infty(\mathbb{T}^3)$ $\mathbb{P}$-a.s. Here, the bars denote the corresponding weak limits. In addition, ϱ satisfies the renormalized equation of continuity.

In order to identify the non-linear density-dependent terms, we keep Remark 7.3.4 in mind and apply the effective viscous flux method in the same way as in Section 4.5, thereby completing the proof. Note that, similarly to Section 7.3, even the limited moment estimates from Proposition 7.4.1 are sufficient for the passage to the limit. □

8 Singular limits

The Navier–Stokes system of partial differential equations describes the entire spectrum of possible motions, ranging from sound waves and cyclone waves in the atmosphere to models of gaseous stars in astrophysics. Its generality constitutes a serious defect of the equations from the point of view of applications. Eliminating unwanted or unimportant modes of motion and building on the essential balances between flow fields allows one to better focus on a particular class of phenomena and to potentially achieve a deeper understanding of the problem. Scaling and asymptotic analysis play an important role in this approach. By scaling the equations, meaning by choosing appropriately the system of the reference units, the parameters determining the behavior of the system become explicit. Asymptotic analysis provides a useful tool in situations when certain of these parameters, called characteristic numbers, vanish or become infinite. As a matter of fact, most, if not all, mathematical models used in fluid mechanics rely on formal asymptotic analysis of more complex systems. The concept of incompressible fluid itself should be viewed as a convenient idealization of a medium in which the speed of sound dominates the characteristic velocity.

We consider the *dimensionless rescaled Navier–Stokes system* in the following form:

$$[\mathrm{Sr}]\mathrm{d}\varrho + \operatorname{div}(\varrho\mathbf{u})\,\mathrm{d}t = 0, \tag{8.1}$$

$$[\mathrm{Sr}]\mathrm{d}(\varrho\mathbf{u}) + \operatorname{div}(\varrho\mathbf{u}\otimes\mathbf{u})\,\mathrm{d}t + \frac{1}{[\mathrm{Ma}]^2}\nabla p(\varrho)\,\mathrm{d}t = \frac{1}{[\mathrm{Re}]}\operatorname{div}\mathbb{S}(\nabla\mathbf{u})\,\mathrm{d}t + \frac{1}{[\mathrm{Fr}]^2}\mathbb{G}(\varrho,\varrho\mathbf{u})\,\mathrm{d}W, \tag{8.2}$$

where

$$p(\varrho) = a\varrho^{\gamma}, \quad \gamma > \frac{N}{2}, \tag{8.3}$$

and

$$\mathbb{S}(\nabla\mathbf{u}) = \mu\Big(\nabla\mathbf{u} + \nabla^t\mathbf{u} - \frac{2}{3}\operatorname{div}\mathbf{u}\mathbb{I}\Big) + \lambda\operatorname{div}\mathbf{u}\mathbb{I}, \tag{8.4}$$

with viscosity coefficients $\mu > 0$ and $\lambda \geq 0$. The equations contain dimensionless parameters – the characteristic numbers: the *Strouhal number* [Sr], the *Mach number* [Ma], the *Reynolds number* [Re], and the *Froude number* [Fr]. We refer to [FN09, Chapter 4] for the derivation of system (8.1)–(8.2) as well as the physical interpretation of the characteristic numbers.

Singular limit processes bridge the gap between fluid motions considered in different geometries, time scales, and/or under different constitutive relations as the case may be. As already pointed out, they describe the situation when one or several of the characteristic numbers vanish or become infinite. In their pioneering paper, Klainerman–Majda [KM81] proposed a general approach to these problems in the con-

https://doi.org/10.1515/9783110492552-008

text of hyperbolic conservation laws. In particular, they examined the passage from compressible to incompressible fluid flow motion via the low Mach number limit. As the problems are typically non-linear, the method applies in general only to short time intervals on which regular solutions are known to exist. A qualitatively new way, at least in the framework of viscous fluids, has been opened by the mathematical theory of weak solutions developed by Lions [Lio98]. In a series of papers, Lions–Masmoudi [LM98, LM99] (see also Desjardins–Grenier [DG99] and Desjardins et al. [DGLM99]) studied various singular limits for the barotropic Navier–Stokes system, among which ssible low Mach number limit.

Another distinguished singular regime corresponds to the inviscid limit characterized by large values of the Reynolds number. There is a vast amount of literature concerning the inviscid limit for the *incompressible* deterministic Navier–Stokes system; see, e.g., Kato [Kat84], Temam–Wang [TW97, TW02], Wang et al. [WXZ12], the survey articles by Weinan [Wei00] and Masmoudi [Mas06], and the references cited therein. On the other hand, much less seems to be known in the context of compressible fluids, even in the deterministic case. Recently several results in this direction have been obtained by Sueur [Sue14] for the barotropic Navier–Stokes system and related issues were discussed by Wang–Williams [WW12].

In this chapter we present a rigorous mathematical approach to the asymptotic analysis of the Navier–Stokes system (8.1)–(8.4) with stochastic perturbations. Notably, we examine the incompressible and the inviscid–incompressible limits. For the readers' convenience we restate Definition 3.4.1, giving the concept of dissipative martingale solution to (8.1)–(8.4), on which our theory is built up.

Definition 8.0.1 (Dissipative martingale solution). Let $\Lambda = \Lambda(\varrho, \mathbf{q})$ be a Borel probability measure on $L^1(\mathbb{T}^N) \times L^1(\mathbb{T}^N)$ such that

$$\Lambda\{\varrho \geq 0\} = 1, \qquad \int_{L^1_x \times L^1_x} \left| \int_{\mathbb{T}^N} \left[\frac{[\mathrm{Sr}]|\mathbf{q}|^2}{\varrho} + \frac{1}{[\mathrm{Ma}]^2} P(\varrho) \right] \mathrm{d}x \right|^r \mathrm{d}\Lambda(\varrho, \mathbf{q}) < \infty, \tag{8.5}$$

where $r \geq 1$ will be determined below.

The quantity $((\Omega, \mathfrak{F}, (\mathfrak{F}_t)_{t\geq 0}, \mathbb{P}), \varrho, \mathbf{u}, W)$ is called a *dissipative martingale solution* to (8.1)–(8.4) with the initial law Λ if:

(1) $(\Omega, \mathfrak{F}, (\mathfrak{F}_t)_{t\geq 0}, \mathbb{P})$ is a stochastic basis with a complete right-continuous filtration;
(2) W is a cylindrical $(\mathfrak{F}_t)$-Wiener process;
(3) the density ϱ and the velocity $\mathbf{u}$ are random distributions adapted to $(\mathfrak{F}_t)_{t\geq 0}$, $\varrho \geq 0$ $\mathbb{P}$-a.s.;
(4) there exists an $\mathfrak{F}_0$-measurable random variable $[\varrho_0, \mathbf{u}_0]$ such that $\Lambda = \mathcal{L}[\varrho_0, \varrho_0 \mathbf{u}_0]$;
(5) the equation of continuity

$$-\int_0^T \partial_t \phi \int_{\mathbb{T}^N} [\mathrm{Sr}] \varrho \psi \, \mathrm{d}x \, \mathrm{d}t = \phi(0) \int_{\mathbb{T}^N} [\mathrm{Sr}] \varrho_0 \psi \, \mathrm{d}x + \int_0^T \phi \int_{\mathbb{T}^N} \varrho \mathbf{u} \cdot \nabla \psi \, \mathrm{d}x \, \mathrm{d}t \tag{8.6}$$

holds for all $\phi \in C_c^\infty([0, T))$ and all $\psi \in C^\infty(\mathbb{T}^N)$ $\mathbb{P}$-a.s.;

(6) the momentum equation

$$
\begin{aligned}
&-\int_0^T \partial_t \phi \int_{\mathbb{T}^N} [\mathrm{Sr}]\varrho\mathbf{u}\cdot\boldsymbol{\varphi}\,\mathrm{d}x\,\mathrm{d}t - \phi(0)\int_{\mathbb{T}^N}[\mathrm{Sr}]\varrho_0\mathbf{u}_0\cdot\boldsymbol{\varphi}\,\mathrm{d}x \\
&= \int_0^T \phi \int_{\mathbb{T}^N}\Big[\varrho\mathbf{u}\otimes\mathbf{u}:\nabla\boldsymbol{\varphi} + \frac{1}{[\mathrm{Ma}]^2}p(\varrho)\operatorname{div}\boldsymbol{\varphi}\Big]\,\mathrm{d}x\,\mathrm{d}t \\
&\quad - \int_0^T \phi \int_{\mathbb{T}^N}\frac{1}{[\mathrm{Re}]}\mathbb{S}(\nabla\mathbf{u}):\nabla\boldsymbol{\varphi}\,\mathrm{d}x\,\mathrm{d}t \\
&\quad + \frac{1}{[\mathrm{Fr}]^2}\sum_{k=1}^{\infty}\int_0^T\phi\int_{\mathbb{T}^N}\mathbf{G}_k(\varrho,\varrho\mathbf{u})\cdot\boldsymbol{\varphi}\,\mathrm{d}x\,\mathrm{d}W_k
\end{aligned}
\tag{8.7}
$$

holds for all $\phi\in C_c^\infty([0,T))$ and all $\boldsymbol{\varphi}\in C^\infty(\mathbb{T}^N)$ $\mathbb{P}$-a.s.;

(7) the energy inequality

$$
\begin{aligned}
&-\int_0^T\partial_t\phi\int_{\mathbb{T}^N}\Big[[\mathrm{Sr}]\frac{1}{2}\varrho|\mathbf{u}|^2+\frac{1}{[\mathrm{Ma}]^2}P(\varrho)\Big]\,\mathrm{d}x\,\mathrm{d}t + \frac{1}{[\mathrm{Re}]}\int_0^T\phi\int_{\mathbb{T}^N}\mathbb{S}(\nabla\mathbf{u}):\nabla\mathbf{u}\,\mathrm{d}x\,\mathrm{d}t \\
&\le \phi(0)\int_{\mathbb{T}^N}\Big[[\mathrm{Sr}]\frac{1}{2}\varrho_0|\mathbf{u}_0|^2+\frac{1}{[\mathrm{Ma}]^2}P(\varrho_0)\Big]\,\mathrm{d}x \\
&\quad + \frac{1}{[\mathrm{Fr}]^4}\frac{1}{2}\sum_{k=1}^{\infty}\int_0^T\phi\int_{\mathbb{T}^N}\varrho^{-1}|\mathbf{G}_k(\varrho,\varrho\mathbf{u})|^2\,\mathrm{d}x\,\mathrm{d}t \\
&\quad + \frac{1}{[\mathrm{Fr}]^2}\sum_{k=1}^{\infty}\int_0^T\phi\int_{\mathbb{T}^N}\mathbf{G}_k(\varrho,\varrho\mathbf{u})\cdot\mathbf{u}\,\mathrm{d}x\,\mathrm{d}W_k
\end{aligned}
\tag{8.8}
$$

holds for all $\phi\in C_c^\infty([0,T))$, $\phi\ge 0$, $\mathbb{P}$-a.s.;

(8) if $b\in C^1(\mathbb{R})$ such that $b'(z)=0$ for all $z\ge M_b$, then, for all $\phi\in C_c^\infty([0,T))$ and all $\psi\in C^\infty(\mathbb{T}^N)$, we have $\mathbb{P}$-a.s.

$$
\begin{aligned}
-\int_0^T\partial_t\phi\int_{\mathbb{T}^N}[\mathrm{Sr}]b(\varrho)\psi\,\mathrm{d}x\,\mathrm{d}t &= \phi(0)\int_{\mathbb{T}^N}[\mathrm{Sr}]b(\varrho_0)\psi\,\mathrm{d}x + \int_0^T\phi\int_{\mathbb{T}^N}b(\varrho)\mathbf{u}\cdot\nabla\psi\,\mathrm{d}x\,\mathrm{d}t \\
&\quad - \int_0^T\phi\int_{\mathbb{T}^N}(b'(\varrho)\varrho-b(\varrho))\operatorname{div}\mathbf{u}\,\psi\,\mathrm{d}x\,\mathrm{d}t.
\end{aligned}
\tag{8.9}
$$

8.1 Incompressible limit

The incompressible regime corresponds to the scaling $[\mathrm{Sr}]=[\mathrm{Re}]=[\mathrm{Fr}]\approx 1$, $[\mathrm{Ma}]\approx\varepsilon\to 0$. Accordingly, when $\varepsilon\to 0$, the speed of the acoustic waves represented by the gradient component of the velocity field becomes infinite. At the same time the fluid density approaches a constant and the velocity becomes solenoidal. The limit behavior is described by the standard *incompressible* Navier–Stokes system. Our first singular limit result is concerned with the *compressible–incompressible* scenario in the context of stochastically driven fluids. Specifically, we consider system (8.1)–(8.4) with $[\mathrm{Sr}]=[\mathrm{Re}]=[\mathrm{Fr}]=1$, $[\mathrm{Ma}]=\varepsilon$ and study the asymptotic behavior of solutions in the low Mach number regime $\varepsilon \to 0$. More specifically, for dimension $N = 2,3$, we study the limit as $\varepsilon \to 0$ in the following system:

$$\mathrm{d}\varrho + \operatorname{div}(\varrho \mathbf{u})\,\mathrm{d}t = 0, \tag{8.10}$$

$$\mathrm{d}(\varrho \mathbf{u}) + \operatorname{div}(\varrho \mathbf{u} \otimes \mathbf{u})\,\mathrm{d}t + \frac{1}{\varepsilon^2}\nabla p(\varrho^\gamma)\,\mathrm{d}t = \operatorname{div}\mathbb{S}(\nabla \mathbf{u})\,\mathrm{d}t + \mathbb{G}(\varrho, \varrho \mathbf{u})\,\mathrm{d}W, \tag{8.11}$$

where

$$p(\varrho) = \varrho^\gamma, \quad \gamma > \frac{N}{2}, \tag{8.12}$$

and

$$\mathbb{S}(\nabla \mathbf{u}) = \mu\left(\nabla \mathbf{u} + \nabla^t \mathbf{u} - \frac{2}{3}\operatorname{div}\mathbf{u}\mathbb{I}\right) + \lambda \operatorname{div}\mathbf{u}\mathbb{I}, \tag{8.13}$$

with viscosity coefficients $\mu > 0$ and $\lambda \geq 0$.

The parameter ε in (8.11) is proportional to the Mach number, i.e., the ratio of the characteristic flow velocity and the speed of sound. From a physical point of view, the fluid should asymptotically behave like an incompressible one if the density is close to a constant, the velocity is small, and we look at large time scales. A suitable scaling of the Navier–Stokes system results in (8.11) with a small parameter ε; see Klein et al. [KBS⁺01]. In the limit of (8.10)–(8.13) we recover the stochastic Navier–Stokes system for incompressible fluids, that is,

$$\mathrm{d}\mathbf{u} + \left[\operatorname{div}(\mathbf{u} \otimes \mathbf{u}) - \mu \Delta \mathbf{u} + \nabla \pi\right]\mathrm{d}t = \Phi(\mathbf{u})\,\mathrm{d}W, \tag{8.14}$$

$$\operatorname{div}(\mathbf{u}) = 0, \tag{8.15}$$

where π denotes the associated pressure and $\Phi(\mathbf{u}) = \mathcal{P}_H \mathbb{G}(1, \mathbf{u})$, with $\mathcal{P}_H$ being the Helmholtz projection onto the space of solenoidal vector fields; cf. Theorem A.1.4. To be more precise, we show that, for a given initial law Λ for (8.14)–(8.15) corresponding to *ill-prepared* initial data for the compressible Navier–Stokes system (8.10)–(8.13), the approximate densities converge to a constant whereas the velocities converge *in law* to a weak martingale solution to the incompressible Navier–Stokes system (8.14)–(8.15) with the initial law Λ. The precise statement is given in Theorem 8.1.6 below. This result is then strengthened in dimension two, where we are able to prove the convergence in probability of the velocities; cf. Theorem 8.1.7.

Remark 8.1.1. In analogy with the deterministic case, the initial data are *ill-prepared* if

$$\int_{\mathbb{T}^N}\left[\frac{1}{2}\frac{|\mathbf{q}_{0,\varepsilon}|^2}{\varrho_{0,\varepsilon}} + \frac{1}{\varepsilon^2}\big(P(\varrho_{0,\varepsilon}) - P'(\overline{\varrho})(\varrho_{0,\varepsilon} - \overline{\varrho}) - P(\overline{\varrho})\big)\right]\mathrm{d}x < \infty \quad \mathbb{P}\text{-a.s.}$$

In other words, the "distance" between the initial density ϱ and its integral mean $\overline{\varrho}$ is controlled in terms of the Mach number. Note that this is a necessary condition for the solutions of the compressible system to remain bounded uniformly for $\varepsilon \to 0$.

The data are *well-prepared* if

$$\mathbf{u}_{0,\varepsilon} = \mathbf{q}_{0,\varepsilon}/\varrho_{0,\varepsilon} \to \mathbf{v}_0, \quad \text{where } \operatorname{div}\mathbf{v}_0 = 0,$$
$$\int_{\mathbb{T}^N} \frac{1}{\varepsilon^2}\big(P(\varrho_{0,\varepsilon}) - P'(\overline{\varrho})(\varrho_{0,\varepsilon} - \overline{\varrho}) - P(\overline{\varrho})\big)\,\mathrm{d}x \to 0,$$

as $\varepsilon \to 0$ $\mathbb{P}$-a.s.

Our approach is based on the concept of dissipative martingale solution (in the sense of Definition 8.0.1) to the compressible Navier–Stokes system (8.10)–(8.13), whose existence was established in Chapter 4. Similarly to its deterministic counterpart, the low Mach number limit problem features two essential difficulties:
- finding suitable uniform bounds independent of the scaling parameter ε;
- analysis of rapidly oscillating *acoustic waves*, at least in the case of ill-prepared initial data.

Here, the necessary uniform bounds follow directly from the associated stochastic analogue of the energy inequality exploiting the basic properties of Itô's stochastic integral; see Section 8.1.3.1. The propagation of acoustic waves is described by a stochastic variant of Lighthill's acoustic analogy: a linear wave equation driven by a stochastic forcing; see Section 8.1.3.2. The desired estimates are obtained via an analogue of the deterministic approach, specifically the so-called local method proposed by Lions–Masmoudi [LM98, LM99], adapted to the stochastic setting.

The proof makes use of the stochastic compactness method presented in Section 2.6. Similarly to the previous chapters of this book, we rely on the Jakubowski–Skorokhod Theorem 2.7.1, which applies to a large class of topological spaces, including separable Banach spaces with weak topology. In the case of two space dimensions pathwise uniqueness for the limit system (8.14)–(8.15) holds true and we gain a stronger convergence result; see Theorem 8.1.7. This is based on the discussion in Section 2.10, particularly the version of the Gyöngy–Krylov characterization of convergence in probability presented in Theorem 2.10.3.

8.1.1 Incompressible Navier–Stokes equations

In this subsection we are concerned with the incompressible stochastic Navier–Stokes system

$$\mathrm{d}\mathbf{u} + \big[\operatorname{div}(\mathbf{u}\otimes\mathbf{u}) - \mu\Delta\mathbf{u} + \nabla\pi\big]\,\mathrm{d}t = \Phi(\mathbf{u})\,\mathrm{d}W, \tag{8.16}$$
$$\operatorname{div}(\mathbf{u}) = 0, \tag{8.17}$$

where $\Phi : L^2(\mathbb{T}^N) \to L_2(\mathfrak{U}; L^2(\mathbb{T}^N))$ is such that

$$\begin{aligned} \|\Phi(\mathbf{v})\|^2_{L_2(\mathfrak{U};L^2_x)} &\le C(1 + \|\mathbf{v}\|^2_{L^2_x}), \\ \|\Phi(\mathbf{v}) - \Phi(\mathbf{w})\|^2_{L_2(\mathfrak{U};L^2_x)} &\le C\|\mathbf{v} - \mathbf{w}\|^2_{L^2_x}, \end{aligned} \tag{8.18}$$

for all $\mathbf{v}, \mathbf{w} \in L^2(\mathbb{T}^N)$. Several notions of solution to (8.16)–(8.17) are typically considered depending on the space dimension. From the PDE point of view, we restrict ourselves to weak solutions (although more can be proved in dimension two), i.e., (8.14)–(8.15) is satisfied in the sense of distributions. From the probabilistic point of view, we consider two concepts, namely, pathwise solutions and martingale solutions. The reader is referred to Section 2.5 for the introduction of these two concepts. Recall that existence of a pathwise solution is stronger and implies existence of a martingale solution. Besides, due to the classical Yamada–Watanabe type argument, existence of a pathwise solution follows from existence of a martingale solution together with pathwise uniqueness. In dimension three, existence of a strong solution, which is closely related to uniqueness, belongs to the celebrated Millennium Prize Problems and remains unsolved. Therefore, we consider weak martingale solutions with initial data gven by a probaility measure on

$$L^2_{\mathrm{div}}(\mathbb{T}^N) = \Big\{ \mathbf{v} \in L^2(\mathbb{T}^N) : \int_{\mathbb{T}^N} \mathbf{v} \cdot \nabla \varphi \, \mathrm{d}x = 0 \ \forall \varphi \in C^\infty(\mathbb{T}^N) \Big\};$$

see for instance Capiński–Gątarek [CG94] or Flandoli–Gątarek [FG95].

Definition 8.1.2. Let Λ be a Borel probability measure on $L^2_{\mathrm{div}}(\mathbb{T}^N)$. Then

$$((\Omega, \mathfrak{F}, (\mathfrak{F}_t)_{t\ge 0}, \mathbb{P}), \mathbf{u}, W)$$

is called a *weak martingale solution* to (8.16)–(8.17) with the initial data Λ, provided:
(1) $(\Omega, \mathfrak{F}, (\mathfrak{F}_t)_{t\ge 0}, \mathbb{P})$ is a stochastic basis with a complete right-continuous filtration;
(2) W is an $(\mathfrak{F}_t)$-cylindrical Wiener process;
(3) the velocity field $\mathbf{u}$ is $(\mathfrak{F}_t)$-adapted and

$$\mathbf{u} \in C_w([0,T]; L^2_{\mathrm{div}}(\mathbb{T}^N)) \cap L^2(0,T; W^{1,2}_{\mathrm{div}}(\mathbb{T}^N)) \quad \mathbb{P}\text{-a.s.};$$

(4) there exists an $\mathfrak{F}_0$-measurable random variable $\mathbf{u}_0$ such that $\Lambda = \mathcal{L}[\mathbf{u}_0]$;
(5) the momentum equation

$$\begin{aligned} &-\int_0^T \partial_t \phi \int_{\mathbb{T}^N} \mathbf{u} \cdot \boldsymbol{\varphi} \, \mathrm{d}x \, \mathrm{d}t - \phi(0) \int_{\mathbb{T}^N} \mathbf{u}_0 \cdot \boldsymbol{\varphi} \, \mathrm{d}x \\ &= \int_0^T \phi \int_{\mathbb{T}^N} \mathbf{u} \otimes \mathbf{u} : \nabla \boldsymbol{\varphi} \, \mathrm{d}x \, \mathrm{d}t - \mu \int_0^T \phi \int_{\mathbb{T}^N} \nabla \mathbf{u} : \nabla \boldsymbol{\varphi} \, \mathrm{d}x \, \mathrm{d}t \\ &\quad + \int_0^T \phi \int_{\mathbb{T}^N} \Phi(\mathbf{u}) \cdot \boldsymbol{\varphi} \, \mathrm{d}x \, \mathrm{d}W \end{aligned}$$

holds for all $\phi \in C_c^\infty([0,T))$ and all $\boldsymbol{\varphi} \in C^\infty_{\mathrm{div}}(\mathbb{T}^N)$ $\mathbb{P}$-a.s.

Under the condition (8.18), the following existence result holds true and can be found for instance in Capiński–Gątarek [CG94] or Flandoli–Gątarek [FG95].

Theorem 8.1.3. *Let $N = 2, 3$ and assume that Φ satisfies* (8.18). *Let Λ be a Borel probability measure on $L^2_{\mathrm{div}}(\mathbb{T}^N)$ such that, for some $r > 2$,*

$$\int_{L^2_x} \|\mathbf{v}\|^r_{L^2_x} \,\mathrm{d}\Lambda(\mathbf{v}) < \infty.$$

Then there exists a weak martingale solution to (8.16)–(8.17) *in the sense of Definition* 8.1.2 *with initial law Λ.*

In dimension two, pathwise uniqueness for weak solutions is known under (8.18); we refer the reader for instance to Capiński–Cutland [CC91] or Capiński [Cap93]. Consequently, we may work with the definition of a weak pathwise solution.

Definition 8.1.4. Let $(\Omega, \mathfrak{F}, (\mathfrak{F}_t)_{t\geq 0}, \mathbb{P})$ be a given stochastic basis with a complete right-continuous filtration and an $(\mathfrak{F}_t)$-cylindrical Wiener process W. Let $\mathbf{u}_0$ be an $\mathfrak{F}_0$-measurable random variable. Then $\mathbf{u}$ is called a *weak pathwise solution* to (8.16)–(8.17) with the initial condition $\mathbf{u}_0$, provided:

(1) the velocity field $\mathbf{u}$ is $(\mathfrak{F}_t)$-adapted and

$$\mathbf{u} \in C_w([0, T]; L^2_{\mathrm{div}}(\mathbb{T}^N)) \cap L^2(0, T; W^{1,2}_{\mathrm{div}}(\mathbb{T}^3)) \quad \mathbb{P}\text{-a.s.};$$

(2) the momentum equation

$$\begin{aligned} &-\int_0^T \partial_t \phi \int_{\mathbb{T}^N} \mathbf{u} \cdot \boldsymbol{\varphi} \,\mathrm{d}x\,\mathrm{d}t - \phi(0) \int_{\mathbb{T}^N} \mathbf{u}_0 \cdot \boldsymbol{\varphi} \,\mathrm{d}x \\ &= \int_0^T \phi \int_{\mathbb{T}^N} \mathbf{u} \otimes \mathbf{u} : \nabla \boldsymbol{\varphi} \,\mathrm{d}x\,\mathrm{d}t - \mu \int_0^T \phi \int_{\mathbb{T}^N} \nabla \mathbf{u} : \nabla \boldsymbol{\varphi} \,\mathrm{d}x\,\mathrm{d}t \\ &\quad + \int_0^T \phi \int_{\mathbb{T}^N} \Phi(\mathbf{u}) \cdot \boldsymbol{\varphi} \,\mathrm{d}x\,\mathrm{d}W \end{aligned}$$

holds for all $\phi \in C^\infty_c([0, T))$ and all $\boldsymbol{\varphi} \in C^\infty_{\mathrm{div}}(\mathbb{T}^N)$ $\mathbb{P}$-a.s.

Theorem 8.1.5. *Let $N = 2$ and assume that Φ satisfies* (8.18). *Let $(\Omega, \mathfrak{F}, (\mathfrak{F}_t)_{t\geq 0}, \mathbb{P})$ be a given stochastic basis with a complete right-continuous filtration and an $(\mathfrak{F}_t)$-cylindrical Wiener process W. Let $\mathbf{u}_0$ be an $\mathfrak{F}_0$-measurable random variable such that $\mathbf{u}_0 \in L^r(\Omega; L^2_{\mathrm{div}}(\mathbb{T}^2))$ for some $r > 2$. Then there exists a unique weak pathwise solution to* (8.16)–(8.17) *in the sense of Definition* 8.1.4 *with the initial condition $\mathbf{u}_0$.*

8.1.2 Main result

Due to the presence of the acoustic waves, the gradient part of the velocity converges only weakly to zero. This problem already occurs at the deterministic level; cf. Lions–Masmoudi [LM98]. Consequently, the limit in the stochastic forcing $\mathbb{G}(\varrho, \varrho\mathbf{u})\,\mathrm{d}W$ can be performed only if $\mathbb{G}$ is *linear* with respect to $\varrho\mathbf{u}$. Unfortunately, a non-linear dependence on the velocity $\mathbf{u}$ cannot be handled by the present method. However, our setting already covers the particular case of

$$\mathbb{G}(\varrho, \varrho\mathbf{u})\,\mathrm{d}W = \varrho\,\mathbb{G}_1\,\mathrm{d}W^1 + \varrho\mathbf{u}\,\mathbb{G}_2\,\mathrm{d}W^2$$

with two independent cylindrical Wiener processes W^1 and W^2 and suitable Hilbert–Schmidt operators $\mathbb{G}_1$ and $\mathbb{G}_2$, which is the main example we have in mind; cf. Section 3.2.2. This result can only be strengthened in the case of unbounded domains, where powerful dispersive estimates are available. We refer to Mensah [Men16], where a variant of Theorem 8.1.6 is shown on the whole space $\mathbb{R}^3$. In this case it is possible to allow the noise to be non-linear in the momentum $\varrho\mathbf{u}$.

To give the precise definition of the diffusion coefficient $\mathbb{G}$, consider $\varrho \in L^\gamma(\mathbb{T}^N)$, $\varrho \geq 0$, and $\mathbf{v} \in L^2(\mathbb{T}^N)$ such that $\sqrt{\varrho}\mathbf{v} \in L^2(\mathbb{T}^N)$. Denote $\mathbf{q} = \varrho\mathbf{v}$ and let $\mathbb{G}(\varrho, \mathbf{q}) : \mathfrak{U} \to L^1(\mathbb{T}^N)$ be defined as follows:

$$\mathbb{G}(\varrho, \mathbf{q})e_k = \mathbf{G}_k(\cdot, \varrho(\cdot), \mathbf{q}(\cdot)) = \mathbf{h}_k(\cdot, \varrho(\cdot)) + \alpha_k \mathbf{q}(\cdot), \tag{8.19}$$

where the coefficients $\alpha_k \in \mathbb{R}$ are constants and $\mathbf{h}_k : \mathbb{T}^N \times [0, \infty) \to \mathbb{R}$ are C^1-functions that satisfy

$$\sum_{k=1}^{\infty} |\mathbf{h}_k(x, 0)|^2 = 0, \quad \sum_{k=1}^{\infty} |\nabla_\varrho \mathbf{h}_k(x, \varrho)|^2 \leq C, \quad \sum_{k=1}^{\infty} |\alpha_k|^2 < \infty. \tag{8.20}$$

Under these assumptions, the existence of a dissipative martingale solution of (8.10)–(8.13) in the sense of Definition 8.0.1 was proved in Theorem 4.0.2.

Note that, due to our assumptions on the operator $\mathbb{G}$, the stochastic perturbation that we obtain in the limit system (8.14)–(8.15) is an affine linear function of the velocity and takes the following form:

$$\Phi(\mathbf{v})e_k\,\mathrm{d}W_k = \mathcal{P}_H \mathbb{G}(1, \mathbf{v})e_k\,\mathrm{d}W_k = (\mathcal{P}_H \mathbf{h}_k(1) + \alpha_k \mathbf{v})\,\mathrm{d}W_k.$$

Besides, due to (8.20) we see that (8.18) holds. Hence Theorems 8.1.3 and 8.1.5 apply.

Our main incompressible limit results are the following.

Theorem 8.1.6. *Let $N = 2, 3$ and let $\mathbb{G}$ satisfy* (8.19)–(8.20). *Let Λ be a given Borel probability measure on $L^2_{\mathrm{div}}(\mathbb{T}^N)$. Let Λ_ε be a Borel probability measure on $L^1(\mathbb{T}^N) \times L^1(\mathbb{T}^N)$ such that, for some constant $M > 0$ independent of ε, we have*

$$\Lambda_\varepsilon \left\{ (\varrho, \mathbf{q}) \in L^1(\mathbb{T}^N) \times L^1(\mathbb{T}^N);\ \frac{1}{M} \leq \varrho \leq M,\ \left| \frac{\varrho - 1}{\varepsilon} \right| \leq M \right\} = 1$$

and, for some $r \geq 4$,

$$\int_{L^1_x \times L^1_x} \left\| \frac{1}{2} \frac{|\mathbf{q}|^2}{\varrho} \right\|^r_{L^1_x} \mathrm{d}\Lambda_\varepsilon(\varrho, \mathbf{q}) \leq C,$$

while the marginal law of Λ_ε *corresponding to the second component converges to* Λ *weakly in the sense of measures on* $L^{\frac{2\gamma}{\gamma+1}}(\mathbb{T}^N)$. *If* $((\Omega^\varepsilon, \mathfrak{F}^\varepsilon, (\mathfrak{F}^\varepsilon_t)_{t\geq 0}, \mathbb{P}^\varepsilon), \varrho_\varepsilon, \mathbf{u}_\varepsilon, W_\varepsilon)$ *is a dissipative martingale solution to* (8.10)–(8.13) *in the sense of Definition* 8.0.1 *with* $[\mathrm{Ma}] = \varepsilon$ *with the initial law* Λ_ε, $\varepsilon \in (0,1)$, *then*

$$\begin{aligned} \varrho_\varepsilon &\to 1 \quad \text{in law on } L^\infty(0,T; L^\gamma(\mathbb{T}^N)), \\ \mathbf{u}_\varepsilon &\to \mathbf{u} \quad \text{in law on } (L^2(0,T; W^{1,2}(\mathbb{T}^N)), w), \end{aligned}$$

where $\mathbf{u}$ *is a weak martingale solution to* (8.16)–(8.17) *in the sense of Definition* 8.1.2 *with the initial law* Λ.

In the two-dimensional case we are able to strengthen the result of Theorem 8.1.6 using the uniqueness of the limit system; cf. Theorem 8.1.5. To be precise, we obtain convergence in probability instead of convergence in law. In order to obtain a pathwise solution with given initial datum $\mathbf{u}_0$, we assume the latter one to be deterministic. Consequently, it is possible to start the evolution of the compressible system close to $\mathbf{u}_0$.

Theorem 8.1.7. *Let* $N = 2$, *let* $\mathbb{G}$ *satisfy* (8.19)–(8.20), *and let* $\mathbf{u}_0 \in L^2_{\mathrm{div}}(\mathbb{T}^2)$ *be a deterministic initial condition. Let* Λ_ε *be a Borel probability measure on* $L^1(\mathbb{T}^2) \times L^1(\mathbb{T}^2)$ *such that, for some constant* $M > 0$ *independent of* ε, *we have*

$$\Lambda_\varepsilon\left\{(\varrho, \mathbf{q}) \in L^1(\mathbb{T}^2) \times L^1(\mathbb{T}^2);\ \frac{1}{M} \leq \varrho \leq M,\ \left|\frac{\varrho - 1}{\varepsilon}\right| < M,\ \left|\frac{\mathbf{q} - \mathbf{u}_0}{\varepsilon}\right| \leq M\right\} = 1.$$

If $((\Omega^\varepsilon, \mathfrak{F}^\varepsilon, (\mathfrak{F}^\varepsilon_t)_{t\geq 0}, \mathbb{P}^\varepsilon), \varrho_\varepsilon, \mathbf{u}_\varepsilon, W_\varepsilon)$ *is a dissipative martingale solution to* (8.10)–(8.13) *in the sense of Definition* 8.0.1 *with* $[\mathrm{Ma}] = \varepsilon$ *with the initial law* Λ_ε, $\varepsilon \in (0,1)$, *then*

$$\begin{aligned} \varrho_\varepsilon &\to 1 \quad \text{in } L^\infty(0,T; L^\gamma(\mathbb{T}^2)) \quad \text{in probability}, \\ \mathbf{u}_\varepsilon &\to \mathbf{u} \quad \text{in } (L^2(0,T; W^{1,2}(\mathbb{T}^2)), w) \quad \text{in probability}, \end{aligned}$$

where $\mathbf{u}$ *is the unique weak pathwise solution to* (8.16)–(8.17) *in the sense of Definition* 8.1.4 *with the initial condition* $\mathbf{u}_0$.

Remark 8.1.8. It turns out that the present problem does not require the full generality of the energy inequality (8.8), that is, its differential form. It suffices to consider an integrated version of (8.8), namely,

$$\mathbb{E}\left[\sup_{t\in[0,T]} \left(\int_{\mathbb{T}^3} \left[\frac{1}{2} \frac{|\varrho \mathbf{u}(t)|^2}{\varrho(t)} + \frac{1}{\varepsilon^2} \frac{\varrho^\gamma(t) - 1 - \gamma(\varrho(t) - 1)}{(\gamma - 1)}\right] \mathrm{d}x\right)^r\right]$$

$$+\mathbb{E}\left[\left(\int_0^T\int_{\mathbb{T}^3}\mathbb{S}(\nabla\mathbf{u}):\nabla\mathbf{u}\,\mathrm{d}x\,\mathrm{d}t\right)^r\right]\le c(r,T),$$

uniformly in ε; cf. Proposition 8.1.9.

8.1.3 Convergence in law – the proof of Theorem 8.1.6

This section is devoted to the study of the limit $\varepsilon\to 0$ in the system (8.10)–(8.13) and the proof of Theorem 8.1.6. To this end, we recall that it was proved in Chapter 4 that, for every $\varepsilon\in(0,1)$, there exists

$$((\Omega^\varepsilon,\mathfrak{F}^\varepsilon,(\mathfrak{F}^\varepsilon_t)_{t\ge 0},\mathbb{P}^\varepsilon),\varrho_\varepsilon,\mathbf{u}_\varepsilon,W_\varepsilon),$$

which is a dissipative martingale solution to (8.10)–(8.13) in the sense of Definition 8.0.1. As already mentioned several times in the limit procedure in Chapter 4, it suffices to consider only one probability space, namely,

$$(\Omega^\varepsilon,\mathfrak{F}^\varepsilon,\mathbb{P}^\varepsilon)=([0,1],\overline{\mathfrak{B}([0,1])},\mathfrak{L})\quad\forall\varepsilon\in(0,1),$$

where $\mathfrak{L}$ denotes the Lebesgue measure on $[0,1]$. Moreover, we assume without loss of generality that there exists one common Wiener process W for all ε.

8.1.3.1 Uniform bounds

We start with an *a priori* estimate which follows from the energy inequality (8.8). It is obtained from (8.8) by integrating and estimating the stochastic integral as well as the correction term by means of the Burkholder–Davis–Gundy inequality, (8.20), and Gronwall's lemma, similarly to the energy bounds in Chapter 4. It is important to note that the constant on the right hand side depends on T and the constants in (8.20) but is independent of ε.

Proposition 8.1.9. *Let $r\ge 2$. Then the following estimate holds true uniformly in $\varepsilon\in(0,1)$:*

$$\begin{aligned}&\mathbb{E}\left[\sup_{t\in[0,T]}\left|\int_{\mathbb{T}^N}\left[\frac{1}{2}\frac{|\varrho_\varepsilon\mathbf{u}_\varepsilon(t)|^2}{\varrho_\varepsilon(t)}+\frac{1}{\varepsilon^2}\frac{\varrho_\varepsilon^\gamma(t)-1-\gamma(\varrho_\varepsilon(t)-1)}{(\gamma-1)}\right]\mathrm{d}x\right|^r\right]\\&\quad+\mathbb{E}\left[\left|\int_0^T\int_{\mathbb{T}^N}\mathbb{S}(\nabla\mathbf{u}_\varepsilon):\nabla\mathbf{u}_\varepsilon\,\mathrm{d}x\,\mathrm{d}t\right|^r\right]\lesssim c(r),\end{aligned}\tag{8.21}$$

where $c(r)$ is given by

$$c(r)\approx 1+\mathbb{E}\left[\left|\int_{\mathbb{T}^N}\left[\frac{1}{2}\frac{|(\varrho_\varepsilon\mathbf{u}_\varepsilon)(0)|^2}{\varrho_\varepsilon(0)}+\frac{1}{\varepsilon^2}\frac{\varrho_\varepsilon^\gamma(0)-1-\gamma(\varrho_\varepsilon(0)-1)}{(\gamma-1)}\right]\mathrm{d}x\right|^r\right]\tag{8.22}$$

and is independent of ε.

Proof. First of all, we observe that the pressure potential $P(\varrho_\varepsilon)$ in (8.8) can be perturbed by a linear function. In particular, one can replace $P(\varrho_\varepsilon)$ by

$$P_{\overline{\varrho}}(\varrho_\varepsilon) = \frac{1}{(\gamma-1)}(\varrho_\varepsilon^\gamma - \gamma\overline{\varrho}^{\gamma-1}(\varrho_\varepsilon - \overline{\varrho}) - \overline{\varrho}^\gamma), \tag{8.23}$$

for any constant $\overline{\varrho} > 0$, and (8.8) remains valid. This is a consequence of the mass conservation

$$\int_{\mathbb{T}^N} \varrho_\varepsilon(t)\,\mathrm{d}x = \int_{\mathbb{T}^N} \varrho_\varepsilon(0)\,\mathrm{d}x, \quad t \in [0,T], \tag{8.24}$$

which holds due to (8.10). So, we have

$$\begin{aligned}
&-\int_0^T \partial_t\phi \int_{\mathbb{T}^N} \Big[\frac{1}{2}\varrho_\varepsilon|\mathbf{u}_\varepsilon|^2 + \frac{1}{\varepsilon^2}P_{\overline{\varrho}}(\varrho_\varepsilon)\Big]\,\mathrm{d}x\,\mathrm{d}t + \int_0^T \phi \int_{\mathbb{T}^N} \mathbb{S}(\nabla\mathbf{u}_\varepsilon) : \nabla\mathbf{u}_\varepsilon\,\mathrm{d}x\,\mathrm{d}t \\
&\le \phi(0)\int_{\mathbb{T}^N} \Big[\frac{1}{2}\varrho_\varepsilon(0)|\mathbf{u}_\varepsilon(0)|^2 + \frac{1}{\varepsilon^2}P_{\overline{\varrho}}(\varrho_\varepsilon(0))\Big]\,\mathrm{d}x \\
&\quad + \frac{1}{2}\sum_{k=1}^\infty \int_0^T \phi \int_{\mathbb{T}^N} \varrho_\varepsilon^{-1}|\mathbf{G}_k(\varrho_\varepsilon,\varrho_\varepsilon\mathbf{u}_\varepsilon)|^2\,\mathrm{d}x\,\mathrm{d}t \\
&\quad + \sum_{k=1}^\infty \int_0^T \phi \int_{\mathbb{T}^N} \mathbf{G}_k(\varrho_\varepsilon,\varrho_\varepsilon\mathbf{u}_\varepsilon)\cdot\mathbf{u}_\varepsilon\,\mathrm{d}x\,\mathrm{d}W_k,
\end{aligned}$$

for all $\varepsilon > 0$. Now we choose ϕ such that it approximates $\mathbf{1}_{[0,t]}$ and gain

$$\begin{aligned}
&\frac{1}{2}\int_{\mathbb{T}^N} \varrho_\varepsilon|\mathbf{u}_\varepsilon|^2\,\mathrm{d}x + \int_0^t \int_{\mathbb{T}^N} \mathbb{S}(\nabla\mathbf{u}_\varepsilon) : \nabla\mathbf{u}_\varepsilon\,\mathrm{d}x\,\mathrm{d}s + \frac{1}{\varepsilon^2}\int_{\mathbb{T}^N} P_{\overline{\varrho}}(\varrho_\varepsilon)\,\mathrm{d}x \\
&\le \frac{1}{2}\int_{\mathbb{T}^N} \frac{|(\varrho_\varepsilon\mathbf{u}_\varepsilon)(0)|^2}{\varrho_\varepsilon(0)}\,\mathrm{d}x + \frac{1}{\varepsilon^2}\int_{\mathbb{T}^N} P_{\overline{\varrho}}(\varrho_\varepsilon(0))\,\mathrm{d}x \\
&\quad + \int_0^t \int_{\mathbb{T}^N} \varrho_\varepsilon^{-1}|\mathbf{G}_k(\varrho_\varepsilon,\varrho_\varepsilon\mathbf{u}_\varepsilon)|^2\,\mathrm{d}x\,\mathrm{d}t + \sum_{k=1}^\infty \int_0^t \phi \int_{\mathbb{T}^N} \mathbf{G}_k(\varrho_\varepsilon,\varrho_\varepsilon\mathbf{u}_\varepsilon)\cdot\mathbf{u}\,\mathrm{d}x\,\mathrm{d}W_k \\
&=: \frac{1}{2}\int_{\mathbb{T}^N} \frac{|(\varrho_\varepsilon\mathbf{u}_\varepsilon)(0)|^2}{\varrho_\varepsilon(0)}\,\mathrm{d}x + \frac{1}{\varepsilon^2}\int_{\mathbb{T}^N} P_{\overline{\varrho}}(\varrho_\varepsilon(0))\,\mathrm{d}x + T_1(t) + T_2(t).
\end{aligned}$$

Using (8.20) and (8.24) we have

$$\begin{aligned}
T_2(t) &\le \frac{1}{2}\sum_{k=1}^\infty \int_0^t \int_{\mathbb{T}^N} \varrho_\varepsilon^{-1}|\mathbf{G}_k(\varrho_\varepsilon,\varrho_\varepsilon\mathbf{u}_\varepsilon)|^2\,\mathrm{d}x\,\mathrm{d}t \lesssim \int_0^t \int_{\mathbb{T}^N} (\varrho_\varepsilon|\mathbf{u}_\varepsilon|^2 + \varrho_\varepsilon^\gamma + 1)\,\mathrm{d}x\,\mathrm{d}t \\
&\lesssim \int_0^t \int_{\mathbb{T}^N} \Big(\varrho_\varepsilon|\mathbf{u}_\varepsilon|^2 + \frac{1}{\varepsilon^2}P_{\overline{\varrho}}(\varrho_\varepsilon) + 1\Big)\,\mathrm{d}x\,\mathrm{d}t.
\end{aligned} \tag{8.25}$$

We apply the rth power on both sides and then take the expectation. As a consequence of the Burkholder–Davis–Gundy inequality, (8.20), and (8.24), we gain

$$\mathbb{E}\Big[\sup_{t\in[0,T]} |T_2(t)|\Big]^r = \mathbb{E}\Big[\sup_{t\in[0,T]} \Big|\int_0^t \int_{\mathbb{T}^N} \mathbf{u}_\varepsilon \cdot \mathbb{G}(\varrho_\varepsilon,\varrho_\varepsilon\mathbf{u}_\varepsilon)\,\mathrm{d}x\,\mathrm{d}W\Big|\Big]^r$$

$$
\begin{aligned}
&= \mathbb{E}\Big[\sup_{t\in[0,T]}\Big|\int_0^t \sum_{k=1}^\infty \int_{\mathbb{T}^N} \mathbf{u}_\varepsilon \cdot \mathbf{G}_k(\varrho_\varepsilon, \varrho_\varepsilon \mathbf{u}_\varepsilon)\,\mathrm{d}x\,\mathrm{d}\beta_k\Big|\Big]^r \\
&\lesssim \mathbb{E}\Big[\int_0^T \sum_{k=1}^\infty \Big(\int_{\mathbb{T}^N} \mathbf{u}_\varepsilon \cdot \mathbf{G}_k(\varrho_\varepsilon, \varrho_\varepsilon \mathbf{u}_\varepsilon)\,\mathrm{d}x\Big)^2\,\mathrm{d}t\Big]^{\frac{r}{2}} \\
&\lesssim \mathbb{E}\Big[\int_0^T \sum_{k=1}^\infty \Big(\int_{\mathbb{T}^N} \varrho_\varepsilon |\mathbf{u}_\varepsilon|^2\,\mathrm{d}x\Big)\Big(\int_{\mathbb{T}^N} \varrho_\varepsilon^{-1} |\mathbf{G}_k(\varrho_\varepsilon, \varrho_\varepsilon \mathbf{u}_\varepsilon)|^2\,\mathrm{d}x\Big)\,\mathrm{d}t\Big]^{\frac{r}{2}}.
\end{aligned}
$$

By Young's inequality and a computation similar to (8.25), we gain, for every $\delta > 0$,

$$
\begin{aligned}
\mathbb{E}\Big[\sup_{t\in[0,T]} |T_2(t)|\Big]^r \le \delta\,\mathbb{E}\Big[\sup_{t\in[0,T]} \int_{\mathbb{T}^N} \varrho_\varepsilon |\mathbf{u}_\varepsilon|^2\,\mathrm{d}x\Big]^r \\
+ c(\delta)\,\mathbb{E}\Big[\int_0^T \int_{\mathbb{T}^N} \Big(\varrho_\varepsilon |\mathbf{u}_\varepsilon|^2 + \frac{1}{\varepsilon^2} P_{\bar{\varrho}}(\varrho_\varepsilon) + 1\Big)\,\mathrm{d}x\,\mathrm{d}t\Big]^r.
\end{aligned}
$$

Finally, taking δ small enough and applying Gronwall's lemma, we obtain

$$
\begin{aligned}
&\mathbb{E}\Big[\Big|\sup_{t\in[0,T]} \int_{\mathbb{T}^N} \Big[\frac{1}{2}\frac{|\varrho_\varepsilon \mathbf{u}_\varepsilon(t)|^2}{\varrho_\varepsilon(t)} + \frac{1}{\varepsilon^2}\frac{\varrho_\varepsilon^\gamma(t) - 1 - \gamma(\varrho_\varepsilon(t) - 1)}{(\gamma-1)}\Big]\,\mathrm{d}x\Big|^r\Big] \\
&\quad + \mathbb{E}\Big[\Big|\int_0^T \int_{\mathbb{T}^N} \mathbb{S}(\nabla \mathbf{u}_\varepsilon) : \nabla \mathbf{u}_\varepsilon\,\mathrm{d}x\,\mathrm{d}t\Big|^r\Big] \\
&\lesssim 1 + \mathbb{E}\Big[\Big|\int_{\mathbb{T}^N} \Big[\frac{1}{2}\frac{|(\varrho_\varepsilon \mathbf{u}_\varepsilon)(0)|^2}{\varrho_\varepsilon(0)} + \frac{1}{\varepsilon^2}\frac{\varrho_\varepsilon^\gamma(0) - 1 - \gamma(\varrho_\varepsilon(0) - 1)}{(\gamma-1)}\Big]\,\mathrm{d}x\Big|^r\Big],
\end{aligned}
$$

choosing $\bar{\varrho} = 1$ in (8.23). In addition, due to the Taylor theorem and our assumptions upon $\Lambda_\varepsilon = \mathcal{L}[\varrho_\varepsilon(0), \varrho_\varepsilon \mathbf{u}_\varepsilon(0)]$, we have

$$
\varrho_\varepsilon^\gamma(0) - 1 - \gamma(\varrho_\varepsilon(0) - 1) \le C\varepsilon^2 \quad \mathbb{P}\text{-a.s.}
$$

Hence (8.21) follows independently of ε. □

We obtain the following corollary.

Corollary 8.1.10. *We have the following uniform bounds:*

$$
\mathbb{E}[\|\nabla \mathbf{u}_\varepsilon\|_{L_t^2 L_x^2}^{2r}] \lesssim c(r), \tag{8.26}
$$

$$
\mathbb{E}\Big[\Big|\sup_{t\in[0,T]} \|\varrho_\varepsilon |\mathbf{u}_\varepsilon|^2\|_{L_x^1}\Big|^r\Big] \lesssim c(r), \tag{8.27}
$$

where $c(r)$ is given by (8.22).

Proof. The estimates (8.26) and (8.27) follow immediately from Proposition 8.1.9. □

Let us now introduce the following essential and residual components of a function h:

$$\begin{aligned}
h &= h_{\text{ess}} + h_{\text{res}},\\
h_{\text{ess}} &= \chi(\varrho_\varepsilon)h, \quad \chi \in C_c^\infty(0,\infty),\ 0 \le \chi \le 1,\ \chi = 1 \text{ on an open interval containing } 1,\\
h_{\text{res}} &= (1-\chi(\varrho_\varepsilon))h.
\end{aligned}$$

The decomposition captures the different behavior of the pressure potential $P_1(\varrho) := \varrho^\gamma - 1 - \gamma(\varrho - 1)$ in different subregions of $[0,\infty)$. This will be described in the following lemma.

Lemma 8.1.11. *There exist constants C_1, C_2, C_3, $C_4 > 0$ such that:*
(i) $C_1|\varrho - 1|^2 \le P_1(\varrho) \le C_2|\varrho - 1|^2$ *if* $\varrho \in \operatorname{supp}\chi$;
(ii) $P_1(\varrho) \ge C_4$ *if* $\varrho \notin \operatorname{supp}\chi$;
(iii) $P_1(\varrho) \ge C_3\varrho^\gamma$ *if* $\varrho \notin \operatorname{supp}\chi$.

Proof. The first statement follows immediately from the Taylor theorem. The second one is a consequence of the fact that P_1 is strictly convex and attains its minimum at $\varrho = 1$. If $\varrho \notin \operatorname{supp}\chi$ and $\varrho \in [0,1)$, then the third statement is a consequence of the second one. Finally, we observe that the function $\frac{P_1(\varrho)}{\varrho^\gamma}$ is increasing for large $\varrho \in [1,\infty)$ and its value at $\varrho = 1$ is zero. This implies the remaining part of (iii) and the proof is complete. □

Accordingly, using Lemma 8.1.11 and the energy estimate (8.21), we obtain the following uniform bounds:

$$\begin{aligned}
\mathbb{E}\Big[\Big|\sup_{t\in[0,T]}\Big\|\Big[\frac{\varrho_\varepsilon - 1}{\varepsilon}\Big]_{\text{ess}}\Big\|_{L_x^2}^2\Big|^r\Big] &\lesssim c(r),\\
\mathbb{E}\Big[\Big|\sup_{t\in[0,T]}\Big\|\Big[\frac{\varrho_\varepsilon - 1}{\varepsilon}\Big]_{\text{res}}\Big\|_{L_x^\gamma}^\gamma\Big|^r\Big] &\lesssim c(r),
\end{aligned}$$

where $c(r)$ is given by (8.22). Therefore, setting $\varphi_\varepsilon := \frac{1}{\varepsilon}(\varrho_\varepsilon - 1)$, we deduce that uniformly in ε

$$\mathbb{E}\Big[\Big|\sup_{t\in[0,T]}\|\varphi_\varepsilon\|_{L_x^{\min(\gamma,2)}}^{\min(\gamma,2)}\Big|^r\Big] \lesssim c(r). \tag{8.28}$$

As the next step, we want to show

$$\lim_{\varepsilon\to 0}\mathbb{E}\Big[\Big|\sup_{t\in[0,T]}\|\varrho_\varepsilon - 1\|_{L_x^\gamma}^\gamma\Big|^r\Big] = 0, \tag{8.29}$$

which in particular leads to

$$\mathbb{E}\Big[\Big|\sup_{t\in[0,T]}\|\varrho_\varepsilon\|_{L_x^\gamma}^\gamma\Big|^r\Big] \le c. \tag{8.30}$$

Let us now verify (8.29). For all $\delta > 0$, there exists $C_\delta > 0$ such that

$$\varrho^\gamma - 1 - \gamma(\varrho - 1) \geq C_\delta |\varrho - 1|^\gamma,$$

if $|\varrho - 1| \geq \delta$ and $\varrho \geq 0$. By (8.21), we obtain

$$\begin{aligned}
\mathbb{E}\Big[\Big|\sup_{t\in[0,T]}\int_{\mathbb{T}^N} |\varrho_\varepsilon - 1|^\gamma \,\mathrm{d}x\Big|^r\Big] &= \mathbb{E}\Big[\Big|\sup_{t\in[0,T]}\int_{\mathbb{T}^N} \mathbf{1}_{\{|\varrho_\varepsilon - 1|\geq\delta\}} |\varrho_\varepsilon - 1|^\gamma \,\mathrm{d}x\Big|^r\Big] \\
&\quad + \mathbb{E}\Big[\Big|\sup_{t\in[0,T]}\int_{\mathbb{T}^N} \mathbf{1}_{\{|\varrho_\varepsilon - 1|<\delta\}} |\varrho_\varepsilon - 1|^\gamma \,\mathrm{d}x\Big|^r\Big] \\
&\leq C_\delta\, \mathbb{E}\Big[\Big|\sup_{t\in[0,T]}\int_{\mathbb{T}^N} (\varrho_\varepsilon^\gamma - 1 - \gamma(\varrho_\varepsilon - 1))\,\mathrm{d}x\Big|^r\Big] + \delta^{\gamma p} \\
&\leq C_\delta \varepsilon^{2p} + \delta^{\gamma p}.
\end{aligned}$$

Letting first $\varepsilon \to 0$ and then $\delta \to 0$ yields the claimed estimate (8.30).

Combining (8.27) and (8.30), respectively, we deduce the uniform bound

$$\mathbb{E}\Big[\Big|\sup_{t\in[0,T]} \|\varrho_\varepsilon \mathbf{u}_\varepsilon\|_{L_x^{\frac{2\gamma}{\gamma+1}}}^{\frac{2\gamma}{\gamma+1}}\Big|^r\Big] \lesssim c(r). \tag{8.31}$$

Moreover, the estimate (8.30) finally allows to show that, for $r \geq 2$, we have

$$\mathbb{E}\big[\|\mathbf{u}_\varepsilon\|_{L_t^2 L_x^2}^{2r}\big] \lesssim c_2(r), \tag{8.32}$$

with $c_2(r)$ given by

$$c_2(r) \approx 1 + \mathbb{E}\Big[\Big|\int_{\mathbb{T}^N}\Big[\frac{1}{2}\frac{|(\varrho_\varepsilon\mathbf{u}_\varepsilon)(0)|^2}{\varrho_\varepsilon(0)} + \frac{1}{\varepsilon^2}\frac{\varrho_\varepsilon^\gamma(0) - 1 - \gamma(\varrho_\varepsilon(0) - 1)}{(\gamma - 1)}\Big]\,\mathrm{d}x\Big|^{2r}\Big]. \tag{8.33}$$

To show (8.32) we argue as in (4.88) and obtain

$$\begin{aligned}
\|\varrho_\varepsilon(0)\|_{L_x^1}^2 \int_0^\tau |(\mathbf{u}_\varepsilon)_{\mathbb{T}^N}|^2 \,\mathrm{d}t &\lesssim \sup_{t\in[0,\tau]} \|\varrho_\varepsilon\|_{L_x^\gamma}^2 \int_0^\tau \|\nabla \mathbf{u}_\varepsilon\|_{L_x^2}^2 \,\mathrm{d}t \\
&\quad + \tau \sup_{t\in[0,\tau]} \big(\|\varrho_\varepsilon\|_{L_x^1}^2 + \|\varrho_\varepsilon|\mathbf{u}_\varepsilon|^2\|_{L_x^1}^2\big).
\end{aligned} \tag{8.34}$$

On account of (8.26), (8.27), (8.30), the mass conservation (8.24), and the assumptions on the initial law, we obtain the desired estimate (8.32).

8.1.3.2 Acoustic equation

In order to proceed we need the Helmholtz projection $\mathcal{P}_H$, which projects $L^2(\mathbb{T}^N)$ onto divergence-free vector fields from

$$L^2_{\mathrm{div}}(\mathbb{T}^N) := \overline{C^\infty_{\mathrm{div}}(\mathbb{T}^N)}^{\|\cdot\|_2}.$$

Moreover, we set $\mathcal{Q} = \mathrm{Id} - \mathcal{P}_H$. Recall that both $\mathcal{P}_H$ and $\mathcal{Q}$ are continuous on all $W^{l,q}(\mathbb{T}^N)$-spaces, $l \in \mathbb{Z}$, $q \in (1,\infty)$; cf. Remark A.1.5.

Let us now project (8.11) onto the space of gradient vector fields. Using the fact that $\mathcal{Q}\nabla f = \nabla f$, system (8.10)–(8.11) rewrites

$$\varepsilon\,\mathrm{d}\varphi_\varepsilon + \operatorname{div}\mathcal{Q}(\varrho_\varepsilon\mathbf{u}_\varepsilon)\,\mathrm{d}t = 0, \tag{8.35}$$

$$\varepsilon\,\mathrm{d}\mathcal{Q}(\varrho_\varepsilon\mathbf{u}_\varepsilon) + \gamma\nabla\varphi_\varepsilon\,\mathrm{d}t = \varepsilon\mathbf{F}_\varepsilon\,\mathrm{d}t + \varepsilon\mathcal{Q}\mathbb{G}(\varrho_\varepsilon,\varrho_\varepsilon\mathbf{u}_\varepsilon)\,\mathrm{d}W, \tag{8.36}$$

where

$$\mathbf{F}_\varepsilon = \nu\Delta\mathcal{Q}\mathbf{u}_\varepsilon + (\lambda+\nu)\nabla\operatorname{div}\mathbf{u}_\varepsilon - \mathcal{Q}[\operatorname{div}(\varrho_\varepsilon\mathbf{u}_\varepsilon\otimes\mathbf{u}_\varepsilon)] - \frac{1}{\varepsilon^2}\nabla[\varrho_\varepsilon^\gamma - 1 - \gamma(\varrho_\varepsilon - 1)].$$

The system (8.35)–(8.36) may be viewed as a stochastic version of Lighthill's acoustic analogy [Lig52, Lig54] associated to the compressible Navier–Stokes system. Note that Proposition 8.1.9 (see also Corollary 8.1.10) yields, for $l > \frac{N}{2} + 1$, using Sobolev's embedding,

$$\mathbb{E}\left[\left|\int_0^T \|\mathbf{F}_\varepsilon\|_{W_x^{-l,2}}\,\mathrm{d}t\right|^r\right] \lesssim c(r), \tag{8.37}$$

with $c(r)$ given by (8.22) uniformly in ε.

8.1.3.3 Compactness

Let us define the path space $\mathcal{X} = \mathcal{X}_\varrho \times \mathcal{X}_\mathbf{u} \times \mathcal{X}_{\varrho\mathbf{u}} \times \mathcal{X}_W$, where

$$\begin{aligned}
\mathcal{X}_\varrho &= C_w([0,T]; L^\gamma(\mathbb{T}^N)),\\
\mathcal{X}_\mathbf{u} &= (L^2(0,T; W^{1,2}(\mathbb{T}^N)), w),\\
\mathcal{X}_{\varrho\mathbf{u}} &= C_w([0,T]; L^{\frac{2\gamma}{\gamma+1}}(\mathbb{T}^N)) \cap C([0,T]; W^{-k,2}(\mathbb{T}^N)), \quad k > \frac{5}{2},\\
\mathcal{X}_W &= C([0,T]; \mathfrak{U}_0).
\end{aligned}$$

To proceed, we have to show tightness of the set

$$\{\mathcal{L}[\varrho_\varepsilon, \mathbf{u}_\varepsilon, \mathcal{P}_H(\varrho_\varepsilon\mathbf{u}_\varepsilon), W];\ \varepsilon \in (0,1)\}.$$

This will be done by the stochastic compactness method, similarly to Chapter 4 (see, in particular, Sections 4.4.3.1 and 4.5.3). We start with tightness of the law of $\mathbf{u}_\varepsilon$.

Proposition 8.1.12. *The set* $\{\mathcal{L}[\mathbf{u}_\varepsilon];\ \varepsilon \in (0,1)\}$ *is tight on* $\mathcal{X}_\mathbf{u}$.

Proof. This is a consequence of (8.32). Indeed, for any $R > 0$, the set

$$B_R = \{\mathbf{u} \in L^2(0,T; W^{1,2}(\mathbb{T}^N));\ \|\mathbf{u}\|_{L_t^2 W_x^{1,2}} \le R\}$$

is relatively compact in $\mathcal{X}_{\mathbf{u}}$ and

$$\mu_{\mathbf{u}_\varepsilon}(B_R^C) = \mathbb{P}(\|\mathbf{u}_\varepsilon\|_{L_t^2 W_x^{1,2}} \geq R) \leq \frac{1}{R}\mathbb{E}\|\mathbf{u}_\varepsilon\|_{L_t^2 W_x^{1,2}} \leq \frac{C}{R},$$

which yields the claim. □

Proposition 8.1.13. *The set* $\{\mathcal{L}[\varrho_\varepsilon];\, \varepsilon \in (0,1)\}$ *is tight on* $\mathcal{X}_\varrho$.

Proof. Due to (8.31), $\{\operatorname{div}(\varrho_\varepsilon \mathbf{u}_\varepsilon)\}$ can be controlled in $W^{-1,\frac{2\gamma}{\gamma+1}}(\mathbb{T}^N)$. In particular, we have

$$\mathbb{E}\Big[\Big|\sup_{t\in[0,T]}\|\operatorname{div}(\varrho_\varepsilon \mathbf{u}_\varepsilon)\|_{W_x^{-1,\frac{2\gamma}{\gamma+1}}}^{\frac{2\gamma}{\gamma+1}}\Big|^r\Big] \lesssim c(r).$$

Therefore the continuity equation (8.11) and (8.30) yield the following uniform bound:

$$\mathbb{E}\Big[\Big|\|\varrho_\varepsilon\|_{C_t^{0,1} W_x^{-1,\frac{2\gamma}{\gamma+1}}}^{\frac{2\gamma}{\gamma+1}}\Big|^r\Big] \lesssim c(r).$$

Now, the required tightness follows by a similar reasoning as in Proposition 8.1.12, together with (8.30) and the compact embedding (see Theorem 1.8.5)

$$L^\infty(0,T;L^\gamma(\mathbb{T}^N)) \cap C^{0,1}([0,T];W^{-1,\frac{2\gamma}{\gamma+1}}(\mathbb{T}^N)) \overset{c}{\hookrightarrow} C_w([0,T];L^\gamma(\mathbb{T}^N)).$$ □

Proposition 8.1.14. *The set* $\{\mathcal{L}[\mathcal{P}_H(\varrho_\varepsilon \mathbf{u}_\varepsilon)];\, \varepsilon \in (0,1)\}$ *is tight on* $\mathcal{X}_{\varrho\mathbf{u}}$.

Proof. We decompose $\mathcal{P}_H(\varrho_\varepsilon \mathbf{u}_\varepsilon)$ into two parts, namely, $\mathcal{P}_H(\varrho_\varepsilon \mathbf{u}_\varepsilon)(t) = Y^\varepsilon(t) + Z^\varepsilon(t)$, where

$$\begin{aligned} Y^\varepsilon(t) &= \mathcal{P}_H \mathbf{q}_\varepsilon(0) - \int_0^t \mathcal{P}_H\big[\operatorname{div}(\varrho_\varepsilon \mathbf{u}_\varepsilon \otimes \mathbf{u}_\varepsilon) - \mu\Delta\mathbf{u}_\varepsilon\big]\,\mathrm{d}s, \\ Z^\varepsilon(t) &= \int_0^t \mathcal{P}_H \mathbb{G}(\varrho_\varepsilon, \varrho_\varepsilon \mathbf{u}_\varepsilon)\,\mathrm{d}W(s). \end{aligned}$$

Hölder continuity of (Y^ε). We show that there exists $l \in \mathbb{N}$ such that, for all $\kappa \in (0,1/2)$, we have

$$\mathbb{E}\|Y^\varepsilon\|_{C_t^\kappa W_x^{-l,2}} \leq C. \tag{8.38}$$

Choose l such that $L^1(\mathbb{T}^N) \hookrightarrow W^{-l+1,2}(\mathbb{T}^N)$. The *a priori* estimates from Corollary 8.1.10 and the continuity of $\mathcal{P}_H$ (cf. Remark A.1.5) yield

$$\begin{aligned} \mathbb{E}\|Y^\varepsilon(t) - Y^\varepsilon(s)\|_{W_x^{-l,2}}^\theta &= \mathbb{E}\Big\|\int_s^t \mathcal{P}_H\big[\operatorname{div}(\varrho_\varepsilon \mathbf{u}_\varepsilon \otimes \mathbf{u}_\varepsilon) + \mu\Delta\mathbf{u}_\varepsilon\big]\,\mathrm{d}s\Big\|_{W_x^{-l,2}}^\theta \\ &\leq C\,\mathbb{E}\Big\|\int_s^t \operatorname{div}(\varrho_\varepsilon \mathbf{u}_\varepsilon \otimes \mathbf{u}_\varepsilon)\,\mathrm{d}s\Big\|_{W_x^{-l,2}}^\theta + C\,\mathbb{E}\Big\|\int_s^t \Delta\mathbf{u}_\varepsilon\,\mathrm{d}s\Big\|_{W_x^{-l,2}}^\theta \\ &\leq C\,\mathbb{E}\Big\|\int_s^t \varrho_\varepsilon \mathbf{u}_\varepsilon \otimes \mathbf{u}_\varepsilon\,\mathrm{d}s\Big\|_{L_x^1}^\theta + C\,\mathbb{E}\Big\|\int_s^t \nabla\mathbf{u}_\varepsilon\,\mathrm{d}s\Big\|_{L_x^1}^\theta \leq C|t-s|^{\theta/2} \end{aligned}$$

and (8.38) follows by the Kolmogorov continuity criterion, Theorem 2.3.11.

Hölder continuity of (Z^ε). Next, we also show

$$\mathbb{E}\|Z^\varepsilon\|_{C_t^\kappa W_x^{-l,2}} \le C,$$

where $l \in \mathbb{N}$ was given by the previous step and $\kappa \in (0,1/2)$. From the embedding $L^1(\mathbb{T}^N) \hookrightarrow W^{-l,2}(\mathbb{T}^N)$, (8.20), the *a priori* estimates, and the continuity of $\mathcal{P}_H$, we get

$$\begin{aligned}
\mathbb{E}\|Z^\varepsilon(t) - Z^\varepsilon(s)\|_{W_x^{-l,2}}^\theta &= \mathbb{E}\left\|\int_s^t \mathcal{P}_H \mathbb{G}(\varrho_\varepsilon, \varrho_\varepsilon \mathbf{u}_\varepsilon)\,\mathrm{d}W\right\|_{W_x^{-l,2}}^\theta \\
&\le C\,\mathbb{E}\left\|\int_s^t \mathbb{G}(\varrho_\varepsilon, \varrho_\varepsilon \mathbf{u}_\varepsilon)\,\mathrm{d}W\right\|_{W_x^{-l,2}}^\theta \le C\,\mathbb{E}\left(\int_s^t \sum_{k=1}^\infty \|\mathbf{G}_k(\varrho_\varepsilon, \varrho_\varepsilon \mathbf{u}_\varepsilon)\|_{W_x^{-l,2}}^2\,\mathrm{d}r\right)^{\frac{\theta}{2}} \\
&\le C\,\mathbb{E}\left(\int_s^t \sum_{k=1}^\infty \|\mathbf{G}_k(\varrho_\varepsilon, \varrho_\varepsilon \mathbf{u}_\varepsilon)\|_{L_x^1}^2\,\mathrm{d}r\right)^{\frac{\theta}{2}} \\
&\le C\,\mathbb{E}\left(\int_s^t \int_{\mathbb{T}^N} (\varrho_\varepsilon + \varrho_\varepsilon |\mathbf{u}_\varepsilon|^2)\,\mathrm{d}x\,\mathrm{d}r\right)^{\frac{\theta}{2}} \\
&\le C|t-s|^{\frac{\theta}{2}}\Big(1 + \mathbb{E}\sup_{0t\in[0,T]} \|\sqrt{\varrho_\varepsilon}\mathbf{u}_\varepsilon\|_{L^2}^\theta + \mathbb{E}\sup_{t\in[0,T]} \|\varrho_\varepsilon\|_{L^\gamma}^{\theta\gamma/2}\Big) \le C|t-s|^{\frac{\theta}{2}}
\end{aligned}$$

and the Kolmogorov continuity criterion, Theorem 2.3.11, applies.

Conclusion. Collecting the above results, we obtain

$$\mathbb{E}\|\mathcal{P}_H(\varrho_\varepsilon \mathbf{u}_\varepsilon)\|_{C_t^\kappa W_x^{-l,2}} \le C,$$

for some $l \in \mathbb{N}$ and all $\kappa \in (0,1/2)$. This implies the desired tightness by making use of (8.31) and the continuity of $\mathcal{P}_H$, together with the following compact embedding (see Theorem 1.8.5):

$$L^\infty(0,T; L^{\frac{2\gamma}{\gamma+1}}(\mathbb{T}^N)) \cap C^\kappa([0,T]; W^{-l,2}(\mathbb{T}^N)) \overset{c}{\hookrightarrow} C_w([0,T]; L^{\frac{2\gamma}{\gamma+1}}(\mathbb{T}^N)). \qquad \square$$

Tightness of μ_W is obvious; cf. Corollary 4.3.9. We conclude with the following Corollary.

Corollary 8.1.15. *The set*

$$\{\mathcal{L}[\varrho_\varepsilon, \mathbf{u}_\varepsilon, \mathcal{P}_H(\varrho_\varepsilon \mathbf{u}_\varepsilon), W];\ \varepsilon \in (0,1)\}$$

is tight on $\mathcal{X}$.

The path space $\mathcal{X}$ is not a Polish space, so our compactness argument is based on the Jakubowski–Skorokhod representation theorem, Theorem 2.7.1. To be more precise, we infer the following result.

Proposition 8.1.16. *There exists a complete probability space $(\tilde{\Omega}, \tilde{\mathfrak{F}}, \tilde{\mathbb{P}})$ with $\mathcal{X}$-valued Borel measurable random variables $[\tilde{\varrho}_\varepsilon, \tilde{\mathbf{u}}_\varepsilon, \tilde{\mathbf{q}}_\varepsilon, \tilde{W}_\varepsilon]$, $\varepsilon \in (0,1)$, as well as $[\tilde{\varrho}, \tilde{\mathbf{u}}, \tilde{\mathbf{q}}, \tilde{W}]$ such that (up to a subsequence):*

(1) *for all $\varepsilon \in (0,1)$, $\mathcal{L}[\tilde{\varrho}_\varepsilon, \tilde{\mathbf{u}}_\varepsilon, \tilde{\mathbf{q}}_\varepsilon, \tilde{W}_\varepsilon]$ and $\mathcal{L}[\varrho_\varepsilon, \mathbf{u}_\varepsilon, \mathcal{P}_H(\varrho_\varepsilon \mathbf{u}_\varepsilon), W]$ coincide on $\mathcal{X}$, in particular,*

$$\tilde{\mathbf{q}}_\varepsilon = \mathcal{P}_H(\tilde{\varrho}_\varepsilon \tilde{\mathbf{u}}_\varepsilon),$$

$\tilde{\mathbb{P}}$-a.s.;

(2) *the law of $[\tilde{\varrho}, \tilde{\mathbf{u}}, \tilde{\mathbf{q}}, \tilde{W}]$ on $\mathcal{X}$ is a Radon measure;*

(3) *$[\tilde{\varrho}_\varepsilon, \tilde{\mathbf{u}}_\varepsilon, \mathcal{P}_H(\tilde{\varrho}_\varepsilon \tilde{\mathbf{u}}_\varepsilon), \tilde{W}_\varepsilon]$ converges in the topology of $\mathcal{X}$ $\tilde{\mathbb{P}}$-a.s. to $[\tilde{\varrho}, \tilde{\mathbf{u}}, \tilde{\mathbf{q}}, \tilde{W}]$, i.e.,*

$$\begin{aligned} \tilde{\varrho}_\varepsilon &\to \tilde{\varrho} \quad \text{in } C_w([0,T]; L^\gamma(\mathbb{T}^N)), \\ \tilde{\mathbf{q}}_\varepsilon &\to \tilde{\mathbf{q}} \quad \text{in } C_w([0,T]; L^{\frac{2\gamma}{\gamma+1}}(\mathbb{T}^N)), \\ \tilde{\mathbf{u}}_\varepsilon &\rightharpoonup \tilde{\mathbf{u}} \quad \text{in } L^2(0,T; W^{1,2}(\mathbb{T}^N)), \\ \tilde{W}_\varepsilon &\to \tilde{W} \quad \text{in } C([0,T]; \mathfrak{U}_0), \end{aligned} \tag{8.39}$$

as $\varepsilon \to 0$ $\tilde{\mathbb{P}}$-a.s.

Similarly to Section 4.3.2 and the subsequent sections within the course of the construction of a dissipative martingale solution in Chapter 4, we observe that it is convenient to work with random distributions as introduced in Section 2.2. This can be seen from the above compactness result as the limit velocity $\tilde{\mathbf{u}}$ is not a stochastic process in the classical sense; cf. Definition 2.1.11. In view of Section 2.3, the stochastic integration theory relies on progressive measurability of the corresponding integrands. Recall that, for random distributions that are adapted in the sense of Definition 2.2.13 and satisfy a suitable integrability assumption, Lemma 2.2.18 guarantees the existence of a progressively measurable stochastic process belonging to the same class of equivalence.

As a consequence, it was discussed in Remark 2.3.7 that the minimal assumption on integrands, under which the stochastic integral is well-defined, is the non-anticipativity of the corresponding joint canonical filtration with respect to the driving Wiener process. In particular, we define

$$\tilde{\mathfrak{F}}_t := \sigma\Big(\sigma_t[\tilde{\varrho}] \cup \sigma_t[\tilde{\mathbf{u}}] \cup \bigcup_{k=1}^{\infty} \sigma_t[\tilde{W}_k]\Big), \quad t \in [0,T],$$

and need to check that $\tilde{\mathfrak{F}}_t$ is independent of $\sigma(\tilde{W}(s) - \tilde{W}(t))$ for all $s > t$.

To this end, we first recall Theorem 2.9.1 and deduce that, for every $\varepsilon \in (0,1)$, $\tilde{W}_\varepsilon = \sum_{k=1}^{\infty} e_k \tilde{W}_{\varepsilon,k}$ is a cylindrical Wiener process with respect to

$$\sigma\Big(\sigma_t[\tilde{\varrho}_\varepsilon] \cup \sigma_t[\tilde{\mathbf{u}}_\varepsilon] \cup \bigcup_{k=1}^{\infty} \sigma_t[\tilde{W}_{\varepsilon,k}]\Big), \quad t \in [0,T].$$

In other words, this filtration is non-anticipative with respect to $\tilde{W}_\varepsilon$. Lemma 2.9.3, together with Proposition 8.1.16, then allows one to pass to the limit as $\varepsilon \to 0$ and the non-anticipativity of $(\tilde{\mathfrak{F}}_t)_{t\geq 0}$ with respect to $\tilde{W}$ follows. Finally, due to Lemma 2.1.35 and Corollary 2.1.36, the process $\tilde{W}$ is a cylindrical Wiener processes with respect to $(\tilde{\mathfrak{F}}_t)_{t\geq 0}$.

8.1.3.4 Identification of the limit

The aim of this section is to identify the limit processes given by Proposition 8.1.16 with a weak martingale solution to (8.14)–(8.15). Namely, we prove the following result, which in turn verifies Theorem 8.1.6.

Theorem 8.1.17. *The multiplet*

$$((\tilde{\Omega}, \tilde{\mathfrak{F}}, (\tilde{\mathfrak{F}}_t)_{t\geq 0}, \tilde{\mathbb{P}}), \tilde{\mathbf{u}}, \tilde{W})$$

is a weak martingale solution to (8.16)–(8.17) *in the sense of Definition* 8.1.2 *with the initial law* Λ.

The proof proceeds in several steps. First of all, we show that, also on the new probability space $(\tilde{\Omega}, \tilde{\mathfrak{F}}, \tilde{\mathbb{P}})$, the approximations $\tilde{\varrho}_\varepsilon$, $\tilde{\mathbf{u}}_\varepsilon$ solve the corresponding compressible Navier–Stokes system (8.10)–(8.13). As a consequence of the equality of laws from Proposition 8.1.16 as well as Theorem 2.9.1 we obtain the following.

Proposition 8.1.18. *Let* $\varepsilon \in (0,1)$. *The multiplet*

$$((\tilde{\Omega}, \tilde{\mathfrak{F}}, (\tilde{\mathfrak{F}}^\varepsilon_t)_{t\geq 0}, \tilde{\mathbb{P}}), \tilde{\varrho}_\varepsilon, \tilde{\mathbf{u}}_\varepsilon, \tilde{W}_\varepsilon)$$

is a dissipative martingale solution to (8.10)–(8.13) *in the sense of Definition* 8.0.1 *with* $[\mathrm{Ma}] = \varepsilon$ *and the initial law* Λ_ε.

Consequently, we recover the result of Proposition 8.1.9 together with all the uniform estimates of the previous subsection. In particular, we find (for a subsequence)

$$\tilde{\varrho}_\varepsilon \to 1 \quad \text{in } L^\infty(0,T;L^\gamma(\mathbb{T}^N)) \ \tilde{\mathbb{P}}\text{-a.s.} \tag{8.40}$$

Due to Corollary 8.1.10 we have the following bounds on the new probability space.

Corollary 8.1.19. *We have the following bounds uniform in* ε, *for all* $l > \frac{N}{2}$:

$$\mathbb{E}\Big[\Big|\sup_{t\in[0,T]} \|\tilde{\varrho}\varepsilon|\tilde{\mathbf{u}}\varepsilon|^2\|_{L^1_x}\Big|^r\Big] \lesssim c(r),$$

$$\mathbb{E}\Big[\Big|\sup_{t\in[0,T]} \|\tilde{\varphi}_\varepsilon\|^{\min(\gamma,2)}_{L^{\min(\gamma,2)}_x}\Big|^r\Big] \lesssim c(r),$$

$$\tilde{\mathbb{E}}\left[\left|\int_0^T \|\tilde{\mathbf{F}}_\varepsilon\|_{W_x^{-l,2}}\,\mathrm{d}t\right|^r\right] \lesssim c(r),$$

where $\tilde{\varphi}_\varepsilon = \frac{\tilde{\varrho}_\varepsilon - 1}{\varepsilon}$ *and*

$$\tilde{\mathbf{F}}_\varepsilon = \nu\Delta\mathcal{Q}\mathbf{u}_\varepsilon + (\lambda+\nu)\nabla\operatorname{div}\tilde{\mathbf{u}}_\varepsilon - \mathcal{Q}[\operatorname{div}(\tilde{\varrho}_\varepsilon\tilde{\mathbf{u}}_\varepsilon\otimes\tilde{\mathbf{u}}_\varepsilon)] - \frac{1}{\varepsilon^2}\nabla[\tilde{\varrho}_\varepsilon^\gamma - 1 - \gamma(\tilde{\varrho}_\varepsilon - 1)].$$

A further consequence of Proposition 8.1.16 is that we can show strong convergence of $\mathcal{P}_H\tilde{\mathbf{u}}_\varepsilon$.

Proposition 8.1.20. *We have the following convergence* $\tilde{\mathbb{P}}$*-a.s.:*

$$\mathcal{P}_H\tilde{\mathbf{u}}_\varepsilon \to \tilde{\mathbf{u}} \quad \text{in } L^2(0,T;L^q(\mathbb{T}^N)) \quad \forall q < \frac{2N}{N-2}. \tag{8.41}$$

Proof. Since the joint laws of $(\varrho_\varepsilon, \mathbf{u}_\varepsilon, \mathcal{P}_H(\varrho_\varepsilon\mathbf{u}_\varepsilon))$ and $(\tilde{\varrho}_\varepsilon, \tilde{\mathbf{u}}_\varepsilon, \tilde{\mathbf{q}}_\varepsilon)$ coincide, we deduce that $\tilde{\mathbf{q}}_\varepsilon = \mathcal{P}_H(\tilde{\varrho}_\varepsilon\tilde{\mathbf{u}}_\varepsilon)$ $\tilde{\mathbb{P}}$-a.s. and consequently it follows from the proof of Proposition 8.1.14 that

$$\tilde{\mathbb{E}}\|\mathcal{P}_H(\tilde{\varrho}_\varepsilon\tilde{\mathbf{u}}_\varepsilon)\|_{C_t^\kappa W_x^{-l,2}} \leq C, \tag{8.42}$$

for all $\kappa \in (0, \frac{1}{2})$ and some $l \in \mathbb{N}$.

Besides, it follows from (8.40) and the convergence of $\tilde{\mathbf{u}}_\varepsilon$ to $\tilde{\mathbf{u}}$ that

$$\tilde{\varrho}_\varepsilon\tilde{\mathbf{u}}_\varepsilon \rightharpoonup \tilde{\mathbf{u}} \quad \text{in } L^2(0,T;L^{\frac{2\gamma}{\gamma+1}}(\mathbb{T}^N))\ \tilde{\mathbb{P}}\text{-a.s.} \tag{8.43}$$

If we pass to the limit in the continuity equation, we see that $\operatorname{div}\tilde{\mathbf{u}} = 0$, which in turn identifies $\tilde{\mathbf{q}}$ with $\tilde{\mathbf{u}}$. Indeed, using continuity of $\mathcal{P}_H$ (cf. Theorem A.1.4), we obtain

$$\mathcal{P}_H(\tilde{\varrho}_\varepsilon\tilde{\mathbf{u}}_\varepsilon) \rightharpoonup \tilde{\mathbf{u}} \quad \text{in } L^2(0,T;L^{\frac{2\gamma}{\gamma+1}}(\mathbb{T}^N))\ \tilde{\mathbb{P}}\text{-a.s.}$$

Thus, with Proposition 8.1.16 and the compact embedding $L^{\frac{2\gamma}{\gamma+1}}(\mathbb{T}^N) \overset{c}{\hookrightarrow} W^{-1,2}(\mathbb{T}^N)$,

$$\mathcal{P}_H(\tilde{\varrho}_\varepsilon\tilde{\mathbf{u}}_\varepsilon) \to \tilde{\mathbf{u}} \quad \text{in } L^2(0,T;W^{-1,2}(\mathbb{T}^N))\ \tilde{\mathbb{P}}\text{-a.s.} \tag{8.44}$$

Since

$$\operatorname{div}\tilde{\mathbf{u}}_\varepsilon \rightharpoonup 0 \quad \text{in } L^2(0,T;L^2(\mathbb{T}^N))\ \tilde{\mathbb{P}}\text{-a.s.}, \tag{8.45}$$

we also have

$$\mathcal{P}_H\tilde{\mathbf{u}}_\varepsilon \rightharpoonup \tilde{\mathbf{u}} \quad \text{in } L^2(0,T;W^{1,2}(\mathbb{T}^N))\ \tilde{\mathbb{P}}\text{-a.s.} \tag{8.46}$$

Note that (8.45) is a consequence of $\operatorname{div}\tilde{\mathbf{u}} = 0$ and the $\tilde{\mathbb{P}}$-a.s. convergence $\tilde{\mathbf{u}}_\varepsilon \rightharpoonup \tilde{\mathbf{u}}$ in $L^2(0,T;W^{1,2}(\mathbb{T}^N))$; cf. Proposition 8.1.16. Combining (8.44) with (8.46), we conclude that

$$\mathcal{P}_H(\tilde{\varrho}_\varepsilon\tilde{\mathbf{u}}_\varepsilon)\cdot\mathcal{P}_H\tilde{\mathbf{u}}_\varepsilon \rightharpoonup |\tilde{\mathbf{u}}|^2 \quad \text{in } L^1((0,T)\times\mathbb{T}^N)\ \tilde{\mathbb{P}}\text{-a.s.}$$

Proposition 8.1.16 yields $\tilde{\mathbb{P}}$-a.s.

$$\left|\int_Q \left(|\mathcal{P}_H \tilde{\mathbf{u}}_\varepsilon|^2 - \mathcal{P}_H(\tilde{\varrho}_\varepsilon \tilde{\mathbf{u}}_\varepsilon)\cdot \mathcal{P}_H \tilde{\mathbf{u}}_\varepsilon\right) \mathrm{d}x\,\mathrm{d}t\right| \le \|\tilde{\varrho}_\varepsilon - 1\|_{L^\infty_t L^\gamma_x} \|\tilde{\mathbf{u}}_\varepsilon\|^2_{L^2_t L^s_x} \to 0,$$

where $s = \frac{2\gamma}{\gamma-1} < \frac{2N}{N-2}$. This implies $\|\mathcal{P}_H \tilde{\mathbf{u}}_\varepsilon\|_{L^2_x} \to \|\tilde{\mathbf{u}}\|_{L^2_x}$ and hence

$$\mathcal{P}_H \tilde{\mathbf{u}}_\varepsilon \to \tilde{\mathbf{u}} \quad \text{in } L^2(0,T; L^2(\mathbb{T}^N))\ \tilde{\mathbb{P}}\text{-a.s.}$$

Combining this with weak convergence in $L^2(0,T;W^{1,2}(\mathbb{T}^N))$ (recall Proposition 8.1.16) yields the claim. □

In the following we aim to identify the limit in the gradient part of the convective term. To this end, we adopt the deterministic approach proposed by Lions–Masmoudi [LM99].

Proposition 8.1.21. *For $l > \frac{N}{2}$, we have*

$$\operatorname{div}(\tilde{\varrho}_\varepsilon \tilde{\mathbf{u}}_\varepsilon \otimes \tilde{\mathbf{u}}_\varepsilon) \rightharpoonup \operatorname{div}(\tilde{\mathbf{u}} \otimes \tilde{\mathbf{u}}) \quad \text{in } L^1(0,T; W^{-l,2}_{\operatorname{div}}(\mathbb{T}^N))\ \tilde{\mathbb{P}}\text{-a.s.}$$

Proof. Following Lions and Masmoudi [LM99], we decompose

$$\begin{aligned}
\tilde{\varrho}_\varepsilon \tilde{\mathbf{u}}_\varepsilon &= \tilde{\mathbf{u}} + \mathcal{P}_H(\tilde{\varrho}_\varepsilon \tilde{\mathbf{u}}_\varepsilon - \tilde{\mathbf{u}}) + \mathcal{Q}(\tilde{\varrho}_\varepsilon \tilde{\mathbf{u}}_\varepsilon - \tilde{\mathbf{u}}),\\
\tilde{\mathbf{u}}_\varepsilon &= \tilde{\mathbf{u}} + \mathcal{P}_H(\tilde{\mathbf{u}}_\varepsilon - \tilde{\mathbf{u}}) + \mathcal{Q}(\tilde{\mathbf{u}}_\varepsilon - \tilde{\mathbf{u}}).
\end{aligned}$$

The claim follows once we have shown that the following convergences hold true weakly in $L^1(0,T; W^{-l,2}_{\operatorname{div}}(\mathbb{T}^N))$ $\tilde{\mathbb{P}}$-a.s.:

$$\operatorname{div}(\tilde{\mathbf{u}} \otimes \mathcal{P}_H(\tilde{\mathbf{u}}_\varepsilon - \tilde{\mathbf{u}})) \rightharpoonup 0, \tag{8.47}$$

$$\operatorname{div}(\tilde{\mathbf{u}} \otimes \mathcal{Q}(\tilde{\mathbf{u}}_\varepsilon - \tilde{\mathbf{u}})) \rightharpoonup 0, \tag{8.48}$$

$$\operatorname{div}(\mathcal{P}_H(\tilde{\varrho}_\varepsilon \tilde{\mathbf{u}}_\varepsilon - \tilde{\mathbf{u}}) \otimes \tilde{\mathbf{u}}) \rightharpoonup 0, \tag{8.49}$$

$$\operatorname{div}(\mathcal{Q}(\tilde{\varrho}_\varepsilon \tilde{\mathbf{u}}_\varepsilon - \tilde{\mathbf{u}}) \otimes \tilde{\mathbf{u}}) \rightharpoonup 0, \tag{8.50}$$

$$\operatorname{div}(P_H(\tilde{\varrho}_\varepsilon \tilde{\mathbf{u}}_\varepsilon - \tilde{\mathbf{u}}) \otimes \mathcal{P}_H(\tilde{\mathbf{u}}_\varepsilon - \tilde{\mathbf{u}})) \rightharpoonup 0, \tag{8.51}$$

$$\operatorname{div}(P_H(\tilde{\varrho}_\varepsilon \tilde{\mathbf{u}}_\varepsilon - \tilde{\mathbf{u}}) \otimes \mathcal{Q}(\tilde{\mathbf{u}}_\varepsilon - \tilde{\mathbf{u}})) \rightharpoonup 0, \tag{8.52}$$

$$\operatorname{div}(\mathcal{Q}(\tilde{\varrho}_\varepsilon \tilde{\mathbf{u}}_\varepsilon - \tilde{\mathbf{u}}) \otimes \mathcal{P}_H(\tilde{\mathbf{u}}_\varepsilon - \tilde{\mathbf{u}})) \rightharpoonup 0, \tag{8.53}$$

$$\operatorname{div}(\mathcal{Q}(\tilde{\varrho}_\varepsilon \tilde{\mathbf{u}}_\varepsilon - \tilde{\mathbf{u}}) \otimes \mathcal{Q}(\tilde{\mathbf{u}}_\varepsilon - \tilde{\mathbf{u}})) \rightharpoonup 0. \tag{8.54}$$

The first four convergences follow from Proposition 8.1.16, (8.43), and the continuity of $\mathcal{P}_H$ and $\mathcal{Q}$, respectively; cf. Theorem A.1.4. The convergences (8.51)–(8.53) are consequences of (8.40) and (8.41). In fact, the only critical part is (8.54). First, we need some improved space regularity. Similarly to Lions and Masmoudi [LM99], we use mollification by means of spatial convolution with a family of regularizing kernels with a parameter $0 < \kappa \ll 1$. As a matter of fact, thanks to the special geometry of the flat torus $\mathbb{T}^N$, the mollified functions can be taken as projections to a finite number (which is the smallest natural number greater than $\frac{1}{\kappa}$) of modes of the trigonometric basis $(e_{\mathbf{m}})_{\mathbf{m}\in\mathbb{Z}^N}$, defined in Section 1.7. Note that this is a different regularization than the mollification from Section 1.7.3. The regularization by projections is more convenient for our purposes here since it commutes with all spatial derivatives as well as with the projections $\mathcal{P}_H$ and $\mathcal{Q}$. For $\delta > 0$ arbitrary, we take $\kappa = \kappa(\delta)$ so small that

$$\tilde{\mathbb{E}}\|(\tilde{\varrho}_\varepsilon\tilde{\mathbf{u}}_\varepsilon)^\kappa - \tilde{\varrho}_\varepsilon^\kappa\tilde{\mathbf{u}}_\varepsilon^\kappa\|_{L_t^2L_x^{\frac{2\gamma}{\gamma+1}}} + \tilde{\mathbb{E}}\|(\tilde{\varrho}_\varepsilon\tilde{\mathbf{u}}_\varepsilon)^\kappa - \tilde{\varrho}_\varepsilon\tilde{\mathbf{u}}_\varepsilon\|_{L_t^2L_x^{\frac{2\gamma}{\gamma+1}}} \leq \delta, \tag{8.55}$$

$$\tilde{\mathbb{E}}\|(\tilde{\varrho}_\varepsilon\tilde{\mathbf{u}}_\varepsilon)^\kappa - \tilde{\mathbf{u}}_\varepsilon^\kappa\|_{L_t^2L_x^{\frac{2\gamma}{\gamma+1}}} + \tilde{\mathbb{E}}\|\tilde{\mathbf{u}}_\varepsilon^\kappa - \tilde{\mathbf{u}}_\varepsilon\|_{L_t^2L_x^{\frac{2N}{N-2}}} + \tilde{\mathbb{E}}\|\tilde{\mathbf{u}}^\kappa - \tilde{\mathbf{u}}\|_{L_t^2L_x^{\frac{2N}{N-2}}} \leq \delta, \tag{8.56}$$

uniformly in ε. We note that the norm $\tilde{\mathbb{E}}\|\tilde{\mathbf{u}}_\varepsilon^\kappa - \tilde{\mathbf{u}}_\varepsilon\|_{L_t^2L_x^{\frac{2N}{N-2}}}$ can be made uniformly small as a consequence of the gradient estimate (8.26). As the mollification commutes with div and $\mathcal{Q}$, it suffices to show that $\tilde{\mathbb{P}}$-a.s.

$$\operatorname{div}(\mathcal{Q}(\tilde{\varrho}_\varepsilon^\kappa\tilde{\mathbf{u}}_\varepsilon^\kappa - \tilde{\mathbf{u}}^\kappa) \otimes \mathcal{Q}(\tilde{\mathbf{u}}_\varepsilon^\kappa - \tilde{\mathbf{u}}^\kappa)) \rightharpoonup 0, \tag{8.57}$$

for fixed κ instead of (8.54). In fact, expectation of the $L^1(0,T;W_{\operatorname{div}}^{-l,2}(\mathbb{T}^N))$-norm of the difference of (8.57) and (8.54) can be estimated in terms of δ using (8.55) and (8.56). To prove (8.57), we write

$$\mathcal{Q}(\tilde{\mathbf{u}}_\varepsilon^\kappa - \tilde{\mathbf{u}}^\kappa) = \mathcal{Q}(\tilde{\varrho}_\varepsilon^\kappa\tilde{\mathbf{u}}_\varepsilon^\kappa - \tilde{\mathbf{u}}^\kappa) + \mathcal{Q}((1 - \tilde{\varrho}_\varepsilon^\kappa)\tilde{\mathbf{u}}_\varepsilon^\kappa).$$

By (8.40), the continuity of $\mathcal{Q}$, and the boundedness of $\tilde{\mathbf{u}}_\varepsilon^\kappa$, we know

$$\mathcal{Q}((1 - \tilde{\varrho}_\varepsilon^\kappa)\tilde{\mathbf{u}}_\varepsilon^\kappa) \to 0 \quad \text{in } L^2((0,T)\times\mathbb{T}^N)\ \tilde{\mathbb{P}}\text{-a.s.}$$

So (8.57) follows from

$$\operatorname{div}(Q(\tilde{\varrho}_\varepsilon^\kappa\tilde{\mathbf{u}}_\varepsilon^\kappa - \tilde{\mathbf{u}}^\kappa) \otimes \mathcal{Q}(\tilde{\varrho}_\varepsilon^\kappa\tilde{\mathbf{u}}_\varepsilon^\kappa - \tilde{\mathbf{u}}^\kappa)) \rightharpoonup 0 \quad \text{in } L^1(0,T;W_{\operatorname{div}}^{-l,2}(\mathbb{T}^N)). \tag{8.58}$$

As $\operatorname{div}(\mathcal{Q}\tilde{\mathbf{u}}^\kappa \otimes \mathcal{Q}\tilde{\mathbf{u}}^\kappa) = \frac{1}{2}\nabla|\mathcal{Q}\tilde{\mathbf{u}}^\kappa|^2$, the convergence (8.58) is a consequence of

$$\operatorname{div}(\mathcal{Q}(\tilde{\varrho}_\varepsilon\tilde{\mathbf{u}}_\varepsilon)^\kappa \otimes \mathcal{Q}(\tilde{\varrho}_\varepsilon\tilde{\mathbf{u}}_\varepsilon)^\kappa) \rightharpoonup 0 \quad \text{in } L^1(0,T;W_{\operatorname{div}}^{-l,2}(\mathbb{T}^N)), \tag{8.59}$$

thanks to (8.43) and (8.55). In order to show (8.59), we need to introduce the function $\tilde{\Psi}_\varepsilon = \Delta^{-1}\operatorname{div}(\tilde{\varrho}_\varepsilon\tilde{\mathbf{u}}_\varepsilon)$, which satisfies $\nabla\tilde{\Psi}_\varepsilon = \mathcal{Q}(\tilde{\varrho}_\varepsilon\tilde{\mathbf{u}}_\varepsilon)$. We have the system of equations

$$\mathrm{d}(\varepsilon\tilde{\varphi}_\varepsilon) = -\nabla\tilde{\Psi}_\varepsilon\,\mathrm{d}t, \quad \mathrm{d}\nabla\tilde{\Psi}_\varepsilon = -\frac{\gamma}{\varepsilon}\nabla\tilde{\varphi}_\varepsilon\,\mathrm{d}t + \tilde{\mathbf{F}}_\varepsilon\,\mathrm{d}t + \mathcal{Q}\mathbb{G}(\tilde{\varrho}_\varepsilon, \tilde{\varrho}_\varepsilon\tilde{\mathbf{u}}_\varepsilon)\,\mathrm{d}\tilde{W}_\varepsilon,$$

where the right hand side only belongs to $W^{-l,2}(\mathbb{T}^N)$. We apply mollification and gain $\tilde{\Psi}_\varepsilon^\kappa = \Delta^{-1}\operatorname{div}(\tilde{\varrho}_\varepsilon\tilde{\mathbf{u}}_\varepsilon)^\kappa$ and $\nabla\tilde{\Psi}_\varepsilon^\kappa = \mathcal{Q}(\tilde{\varrho}_\varepsilon\tilde{\mathbf{u}}_\varepsilon)^\kappa$. The system of equations for $\tilde{\varphi}_\varepsilon^\kappa$ and $\tilde{\Psi}_\varepsilon^\kappa$ reads

$$\mathrm{d}(\varepsilon\tilde{\varphi}_\varepsilon^\kappa) = -\Delta\tilde{\Psi}_\varepsilon^\kappa\,\mathrm{d}t, \quad \mathrm{d}\nabla\tilde{\Psi}_\varepsilon^\kappa = -\frac{\gamma}{\varepsilon}\nabla\tilde{\varphi}_\varepsilon^\kappa\,\mathrm{d}t + \tilde{\mathbf{F}}_\varepsilon^\kappa\,\mathrm{d}t + \mathcal{Q}\mathbb{G}(\tilde{\varrho}_\varepsilon, \tilde{\varrho}_\varepsilon\tilde{\mathbf{u}}_\varepsilon)^\kappa\,\mathrm{d}\tilde{W}_\varepsilon. \tag{8.60}$$

We note that, for the special choice where the mollification is taken as the projection onto a finite number of Fourier modes, the system (8.60) reduces to a *finite number* of equations. Now, we apply Itô's formula to the function

$$f(\varepsilon\tilde{\varphi}_\varepsilon^\kappa, \nabla\tilde{\Psi}_\varepsilon^\kappa) = \int_{\mathbb{T}^N} \varepsilon\tilde{\varphi}_\varepsilon^\kappa \nabla\tilde{\Psi}_\varepsilon^\kappa \cdot \boldsymbol{\varphi}\,\mathrm{d}x,$$

with $\boldsymbol{\varphi} \in C^\infty_{\mathrm{div}}(\mathbb{T}^N)$ arbitrary, and gain

$$\begin{aligned}
&\int_{\mathbb{T}^N} \varepsilon\tilde{\varphi}_\varepsilon^\kappa(t)\nabla\tilde{\Psi}_\varepsilon^\kappa(t)\cdot\boldsymbol{\varphi}\,\mathrm{d}x \\
&\quad = -\int_0^t\int_{\mathbb{T}^N} \Delta\tilde{\Psi}_\varepsilon^\kappa\nabla\tilde{\Psi}_\varepsilon^\kappa\cdot\boldsymbol{\varphi}\,\mathrm{d}x\,\mathrm{d}s - \gamma\int_0^t\int_{\mathbb{T}^N}\tilde{\varphi}_\varepsilon^\kappa\nabla\tilde{\varphi}_\varepsilon^\kappa\cdot\boldsymbol{\varphi}\,\mathrm{d}x\,\mathrm{d}s \\
&\qquad + \varepsilon\int_0^t\int_{\mathbb{T}^N}\tilde{\varphi}_\varepsilon^\kappa\tilde{\mathbf{F}}_\varepsilon^\kappa\cdot\boldsymbol{\varphi}\,\mathrm{d}x\,\mathrm{d}s + \varepsilon\int_{\mathbb{T}^N}\int_0^t\tilde{\varphi}_\varepsilon^\kappa\boldsymbol{\varphi}\cdot\mathcal{Q}\mathbb{G}(\tilde{\varrho}_\varepsilon,\tilde{\varrho}_\varepsilon\tilde{\mathbf{u}}_\varepsilon)^\kappa\,\mathrm{d}\tilde{W}_\varepsilon\,\mathrm{d}x.
\end{aligned}$$

Note that we have

$$\begin{aligned}
&\int_0^t\int_{\mathbb{T}^N}\Delta\tilde{\Psi}_\varepsilon^\kappa\nabla\tilde{\Psi}_\varepsilon^\kappa\cdot\boldsymbol{\varphi}\,\mathrm{d}x\,\mathrm{d}s \\
&\quad = \frac{1}{2}\int_0^t\int_{\mathbb{T}^N}\nabla|\nabla\tilde{\Psi}_\varepsilon^\kappa|^2\cdot\boldsymbol{\varphi}\,\mathrm{d}x\,\mathrm{d}s - \int_0^t\int_{\mathbb{T}^N}\nabla\tilde{\Psi}_\varepsilon^\kappa\otimes\nabla\tilde{\Psi}_\varepsilon^\kappa : \nabla\boldsymbol{\varphi}\,\mathrm{d}x\,\mathrm{d}s \\
&\quad = -\int_0^t\int_{\mathbb{T}^N}\nabla\tilde{\Psi}_\varepsilon^\kappa\nabla\tilde{\Psi}_\varepsilon^\kappa : \nabla\boldsymbol{\varphi}\,\mathrm{d}x\,\mathrm{d}s, \\
&\int_0^t\int_{\mathbb{T}^N}\tilde{\varphi}_\varepsilon^\kappa\nabla\tilde{\varphi}_\varepsilon^\kappa\cdot\boldsymbol{\varphi}\,\mathrm{d}x\,\mathrm{d}s = \frac{1}{2}\int_0^t\int_{\mathbb{T}^N}\nabla|\tilde{\varphi}_\varepsilon^\kappa|^2\cdot\boldsymbol{\varphi}\,\mathrm{d}x\,\mathrm{d}s = 0,
\end{aligned}$$

due to $\mathrm{div}\,\boldsymbol{\varphi} = 0$, so we obtain

$$\begin{aligned}
&\int_0^t\int_{\mathbb{T}^N}\nabla\tilde{\Psi}_\varepsilon^\kappa\otimes\nabla\tilde{\Psi}_\varepsilon^\kappa : \nabla\boldsymbol{\varphi}\,\mathrm{d}x\,\mathrm{d}s \\
&\quad = -\varepsilon\int_{\mathbb{T}^N}\tilde{\varphi}_\varepsilon^\kappa(t)\nabla\tilde{\Psi}_\varepsilon^\kappa(t)\cdot\boldsymbol{\varphi}\,\mathrm{d}x \\
&\qquad + \varepsilon\int_0^t\int_{\mathbb{T}^N}\tilde{\varphi}_\varepsilon^\kappa\tilde{\mathbf{F}}_\varepsilon^\kappa\cdot\boldsymbol{\varphi}\,\mathrm{d}x\,\mathrm{d}s + \varepsilon\int_{\mathbb{T}^N}\int_0^t\tilde{\varphi}_\varepsilon^\kappa\boldsymbol{\varphi}\cdot\mathcal{Q}\mathbb{G}(\tilde{\varrho}_\varepsilon,\tilde{\varrho}_\varepsilon\tilde{\mathbf{u}}_\varepsilon)^\kappa\,\mathrm{d}\tilde{W}_\varepsilon\,\mathrm{d}x.
\end{aligned}$$

For fixed $\kappa > 0$, the right hand side vanishes $\tilde{\mathbb{P}}$-a.s. for $\varepsilon \to 0$ at least after taking a subsequence due to Corollary 8.1.19, Proposition 8.1.16, and the properties of the mollification. Finally we conclude with (8.59), which implies the last missing convergence (8.54) as explained above. □

Now, everything is in hand to complete the proof of Theorem 8.1.17, which implies the proof of our main result, Theorem 8.1.6.

Proof of Theorem 8.1.17. In order to show that (8.14)–(8.15) is satisfied in the sense of Definition 8.1.2, let us take a divergence-free test function $\boldsymbol{\varphi} \in C^\infty_{\mathrm{div}}(\mathbb{T}^N)$. This way we only study the approximate equation (8.11) projected by $\mathcal{P}_H$ and the pressure term drops out. By Proposition 8.1.16, we can pass to the limit in the deterministic parts of the equation. We comment on the passage to the limit in the terms coming from the stochastic integral in detail. To this end, we write

$$\begin{aligned}
&\|\mathbb{G}(\tilde{\varrho}_\varepsilon, \tilde{\varrho}_\varepsilon \tilde{\mathbf{u}}_\varepsilon) - \mathbb{G}(1, \tilde{\mathbf{u}})\|_{L_2(\mathfrak{U}; W_x^{-l,2})} \\
&\leq \Big(\sum_{k\geq 1} \|\mathbf{h}_k(\tilde{\varrho}_\varepsilon) - \mathbf{h}_k(1)\|^2_{W_x^{-l,2}} \Big)^{\frac{1}{2}} + \Big(\sum_{k\geq 1} |\alpha_k|^2 \|\tilde{\varrho}_\varepsilon \tilde{\mathbf{u}}_\varepsilon - \tilde{\mathbf{u}}\|^2_{W_x^{-l,2}} \Big)^{\frac{1}{2}} \\
&= I_1 + I_2.
\end{aligned}$$

For I_2, we use (8.20) together with (8.44) to obtain $I_2 \to 0$ for a.e. (ω, t). For I_1, we apply the Minkowski integral inequality, the mean value theorem, and (8.20), to obtain

$$\begin{aligned}
I_1 &\leq C\Big(\sum_{k\geq 1} \|\mathbf{h}_k(\tilde{\varrho}_\varepsilon) - \mathbf{h}_k(1)\|^2_{L^1_x} \Big)^{\frac{1}{2}} \leq C \int_{\mathbb{T}^N} \Big(\sum_{k\geq 1} |\mathbf{h}_k(\tilde{\varrho}_\varepsilon) - \mathbf{h}_k(1)|^2 \Big)^{\frac{1}{2}} \mathrm{d}x \\
&\leq C \int_{\mathbb{T}^N} (1 + \tilde{\varrho}_\varepsilon^{\frac{\gamma-1}{2}}) |\tilde{\varrho}_\varepsilon - 1| \, \mathrm{d}x \leq C \Big[\int_{\mathbb{T}^N} (1 + \tilde{\varrho}_\varepsilon^{\frac{\gamma-1}{2}})^p \, \mathrm{d}x \Big]^{\frac{1}{p}} \Big[\int_{\mathbb{T}^N} |\tilde{\varrho}_\varepsilon - 1|^q \, \mathrm{d}x \Big]^{\frac{1}{q}},
\end{aligned}$$

where the conjugate exponents $p, q \in (1, \infty)$ are chosen in such a way that

$$p\frac{\gamma - 1}{2} < \gamma + 1 \quad \text{and} \quad q < \gamma.$$

Therefore, using (8.30) and (8.40) we deduce

$$\int_0^T I_1 \, \mathrm{d}t \to 0$$

$\tilde{\mathbb{P}}$-a.s. and we obtain

$$\mathbb{G}(\tilde{\varrho}_\varepsilon, \tilde{\varrho}_\varepsilon \tilde{\mathbf{u}}_\varepsilon) \to \mathbb{G}(1, \tilde{\mathbf{u}}) \quad \text{in } L^1(0, T; L_2(\mathfrak{U}; W^{-l,2}(\mathbb{T}^N))) \tag{8.61}$$

$\tilde{\mathbb{P}}$-a.s. Besides, since, for some $r > 2$,

$$\begin{aligned}
&\tilde{\mathbb{E}} \int_s^t \|\mathbb{G}(\tilde{\varrho}_\varepsilon, \tilde{\varrho}_\varepsilon \tilde{\mathbf{u}}_\varepsilon)\|^p_{L_2(\mathfrak{U}; W_x^{-l,2})} \, \mathrm{d}\sigma \\
&\leq C \tilde{\mathbb{E}} \int_s^t \|\tilde{\varrho}_\varepsilon\|^{\frac{r}{2}}_{L^2_x} (1 + \|\tilde{\varrho}_\varepsilon\|^\gamma_{L^\gamma_x} + \|\sqrt{\tilde{\varrho}_\varepsilon} \tilde{\mathbf{u}}_\varepsilon\|^2_{L^2_x})^{\frac{r}{2}} \, \mathrm{d}\sigma \\
&\leq C\Big(1 + \tilde{\mathbb{E}} \sup_{0\leq t\leq T} \|\tilde{\varrho}_\varepsilon\|^{\gamma r}_{L^\gamma_x} + \tilde{\mathbb{E}} \sup_{0\leq t\leq T} \|\sqrt{\tilde{\varrho}_\varepsilon} \tilde{\mathbf{u}}_\varepsilon\|^{2r}_{L^2_x} \Big) \leq C,
\end{aligned}$$

using (8.27) and (8.30), we obtain

$$\mathbb{G}(\tilde{\varrho}_\varepsilon, \tilde{\varrho}_\varepsilon \tilde{\mathbf{u}}_\varepsilon) \to \mathbb{G}(1, \tilde{\mathbf{u}}) \quad \text{in } L^2(0, T; L_2(\mathfrak{U}; W^{-l,2}(\mathbb{T}^N))).$$

Combining this with the convergence of $\tilde{W}_h$ from Proposition 8.1.16, we may apply Lemma 2.6.6 to pass to the limit in the stochastic integral. Hence $\tilde{\mathbf{u}}$ solves (8.14)–(8.15). It follows immediately from our construction that $\tilde{\mathbb{P}}$-a.s.

$$\tilde{\mathbf{u}} \in L^2(0,T;W^{1,2}_{\mathrm{div}}(\mathbb{T}^N)).$$

Besides, since we have $\tilde{\mathbb{P}}$-a.s. (due to Proposition 8.1.16 and (8.40))

$$\sqrt{\tilde{\varrho}_\varepsilon}\tilde{\mathbf{u}}_\varepsilon \rightharpoonup \tilde{\mathbf{u}} \quad \text{in } L^1(\Omega;L^1(Q)),$$

the lower semi-continuity of the functional

$$\tilde{\mathbf{w}} \mapsto \tilde{\mathbb{E}}\Big[\Big|\sup_{t\in(0,T)}\int_{\mathbb{T}^N}|\tilde{\mathbf{w}}|^2\,\mathrm{d}x\Big|^r\Big]$$

yields

$$\tilde{\mathbb{E}}\Big[\Big|\sup_{t\in(0,T)}\int_{\mathbb{T}^N}|\tilde{\mathbf{u}}|^2\,\mathrm{d}x\Big|^r\Big] \lesssim c(r), \tag{8.62}$$

on account of Corollary 8.1.19. The usual argument about the fractional time derivative in the distributional sense implies $\tilde{\mathbf{u}} \in C_w([0,T];L^2_{\mathrm{div}}(\mathbb{T}^N))$ $\tilde{\mathbb{P}}$-a.s. In fact we can argue as in the proof of Proposition 8.1.14 and decompose $\tilde{\mathbf{u}}$ into the sum of

$$\tilde{Y}(t) = \tilde{\mathbf{u}}(0) - \int_0^t \mathcal{P}_H[\mathrm{div}(\tilde{\mathbf{u}}\otimes\tilde{\mathbf{u}}) - \mu\Delta\tilde{\mathbf{u}}]\,\mathrm{d}s,$$
$$\tilde{Z}(t) = \int_0^t \mathcal{P}_H\mathbb{G}(1,\tilde{\mathbf{u}})\,\mathrm{d}\tilde{W}(s),$$

in order to show

$$\mathbb{E}\|\tilde{\mathbf{u}}\|_{C^\kappa_t W^{-l,2}_x} \leq C,$$

for all $\kappa \in (0,\frac{1}{2})$ and some $l \in \mathbb{N}$. Due to (8.62), this implies $\tilde{\mathbf{u}} \in C_w([0,T];L^2_{\mathrm{div}}(\mathbb{T}^N))$ $\tilde{\mathbb{P}}$-a.s. The proof is hereby complete. □

8.1.4 Convergence in probability – the proof of Theorem 8.1.7

In order to complete the proof of Theorem 8.1.7, we make use of Theorem 2.10.3, which is a generalization of the Gyöngy–Krylov characterization of convergence in probability introduced in [GK96], adapted to the case of sub-Polish spaces. It applies to situations when pathwise uniqueness and existence of a martingale solution are valid and allows one to establish existence of a pathwise solution. We recall that, in the case of $N = 2$, pathwise uniqueness for (8.16)–(8.17) is known; cf. Theorem 8.1.5. We consider two null sequences (ε_n) and (ε_m). Let $(\varrho_n, \mathbf{u}_n) = (\varrho_{\varepsilon_n}, \mathbf{u}_{\varepsilon_n})$ and $(\varrho_m, \mathbf{u}_m) = (\varrho_{\varepsilon_m}, \mathbf{u}_{\varepsilon_m})$ be density and velocity corresponding to a dissipative martingale solution to (8.10)–(8.13) in the sense of Definition 8.0.1 with $a = \frac{1}{\varepsilon_n^2}$ and $a = \frac{1}{\varepsilon_m^2}$, respectively. As explained at the beginning of Section 8.1.3, we can assume that both sequences of dissipative martingale solutions are defined on the same probability space and with the same Wiener process W. We consider the collection of joint laws of $(\mathbf{X}_n, \mathbf{X}_m, W)$, where

$$\mathbf{X}_n = (\varrho_n, \mathbf{u}_n, \mathcal{P}_H(\varrho_n \mathbf{u}_n)),$$
$$\mathbf{X}_m = (\varrho_m, \mathbf{u}_m, \mathcal{P}_H(\varrho_m \mathbf{u}_m)),$$

on the extended path space

$$\mathcal{X}^J = (\mathcal{X}_\varrho \times \mathcal{X}_\mathbf{u} \times \mathcal{X}_{\varrho\mathbf{u}})^2 \times \mathcal{X}_W,$$

where

$$\mathcal{X}_\varrho = C_w([0,T]; L^\gamma(\mathbb{T}^2)),$$
$$\mathcal{X}_\mathbf{u} = (L^2(0,T; W^{1,2}(\mathbb{T}^2)), w),$$
$$\mathcal{X}_{\varrho\mathbf{u}} = C_w([0,T]; L^{\frac{2\gamma}{\gamma+1}}(\mathbb{T}^2)) \cap C([0,T]; W^{-k,2}(\mathbb{T}^2)), \quad k > \frac{5}{2},$$
$$\mathcal{X}_W = C([0,T]; \mathfrak{U}_0).$$

Similarly to Corollary 8.1.15, the following fact holds true. The proof is nearly identical and will therefore be left to the reader.

Proposition 8.1.22. *The set*

$$\{\mathcal{L}[\mathbf{X}_n, \mathbf{X}_m, W];\, n, m \in \mathbb{N}\}$$

is tight on $\mathcal{X}^J$.

Let us take any subsequence $(\mathbf{X}_{n_k}, \mathbf{X}_{m_k}, W)$. By the Jakubowski–Skorokhod theorem, Theorem 2.7.1, we infer (for a further subsequence but without loss of generality we keep the same notation) the existence of a probability space $(\bar{\Omega}, \bar{\mathfrak{F}}, \bar{\mathbb{P}})$ with a sequence of random variables $(\hat{\mathbf{X}}_{n_k}, \check{\mathbf{X}}_{m_k}, \bar{W}_k)$ with

$$\hat{\mathbf{X}}_{n_k} = (\hat{\varrho}_{n_k}, \hat{\mathbf{u}}_{n_k}, \hat{\mathbf{q}}_{n_k}), \quad k \in \mathbb{N},$$
$$\check{\mathbf{X}}_{m_k} = (\check{\varrho}_{m_k}, \check{\mathbf{u}}_{m_k}, \check{\mathbf{q}}_{m_k}), \quad k \in \mathbb{N},$$

converging almost surely in $\mathcal{X}^J$ to a random variable $(\hat{\mathbf{X}}, \check{\mathbf{X}}, \bar{W})$, with

$$\hat{\mathbf{X}} = (\hat{\varrho}, \hat{\mathbf{u}}, \hat{\mathbf{q}}),$$
$$\check{\mathbf{X}} = (\check{\varrho}, \check{\mathbf{u}}, \check{\mathbf{q}}).$$

Moreover,

$$\mathcal{L}[\hat{\mathbf{X}}_{n_k}, \check{\mathbf{X}}_{m_k}, \bar{W}_k] = \mathcal{L}[\mathbf{X}_{n_k}, \mathbf{X}_{m_k}, W]$$

on $\mathcal{X}^J$ for all $k \in \mathbb{N}$. Observe that, in particular, $\mathcal{L}[\mathbf{X}_{n_k}, \mathbf{X}_{m_k}, W]$ converges weakly to the measure $\mathcal{L}[\hat{\mathbf{X}}, \check{\mathbf{X}}, \bar{W}]$. As the next step, we should recall the technique established in Section 8.1.3.4. Analogously, it can be applied to both

$$(\hat{\varrho}_{n_k}, \hat{\mathbf{u}}_{n_k}, \hat{\mathbf{q}}_{n_k}, \bar{W}_k), \quad (\hat{\varrho}, \hat{\mathbf{u}}, \hat{\mathbf{q}}, \bar{W})$$

and

$$(\check{\varrho}_{m_k}, \check{\mathbf{u}}_{m_k}, \check{\mathbf{q}}_{m_k}, \bar{W}_k), \quad (\check{\varrho}, \check{\mathbf{u}}, \check{\mathbf{q}}, \bar{W})$$

in order to show that $(\hat{\mathbf{u}}, \bar{W})$ and $(\check{\mathbf{u}}, \bar{W})$ are weak martingale solutions to (8.14)–(8.15) defined on the same stochastic basis $(\bar{\Omega}, \bar{\mathfrak{F}}, (\bar{\mathfrak{F}}_t), \bar{\mathbb{P}})$, where $(\bar{\mathfrak{F}}_t)_{t\geq 0}$ is the $\bar{\mathbb{P}}$-augmented canonical filtration of $(\hat{\mathbf{u}}, \check{\mathbf{u}}, \bar{W})$. Besides, we obtain

$$\hat{\varrho} = \check{\varrho} = 1, \quad \hat{\mathbf{q}} = \hat{\mathbf{u}}, \quad \check{\mathbf{q}} = \check{\mathbf{u}} \quad \bar{\mathbb{P}}\text{-a.s.}$$

In order to verify the assumption of Theorem 2.10.3, we employ the pathwise uniqueness result for (8.14)–(8.15) in two dimensions; cf. Theorem 8.1.5. Indeed, it follows from our assumptions on the approximate initial laws Λ_ε that $\hat{\varrho}(0) = \check{\varrho}(0) = 1$ as well as $\hat{\mathbf{u}}(0) = \check{\mathbf{u}}(0) = \mathbf{u}_0$ $\bar{\mathbb{P}}$-a.s. Therefore, according to Theorem 8.1.5, the solutions $\hat{\mathbf{u}}$ and $\check{\mathbf{u}}$ coincide $\bar{\mathbb{P}}$-a.s. and

$$\begin{aligned}
&\mathcal{L}[\hat{\mathbf{X}}, \check{\mathbf{X}}]((\mathbf{X}_1, \mathbf{X}_2) \in \mathcal{X}^J : \mathbf{X}_1 = \mathbf{X}_2) \\
&= \bar{\mathbb{P}}((\hat{\varrho}, \hat{\mathbf{u}}, \hat{\mathbf{q}}) = (\check{\varrho}, \check{\mathbf{u}}, \check{\mathbf{q}})) = \bar{\mathbb{P}}(\hat{\mathbf{u}} = \check{\mathbf{u}}) = 1.
\end{aligned}$$

Now, everything is in hand to apply Theorem 2.10.3. It implies that the original sequence $(\varrho_\varepsilon, \mathbf{u}_\varepsilon, \mathcal{P}_H(\varrho_\varepsilon \mathbf{u}_\varepsilon))$, defined on the initial probability space $(\Omega, \mathfrak{F}, \mathbb{P})$, converges in probability in the topology of $\mathcal{X}_\varrho \times \mathcal{X}_{\mathbf{u}} \times \mathcal{X}_{\varrho\mathbf{u}}$ to a random variable $(\varrho, \mathbf{u}, \mathbf{q})$. Therefore, by the method from Section 8.1.3.4, we finally deduce that $\mathbf{u}$ is a pathwise weak solution to (8.14)–(8.15). Actually, identification of the limit is more straightforward here since in this case all the work is done for the initial setting and only one fixed driving Wiener process W is considered. The proof of Theorem 8.1.7 is complete. □

8.2 Inviscid–incompressible limit

In the inviscid limit, the Reynolds number (essentially the reciprocal of the viscosity) tends to infinity. Accordingly, the viscous friction in the flow can be neglected. Although inviscid fluids such as superfluids are rare to find in reality, inviscid fluid models have many applications in physics and engineering. More precisely, there are many limit regimes in which the fluid motion is “inviscid” although the concrete fluid is "viscous". From the physical point of view, if the Mach number tends to zero and the Reynolds number becomes infinite, the fluid flow becomes incompressible and inviscid. The limit behavior is therefore described by the *incompressible* Euler system. We rigorously describe the asymptotic behavior of the system (8.1)–(8.4) in the low Mach number-low viscosity regime, meaning [Sr] = [Fr] = 1, [Ma] = ε, [Re] $\approx \frac{1}{\varepsilon}$. That is, for dimension $N = 2, 3$, we study the system

$$d\varrho + \operatorname{div}(\varrho\mathbf{u})\,dt = 0 \tag{8.63}$$

$$d(\varrho\mathbf{u}) + \operatorname{div}(\varrho\mathbf{u}\otimes\mathbf{u})\,dt + \frac{1}{\varepsilon^2}\nabla p(\varrho)\,dt = \operatorname{div}\mathbb{S}_\varepsilon(\nabla\mathbf{u})\,dt + \mathbb{G}(\varrho,\varrho\mathbf{u})\,dW, \tag{8.64}$$

$$\mathbb{S}_\varepsilon(\nabla\mathbf{u}) = \mu_\varepsilon\left(\nabla\mathbf{u} + \nabla^t\mathbf{u} - \frac{2}{3}\operatorname{div}\mathbf{u}\mathbb{I}\right) + \lambda_\varepsilon\operatorname{div}\mathbf{u}\mathbb{I}, \tag{8.65}$$

where

$$\mu_\varepsilon,\ \lambda_\varepsilon \to 0 \quad \text{as}\ \varepsilon \to 0.$$

Similarly to the previous part, the underlying spatial domain is the flat torus $\mathbb{T}^N$. In particular, we avoid the highly non-trivial and still not fully understood problem of boundary layer occurring in the case of physically relevant no-slip conditions; see, e.g., Weinan [Wei00].

The scaling in (8.63)–(8.65) reflects the situation when the Mach number is low and the Reynolds number is high, meaning the fluid is in a highly turbulent almost incompressible regime; see, e.g., Klein et al. [KBS$^+$01]. Under these circumstances, the motion is expected to be governed by the incompressible Euler system

$$\operatorname{div}\mathbf{v} = 0 \tag{8.66}$$

$$d\mathbf{v} + \mathbf{v}\cdot\nabla\mathbf{v}\,dt + \nabla\Pi\,dt = \mathbb{G}(1,\mathbf{v})\,dW, \tag{8.67}$$

where Π is the associated pressure. To compare the primitive and limit systems, we require the following:

- The Navier–Stokes system (8.63)–(8.65) possesses a dissipative martingale solution $((\Omega, \mathfrak{F}, (\mathfrak{F}_t)_{t\geq 0}, \mathbb{P}), \varrho, \mathbf{u}, W)$ and the Euler system (8.68)–(8.69) a strong solution $(\mathbf{v}, \Pi)$, defined on the same probability space $(\Omega, \mathfrak{F}, (\mathfrak{F}_t)_{t\geq 0}, \mathbb{P})$ and with the same Wiener process W.
- Both $\mathbf{v}$ and the pressure $\nabla\Pi$ are smooth enough in the x-variable, so that $r = 1$, $\mathbf{U} = \mathbf{v}$ can be taken as test functions in the relative energy inequality (6.6).

8.2.1 Solutions of the Euler system

In this subsection we are concerned with the incompressible stochastic Euler system

$$\operatorname{div}\mathbf{v} = 0 \tag{8.68}$$

$$\mathrm{d}\mathbf{v} + \mathbf{v} \cdot \nabla \mathbf{v} \, \mathrm{d}t + \nabla \Pi \, \mathrm{d}t = \Phi(\mathbf{v}) \, \mathrm{d}W, \tag{8.69}$$

where $\Phi : W^{s,2}(\mathbb{T}^N) \to L_2(\mathfrak{U}; W^{s,2}(\mathbb{T}^N))$ such that

$$\begin{aligned} \|\Phi(\mathbf{v})\|^2_{L_2(\mathfrak{U}; W^{s,2}_x)} &\leq C(1 + \|\mathbf{v}\|^2_{W^{s,2}_x}), \\ \|\Phi(\mathbf{v}) - \Phi(\mathbf{w})\|^2_{L_2(\mathfrak{U}; W^{s,2}_x)} &\leq C\|\mathbf{v} - \mathbf{w}\|^2_{W^{s,2}_x}, \end{aligned} \tag{8.70}$$

for all $\mathbf{v}, \mathbf{w} \in W^{s,2}(\mathbb{T}^N)$ with $s > \frac{N}{2} + 1$. Different from the incompressible case, the only available framework for three-dimensional flows is the concept of local strong solutions. This is analogous to the deterministic theory. A first stochastic result was given by Mikulevicius–Valiukevicius [MV00], who studied (8.68)–(8.69) with additive noise via a semi-deterministic approach using the flow transformation in the spirit of Bensoussan–Temam [BT73]. A first fully stochastic result was shown by Kim [Kim09], where the existence of local strong solutions to (8.68)–(8.69) was shown on the whole space. Finally, the case of bounded domains has been studied by Glatt-Holtz and Vicol [GHV14]. Inspired by [GHV14], we introduce the notion of local strong solutions of the Euler system (8.68)–(8.69) similarly to Definition 5.0.1.

Definition 8.2.1 (Local strong solution). Let $(\Omega, \mathfrak{F}, (\mathfrak{F}_t)_{t\geq 0}, \mathbb{P})$ be a stochastic basis with a complete right-continuous filtration and let W be a cylindrical $(\mathfrak{F}_t)$-Wiener process. Let $\mathbf{v}_0$ be an $\mathfrak{F}_0$-measurable random variable in the space $W^{s,2}_{\mathrm{div}}(\mathbb{T}^N)$. A tuple $(\mathbf{v}, \mathfrak{t})$ is called a *local strong pathwise solution* to system (8.68)–(8.69), provided:

(1) $\mathfrak{t}$ is an a.s. strictly positive $(\mathfrak{F}_t)$-stopping time;

(2) the velocity $\mathbf{v}$ is a $W^{s,2}(\mathbb{T}^N)$-valued $(\mathfrak{F}_t)$-progressively measurable stochastic process such that

$$\mathbf{v}(\cdot \wedge \mathfrak{t}) \in C([0,T]; W^{s,2}_{\mathrm{div}}(\mathbb{T}^N)) \quad \mathbb{P}\text{-a.s.};$$

(3) the equations

$$\begin{aligned} &\mathrm{div}\, \mathbf{v} = 0, \\ &\mathbf{v}(t \wedge \mathfrak{t}) = \mathbf{v}(0) - \int_0^{t\wedge\mathfrak{t}} \mathcal{P}_H[\mathbf{v} \cdot \nabla \mathbf{v}] \, \mathrm{d}t + \int_0^{t\wedge\mathfrak{t}} \mathcal{P}_H[\Phi(\mathbf{v})] \, \mathrm{d}W \end{aligned} \tag{8.71}$$

hold $\mathbb{P}$-a.s. for all $t \in [0,T]$.

Recall that $\mathcal{P}_H$ denotes the standard Helmholtz projection onto the space of solenoidal vector fields; cf. Theorem A.1.4.

The existence of local strong pathwise solutions to the stochastic Euler system was established by Glatt-Holtz and Vicol [GHV14, Theorem 4.3]. It reads, with minor modifications, as follows.

Theorem 8.2.2. *Let $N = 2,3$ and let $s \in \mathbb{N}$ satisfy $s > \frac{N}{2} + 1$. Let the coefficients Φ satisfy hypothesis* (8.70) *and let $\mathbf{v}_0$ be an $\mathfrak{F}_0$-measurable $W^{s,2}_{\mathrm{div}}(\mathbb{T}^N)$-valued random variable. Then there exists a unique local strong pathwise solution $(\mathbf{v}, \mathfrak{t})$ to problem* (8.68)–(8.69) *in the sense of Definition* 8.2.1 *with the initial condition $\mathbf{v}_0$.*

Remark 8.2.3. The solution constructed by Glatt-Holtz and Vicol [GHV14, Theorem 4.3] is in fact a maximal strong pathwise solution; see also Definition 5.0.2. However, the concept of Definition 8.2.1 suffices for our purposes.

In the following we assume a very simple form of Φ, which is

$$\Phi(\mathbf{v}) = \mathbb{F} + \mathbf{v}\mathbb{H}, \quad \text{where } \mathbb{F} = (F_k)_{k\in\mathbb{N}},\ \mathbb{H} = (H_k)_{k\in\mathbb{N}}. \tag{8.72}$$

Here F_k, H_k are real numbers such that $\sum_{k=1}^{\infty} |F_k| < \infty$ and $\sum_{k=1}^{\infty} |H_k| < \infty$. The advantage of such a choice is that the pressure Π can be computed explicitly from (8.71) and does not contain a stochastic component (in the general case an additional stochastic integral is part of the pressure; see [Bre15, Section 2]). Indeed, seeing that

$$\mathcal{P}_H[\Phi(\mathbf{v})] = \Phi(\mathbf{v}),$$

we obtain

$$\nabla\Pi = -\mathcal{Q}[\mathbf{v}\cdot\nabla\mathbf{v}] = -\nabla\Delta^{-1}\mathrm{div}\,(\mathbf{v}\otimes\mathbf{v}), \tag{8.73}$$

where $\mathcal{Q} = \mathrm{Id} - \mathcal{P}_H$. Accordingly, the second equation in (8.71) reads

$$\mathbf{v}(t\wedge\mathfrak{t}) = \mathbf{v}(0) - \int_0^{t\wedge\mathfrak{t}} [\mathbf{v}\cdot\nabla\mathbf{v}]\,\mathrm{d}t - \int_0^{t\wedge\mathfrak{t}} \nabla\Pi\,\mathrm{d}t + \int_0^{t\wedge\mathfrak{t}} \Phi(\mathbf{v})\,\mathrm{d}W. \tag{8.74}$$

8.2.2 Main result

In accordance with (8.72), we assume a very simple form of the diffusion coefficient $\mathbb{G}$ in (8.63)–(8.65), namely that it is an affine function of density and momentum

$$\mathbb{G}(\varrho,\mathbf{q}) = \varrho\mathbb{F} + \mathbf{q}\mathbb{H}, \quad \text{where } \mathbb{F} = (F_k)_{k\in\mathbb{N}},\ \mathbb{H} = (H_k)_{k\in\mathbb{N}}. \tag{8.75}$$

Here F_k, H_k are real numbers such that $\sum_{k=1}^{\infty} |F_k| < \infty$ and $\sum_{k=1}^{\infty} |H_k| < \infty$. Now we have a special case of the assumptions (3.13) and (3.14) supposed in the existence theory from Chapter 4. The advantage of (8.75) can be found in (8.73) above. Our main inviscid–incompressible limit result is the following.

Theorem 8.2.4. *Let $N = 2,3$ and let $\mathbb{G}$ satisfy* (8.75). *Let $(\Omega, \mathfrak{F}, \mathbb{P})$ be a complete probability space and W a cylindrical Wiener process on $(\Omega, \mathfrak{F}, \mathbb{P})$. Assume that*

$$((\Omega, \mathfrak{F}, (\mathfrak{F}_t^{\varepsilon})_{t\geq 0}, \mathbb{P}), \varrho_\varepsilon, \mathbf{u}_\varepsilon, W)$$

is a dissipative martingale solution to (8.63)–(8.65) *in the sense of Definition* 8.0.1, *and* $(\mathbf{v}, \mathfrak{t})$ *defined on the same probability space* $(\Omega, \mathfrak{F}, \mathbb{P})$ *is a unique local strong solution of the Euler system* (8.68)–(8.69) *driven by the same cylindrical Wiener process W in the sense of Definition* 8.2.1. *Assume that the initial data* $[\varrho_{0,\varepsilon}, (\varrho\mathbf{u})_{0,\varepsilon}]$ *and* $\mathbf{v}_0$ *satisfy*

$$[\varrho_{0,\varepsilon}, (\varrho\mathbf{u})_{0,\varepsilon}] \in L^\gamma(\mathbb{T}^N) \times L^{\frac{2\gamma}{\gamma+1}}(\mathbb{T}^N), \quad \mathbb{P}\text{-}a.s.,$$
$$\mathbf{v}_0 \in W^{s,2}(\mathbb{T}^N), \quad \operatorname{div} \mathbf{v}_0 = 0, \quad \mathbb{P}\text{-}a.s.,$$
$$\mathbb{E}[\|\mathbf{v}_0\|^q_{W^{s,2}_x}] < \infty \quad \textit{for all } 1 \le q < \infty$$

and

$$\varrho_{0,\varepsilon} \ge \underline{\varrho} > 0, \quad \frac{|\varrho_{0,\varepsilon} - 1|}{\varepsilon} \le \delta(\varepsilon), \quad |(\varrho\mathbf{u})_{0,\varepsilon} - \mathbf{v}_0| \le \delta(\varepsilon) \quad \mathbb{P}\text{-}a.s.,$$

where

$$\delta(\varepsilon) \to 0 \quad \textit{as } \varepsilon \to 0.$$

Then

$$\sup_{t\in[0,T]} \mathbb{E}\left[\int_{\mathbb{T}^N} \left[\frac{1}{2}\varrho_\varepsilon|\mathbf{u}_\varepsilon - \mathbf{v}|^2 + \frac{1}{\varepsilon^2}(P(\varrho_\varepsilon) - P'(1)(\varrho_\varepsilon - 1) - P(1))\right] \mathrm{d}x(t \wedge \mathfrak{t})\right] \to 0 \tag{8.76}$$

as $\varepsilon \to 0$.

Remark 8.2.5. It follows from (8.76) that

$$\mathbb{E}\left[\int_0^{T\wedge\mathfrak{t}} \|\mathbf{u}_\varepsilon - \mathbf{v}\|^2_{L^2_x} \, \mathrm{d}t\right] \to 0 \quad \text{as } \varepsilon \to 0.$$

Moreover, we have

$$\varrho_\varepsilon(\cdot \wedge \mathfrak{t}) \to 1 \quad \text{strongly in } L^{\gamma_*}(\Omega \times (0,T) \times \mathbb{T}^N),$$

where $\gamma_* = \min\{2, \gamma\}$.

Remark 8.2.6. The situation considered in Theorem 8.2.4 corresponds to the so-called well-prepared data. With some additional technical effort, our approach can be extended to the case of ill-prepared data; cf. Masmoudi [Mas01] and [FN14] for the related deterministic results.

Remark 8.2.7. Note that the inviscid limit in the purely *incompressible* setting was studied by Glatt-Holtz et al. [GHŠV15] in the two-dimensional setting.

8.2.3 Proof of Theorem 8.2.4

As compactness is lost in the inviscid limit, our main tool is the relative energy inequality from Chapter 6 which we recall in the following. Given a dissipative martingale solution $[\varrho_\varepsilon, \mathbf{u}_\varepsilon]$ of system (8.63)–(8.65) and a pair of arbitrary smooth processes $[r, \mathbf{U}]$, the *relative energy* functional is given by

$$\mathcal{E}(\varrho_\varepsilon, \mathbf{u}_\varepsilon | r, \mathbf{U}) = \int_{\mathbb{T}^N} \left[\frac{1}{2}\varrho_\varepsilon |\mathbf{u}_\varepsilon - \mathbf{U}|^2 + \frac{1}{\varepsilon^2}(P(\varrho_\varepsilon) - P'(r)(\varrho_\varepsilon - r) - P(r)) \right] \mathrm{d}x. \tag{8.77}$$

Here, r and $\mathbf{U}$ are stochastic processes adapted to $(\mathfrak{F}_t)_{t\geq 0}$ such that

$$\mathrm{d}r = \mathrm{d}^\mathrm{d} r\, dt + \mathrm{d}^\mathrm{s} r\, dW, \quad \mathrm{d}\mathbf{U} = \mathrm{d}^\mathrm{d}\mathbf{U}\, dt + \mathrm{d}^\mathrm{s}\mathbf{U}\, dW, \tag{8.78}$$

where the processes $\mathrm{d}^\mathrm{d} r$, $\mathrm{d}^\mathrm{s} r$, $\mathrm{d}^\mathrm{d}\mathbf{U}$, and $\mathrm{d}^\mathrm{s}\mathbf{U}$ are sufficiently regular with respect to the spatial variable. It is shown in Theorem 6.1.1 that under these assumptions the relative energy inequality

$$\begin{aligned} &- \int_0^T \partial_t \phi\, \mathcal{E}(\varrho_\varepsilon, \mathbf{u}_\varepsilon | r, \mathbf{U})\, \mathrm{d}t + \int_0^T \phi \int_{\mathbb{T}^N} (\mathbb{S}_\varepsilon(\nabla \mathbf{u}_\varepsilon) - \mathbb{S}_\varepsilon(\nabla \mathbf{U})) : \nabla(\mathbf{u}_\varepsilon - \mathbf{U})\, \mathrm{d}x\, \mathrm{d}t \\ &\quad \leq \phi(0)\mathcal{E}(\varrho_\varepsilon, \mathbf{u}_\varepsilon | r, \mathbf{U})(0) + \int_0^T \phi \mathcal{R}(\varrho_\varepsilon, \mathbf{u}_\varepsilon | r, \mathbf{U})\, \mathrm{d}t + \int_0^T \phi\, \mathrm{d}M_{\mathrm{RE}} \end{aligned} \tag{8.79}$$

holds for all $\phi \in C_c^\infty([0,T))$, $\phi \geq 0$, $\mathbb{P}$-a.s. Here, M_{RE} is a real-valued square integrable martingale. The remaining term reads

$$\begin{aligned} &\mathcal{R}(\varrho_\varepsilon, \mathbf{u}_\varepsilon | r, \mathbf{U}) \\ &\quad = \int_{\mathbb{T}^N} \mathbb{S}(\nabla \mathbf{U}) : (\nabla \mathbf{U} - \nabla \mathbf{u}_\varepsilon)\, \mathrm{d}x + \int_{\mathbb{T}^N} \varrho(\mathrm{d}^\mathrm{d}\mathbf{U} + \mathbf{u}_\varepsilon \cdot \nabla \mathbf{U})(\mathbf{U} - \mathbf{u}_\varepsilon)\, \mathrm{d}x \\ &\qquad + \int_{\mathbb{T}^N} ((r - \varrho_\varepsilon)P''(r)\mathrm{d}^\mathrm{d} r + \nabla P'(r)(r\mathbf{U} - \varrho_\varepsilon \mathbf{u}_\varepsilon))\, \mathrm{d}x - \int_{\mathbb{T}^N} \operatorname{div}\mathbf{U}(p(\varrho_\varepsilon) - p(r))\, \mathrm{d}x \\ &\qquad + \frac{1}{2}\sum_{k=1}^\infty \int_{\mathbb{T}^N} \varrho \left| \frac{\mathbf{G}_k(\varrho_\varepsilon, \varrho_\varepsilon \mathbf{u}_\varepsilon)}{\varrho_\varepsilon} - \mathrm{d}^\mathrm{s}\mathbf{U}(e_k) \right|^2 \mathrm{d}x - \frac{1}{2}\sum_{k=1}^\infty \int_{\mathbb{T}^N} \varrho_\varepsilon P'''(r) |\mathrm{d}^\mathrm{s} r(e_k)|^2\, \mathrm{d}x \\ &\qquad + \frac{1}{2}\sum_{k=1}^\infty \int_{\mathbb{T}^N} p''(r) |\mathrm{d}^\mathrm{s} r(e_k)|^2\, \mathrm{d}x. \end{aligned} \tag{8.80}$$

Suppose that $\mathbf{v}$, with a stopping time $\mathfrak{t}$, is a local strong solution of the Euler system (8.68)–(8.69). For each $L > 0$, let

$$\tau_L = \inf\{t \in [0,T];\ \|\nabla \mathbf{v}(t)\|_{L_x^\infty} > L\}$$

be another stopping time. In view of the existence result Theorem 8.2.2, we assume, without loss of generality, that $\tau_L \leq \mathfrak{t}$. Choosing the functions $r = 1$, $\mathbf{U}(t) = \mathbf{v}(t \wedge \tau_L)$ in (8.79), together with the choice $\phi = \chi_{[0,t]}$, yields

$$\mathcal{E}(\varrho_\varepsilon, \mathbf{u}_\varepsilon | 1, \mathbf{v})(\tau \wedge \tau_L) + \int_0^{\tau\wedge\tau_L} \int_{\mathbb{T}^N} (\mathbb{S}_\varepsilon(\nabla \mathbf{v}) - \mathbb{S}_\varepsilon(\nabla \mathbf{u}_\varepsilon)) : (\nabla \mathbf{v} - \nabla \mathbf{u}_\varepsilon)\, \mathrm{d}x\, \mathrm{d}t$$

$$
\begin{aligned}
&\leq \mathcal{E}(\varrho_\varepsilon, \mathbf{u}_\varepsilon | 1, \mathbf{v})(0) + M_{\mathrm{RE}}(\tau \wedge \tau_L) - M_{\mathrm{RE}}(0) \\
&\quad - \int_0^{\tau\wedge\tau_L} \int_{\mathbb{T}^N} \varrho_\varepsilon (\mathbf{u}_\varepsilon - \mathbf{v}) \cdot \nabla \mathbf{v} \cdot (\mathbf{u}_\varepsilon - \mathbf{v}) \,\mathrm{d}x \,\mathrm{d}t + \int_0^{\tau\wedge\tau_L} \int_{\mathbb{T}^N} \mathbb{S}_\varepsilon(\nabla \mathbf{v}) : (\nabla \mathbf{v} - \nabla \mathbf{u}_\varepsilon) \,\mathrm{d}x \,\mathrm{d}t \\
&\quad - \int_0^{\tau\wedge\tau_L} \int_{\mathbb{T}^N} \varrho_\varepsilon \nabla \Pi \cdot (\mathbf{v} - \mathbf{u}_\varepsilon) \,\mathrm{d}x \,\mathrm{d}t \\
&\quad + \frac{1}{2} \sum_{k=1}^\infty \int_0^{\tau\wedge\tau_L} \int_{\mathbb{T}^N} \varrho_\varepsilon \left| \frac{1}{\varrho} \mathbf{G}_k(\varrho_\varepsilon, \varrho_\varepsilon \mathbf{u}_\varepsilon) - \mathbf{G}_k(1, \mathbf{v}) \right|^2 \mathrm{d}x \,\mathrm{d}t.
\end{aligned} \tag{8.81}
$$

Note that the terms involving $\mathrm{d}^d r$, $\mathrm{d}^s r$, $\nabla_x P'(r)$ vanish since r is constant. We also used $\operatorname{div} \mathbf{v} = 0$. We show that, similarly to the proof of Theorem 6.2.2, the terms on the right hand side of (8.81) can be absorbed by means of a Gronwall type argument. To see this, we first observe

$$
\begin{aligned}
\left| \int_0^{\tau\wedge\tau_L} \int_{\mathbb{T}^N} \varrho_\varepsilon (\mathbf{u}_\varepsilon - \mathbf{v}) \cdot \nabla \mathbf{v} \cdot (\mathbf{u}_\varepsilon - \mathbf{v}) \,\mathrm{d}x \,\mathrm{d}t \right| &\leq c \sup_{t \in [0,\tau_L]} \|\nabla \mathbf{v}\|_{L^\infty_x} \int_0^{\tau\wedge\tau_L} \mathcal{E}(\varrho_\varepsilon, \mathbf{u}_\varepsilon | 1, \mathbf{v}) \,\mathrm{d}t \\
&\leq cL \int_0^{\tau\wedge\tau_L} \mathcal{E}(\varrho_\varepsilon, \mathbf{u}_\varepsilon | 1, \mathbf{v}) \,\mathrm{d}t.
\end{aligned} \tag{8.82}
$$

Similarly,

$$
\begin{aligned}
&\left| \int_0^{\tau\wedge\tau_L} \int_{\mathbb{T}^N} \mathbb{S}_\varepsilon(\nabla \mathbf{v}) : (\nabla \mathbf{v} - \nabla \mathbf{u}_\varepsilon) \,\mathrm{d}x \,\mathrm{d}t \right| \\
&\quad \leq \frac{1}{2} \int_0^{\tau\wedge\tau_L} \int_{\mathbb{T}^N} (\mathbb{S}_\varepsilon(\nabla \mathbf{v}) - \mathbb{S}_\varepsilon(\nabla \mathbf{u}_\varepsilon)) : (\nabla \mathbf{v} - \nabla \mathbf{u}_\varepsilon) \,\mathrm{d}x \,\mathrm{d}t + c \int_0^{\tau\wedge\tau_L} \int_{\mathbb{T}^N} |\mathbb{S}(\nabla \mathbf{v})|^2 \,\mathrm{d}x \,\mathrm{d}t \\
&\quad \leq \frac{1}{2} \int_0^{\tau\wedge\tau_L} \int_{\mathbb{T}^N} (\mathbb{S}_\varepsilon(\nabla \mathbf{v}) - \mathbb{S}_\varepsilon(\nabla \mathbf{u}_\varepsilon)) : (\nabla \mathbf{v} - \nabla \mathbf{u}_\varepsilon) \,\mathrm{d}x \,\mathrm{d}t + (\mu_\varepsilon + \lambda_\varepsilon) cTL^2,
\end{aligned} \tag{8.83}
$$

whence (8.81) reduces to

$$
\begin{aligned}
&\mathcal{E}(\varrho_\varepsilon, \mathbf{u}_\varepsilon | 1, \mathbf{v})(\tau \wedge \tau_L) + \frac{1}{2} \int_0^{\tau\wedge\tau_L} \int_{\mathbb{T}^N} (\mathbb{S}_\varepsilon(\nabla \mathbf{v}) - \mathbb{S}_\varepsilon(\nabla \mathbf{u}_\varepsilon)) : (\nabla \mathbf{v} - \nabla \mathbf{u}_\varepsilon) \,\mathrm{d}x \,\mathrm{d}t \\
&\quad \leq \mathcal{E}(\varrho_\varepsilon, \mathbf{u}_\varepsilon | 1, \mathbf{v})(0) + M_{\mathrm{RE}}(\tau \wedge \tau_L) - M_{\mathrm{RE}}(0) + cL \int_0^{\tau\wedge\tau_L} \mathcal{E}(\varrho_\varepsilon, \mathbf{u}_\varepsilon | 1, \mathbf{v}) \,\mathrm{d}t \\
&\qquad + (\mu_\varepsilon + \lambda_\varepsilon) cTL^2 - \int_0^{\tau\wedge\tau_L} \int_{\mathbb{T}^N} \varrho_\varepsilon \nabla \Pi \cdot (\mathbf{v} - \mathbf{u}_\varepsilon) \,\mathrm{d}x \,\mathrm{d}t \\
&\qquad + \frac{1}{2} \sum_{k=1}^\infty \int_0^{\tau\wedge\tau_L} \int_{\mathbb{T}^N} \varrho_\varepsilon \left| \frac{1}{\varrho_\varepsilon} \mathbf{G}_k(\varrho_\varepsilon, \varrho_\varepsilon \mathbf{u}_\varepsilon) - \mathbf{G}_k(1, \mathbf{v}) \right|^2 \mathrm{d}x \,\mathrm{d}t.
\end{aligned} \tag{8.84}
$$

Next, using $\operatorname{div} \mathbf{v} = 0$, the integral containing the pressure can be written

$$
\begin{aligned}
&\int_0^{\tau\wedge\tau_L} \int_{\mathbb{T}^N} \varrho_\varepsilon \nabla \Pi \cdot (\mathbf{v} - \mathbf{u}_\varepsilon) \,\mathrm{d}x \,\mathrm{d}t \\
&\quad = \int_0^{\tau\wedge\tau_L} \int_{\mathbb{T}^N} \varrho_\varepsilon \nabla \Pi \cdot \mathbf{v} \,\mathrm{d}x \,\mathrm{d}t - \int_0^{\tau\wedge\tau_L} \int_{\mathbb{T}^N} \varrho_\varepsilon \nabla \Pi \cdot \mathbf{u}_\varepsilon \,\mathrm{d}x \,\mathrm{d}t \\
&\quad = \varepsilon \int_0^{\tau\wedge\tau_L} \int_{\mathbb{T}^N} \frac{\varrho_\varepsilon - 1}{\varepsilon} \nabla \Pi \cdot \mathbf{v} \,\mathrm{d}x \,\mathrm{d}t - \int_0^{\tau\wedge\tau_L} \int_{\mathbb{T}^N} \varrho_\varepsilon \nabla \Pi \cdot \mathbf{u}_\varepsilon \,\mathrm{d}x \,\mathrm{d}t.
\end{aligned}
$$

Finally, we handle the integral

$$\sum_{k=1}^{\infty}\int_{\mathbb{T}^N}\varrho_\varepsilon\left|\frac{1}{\varrho_\varepsilon}\mathbf{G}_k(\varrho_\varepsilon,\varrho_\varepsilon\mathbf{u}_\varepsilon)-\mathbf{G}_k(1,\mathbf{v})\right|^2\mathrm{d}x.$$

Applying our assumption (8.75), we obtain

$$\begin{aligned}\sum_{k=1}^{\infty}\int_{\mathbb{T}^N}\varrho_\varepsilon\left|\frac{1}{\varrho_\varepsilon}\mathbf{G}_k(\varrho_\varepsilon,\varrho_\varepsilon\mathbf{u}_\varepsilon)-\mathbf{G}_k(1,\mathbf{v})\right|^2\mathrm{d}x\\ =\sum_{k=1}^{\infty}\int_{\mathbb{T}^N}\varrho_\varepsilon|(\mathbf{u}_\varepsilon-\mathbf{v})H_k|^2\,\mathrm{d}x\le c\,\mathcal{E}(\varrho_\varepsilon,\mathbf{u}_\varepsilon|1,\mathbf{v}),\end{aligned}$$

using $\sum_{k=1}^{\infty}|H_k|^2<\infty$ (which is a consequence of $\sum_{k=1}^{\infty}|H_k|<\infty$). After integrating in time and applying Gronwall's lemma, equation (8.84) gives rise to

$$\begin{aligned}\mathbb{E}[\mathcal{E}(\varrho_\varepsilon,\mathbf{u}_\varepsilon|1,\mathbf{v})(\tau\wedge\tau_L)]\le c(L,T)\big(\mathbb{E}[\mathcal{E}(\varrho_\varepsilon,\mathbf{u}_\varepsilon|1,\mathbf{v})(0)]+\mu_\varepsilon+\lambda_\varepsilon\big)\\ +\varepsilon\mathbb{E}\bigg[\int_0^{\tau\wedge\tau_L}\int_{\mathbb{T}^N}\frac{\varrho_\varepsilon-1}{\varepsilon}\nabla\Pi\cdot\mathbf{v}\,\mathrm{d}x\,\mathrm{d}t\bigg]-\mathbb{E}\bigg[\int_0^{\tau\wedge\tau_L}\int_{\mathbb{T}^N}\varrho_\varepsilon\nabla\Pi\cdot\mathbf{u}_\varepsilon\,\mathrm{d}x\,\mathrm{d}t\bigg].\end{aligned}\tag{8.85}$$

In order to control the last two terms in (8.85), we use again (8.79), this time for $r=1$, $\mathbf{U}=0$, obtaining

$$\begin{aligned}\mathbb{E}\bigg[\int_{\mathbb{T}^N}\Big[\frac{1}{2}\varrho_\varepsilon|\mathbf{u}_\varepsilon|^2+\frac{1}{\varepsilon^2}(P(\varrho_\varepsilon)-P'(1)(\varrho_\varepsilon-1)-P(1))\Big]\,\mathrm{d}x(\tau\wedge\tau_L)\bigg]\\ \le\mathbb{E}\bigg[\int_{\mathbb{T}^N}\Big[\frac{1}{2}\varrho_\varepsilon|\mathbf{u}_\varepsilon|^2+\frac{1}{\varepsilon^2}(P(\varrho_\varepsilon)-P'(1)(\varrho_\varepsilon-1)-P(1))\Big]\,\mathrm{d}x(0)\bigg].\end{aligned}$$

Thus, since the right hand side of the above inequality is bounded uniformly for $\varepsilon\to0$, we deduce the following uniform bounds (recall (6.19) and set $\gamma_*=\min\{\gamma,2\}$):

$$\mathbb{E}\bigg[\int_{\mathbb{T}^N}\frac{1}{2}\varrho_\varepsilon|\mathbf{u}_\varepsilon|^2\,\mathrm{d}x(\tau\wedge\tau_L)\bigg]\le c,\quad\mathbb{E}\bigg[\int_{\mathbb{T}^N}\frac{|\varrho_\varepsilon-1|^{\gamma_*}}{\varepsilon^2}\,\mathrm{d}x(\tau\wedge\tau_L)\bigg]\le c.\tag{8.86}$$

Using also (8.73), the continuity of $\nabla\Delta^{-1}\mathrm{div}$, and regularity of $\mathbf{v}$, we obtain

$$\begin{aligned}&\bigg|\mathbb{E}\bigg[\int_0^{\tau\wedge\tau_L}\int_{\mathbb{T}^N}\frac{\varrho_\varepsilon-1}{\varepsilon}\nabla\Pi\cdot\mathbf{v}\,\mathrm{d}x\,\mathrm{d}t\bigg]\bigg|\\ &\le\bigg\|\frac{\varrho_\varepsilon-1}{\varepsilon}\bigg\|_{L_x^{\gamma_*}}\|\nabla_x\Pi\|_{L_x^{2\gamma_*'}}\|\mathbf{v}\|_{L_x^{2\gamma_*'}}\le\bigg\|\frac{\varrho_\varepsilon-1}{\varepsilon}\bigg\|_{L_x^{\gamma_*}}\|\mathbf{v}\otimes\mathbf{v}\|_{L_x^{2\gamma_*'}}\|\mathbf{v}\|_{L_x^{2\gamma_*'}}\le c\end{aligned}$$

uniformly for $\varepsilon\to0$. Additionally, (8.86) implies

$$\begin{aligned}\varrho_\varepsilon\mathbf{u}_\varepsilon(\cdot\wedge\tau_L)&\to\mathbf{v}(\cdot\wedge\tau_L)\quad\text{weakly in }L^{\frac{2\gamma_*}{\gamma_*+1}}(\Omega\times(0,T)\times\mathbb{T}^N),\\ \varrho_\varepsilon(\cdot\wedge\tau_L)&\to1\quad\text{strongly in }L^{\gamma_*}(\Omega\times(0,T)\times\mathbb{T}^N).\end{aligned}$$

Passing to the limit in the continuity equation (8.63) shows that $\mathrm{div}\,\mathbf{v}=0$. In particular, the last two terms on the right hand side of (8.85) vanish for $\varepsilon\to0$. We have proved Theorem 8.2.4.

A Appendix

For the convenience of the reader, a number of standard results used in the preceding text is summarized in this chapter.

A.1 Elliptic equations and related problems

In this section we review important consequences of the theory of *elliptic* equations. The standard reference material are the monographs by Gilbarg–Trudinger [GT83], Ladyzhenskaya–Ural'ceva [LU68], or the nowadays classical paper by Agmon et al. [ADN59]. As the underlying spatial domain is the flat torus $\mathbb{T}^N$, the problem simplifies considerably since the boundary behavior does not play any role in the analysis.

We start by considering the problem

$$\Delta u = f \quad \text{in } \mathbb{T}^N. \tag{A.1}$$

A weak solution to (A.1) can already be defined if $f = \operatorname{div} \mathbf{f}$. We call a function u a weak solution to (A.1) with $f = \operatorname{div} \mathbf{f}$ if

$$\int_{\mathbb{T}^N} \nabla u \cdot \nabla \psi \, \mathrm{d}x = \int_{\mathbb{T}^N} \mathbf{f} \cdot \nabla \psi \, \mathrm{d}x$$

for all $\psi \in C^\infty(\mathbb{T}^N)$. The following is the fundamental statement on the L^q-regularity that may be seen as a direct consequence of the discrete variant of the Hörmander–Mikhlin theorem (Theorem 1.7.1).

Theorem A.1.1. *Let* $1 < q < \infty$ *and suppose that* $\mathbf{f} \in L^q(\mathbb{T}^N)$. *Then problem* (A.1) *with* $f = \operatorname{div} \mathbf{f}$ *admits a weak solution u, denoted by* $\Delta^{-1} \operatorname{div} \mathbf{f}$, *unique in the class*

$$u \in W^{1,q}(\mathbb{T}^N), \quad \int_{\mathbb{T}^N} u \, \mathrm{d}x = 0.$$

Moreover, there exists a positive constant $c = c(q)$ *such that*

$$\|u\|_{W^{1,q}_x} \le c \|\mathbf{f}\|_{L^q_x}.$$

In the context of the so-called *strong solutions*, equation (A.1) is satisfied a.e. in $\mathbb{T}^N$. Iterating the result stated in Theorem A.1.1 we easily obtain the following.

Theorem A.1.2. *Let* $1 < q < \infty$ *and suppose that* $f \in L^q(\mathbb{T}^N)$ *with* $\int_{\mathbb{T}^N} f \, \mathrm{d}x = 0$. *Then problem* (A.1) *admits a solution u, denoted by* $\Delta^{-1} f$, *unique in the class*

$$u \in W^{1,q}(\mathbb{T}^N), \quad \int_{\mathbb{T}^N} u \, \mathrm{d}x = 0.$$

Moreover, there exists a positive constant $c = c(q)$ *such that*

$$\|u\|_{W^{2,q}_x} \le c \|f\|_{L^q_x}.$$

https://doi.org/10.1515/9783110492552-009

The above results can be generalized to higher order derivatives by iteration. We may go even further and consider the elliptic problem (A.1) in the following class of periodic distributions:

$$\langle u, \Delta\psi\rangle = \langle f, \psi\rangle, \quad \langle f, 1\rangle = 0$$

for all $\psi \in C^\infty(\mathbb{T}^N)$. The driving force f may be taken in the scale of Hilbert spaces $W^{\ell,2}(\mathbb{T}^N)$, $\ell \in \mathbb{R}$, introduced in Section 1.7, or even $W^{\ell,q}(\mathbb{T}^N)$, $1 < q < \infty$. The following statement is again a straightforward consequence of Theorem 1.7.1 and contains Theorem A.1.1 and Theorem A.1.2 as special cases.

Theorem A.1.3. *Let $1 < q < \infty$ and $\ell \in \mathbb{R}$ and suppose that $f \in W^{\ell,q}(\mathbb{T}^N)$ with $\langle f, 1\rangle = 0$. Then problem* (A.1) *admits a solution u, denoted by $\Delta^{-1} f$, unique in the class*

$$u \in W^{\ell+2,q}(\mathbb{T}^N), \quad \langle u, 1\rangle = 0.$$

Moreover, there exists a positive constant $c = c(q)$ such that

$$\|u\|_{W_x^{\ell+2,q}} \le c\|f\|_{W_x^{\ell,q}}.$$

A first important consequence of the elliptic theory is the existence of the *Helmholtz decomposition*. It allows to decompose any vector-valued function into a divergence-free part and a gradient part. Set

$$L^p_{\mathrm{div}}(\mathbb{T}^N) := \{\mathbf{v} \in L^p(\mathbb{T}^N, \mathbb{R}^N) \mid \operatorname{div} \mathbf{v} = 0\},$$
$$L^p_{\mathrm{g}}(\mathbb{T}^N) := \{\mathbf{v} \in L^p(\mathbb{T}^N) \mid \mathbf{v} = \nabla\Psi,\ \Psi \in W^{1,p}(\mathbb{T}^N)\}.$$

The *Helmholtz decomposition* is defined by

$$\mathbf{v} = \mathcal{P}_H \mathbf{v} + \mathcal{Q}\mathbf{v} \quad \text{for any } \mathbf{v} \in L^p(\mathbb{T}^N; \mathbb{R}^N),$$
$$\mathcal{P}_H \mathbf{v} = \mathbf{v} - \nabla\Delta^{-1} \operatorname{div} \mathbf{v}, \quad \mathcal{Q}\mathbf{v} = \nabla\Delta^{-1} \operatorname{div} \mathbf{v}.$$

Basic properties of Helmholtz decomposition follow directly from Theorem A.1.3.

Theorem A.1.4. *The Lebesgue space $L^p(\mathbb{T}^N; \mathbb{R}^N)$ admits a decomposition*

$$L^p(\mathbb{T}^N) = L^p_{\mathrm{div}}(\mathbb{T}^N) \oplus L^p_{\mathrm{g}}(\mathbb{T}^N), \quad 1 < p < \infty.$$

More precisely,

$$\mathbf{v} = \mathcal{P}_H \mathbf{v} + \mathcal{Q}\mathbf{v} \quad \textit{for any } \mathbf{v} \in L^q(\mathbb{T}^N),$$

with

$$\mathcal{Q}\mathbf{v} = \nabla\Psi, \quad \Psi \in W^{1,p}(\mathbb{T}^N), \quad \Psi = \Delta^{-1} \operatorname{div} \mathbf{v}.$$

In the particular case $p = 2$, the decomposition is orthogonal with respect to the $L^2(\mathbb{T}^N; \mathbb{R}^N)$-inner product.

Remark A.1.5. In accordance with the regularity properties of the elliptic operators reviewed in Theorem A.1.3, both $\mathcal{P}_H$ and $\mathcal{Q}$ are continuous linear operators on $W^{l,q}(\mathbb{T}^N)$ for any $1 < q < \infty$ and any $l \in \mathbb{Z}$. In particular, they are continuous on $L^q(\mathbb{T}^N)$ for any $1 < q < \infty$.

Remark A.1.6. The fact that the functions are space periodic makes all statements about Helmholtz decomposition and related problems quite easy in view of the Hörmander–Mikhlin result stated in Theorem 1.7.1. In particular, the linear operators $\nabla\Delta\mathrm{div}$ are represented by discrete Fourier symbols $m_i m_j \frac{1}{|\mathbf{m}|^2}$, $i = 1, \ldots, N$, that are L^p-multipliers.

It is standard in fluid mechanics to define Helmholtz decomposition on general even unbounded spatial domains; see e.g. Farwig et al. [FKS05].

Another consequence of Theorem 1.7.1 is a type of Poincaré inequality that concerns norms in the negative Sobolev spaces in the spirit of Nečas [Nec67].

Lemma A.1.7. *Let $1 < q < \infty$ and $\ell \in \mathbb{R}$. Then there is $c = c(q)$ such that*

$$\left\| f - \frac{1}{|\mathbb{T}^N|}\langle f, 1\rangle \right\|_{W_x^{\ell,q}} \le c\|\nabla f\|_{W_x^{\ell-1,q}},$$

for all $f \in W^{\ell,q}(\mathbb{T}^N)$.

Proof. We write f in terms of the corresponding Fourier series

$$f = \sum_{\mathbf{m}\in\mathbb{Z}^N} a_{\mathbf{m}} e_{\mathbf{m}};$$

cf. Section 1.7. Then the desired result follows from Theorem 1.7.1, as the operator

$$\nabla\Delta^{-1/2} \quad \text{represented by the symbol } \frac{\mathbf{m}}{|\mathbf{m}|}$$

is an L^q-multiplier for any $1 < q < \infty$; cf. Theorem 1.7.1. □

Next we report Korn's inequality; see also [FN09, Theorem 10.22].

Theorem A.1.8. *Let $1 < q < \infty$ and $N \ge 2$. There exists a positive constant $c = c(q, N)$ such that*

$$\|\nabla \mathbf{v}\|_{L_x^q} \le c\left\| \nabla\mathbf{v} + \nabla^T\mathbf{v} - \frac{2}{N}\mathrm{div}\,\mathbf{v}\mathbb{I}\right\|_{L_x^q}$$

for any $\mathbf{v} \in W^{1,q}(\mathbb{T}^N)$, where $\mathbb{I} = (\delta_{i,j})_{i,j=1}^N$ is the identity matrix.

Proof. In view of Theorem 1.7.1, it suffices to observe that the kernel of the linear operator

$$\mathbf{v} \mapsto \nabla \mathbf{v} + \nabla^T \mathbf{v} - \frac{2}{N} \operatorname{div} \mathbf{v} \mathbb{I}$$

consists of constant functions. Let

$$\nabla \mathbf{v} + \nabla^T \mathbf{v} - \frac{2}{N} \operatorname{div} \mathbf{v} \mathbb{I} = 0.$$

Without loss of generality, we assume that **v** is smooth. Integrating by parts, we easily obtain

$$0 = \int_{\mathbb{T}^N} \left| \nabla \mathbf{v} + \nabla^T \mathbf{v} - \frac{2}{N} \operatorname{div} \mathbf{v} \mathbb{I} \right|^2 \, dx = 2 \|\nabla \mathbf{v}\|_{L_x^2}^2 + 2\left(1 - \frac{2}{N}\right) \|\operatorname{div} \mathbf{v}\|_{L_x^2}^2;$$

whence $\nabla \mathbf{v} = 0$. □

Remark A.1.9. Under the assumptions of Theorem A.1.8 it is not possible to obtain the inequality

$$\|\mathbf{v}\|_{W_x^{1,q}} \leq c \left\| \nabla \mathbf{v} + \nabla^T \mathbf{v} - \frac{2}{N} \operatorname{div} \mathbf{v} \mathbb{I} \right\|_{L_x^q} \tag{A.2}$$

for any $\mathbf{v} \in W^{1,q}(\mathbb{T}^N)$, as the right hand side obviously vanishes on the set of constant functions. Inequality (A.2) holds, however, if we suppose in addition symmetry conditions

$$u^i(t, \cdot, -x_i, \cdot) = -u^i(t, \cdot, x_i, \cdot) \quad i = 1, \dots, N,$$
$$u^i(t, \cdot, -x_j, \cdot) = u^i(t, \cdot, x_j, \cdot) \quad i \neq j, \; i, j = 1, \dots, N,$$

on **v**; cf. [Ebi83]. Clearly, the symmetry conditions reduce the kernel of the operator

$$\mathbf{v} \mapsto \nabla \mathbf{v} + \nabla^T \mathbf{v} - \frac{2}{N} \operatorname{div} \mathbf{v} \mathbb{I}$$

to zero.

Remark A.1.10. Following the arguments from the proof of Theorem A.1.8, one can verify the standard Korn inequality

$$\|\nabla \mathbf{v}\|_{L_x^q} \leq c \|\nabla \mathbf{v} + \nabla^T \mathbf{v}\|_{L_x^q}$$

for any $\mathbf{v} \in W^{1,q}(\mathbb{T}^N)$ in the same way.

We conclude this section by reporting a variant of the celebrated Div–Curl lemma. A proof can be found in [FN09, Theorem 10.27].

Lemma A.1.11. *Let*

$$\mathbf{u}_\varepsilon \to \mathbf{u} \quad \text{weakly in } L^p(\mathbb{T}^N), \quad \mathbf{v}_\varepsilon \to \mathbf{v} \quad \text{weakly in } L^q(\mathbb{T}^N),$$

where

$$\frac{1}{p} + \frac{1}{q} = \frac{1}{r} < 1.$$

Then

$$\mathbf{u}_\varepsilon \cdot \nabla\Delta^{-1}\mathrm{div}\,[\mathbf{v}_\varepsilon] - \mathbf{v}_\varepsilon \cdot \nabla\Delta^{-1}\mathrm{div}\,[\mathbf{u}_\varepsilon] \to \mathbf{u} \cdot \nabla\Delta^{-1}\mathrm{div}\,[\mathbf{v}] - \mathbf{v} \cdot \nabla\Delta^{-1}\mathrm{div}\,[\mathbf{u}]$$

weakly in $L^r(\mathbb{T}^N)$.

A.2 Regularity for parabolic equations

We consider the parabolic problem

$$\left\{\begin{array}{ll} \partial_t u - \varepsilon\Delta u = f & \text{in } (0,T)\times\mathbb{T}^N, \\ u(0,x) = u_0(x), & x \in \mathbb{T}^N, \end{array}\right\} \tag{A.3}$$

where $\varepsilon > 0$ is a positive constant. We call a function u a weak solution to (A.3) if

$$-\int_0^T \partial_t\phi \int_{\mathbb{T}^N} u\psi\,\mathrm{d}x\,\mathrm{d}t = \phi(0)\int_{\mathbb{T}^N} u_0\,\psi\,\mathrm{d}x - \varepsilon\int_0^T \phi\int_{\mathbb{T}^N} \nabla u\cdot\nabla\psi\,\mathrm{d}x\,\mathrm{d}t + \int_0^T \phi\int_{\mathbb{T}^N} f\,\psi\,\mathrm{d}x\,\mathrm{d}t$$

for all $\phi \in C_c^\infty([0,T))$ and all $\psi \in C^\infty(\mathbb{T}^N)$. The following statement is the maximal regularity property (see Amann [Ama93, Ama95] and Ladyshenskaya et al. [LSU68]).

Theorem A.2.1. *Let* $1 < p, q < \infty$ *and suppose that*

$$f \in L^p(0,T;L^q(\mathbb{T}^N)), \quad u_0 \in X_{p,q} = \{L^q(\mathbb{T}^N); W^{2,q}(\mathbb{T}^N)\}_{1-1/p,p},$$

where $\{\cdot;\cdot\}$ *denotes the real interpolation space.*

Then problem (A.3) *admits a weak solution u, unique in the class*

$$u \in L^p(0,T;W^{2,q}(\mathbb{T}^N)), \quad \partial_t u \in L^p(0,T;L^q(\mathbb{T}^N)).$$

Moreover, there exists a positive constant $c = c(p,q,\varepsilon,T)$ *such that*

$$\sup_{t\in[0,T]} \|u(t)\|_{X_{p,q}} + \|u\|_{L^p_t W^{2,q}_x} + \|\partial_t u\|_{L^p_t L^q_x} \le c(\|\mathbf{f}\|_{L^p_t L^q_x} + \|u_0\|_{X_{p,q}}).$$

Note that the solution obtained in Theorem A.2.1 is in fact a *strong solution*, the first equation in (A.3) is satisfied a.e. in $(0,T)\times\mathbb{T}^N$ and the initial condition holds a.e. in $\mathbb{T}^N$.

We refer to Bergh–Löfström [BL76] for the definition of the interpolation spaces. In particular, we have $X_{q,q}(\mathbb{T}^N)=W^{2-\frac{2}{q},q}(\mathbb{T}^N)$.

The periodic structure makes it particularly easy to extend the above result to a more general class of data. Note that the periodic Laplacian commutes with all fractional derivatives with respect to the space variables. The following statement holds (cf. also Hieber–Prüss [HP97]).

Theorem A.2.2. *Let $1<p,q<\infty$, $\ell\in\mathbb{R}$ and suppose that*

$$f\in L^p(0,T;W^{\ell,q}(\mathbb{T}^N)),\quad u_0\in X^{\ell}_{p,q}=\{W^{\ell,q}(\mathbb{T}^N);W^{2+\ell,q}(\mathbb{T}^N)\}_{1-1/p,p}.$$

Then problem (A.3) *admits a weak solution u, unique in the class*

$$u\in L^p(0,T;W^{2+\ell,q}(\mathbb{T}^N)),\quad \partial_t u\in L^p(0,T;W^{\ell,q}(\mathbb{T}^N)).$$

Moreover, there exists a positive constant $c=c(p,q,\ell,\varepsilon,T)$ such that

$$\sup_{t\in[0,T]}\|u(t)\|_{X^{\ell}_{p,q}}+\|u\|_{L^p_tW^{2+\ell,q}_x}+\|\partial_t u\|_{L^p_tW^{\ell,q}_x}\le c(\|\mathbf{f}\|_{L^p_tW^{\ell,q}_x}+\|u_0\|_{X^{\ell}_{p,q}}).$$

Maximal regularity in the classes of smooth functions relies on a classical argument. A result in this direction reads as follows (see Lunardi [Lun12, Theorem 5.1.2]).

Theorem A.2.3. *Let $\nu\in(0,1)$ and suppose*

$$f\in C([0,T];C^{0,\nu}(\mathbb{T}^N)),\quad u_0\in C^{2,\nu}(\mathbb{T}^N).$$

Then problem (A.3) *admits a solution u, unique in the class*

$$u\in C([0,T];C^{2,\nu}(\mathbb{T}^N)),\quad \partial_t u\in C([0,T];C^{0,\nu}(\mathbb{T}^N)).$$

Moreover, there exists a positive constant $c=c(p,q,\varepsilon,T)$ such that

$$\|\partial_t u\|_{C_tC^{0,\nu}_x}+\|u\|_{C_tC^{2,\nu}_x}\le c(\|u_0\|_{C^{2,\nu}_x}+\|f\|_{C_tC^{0,\nu}_x}).$$

Similarly to the above, we may use the periodic structure to generalize Theorem A.2.3.

Theorem A.2.4. *Let $\nu\in(0,1)$ $k\ge 0$ be a non-negative integer and suppose that*

$$f\in C([0,T];C^{k,\nu}(\mathbb{T}^N)),\quad u_0\in C^{2+k,\nu}(\mathbb{T}^N).$$

Then problem (A.3) *admits a solution u, unique in the class*

$$u \in C([0,T];C^{2,k+\nu}(\mathbb{T}^N)), \quad \partial_t u \in C([0,T];C^{0,k+\nu}(\mathbb{T}^N)).$$

Moreover, there exists a positive constant $c = c(p,q,\varepsilon,kT)$ *such that*

$$\|\partial_t u\|_{C_t C_x^{k,\nu}} + \|u\|_{C_t C_x^{2+k,\nu}} \leq c(\|u_0\|_{C_x^{2+k,\nu}} + \|f\|_{C_t C_x^{k,\nu}}).$$

In the following, we consider a more particular parabolic equation which arises when regularizing the continuity equation with an artificial viscosity. We consider the parabolic problem

$$\begin{Bmatrix} \partial_t u - \varepsilon \Delta u = \operatorname{div}(u\mathbf{F}) & \text{in } (0,T)\times\mathbb{T}^N, \\ u(0,x) = u_0(x), & x \in \mathbb{T}^N, \end{Bmatrix} \tag{A.4}$$

where $\mathbf{F} : \mathbb{T}^N \to \mathbb{R}^N$ is given. The following Theorem is concerned with the existence and regularity of solutions to (A.4) (see Lunardi [Lun12, Theorem 5.1.21] as well as [FNP01, Lemma 2.2] and Protter–Weinberger [PW67] for the comparison principle).

Theorem A.2.5. *Let* $\nu > 0$ *and suppose*

$$\mathbf{F} \in C([0,T];C^2(\mathbb{T}^N)), \quad u_0 \in C^{2,\nu}(\mathbb{T}^N), \quad \underline{u} \leq u_0 \leq \overline{u},$$

where $\underline{u}, \overline{u} > 0$.

Then problem (A.4) *admits a solution u, unique in* $C([0,T];C^{2,\nu}(\mathbb{T}^N))$. *Moreover, we have*

$$\underline{u}\exp\left(\int_0^t \|\operatorname{div}\mathbf{F}\|_{L_x^\infty}\,\mathrm{ds}\right) \leq u(t,x) \leq \overline{u}\exp\left(\int_0^t \|\operatorname{div}\mathbf{F}\|_{L_x^\infty}\,\mathrm{ds}\right),$$

for all $t \in [0,T]$ *and all* $x \in \mathbb{T}^N$.

Remark A.2.6. It is easy to see that the maximum principle from Theorem A.2.5 continues to hold for solutions to (A.4) with $\varepsilon = 0$ provided $\mathbf{F}$ satisfies appropriate regularity assumptions.

Let us finally present a result concerning the stability of solutions to (A.4); cf. [FNP01, Lemma 2.2].

Lemma A.2.7. *Under the assumptions of Theorem* A.2.5 *let* u_1 *and* u_2 *be two solutions to* (A.4) *with right hand sides* $\mathbf{F}_1$ *and* $\mathbf{F}_2$, *respectively, and the same initial datum* u_0. *Then we have*

$$\|u_1 - u_2\|_{C_t W_x^{1,2}} \leq Tc(T,K)\|\mathbf{F}_1 - \mathbf{F}_2\|_{C_t W_x^{1,2}},$$

provided $\mathbf{F}_1$ *and* $\mathbf{F}_2$ *belong to the set*

$$\{\mathbf{F} \in C([0,T];C^2(\mathbb{T}^N)) : \|\mathbf{F}\|_{L_t^\infty W_x^{1,\infty}} \leq K\}.$$

A.3 Renormalized solutions of the continuity equation

In this section we explain the main ideas of the regularization technique developed by DiPerna–Lions [DL89] and discuss the basic properties of the renormalized solutions to the equation of continuity. Detailed proofs of the statements for the corresponding problem on the whole space $\mathbb{R}^N$ can be found in [FN09, Appendix 10.18]. The straightforward modifications for the periodic setting are left to the reader.

Theorem A.3.1. *Let $N \geq 2$, $\beta \in [1,\infty)$, $q \in [1,\infty]$, $\frac{1}{q} + \frac{1}{\beta} \in (0,1]$. Suppose that the functions $(\varrho, \mathbf{u}) \in L^\beta((0,T) \times \mathbb{T}^N) \times L^q(0,T; W^{1,q}(\mathbb{T}^N))$, where $\varrho \geq 0$ a.e. in $(0,T) \times \mathbb{T}^N$, satisfy the transport equation*

$$\partial_t \varrho + \operatorname{div}(\varrho \mathbf{u}) = 0 \quad \text{in } \mathcal{D}'((0,T) \times \mathbb{T}^N). \tag{A.5}$$

Then

$$\partial_t b(\varrho) + \operatorname{div}((b(\varrho)\mathbf{u}) + (\varrho b'(\varrho) - b(\varrho))\operatorname{div}\mathbf{u} = 0 \quad \text{in } \mathcal{D}'((0,T) \times \mathbb{T}^N) \tag{A.6}$$

for any

$$b \in C^1([0,\infty)) \cap W^{1,\infty}((0,\infty)). \tag{A.7}$$

Once the renormalized continuity equation is established for any b belonging to (A.7), it is satisfied for any "renormalizing" function b belonging to a larger class. This is clarified in the following lemma.

Lemma A.3.2. *Let $N \geq 2$, $\beta \in [1,\infty)$, $q \in [1,\infty]$, $\frac{1}{q} + \frac{1}{\beta} \in (0,1]$. Suppose that the functions $(\varrho, \mathbf{u}) \in L^\beta((0,T) \times \mathbb{T}^N) \times L^q(0,T; W^{1,q}(\mathbb{T}^N))$, where $\varrho \geq 0$ a.e. in $(0,T) \times \mathbb{T}^N$, satisfy the renormalized continuity equation* (A.6) *for any b belonging to the class* (A.7)*. Then we have:*

(i) *Equation* (A.6) *holds for any*

$$\begin{aligned} &b \in C([0,\infty)) \cap C^1((0,\infty)), \\ &\lim_{s \to 0+} (sb'(s) - b(s)) \in \mathbb{R}, \\ &|b'(s)| \leq cs^\lambda \quad \text{if } s \in (1,\infty) \text{ for a certain } \lambda \leq \tfrac{\beta}{2} - 1 \end{aligned} \tag{A.8}$$

(ii) *The function $z \to b(z)$ in statements* (i) *can be replaced by $z \to cz + b(z)$, $c \in \mathbb{R}$, where b satisfies either*

$$b \in C^1([0,\infty)), \quad |b'(s)| \leq cs^\lambda, \quad \text{for } s > 1, \text{ where } \lambda \leq \tfrac{\beta}{2} - 1, \tag{A.9}$$

or (A.8) *as the case may be.*

(iii) *We have*

$$\partial_t(\varrho B(\varrho)) + \operatorname{div}_x(\varrho B(\varrho)\mathbf{u}) + b(\varrho)\operatorname{div}_x\mathbf{u} = 0 \quad \text{in } \mathcal{D}'((0,T)\times\mathbb{T}^N)$$

for any

$$b \in C([0,\infty)) \cap L^\infty(0,\infty), \quad B(\varrho) = B(1) + \int_1^\varrho \frac{b(z)}{z^2}\,\mathrm{d}z.$$

Next, we shall investigate the pointwise behavior of renormalized solutions with respect to time.

Lemma A.3.3. *Let $N \geq 2$, $\beta, q \in (1,\infty)$, $\frac{1}{q} + \frac{1}{\beta} \in (0,1]$. Suppose that the functions $(\varrho, \mathbf{u}) \in L^\infty(0,T;L^\beta(\mathbb{T}^N)) \times L^q(0,T;W^{1,q}(\mathbb{T}^N))$, $\varrho \geq 0$ a.e. in $(0,T)\times\mathbb{T}^N$, satisfy the continuity equation* (A.5) *and the renormalized continuity equation* (A.6) *for any b belonging to class* (A.7). *Then*

$$\varrho \in C_w([0,T];L^\beta(\mathbb{T}^N)) \cap C([0,T],L^p(\mathbb{T}^N))$$

for any $1 \leq p < \beta$.

A.4 A generalized Itô formula

We conclude with a generalized version of Itô's formula, Theorem 2.4.1.

Theorem A.4.1. *Let W be a cylindrical Wiener process in $\mathfrak{U}$ defined on a stochastic basis $(\Omega, \mathfrak{F}, (\mathfrak{F}_t)_{t\geq0}, \mathbb{P})$. Let s be a stochastic process on $(\Omega, \mathfrak{F}, (\mathfrak{F}_t)_{t\geq0}, \mathbb{P})$ such that, for some $\lambda > 0$,*

$$s \in C_w([0,T];W^{-\lambda,2}(\mathbb{T}^N)) \cap L^\infty(0,T;L^1(\mathbb{T}^N)) \quad \mathbb{P}\text{-a.s.},$$

$$\mathbb{E}\Big[\sup_{t\in[0,T]} \|s\|_{L^1_x}^2\Big] < \infty, \tag{A.10}$$

$$\mathrm{d}s = \mathrm{d}^\mathrm{d}s\,\mathrm{d}t + \mathrm{d}^\mathrm{s}s\,\mathrm{d}W. \tag{A.11}$$

Here $\mathrm{d}^\mathrm{d}s$, $\mathrm{d}^\mathrm{s}s$ are progressively measurable with

$$\begin{aligned}
&\mathrm{d}^\mathrm{d}s \in L^2(\Omega;L^1(0,T;W^{-\lambda,q}(\mathbb{T}^N))),\\
&\mathrm{d}^\mathrm{s}s \in L^2(\Omega;L^2(0,T;L_2(\mathfrak{U};W^{-m,2}(\mathbb{T}^N)))),\\
&\sum_{k=1}^\infty \int_0^T \|\mathrm{d}^\mathrm{s}s(e_k)\|_{L^1_x}^2\,\mathrm{d}t \in L^1(\Omega),
\end{aligned} \tag{A.12}$$

for some $q > 1$ and some $m \in \mathbb{N}$.

Let r be a stochastic process on $(\Omega, \mathfrak{F}, (\mathfrak{F}_t)_{t\geq 0}, \mathbb{P})$ *satisfying*

$$r \in C([0,T]; W^{\lambda,q'} \cap C(\mathbb{T}^N)) \quad \mathbb{P}\text{-a.s.},$$
$$\mathbb{E}\Big[\sup_{t\in[0,T]} \|r\|^2_{W^{\lambda,q'}_x \cap C_x}\Big] < \infty, \tag{A.13}$$
$$\mathrm{d}r = \mathrm{d}^{\mathrm{d}} r + \mathrm{d}^{\mathrm{s}} r \, \mathrm{d}W, \tag{A.14}$$

where $q' = \frac{q}{q-1}$. *Here* $\mathrm{d}^{\mathrm{d}} r$, $\mathrm{d}^{\mathrm{s}} r$ *are progressively measurable with*

$$\begin{aligned} &\mathrm{d}^{\mathrm{d}} r \in L^2(\Omega; L^1(0,T; W^{\lambda,q'} \cap C(\mathbb{T}^N))), \\ &\mathrm{d}^{\mathrm{s}} r \in L^2(\Omega; L^2(0,T; L_2(\mathfrak{U}; W^{-m,2}(\mathbb{T}^N)))), \\ &\sum_{k=1}^{\infty} \int_0^T \|\mathrm{d}^{\mathrm{s}} r(e_k)\|^2_{W^{\lambda,q'}_x \cap C_x} \, \mathrm{d}t \in L^1(\Omega). \end{aligned} \tag{A.15}$$

Let Q be $[\lambda + 2]$*-continuously differentiable function satisfying*

$$\mathbb{E}\Big[\sup_{t\in[0,T]} \|Q^{(j)}(r)\|^2_{W^{\lambda,q'}_x \cap C_x}\Big] < \infty \quad j = 0,1,2. \tag{A.16}$$

Then

$$\begin{aligned} \mathrm{d}\Big(\int_{\mathbb{T}^3} sQ(r)\,\mathrm{d}x\Big) = \int_{\mathbb{T}^3} \Big[s\Big(Q'(r)\mathrm{d}^{\mathrm{d}} r + \frac{1}{2}\sum_{k=1}^{\infty} Q''(r)|\mathrm{d}^{\mathrm{s}} r(e_k)|^2 \Big)\Big] \,\mathrm{d}x\,\mathrm{d}t \\ + \langle Q(r), \mathrm{d}^{\mathrm{d}} s\rangle \,\mathrm{d}t + \Big(\sum_{k=1}^{\infty} \int_{\mathbb{T}^3} \mathrm{d}^{\mathrm{s}} s(e_k)\,\mathrm{d}^{\mathrm{s}} r(e_k)\,\mathrm{d}x \Big)\,\mathrm{d}t + \mathrm{d}\mathbb{M}, \end{aligned} \tag{A.17}$$

where

$$\mathbb{M} = \sum_{k=1}^{\infty} \int_0^t \int_{\mathbb{T}^3} [sQ'(r)\,\mathrm{d}^{\mathrm{s}} r(e_k) + Q(r)\mathrm{d}^{\mathrm{s}} s(e_k)]\,\mathrm{d}x\,\mathrm{d}W_k. \tag{A.18}$$

Proof. In accordance with hypothesis (A.13), relation (A.14) holds pointwise in $\mathbb{T}^N$. Consequently, we may apply Itô's formula, Theorem 2.4.1, to obtain

$$\mathrm{d}Q(r) = Q'(r)[\mathrm{d}^{\mathrm{d}} r\,\mathrm{d}t + \mathrm{d}^{\mathrm{s}} r\,\mathrm{d}W] + \frac{1}{2}\sum_{k=1}^{\infty} Q''(r)|\mathrm{d}^{\mathrm{s}} r(e_k)|^2\,\mathrm{d}t \tag{A.19}$$

pointwise in $\mathbb{T}^N$.

Next, we regularize (A.11) by taking a spatial convolution with a suitable family of regularizing kernels; cf. Section 1.7.3. Denoting by $[v]_\delta = [v]_{x,\delta}$ the regularization of v, we write

$$\mathrm{d}[s]_\delta = [\mathrm{d}^{\mathrm{d}} s]_\delta\,\mathrm{d}t + [\mathrm{d}^{\mathrm{s}} s]_\delta\,\mathrm{d}W$$

pointwise in $\mathbb{T}^N$. Thus, by Itô's product rule, Proposition 2.4.2,

$$\begin{aligned} d([s]_\delta Q(r)) &= [s]_\delta \, dQ(r) + Q(r)\, d[s]_\delta + \sum_{k=1}^{\infty} [d^s s]_\delta(e_k)\, d^s r(e_k)\, dt \\ &= [s]_\delta \left(Q'(r)\, d^d r + \frac{1}{2} \sum_{k=1}^{\infty} Q''(r) |d^s r(e_k)|^2 \right) dt \\ &\quad + Q(r)[d^d s]_\delta \, dt + [[s]_\delta Q'(r)\, d^s r + Q(r)[d^s s]_\delta]\, dW \\ &\quad + \sum_{k=1}^{\infty} [d^s s]_\delta(e_k)\, d^s r(e_k)\, dt \end{aligned} \tag{A.20}$$

pointwise in $\mathbb{T}^N$. Integrating (A.20), we therefore obtain

$$\begin{aligned} d \int_{\mathbb{T}^3} [s]_\delta Q(r)\, dx &= \int_{\mathbb{T}^3} \left[[s]_\delta \left(Q'(r) d^d r + \frac{1}{2} \sum_{k=1}^{\infty} Q''(r) |d^s r(e_k)|^2 \right) \right] dx\, dt \\ &\quad + Q(r)[d^d s]_\delta \, dt + \int_{\mathbb{T}^3} [[s]_\delta Q'(r)\, d^s r + Q(r)[d^s s]_\delta]\, dx\, dW \\ &\quad + \sum_{k=1}^{\infty} \int_{\mathbb{T}^3} [d^s s]_\delta(e_k)\, d^s r(e_k)\, dx\, dt. \end{aligned} \tag{A.21}$$

Finally, using hypotheses (A.10), (A.12), (A.13), (A.15), and (A.16) we are able to perform the limit $\delta \to 0$ in (A.21), completing the proof. □

Remark A.4.2. The result stated in Lemma A.4.1 is not optimal with respect to the regularity properties of the processes r and s. As a matter of fact, we could regularize both r and s in the above proof to conclude that (A.17) holds as long as all expressions in (A.17) and (A.18) are well-defined.

B Bibliographical remarks

The results collected in Chapter 1 are standard and can be found in the literature. The precise references are provided directly in the text.

Chapter 2 presents an introduction to the principal concepts from probability theory and stochastic analysis with applications to stochastic PDEs. The standard references include for instance the monographs by Karatzas–Shreve [KS91] and Da Prato–Zabczyk [DPZ92]. The novelty of our presentation is twofold. First, various concepts are discussed in the context of more general topological spaces, such as locally convex topological vector spaces, rather than Polish spaces. Second, we introduce the concept of random distribution (see Section 2.2). It is a generalization of the notion of stochastic process, which allows one to treat random elements as space-time distributions rather than functions of time taking values in some abstract function space. In Section 2.2 and the sequel we establish various original results that explain how random distributions naturally fit into the stochastic Itô integration theory and provide simplifications in the analysis of stochastic PDEs.

The main part of this book starts with Chapter 3. Here, we introduce our model system and the main questions of interest. A particular emphasis is put on various notions of solutions that are treated later on.

As the first step in our analysis of the compressible Navier–Stokes system driven by stochastic forces we establish existence of a dissipative martingale solution in Chapter 4. This result is motivated by [BH16] and [BFH17]. In the former, existence of the so-called finite energy weak martingale solution was proved. In the latter, we show that solutions constructed in [BH16] satisfy a more general version of an energy inequality. In Chapter 4 we put forward a direct and more refined construction based on a different approximation procedure.

Chapter 5 is devoted to the existence of strong solutions. These solutions are strong in the PDE and probabilistic sense and can only be showed to exist locally in time, that is, up to a positive stopping time. The results of this Chapter can be found in [BFH16a].

In Chapter 6 we continue our discussion of dissipative martingale solutions and show that they satisfy a relative energy inequality. This is a kind of distance between a dissipative martingale solution and a pair of arbitrary smooth stochastic processes. The results of this Chapter are based on [BFH17].

As the next step, we focus on the long time behavior of dissipative martingale solutions in Chapter 7. Namely, we prove the existence of a stationary dissipative martingale solution. This result relies on [BFHM17].

Finally, in Chapter 8 we present two singular limit results, an incompressible limit (Section 8.1) and an inviscid–incompressible limit (Section 8.2). The former result is motivated by [BFH16b] but the discussion relies on dissipative martingale solutions rather than finite energy martingale solutions employed in [BFH16b]. The inviscid–

https://doi.org/10.1515/9783110492552-010

incompressible limit is based on the relative energy inequality and can be found in [BFH17].

The results in the appendix are again standard and references are provided directly in the text.

Bibliography

[Ada75] R. A. Adams, *Sobolev Spaces*, Academic Press, New York, 1975.

[ADN59] S. Agmon, A. Douglis, and L. Nirenberg, Estimates near the boundary for solutions of elliptic partial differential equations, *Commun. Pure Appl. Math.*, 12:623–727, 1959.

[Ama93] H. Amann, Nonhomogeneous linear and quasilinear elliptic and parabolic boundary value problems, In *Function Spaces, Differential Operators and Nonlinear Analysis (Friedrichroda, 1992)*, volume 133 of *Teubner-Texte Math.*, pages 9–126, Teubner, Stuttgart, 1993.

[Ama95] H. Amann, *Linear and Quasilinear Parabolic Problems: Volume I: Abstract Linear Theory*, volume 1, Springer Science & Business Media, Berlin, 1995.

[ABF99] P. Angot, C.-H. Bruneau, and P. Fabrie, A penalization method to take into account obstacles in incompressible viscous flows, *Numer. Math.*, 81(4):497–520, 1999.

[AKM83] S. N. Antontsev, A. V. Kazhikhov, and V. N. Monakhov, *Kraevye Zadachi Mekhaniki Neodnorodnykh Zhidkostei*, "Nauka" Sibirsk. Otdel., Novosibirsk, 1983.

[Aub63] J.-P. Aubin, Un théorème de compacité. *C. R. Acad. Sci. Paris*, 256:5042–5044, 1963.

[Bal89] J. M. Ball, A version of the fundamental theorem for Young measures, volume 344 of *Lect. Notes in Physics*, pages 207–215, Springer-Verlag, Berlin, 1989.

[Bar82] M. T. Barlow, One-dimensional stochastic differential equations with no strong solution, *J. Lond. Math. Soc. (2)*, 26(2):335–347, 1982.

[Bec66] E. Becker, *Gasdynamik*, Teubner-Verlag, Stuttgart, 1966.

[Ben95] A. Bensoussan. Stochastic Navier–Stokes equations, *Acta Appl. Math.*, 38(3):267–304, 1995.

[BF00] A. Bensoussan and J. Frehse, Local solutions for stochastic Navier–Stokes equations, *ESAIM: Math. Model. Numer. Anal.*, 34(2):241–273, 2000.

[BT73] A. Bensoussan and R. Temam, Équations stochastiques du type Navier–Stokes, *J. Funct. Anal.*, 13:195–222, 1973.

[BL76] J. Bergh and J. Löfström, *Interpolation Spaces. An introduction*, volume 223 of Grundlehren der Mathematischen Wissenschaften, Springer-Verlag, Berlin, 1976.

[Bil99] P. Billingsley, *Convergence of Probability Measures*, Wiley Series in Probability and Statistics: Probability and Statistics, John Wiley & Sons, Inc., New York, second edition, 1999. A Wiley-Interscience Publication.

[BD83] D. Blackwell and L. E. Dubins, An extension of Skorohod's almost sure representation theorem, *Proc. Am. Math. Soc.*, 89(4):691–692, 1983.

[Bog07] V. I. Bogachev, *Measure Theory, Vol. I, II*, Springer-Verlag, Berlin, 2007.

[Bre15] D. Breit, Existence theory for stochastic power law fluids, *J. Math. Fluid Mech.*, 17(2):295–326, 2015.

[BH16] D. Breit and M. Hofmanová, Stochastic Navier–Stokes equations for compressible fluids, *Indiana Univ. Math. J.*, 65(4):1183–1250, 2016.

[BFH16a] D. Breit, E. Feireisl, and M. Hofmanova, Local strong solutions to the stochastic compressible Navier–Stokes system, *ArXiv e-prints*, June 2016. arXiv:1606.05441v1.

[BFH16b] D. Breit, E. Feireisl, and M. Hofmanová, Incompressible limit for compressible fluids with stochastic forcing, *Arch. Ration. Mech. Anal.*, 222(2):895–926, 2016.

[BFH17] D. Breit, E. Feireisl, and M. Hofmanová, Compressible fluids driven by stochastic forcing: the relative energy inequality and applications, *Commun. Math. Phys.*, 350(2):443–473, 2017.

[BFHM17] D. Breit, E. Feireisl, M. Hofmanova, and B. Maslowski, Stationary solutions to the compressible Navier–Stokes system driven by stochastic forces, *ArXiv e-prints*, March 2017. arXiv:1703.03177v1.

https://doi.org/10.1515/9783110492552-011

[BD07] D. Bresch and B. Desjardins, On the existence of global weak solutions to the Navier–Stokes equations for viscous compressible and heat conducting fluids, *J. Math. Pures Appl.*, 87:57–90, 2007.

[BDGV07] D. Bresch, B. Desjardins, and D. Gérard-Varet, On compressible Navier–Stokes equations with density dependent viscosities in bounded domains, *J. Math. Pures Appl.*, 87:227–235, 2007.

[Bre83] H. Brezis, Analyse fonctionnelle. Collection mathématiques appliquées pour la maıtrise. [Collection of Applied Mathematics for the Master'S Degree], Masson, Paris, 1983.

[Brz95] Z. Brzeźniak, Stochastic partial differential equations in M-type 2 Banach spaces, *Potential Anal.*, 4(1):1–45, 1995.

[BF17] Z. Brzeźniak and B. Ferrario, Stationary solutions for stochastic damped Navier–Stokes equations in $\mathbb{R}^d$, *ArXiv e-prints*, February 2017. arXiv:1702.00697.

[BO07] Z. Brzeźniak and M. Ondreját, Strong solutions to stochastic wave equations with values in Riemannian manifolds, *J. Funct. Anal.*, 253(2):449–481, 2007.

[BP99] Z. Brzeźniak and S. Peszat, Strong local and global solutions for stochastic Navier–Stokes equations, In *Infinite Dimensional Stochastic Analysis*, pages 85–98 R. Neth. Acad. Arts Sci., Amsterdam, 1999.

[BOS16] Z. Brzeźniak, M. Ondreját, and J. Seidler, Invariant measures for stochastic nonlinear beam and wave equations, *J. Differ. Equ.*, 260(5):4157–4179, 2016.

[BMO15] Z. Brzeźniak, E. Motyl, and M. Ondrejat, Invariant measure for the stochastic Navier–Stokes equations in unbounded 2D domains, *ArXiv e-prints*, February 2015. arXiv:1502.02637.

[Cap93] M. Capiński, A note on uniqueness of stochastic Navier–Stokes equations, *Univ. Iagel. Acta Math.*, (30):219–228, 1993.

[CC91] M. Capiński and N. Cutland, Stochastic Navier–Stokes equations, *Acta Appl. Math.*, 25(1):59–85, 1991.

[CG94] M. Capiński and D. Gątarek, Stochastic equations in Hilbert space with application to Navier–Stokes equations in any dimension, *J. Funct. Anal.*, 126(1):26–35, 1994.

[Car13] R. W. Carroll, *Abstract Methods in Partial Differential Equations*, Courier Corporation, North Chelmsford, 2013.

[CdFV04] C. Castaing, P. R. de Fitte, and M. Valadier, *Young Measures on Topological Spaces: With Applications in Control Theory and Probability Theory*, volume 571, Springer Science & Business Media, Berlin, 2004.

[CCK04] Y. Cho, H. J. Choe, and H. Kim, Unique solvability of the initial boundary value problems for compressible viscous fluids, *J. Math. Pures Appl.*, 83(2):243–275, 2004.

[CM90] A. J. Chorin and J. E. Marsden, *A Mathematical Introduction to Fluid Mechanics*, volume 4 of *Texts in Applied Mathematics*, Springer-Verlag, New York, second edition, 1990.

[Coh13] D. L. Cohn, *Measure Theory*, Springer, Berlin, 2013.

[Cul68] H. F. Cullen, The Stone-Weierstrass theorem and complex Stone-Weierstrass theorem, In *Introduction to General Topology*, pages 286–293, Heath, Boston, 1968.

[DPD03] G. Da Prato and A. Debussche, Ergodicity for the 3D stochastic Navier–Stokes equations, *J. Math. Pures Appl. (9)*, 82(8):877–947, 2003.

[DPD08] G. Da Prato and A. Debussche, On the martingale problem associated to the 2D and 3D stochastic Navier–Stokes equations, *Atti Accad. Naz. Lincei Cl. Sci. Fis. Mat. Natur. Rend. Lincei (9) Mat. Appl.*, 19(3):247–264, 2008.

[DPZ92] G. Da Prato and J. Zabczyk, *Stochastic Equations in Infinite Dimensions*, volume 44 of *Encyclopedia of Mathematics and its Applications*, Cambridge University Press, Cambridge, 1992.

[DPZ96] G. Da Prato and J. Zabczyk, *Ergodicity for Infinite-Dimensional Systems*, volume 229 of *London Mathematical Society Lecture Note Series*, Cambridge University Press, Cambridge, 1996.

[Daf79] C. M. Dafermos, The second law of thermodynamics and stability, *Arch. Ration. Mech. Anal.*, 70(2):167–179, 1979.

[dB59] L. de Branges, The Stone–Weierstrass theorem, *Proc. Am. Math. Soc.*, 10:822–824, 1959.

[DLS10] C. De Lellis and L. Székelyhidi, On admissibility criteria for weak solutions of the Euler equations, *Arch. Ration. Mech. Anal.*, 195(1):225–260, 2010.

[Deb13] A. Debussche, Ergodicity results for the stochastic Navier–Stokes equations: an introduction, In *Topics in Mathematical Fluid Mechanics*, volume 2073 of *Lecture Notes in Math.*, pages 23–108, Springer, Heidelberg, 2013.

[DR14] A. Debussche and M. Romito, Existence of densities for the 3D Navier–Stokes equations driven by gaussian noise, *Probab. Theory Relat. Fields*, 158(3–4):575–596, 2014.

[DV10] A. Debussche and J. Vovelle, Scalar conservation laws with stochastic forcing, *J. Funct. Anal.*, 259(4):1014–1042, 2010.

[DGHT11] A. Debussche, N. Glatt-Holtz, and R. Temam, Local martingale and pathwise solutions for an abstract fluids model, *Physica D*, 240(14–15):1123–1144, 2011.

[DM75] C. Dellacherie and P.-A. Meyer, *Probabilités et potentiel*, Hermann, Paris, 1975. Chapitres I à IV, Édition entièrement refondue, Publications de l'Institut de Mathématique de l'Université de Strasbourg, No. XV, Actualités Scientifiques et Industrielles, No. 1372.

[DG99] B. Desjardins and E. Grenier, Low Mach number limit of viscous compressible flows in the whole space, *R. Soc. Lond. Proc. Ser. A Math. Phys. Eng. Sci.*, 455(1986):2271–2279, 1999.

[DGLM99] B. Desjardins, E. Grenier, P.-L. Lions, and N. Masmoudi, Incompressible limit for solutions of the isentropic Navier–Stokes equations with Dirichlet boundary conditions, *J. Math. Pures Appl. (9)*, 78(5):461–471, 1999.

[Die70] J. Dieudonné. *Treatise on Analysis. Vol. II*, Translated from French by I. G. Macdonald, volume 10–II of *Pure and Applied Mathematics*, Academic Press, New York–London, 1970.

[DL89] R. J. DiPerna and P.-L. Lions, Ordinary differential equations, transport theory and Sobolev spaces, *Invent. Math.*, 98(3):511–547, 1989.

[Dud02] R. M. Dudley, *Real Analysis and Probability*, volume 74 of *Cambridge Studies in Advanced Mathematics*, Cambridge University Press, Cambridge, 2002. Revised reprint of the 1989 original.

[DK10] J. J. Duistermaat and J. A. C. Kolk, *Distributions*, Springer, Berlin, 2010.

[Ebi83] D. G. Ebin, Viscous fluids in a domain with frictionless boundary, In *Global Analysis–Analysis on Manifolds*, volume 57 of *Teubner-Texte Math.*, pages 93–110, Teubner, Leipzig, 1983.

[Edw94] R. E. Edwards, *Functional Analysis: Theory and Applications, Holt, Rinehart and Winston, New York, 1965*, volume 36, 1994.

[ET99] I. Ekeland and R. Temam, *Convex Analysis and Variational Problems*, SIAM, Philadelphia, 1999.

[FKS05] R. Farwig, H. Kozono, and H. Sohr, An L^q-approach to Stokes and Navier–Stokes equations in general domains, *Acta Math.*, 195:21–53, 2005.

[Fei00] E. Feireisl, Global attractors for the Navier–Stokes equations of three-dimensional compressible flow, *C. R. Acad. Sci. Paris Sér. I Math.*, 331(1):35–39, 2000.

[Fei02] E. Feireisl, Compressible Navier–Stokes equations with a non-monotone pressure law, *J. Differ. Equ.*, 184:97–108, 2002.

[Fei04] E. Feireisl, *Dynamics of Viscous Compressible Fluids*, volume 26 of *Oxford Lecture Series in Mathematics and its Applications*, Oxford University Press, Oxford, 2004.

[FN09] E. Feireisl and A. Novotný, *Singular Limits in Thermodynamics of Viscous Fluids*, Advances in Mathematical Fluid Mechanics, Birkhäuser Verlag, Basel, 2009.

[FN12] E. Feireisl and A. Novotný, Weak-strong uniqueness property for the full Navier–Stokes–Fourier system, *Arch. Ration. Mech. Anal.*, 204(2):683–706, 2012.

[FN14] E. Feireisl and A. Novotný, Inviscid incompressible limits under mild stratification: a rigorous derivation of the Euler-Boussinesq system, *Appl. Math. Optim.*, 70:279–307, 2014.

[FP01] E. Feireisl and H. Petzeltová, Bounded absorbing sets for the Navier–Stokes equations of compressible fluid, *Commun. Partial Differ. Equ.*, 26(7–8):1133–1144, 2001.

[FP10] E. Feireisl and D. Pražák, *Asymptotic Behavior of Dynamical Systems in Fluid Mechanics*, volume 4 of *AIMS Series on Applied Mathematics*, American Institute of Mathematical Sciences (AIMS), Springfield, MO, 2010.

[FNP01] E. Feireisl, A. Novotný, and H. Petzeltová, On the existence of globally defined weak solutions to the Navier–Stokes equations, *J. Math. Fluid Mech.*, 3(4):358–392, 2001.

[FNS11] E. Feireisl, A. Novotný, and Y. Sun, Suitable weak solutions to the Navier–Stokes equations of compressible viscous fluids, *Indiana Univ. Math. J.*, 60(2):611–631, 2011.

[FJN12] E. Feireisl, B. J. Jin, and A. Novotný, Relative entropies, suitable weak solutions, and weak-strong uniqueness for the compressible Navier–Stokes system, *J. Math. Fluid Mech.*, 14(4):717–730, 2012.

[FMN13] E. Feireisl, B. Maslowski, and A. Novotný, Compressible fluid flows driven by stochastic forcing, *J. Differ. Equ.*, 254(3):1342–1358, 2013.

[Fer88] X. Fernique, Un modele presque sûr pour la convergence en loi, *C. R. Acad. Sci. Paris, Sér.*, 1(306):335–338, 1988.

[Fla94] F. Flandoli. Dissipativity and invariant measures for stochastic Navier–Stokes equations, *Nonlinear Differ. Equ. Appl.*, 1(4):403–423, 1994.

[Fla08] F. Flandoli, An introduction to 3D stochastic fluid dynamics, In *SPDE in Hydrodynamic: Recent Progress and Prospects*, volume 1942 of *Lecture Notes in Math.*, pages 51–150, Springer, Berlin, 2008.

[FG95] F. Flandoli and D. Gątarek, Martingale and stationary solutions for stochastic Navier–Stokes equations, *Probab. Theory Relat. Fields*, 102(3):367–391, 1995.

[FM12] F. Flandoli and A. Mahalov, Stochastic three-dimensional rotating Navier–Stokes equations: averaging, convergence and regularity, *Arch. Ration. Mech. Anal.*, 205(1):195–237, 2012.

[FR08] F. Flandoli and M. Romito, Markov selections for the 3D stochastic Navier–Stokes equations, *Probab. Theory Relat. Fields*, 140(3–4):407–458, 2008.

[Foi72a] C. Foiaş, Statistical study of Navier–Stokes equations, I, *Rend. Semin. Mat. Univ. Padova*, 48:219–348 (1972).

[Foi72b] C. Foiaş, Statistical study of Navier–Stokes equations, II, *Rend. Semin. Mat. Univ. Padova*, 49:9–123 (1973).

[GGZ75] H. Gajewski, K. Gröger, and K. Zacharias, Nichtlineare operatorgleichungen und operatordifferentialgleichungen, *Math. Nachr.*, 67(22), 1975.

[Gal00] I. Gallagher, A remark on smooth solutions of the weakly compressible periodic Navier–Stokes equations, *J. Math. Kyoto Univ.*, 40(3):525–540, 2000.

[Ger11] P. Germain, Weak–strong uniqueness for the isentropic compressible Navier–Stokes system, *J. Math. Fluid Mech.*, 13(1):137–146, 2011.

[GhS80] Ĭ. Ī. Gīhman and A. V. Skorohod, *The Theory of Stochastic Processes, I*, volume 210 of *Grundlehren der Mathematischen Wissenschaften [Fundamental Principles of*

Mathematical Sciences], Springer-Verlag, Berlin–New York, English edition, 1980. Translated from Russian by Samuel Kotz .

[GT83] D. Gilbarg and N. Trudinger, *Elliptic Partial Differential Equations of Second Order*, Springer-Verlag, Berlin, 1983.

[GHV14] N. E. Glatt-Holtz and V. C. Vicol, Local and global existence of smooth solutions for the stochastic Euler equations with multiplicative noise, *Ann. Probab.*, 42(1):80–145, 2014.

[GHŠV15] N. Glatt-Holtz, V. Šverák, and V. Vicol, On inviscid limits for the stochastic Navier–Stokes equations and related models, *Arch. Ration. Mech. Anal.*, 217(2):619–649, 2015.

[GM05] B. Goldys and B. Maslowski, Exponential ergodicity for stochastic Burgers and 2D Navier–Stokes equations, *J. Funct. Anal.*, 226(1):230–255, 2005.

[GM06] B. Goldys and B. Maslowski, Lower estimates of transition densities and bounds on exponential ergodicity for stochastic PDE's, *Ann. Probab.*, 34(4):1451–1496, 2006.

[Gra08] L. Grafakos, *Classical Fourier Analysis*, volume 249 of *Graduate Texts in Mathematics*, Springer, New York, second edition, 2008.

[GK96] I. Gyöngy and N. Krylov, Existence of strong solutions for Itô's stochastic equations via approximations, *Probab. Theory Relat. Fields*, 105(2):143–158, 1996.

[HM06] M. Hairer and J. C. Mattingly, Ergodicity of the 2D Navier–Stokes equations with degenerate stochastic forcing, *Ann. of Math. (2)*, 164(3):993–1032, 2006.

[HP97] M. Hieber and J. Pruss, Heat kernels and maximal lp—lq estimates for parabolic evolution equations, *Commun. Partial Differ. Equ.*, 22(9–10):1647–1669, 1997.

[Hof13] M. Hofmanová, Degenerate parabolic stochastic partial differential equations, *Stoch. Process. Appl.*, 123(12):4294–4336, 2013.

[Itô46] K. Itô, On a stochastic integral equation, *Proc. Jpn. Acad.*, 22(1–4):32–35, 1946.

[Itô51] K. Itô, On stochastic differential equations, *Mem. Am. Math. Soc.*, 4:1–51, 1951.

[IN64] K. Itô and M. Nisio, On stationary solutions of a stochastic differential equation, *J. Math. Kyoto Univ.*, 4:1–75, 1964.

[Jak97] A. Jakubowski, The almost sure Skorokhod representation for subsequences in nonmetric spaces, *Teor. Veroâtn. Primen.*, 42(1):209–216, 1997.

[KS91] I. Karatzas and S. E. Shreve, *Brownian Motion and Stochastic Calculus*, volume 113 of *Graduate Texts in Mathematics*, Springer-Verlag, New York, second edition, 1991.

[Kat84] T. Kato, Remarks on the zero viscosity limit for nonstationary Navier–Stokes flows with boundary, In *Seminar on PDE's*, S. S. Chern (ed.), Springer, New York, 1984.

[Kel55] J. L. Kelley, *General Topology*, D. Van Nostrand Company, Inc., Toronto–New York–London, 1955.

[Kim09] J. U. Kim, Existence of a local smooth solution in probability to the stochastic Euler equations in $\mathbf{R}^3$, *J. Funct. Anal.*, 256(11):3660–3687, 2009.

[Kim11] J. U. Kim, On the stochastic quasi-linear symmetric hyperbolic system, *J. Differ. Equ.*, 250(3):1650–1684, 2011.

[KM81] S. Klainerman and A. Majda, Singular limits of quasilinear hyperbolic systems with large parameters and the incompressible limit of compressible fluids, *Commun. Pure Appl. Math.*, 34(4):481–524, 1981.

[KBS+01] R. Klein, N. Botta, T. Schneider, C.-D. Munz, S. Roller, A. Meister, L. Hoffmann, and T. Sonar, Asymptotic adaptive methods for multi-scale problems in fluid mechanics, In *Practical Asymptotics*, pages 261–343, Springer, Berlin, 2001.

[KK83] G. Köthe and G. Köthe, *Topological Vector Spaces*, Springer, Berlin, 1983.

[KR79] N. V. Krylov and B. L. Rozovskii, Stochastic evolution equations, *Itogi Nauki i Tekhniki, Ser. Sovremen. Prob. Mat. Noveishie Dostizheniya*, 14:71–146, 1979.

[KJF77] A. Kufner, O. John, and S. Fucik, *Function Spaces*, volume 3, Springer Science & Business Media, Berlin, 1977.

[KS12] S. Kuksin and A. Shirikyan, *Mathematics of Two-Dimensional Turbulence*, volume 194, Cambridge University Press, Cambridge, 2012.

[LSU68] O. A. Ladyzenskaa, V. A. Solonnikov, and N. N. Uraltseva, *Linear and Quasi-Linear Equations of Parabolic Type*, American Mathematical Society, Providence, 1968.

[LU68] O. A. Ladyzhenskaya and N. N. Uralceva, *Equations aux dérivées partielles de type elliptique*, Dunod, Paris, 1968.

[LJFR99] D. Leão Jr, M. D. Fragoso, and P. R. C. Ruffino, Characterizations of radon spaces, *Stat. Probab. Lett.*, 42(4):409–413, 1999.

[LV11] N. Leger and A. Vasseur, Relative entropy and the stability of shocks and contact discontinuities for systems of conservation laws with non-bv perturbations, *Arch. Ration. Mech. Anal.*, 201(1):271–302, 2011.

[Lig52] M. J. Lighthill, On sound generated aerodynamically, I, General theory, *Proc. R. Soc. Lond. Ser. A*, 211:564–587, 1952.

[Lig54] M. J. Lighthill, On sound generated aerodynamically, II, Turbulence as a source of sound, *Proc. R. Soc. Lond. Ser. A*, 222:1–32, 1954.

[Lio69] J.-L. Lions, *Quelques méthodes de résolution des problemes aux limites non linéaires*, volume 31, Dunod, Paris, 1969.

[Lio98] P.-L. Lions, *Mathematical Topics in Fluid Mechanics, Vol. 2: Compressible Models*, volume 10 of *Oxford Lecture Series in Mathematics and its Applications*, The Clarendon Press, Oxford University Press, New York, 1998, Oxford Science Publications.

[LM98] P.-L. Lions and N. Masmoudi, Incompressible limit for a viscous compressible fluid, *J. Math. Pures Appl. (9)*, 77(6):585–627, 1998.

[LM99] P.-L. Lions and N. Masmoudi, Une approche locale de la limite incompressible, *C. R. Acad. Sci., Ser. 1 Math.*, 329(5):387–392, 1999.

[Lun12] A. Lunardi, *Analytic Semigroups and Optimal Regularity in Parabolic Problems*, Springer Science & Business Media, Berlin, 2012.

[Maj12] A. Majda, *Compressible Fluid Flow and Systems of Conservation Laws in Several Space Variables*, volume 53, Springer Science & Business Media, Berlin, 2012.

[MNRR96] J. Málek, J. Necas, M. Rokyta, and M. Ruzicka, *Weak and Measure-Valued Solutions to Evolutionary PDEs*, volume 13, CRC Press, Boca Raton, 1996.

[Mas01] N. Masmoudi, Incompressible, inviscid limit of the compressible Navier–Stokes system, *Ann. Inst. Henri Poincaré, Anal. Non Linéaire*, 18:199–224, 2001.

[Mas06] N. Masmoudi, Examples of singular limits in hydrodynamics, In *Handbook of Differential Equations, III*, C. Dafermos, E. Feireisl (eds.), Elsevier, Amsterdam, 2006.

[MN79] A. Matsumura and T. Nishida, The initial value problem for the equations of motion of compressible viscous and heat-conductive fluids, *Proc. Jpn. Acad., Ser. A, Math. Sci.*, 55(9):337–342, 1979.

[MN83] A. Matsumura and T. Nishida, Initial-boundary value problems for the equations of motion of compressible viscous and heat-conductive fluids, *Commun. Math. Phys.*, 89(4):445–464, 1983.

[Maz13] V. Maz'ya, *Sobolev Spaces*, Springer, Berlin, 2013.

[MV08] A. Mellet and A. Vasseur, Existence and uniqueness of global strong solutions for one-dimensional compressible Navier–Stokes equations, *SIAM J. Math. Anal.*, 39(4):1344–1365, 2008.

[Men16] P. R. Mensah, Existence of martingale solutions and the incompressible limit for stochastic compressible flows on the whole space, *Ann. Mat. Pura Appl.*, 196(6):2105-–2133, 2017.

[MR04] R. Mikulevicius and B. L. Rozovskii, Stochastic Navier–Stokes equations for turbulent flows, *SIAM J. Math. Anal.*, 35(5):1250–1310, 2004.

[MR05] R. Mikulevicius and B. L. Rozovskii, Global l2-solutions of stochastic Navier–Stokes equations, *Ann. Probab.*, 33(1):137–176, 2005.

[MV00] R. Mikulevicius and G. Valiukevicius, On stochastic Euler equation in rd, *Electron. J. Probab.*, 5(6):20, 2000.

[Nec67] J. Necas, *Les méthodes directes en théorie des équations elliptiques*, Academia, Prague, 1967.

[Ond04] M. Ondreját, Uniqueness for stochastic evolution equations in Banach spaces, *Diss. Math. (Rozprawy Mat.)*, 426:63, 2004.

[Ped97] P. Pedregal, *Parametrized Measures and Variational Principles*, volume 30 of *Progress in Nonlinear Differential Equations and Their Applications*, Birkhäuser, Basel, 1997.

[PW15] P. I. Plotnikov and W. Weigant, Isothermal Navier–Stokes equations and Radon transform, *SIAM J. Math. Anal.*, 47(1):626–653, 2015.

[PR07] C. Prévôt and M. Röckner, *A Concise Course on Stochastic Partial Differential Equations*, volume 1905 of *Lecture Notes in Mathematics*, Springer, Berlin, 2007.

[PW67] M. H. Protter and H. F. Weinberger, *Maximum Principles in Differential Equations*, Prentice Hall, Inc., London, 1967.

[Rom10] M. Romito, Existence of martingale and stationary suitable weak solutions for a stochastic Navier–Stokes system, *Stochastics*, 82(1–3):327–337, 2010.

[Rud87] W. Rudin, *Real and Complex Analysis*, Tata McGraw-Hill Education, New Delhi, 1987.

[SR09] L. Saint-Raymond, Hydrodynamic limits: some improvements of the relative entropy method, *Ann. Inst. Henri Poincaré, Anal. Non Linéaire*, 26(3):705–744, 2009.

[Ser86] D. Serre, Solutions faibles globales des équations de Navier–Stokes pour un fluide compressible, *C. R. Acad. Sci. Paris Sér. I Math.*, 303(13):639–642, 1986.

[Sim86] J. Simon, Compact sets in the space lp (o, t; b), *Ann. Mat. Pura Appl.*, 146(1):65–96, 1986.

[Sko61] A. V. Skorohod, On the existence and uniqueness of solutions of stochastic differential equations, *Sib. Mat. Zh.*, 2:129–137, 1961.

[Sko62] A. V. Skorohod, On stochastic differential equations, In *Proc. Sixth All-Union Conf. Theory Prob. and Math. Statist.* (Vilnius (Russian) 1960), pages 159–168, Gosudarstv. Izdat. Političesk. i Navčn. Lit. Litovsk. SSR, Vilnius, 1962.

[Smi15] S. Smith, Random perturbations of viscous compressible fluids: global existence of weak solutions, *ArXiv e-prints*, April 2015.

[Ste70] E. M. Stein, *Singular Integrals and Differentiability Properties of Functions*, volume 30 of *Princeton Math. Series*, Princeton University Press, Princeton, 1970.

[Sue14] F. Sueur, On the inviscid limit for the compressible Navier–Stokes system in an impermeable bounded domain, *J. Math. Fluid Mech.*, 16(1):163–178, 2014.

[SWZ11] Y. Sun, C. Wang, and Z. Zhang, A Beale-Kato-Majda criterion for the 3-D compressible Navier–Stokes equations, *J. Math. Pures Appl.*, 95(1):36–47, 2011.

[TW97] R. Temam and X. Wang, On the behavior of the solutions of the Navier–Stokes equations at vanishing viscosity, *Ann. Sc. Norm. Pisa*, 25:807–828, 1997.

[TW02] R. Temam and X. Wang, Boundary layers associated with incompressible Navier–Stokes equations: the noncharacteristic boundary case, *J. Differ. Equ.*, 179:647–686, 2002.

[Tor00] E. Tornatore, Global solution of bi-dimensional stochastic equation for a viscous gas, *Nonlinear Differ. Equ. Appl.*, 7(4):343–360, 2000.

[TFY97] E. Tornatore and H. F. Yashima, One-dimensional stochastic equations for a viscous barotropic gas, *Ric. Mat.*, 46(2):255–283 (1998), 1997.

[VK95] V. A. Vaĭgant and A. V. Kazhikhov, On the existence of global solutions of two-dimensional Navier–Stokes equations of a compressible viscous fluid, *Sib. Mat. Zh.*, 36(6):1283–1316, 1995.

[VZ86] A. Valli and W. M. Zajączkowski, Navier–Stokes equations for compressible fluids: global existence and qualitative properties of the solutions in the general case, *Commun. Math. Phys.*, 103(2):259–296, 1986.

[vNVW07] J. M. A. M. van Neerven, M. C. Veraar, and L.Weis, Stochastic integration in UMD Banach spaces, *Ann. Probab.*, 35(4):1438–1478, 2007.

[WW15] D. Wang and H. Wang, Global existence of martingale solutions to the three-dimensional stochastic compressible Navier–Stokes equations, *Differ. Integral Equ.*, 28(11–12):1105–1154, 2015.

[WW12] Y.-G. Wang and M. Williams, The inviscid limit and stability of characteristic boundary layers for the compressible Navier–Stokes equations with Navier-friction boundary conditions, *Ann. Inst. Fourier (Grenoble)*, 62(6):2257–2314 (2013), 2012.

[WXZ12] L. Wang, Z. Xin, and A. Zang, Vanishing viscous limits for 3D Navier–Stokes equations with a Navier-slip boundary condition, *J. Math. Fluid Mech.*, 14(4):791–825, 2012.

[Wei00] E. Weinan, Boundary layer theory and the zero-viscosity limit of the Navier–Stokes equation, *Acta Math. Sin. Engl. Ser.*, 16(2):207–218, 2000.

[Wei01] E. Weinan, Stochastic hydrodynamics, In *Current Developments in Mathematics, 2000*, pages 109–147, Int. Press, Somerville, MA, 2001.

[Wei12] F. Weisz, Summability of multi-dimensional trigonometric Fourier series, *Surv. Approx. Theory*, 7:1–179, 2012.

[You36] L. C. Young, An inequality of the Hölder type, connected with Stieltjes integration, *Acta Math.*, 67(1):251–282, 1936.

[You69] L. C.Young, *Lectures on the Calculus of Variations and Optimal Control Theory*, Saunders, Philadelphia, 1969.

[Zie89] W. P. Ziemer, *Weakly Differentiable Functions. Sobolev Spaces And Functions of Bounded Variation*, volume 120 of *Graduate Texts in Mathematics*, Springer-Verlag, New York, 1989.

Index

De Gruyter Series in Applied and Numerical Mathematics

Volume 2
Zahari Zlatev, Ivan Dimov, István Faragó, Ágnes Havasi
Richardson Extrapolation. Practical Aspects and Applications
ISBN 978-3-11-051649-4, e-ISBN 978-3-11-053300-2, Set-ISBN 978-3-11-053301-9

Volume 1
Anvarbek Meirmanov, Oleg V. Galtsev, Reshat N. Zimin
Free Boundaries in Rock Mechanics
ISBN 978-3-11-054490-9, e-ISBN 978-3-11-054616-3, Set-ISBN 978-3-11-054617-0

www.ingramcontent.com/pod-product-compliance
Ingram Content Group UK Ltd.
Pitfield, Milton Keynes, MK11 3LW, UK
UKHW050954290926
15470UKWH00006B/1807

9 783110 490503